Food and Culture

A Reader

edited by

Carole Counihan
and Penny Van Esterik

ROUTLEDGE
New York and London

Published in 1997 by

Routledge
29 West 35th Street
New York, NY 10001

Published in Great Britain in 1997 by

Routledge
11 New Fetter Lane
London EC4P 4EE

Printed in the United States of America
Design: Jack Donner

Library of Congress Cataloging-in-Publication Data

Food and culture : a reader / edited by Carole Counihan and Penny Van Esterik.
p. cm.
Includes bibliographical references and index.
ISBN 0-415-91709-3 (hb: alk. paper). — ISBN 0-415-91710-7 (pb: alk. paper)
1. Food—Social aspects. 2. Food habits. I. Counihan, Carole, 1948– . II. Van Esterik, Penny.
GT2850.F64 1997
394.1—dc21 96-46430
CIP

Contents

Foreword

People ask me: Why do you write about food, and eating and drinking? Why don't you write about the struggle for power and security, and about love, the way others do?

They ask it accusingly, as it I were somehow gross, unfaithful to the honor of my craft.

The easiest answer is to say that, like most other humans, I am hungry. But there is more than that. It seems to me that our three basic needs, for food and security and love, are so mixed and mingled and entwined that we cannot straightly think of one without the others. So it happens that when I write of hunger, I am really writing about love and the hunger for it . . . and warmth and richness and fine reality of hunger satisfied . . . and it is all one.

I tell about myself, and how I ate bread on a lasting hillside, or drank red wine in a room now blown to bits, and it happens without my willing it that I am telling too about the people with me then, and their other deeper needs for love and happiness.

There is food in the bowl, and more often than not, because of what honesty I have, there is nourishment in the heart, to feed the wilder, more insistent hungers. We must eat. If, in the face of that dread fact, we can find other nourishment, and tolerance and compassion for it, we'll be no less full of human dignity.

There is a communion of more than our bodies when bread is broken and wine drunk. And that is my answer, when people ask me: Why do you write about hunger, and not wars or love?

M. F. K. Fisher

Introduction

CAROLE COUNIHAN AND PENNY VAN ESTERIK

We have begun this reader with the foreword from M. F. K. Fisher's *The Gastronomical Me* because no one has written about food with more passion, verve, or appreciation.[1] Fisher begins with these words: "People ask me: Why do you write about food, and eating and drinking?" Why indeed? We believe with her that "[t]here is communion of more than our bodies when bread is broken and wine drunk." We believe that worlds pass between one bite and another, and we hope this collection gives a taste of those worlds.

Food touches everything. Food is the foundation of every economy. It is a central pawn in political strategies of states and households. Food marks social differences, boundaries, bonds, and contradictions. Eating is an endlessly evolving enactment of gender, family, and community relationships. In this volume we see how food-sharing creates solidarity, and how food scarcity damages the human community and the human spirit. We see how men and women define themselves differently through their foodways, and how women across cultures so often speak through food and appetite. We examine some of the meanings of eating, fasting, being fat, and being thin, and we show their links to cultural images of masculinity and femininity.

Food is life, and life can be studied and understood through food. The articles we present here are designed to aid in such understanding. They are rich with detailed observation and creative thinking, offering diverse perspectives about other cultures. Food is both a scholarly concern and a real-life concern.

Because food crosses so many conceptual boundaries, it must be interpreted from a wide range of disciplinary perspectives. In fact, the study of food can be used to question the limitations of academic disciplinary boundaries. In spite of the interdisciplinary nature of food studies, however, anthropology dominates the field and this reader. Why? Because anthropology is holistic by definition. Anthropology integrates meaning-centered interpretive approaches, evolutionary and materialist perspectives, and advocacy-oriented critiques, all important to the study of food.

The development of research interests in food is as old as anthropology. Early anthropologists studied food because of its central role in many cultures, and several wrote pointed pieces on foodways, most notably Audrey Richards (1932, 1939) but also Raymond Firth (1934), Cora Du Bois (1941), and M. and S. L. Fortes (1936).[2] Nutritional anthropology grew out of the subfield of medical anthropology in the mid-seventies. Key works over the last few years illustrate the range of symbolic, materialist, and ecological perspectives used to explain patterns of food selection and their nutritional consequences.[3] The field has produced a number of recent food-oriented ethnographies that demonstrate how food is integrated into all aspects of life, such as Meigs's *Food, Sex, and Pollution* (1984), Kahn's *Always Hungry, Never Greedy: Food and the Expression of Gender in a Melanesian Society* (1986), Weismantel's *Food, Gender and Poverty in the Ecuadorian Andes* (1988), Pollock's *These Roots Remain: Food Habits in Islands of the Central and Eastern Pacific since Western Contact* (1992), Ohnuki-Tierney's *Rice As Self: Japanese Identities through Time* (1993), and Dettwyler's *Dancing Skeletons: Life and Death in West Africa* (1994). Furthermore an interest in food studies has recently burgeoned in other scholarly disciplines, such as history, sociology, philosophy, and literary criticism.[4] And food, of course, has always been a topic of fascination to the general public, as evidenced by the huge numbers of published cookbooks as well as the centrality of food in many marvelous works of literature.[5] We hope that as readers learn about food, M.F.K. Fisher's optimistic prediction will come true: "with our gastronomical growth will come, inevitably, knowledge and perception of a hundred other things, but mainly of ourselves."[6]

We open this volume with the section "Food, Meaning, and Voice" because food's extraordinary ability to convey meaning as well as nourish bodies constitutes one source of its power. Because everyone eats and many people cook, the meanings attached to food speak to many more people than do the meanings attached to more esoteric objects and practices. "It is because they are ordinarily immersed in everyday practice in a material way that foods, abstracted as symbols from this material process, can condense in themselves a wealth of ideological meanings" (Weismantel 1988:7–8). Scholars have noted how food presents a rich symbolic alphabet through its diversity of color, texture, smell, and taste; its ability to be elaborated and combined in infinite ways; and its immersion in norms of manners and cuisine. We present some of the classic articles on how food communicates, and we introduce readers to a variety of the theoretical approaches that have been applied to interpreting foodways—semiotic, structuralist, materialist, and cultural.

Together, the contributors presented in "Food, Meaning, and Voice" convey a sense of the power and reach of food's symbolic resonance. They show how food holds the keys to any culture and presents manifold channels for analysis. We learn much from our authors about food consumption and health, food choice and rules, and food symbolism. The remaining sections and articles explore more specific dimensions of the social and cultural uses of human foodways.

In the second section, contributors focus on practices and meanings of giving, receiving, and refusing food. Commensality, or food-sharing, has been a dominant concern of social scientists for decades. Food scholars owe a particular debt of gratitude to the brilliant French sociologist Marcel Mauss (1967), who made food exchanges central in his cross-cultural study, *The Gift*. Sahlins's thoughtful study, *Stone Age Economics* (1972), built on Mauss and underlined the important role of food exchanges in sociopolitical alliances. Gregory pointed out the important distinction between consumption in a gift economy and consumption in a commodity economy: "Consumption in a gift economy, then, is not simply the act of eating food. It is primarily concerned with the regulation of relations between people in the process of social and biological reproduction" (Gregory 1982:79). Food events encode and regulate key social relations. Other anthropologists have noted the key role of food in feasts of communal solidarity and political ranking.[7] Food-sharing is the medium for creating and maintaining social relations both within and beyond the household. Because of the mandatory nature of food-sharing, food refusal and fasting have powerful social and symbolic weight. As Brumberg shows, appetite can be a poweful voice. While in many cultures across the globe both men and women fast for political and religious reasons, in Western cultures, abstention from food has been a particularly salient and meaningful activity for women—both in a religious context as Bynum demonstrates for medieval women, and in pursuit of power and control in a secular context, as Bordo shows for contemporary women in her piece on anorexia nervosa.

Across history and cultures, women have a special relationship to food and a particularly vivid experience of their bodies. We explore the relationships between food, body, and culture and give particular attention to women's perspectives on these connections in the third section. In understanding the social construction of bodies, we learn of the relation between body image, eating, and sexuality in different cultures. Cultures that link female corporeal expansiveness to fertility often accord women status and respect. The cultural disassociation of women's body from reproduction often coexists with the definition of women's bodies as aesthetic objects and the focus on thinness that are so oppressive to Europeanized women. Cultural acceptance of diverse bodies is one key step in female empowerment.[8]

Control of food across history and cultures has often been a key source of power for women. Several authors have noted that women's ability to prepare and serve food gives them direct influence over others, both material and magical. Weismantel (1988) underscores how serving the daily soup gives the female head of an Ecuadorian Indian household the ability to assign relative status distinctions within the family based on how much meat she serves each person. Behar (1989) uses records from the Mexican Inquisition to show how "women made men 'eat' their witchcraft, using their power over the domain of food preparation for subversive ends . . . in eating, the pollution was introduced directly and effectively into the body. . . . The belief that food could be used to harm rather than to nurture gave women a very specific and real power that could serve as an important defense against abusive male dominance" (180).

Not only is food a medium for gender definitions, but it is also linked to overall social hierarchies and power relations. Access to food might be called the most basic human right, yet with the development of capitalism and its handmaidens of colonialism, imperialism, and food commodification, access to food has become a key measure of power and powerlessness, as Lappé and Collins (1986) so powerfully demonstrate. Culturally and economically marginal people often suffer hunger and malnutrition and rarely eat meat, or they only eat despised cuts or innards, some of which may serve as the foundations of entire cultural cuisines, as pigs' feet, chitterlings, and cracklings do in the African-American culture of the southern United States. Contributors in the fourth section explore the relationship between food commodification and scarcity, and show their link to power. The international hierarchy of "First World" vs. "Third World" is clearly expressed through differential access to food, especially meat, which is often raised in the Third World at the expense of traditional cereal crops and exported to rich First World consumers. Contributors explore how processes such as colonization and the delocalization caused by global food industries are implicated in changing conditions of food security at home and abroad. Articles also explore cultural myths surrounding hunger and the ways in which people combat hunger through advocacy and political organization.

HOW TO USE THIS BOOK

"Write a book with legs," our editor urged us—a book with staying power that will answer the needs of readers, scholars, and teachers for years to come. This volume is interdisciplinary and ranges across the social sciences and humanities. Perspectives from anthropology, sociology, psychology, history, literary criticism, philosophy, and social activism are all represented. Articles complement those included in *Cooking, Eating, Thinking: Transformative Philosophies of Food*, edited by Deane Curtin and Lisa Heldke (1992), which focuses on philosophical musings about foodways. Our articles do not duplicate any of the biocultural articles included in Marvin Harris's and Eric Ross's edited volume, *Food and Evolution* (1987), which contains studies of primate diet, hominid evolution, and materialist theories of foodways.[9] While cultural materialism has an appealing simplicity to it, Harris and Ross do not engage with current approaches to social, symbolic, or meaning-centered interpretations of food use. We consider the political and economic contexts of food use in addition to the meaning-centered contexts without prioritizing one over the other.

The social, symbolic, and political-economic perspectives included here make a valuable contribution to courses in anthropology, history, sociology, social work and political science as well as community health, nutrition, and nursing. They can teach students about connections between beliefs and practices, economy and ideology, and relationships and roles. They present a range of theoretical perspectives—structuralism, semiotics, psychoanalytic theory, Marxism, and feminism—and can be used to teach students theory through concrete applications.

CROSS-CUTTING THEMES

The four sections of the reader represent four interconnected themes key to the understanding of food and culture. In addition to the themes represented by these sections, a number of possible other themes could be developed through these readings.

Gender is a primary theme cross-cuts all sections of the book. Readers could move through the articles on gender for a different reading of food, as the following selections illustrate. Meigs explores women's food work and gender identity among the Hua of Papua New Guinea; De Vault, Hughes, and Massara among diverse modern North Americans. Several other articles explore gender in other cultural groups: Shack in the Gurage of Ethiopia, Sobo in rural Jamaicans, Bynum in medieval holy women, Allison in urban Japan. Anna Freud considers women's role in infant feeding as a key channel of psychosocial development. Brumberg explores the complex meanings of food refusal to Victorian anorexic girls. De Vault examines women's roles in feeding families and finds that it is usually women who mentally construct and physically carry out the myriad activities ensuring that men and children are fed culturally appropriate meals. Van Esterik focuses on breast-feeding as a key to women's cultural place. Control over food production and distribution is at the heart of women's status and power in many subsistence cultures, as Counihan's article about Sardinia shows. With industrialization and colonization, women's social position has declined as their important role in subsistence production has decreased and wage labor has risen. In cultures where women's social position is low, they may have difficulty getting enough to eat, or even claiming their legitimate right to eat, as Fitchen shows. In fact, women are more often malnourished than men and suffer more severely in times of famine, a particularly dangerous problem for pregnant and lactating women whose needs for calories and nutrients are unusually high.[10] Issues of men's and women's power are continually expressed through their control over food resources, their complementarity or opposition in food roles, and their symbolic depiction in food imagery.[11]

The need to understand the link between behavior and meaning is a second important theme in this book. Understanding the complex ways in which social norms, cultural meanings, and economic realities underlie food habits is essential for making successful policy recommendations and for integrating indigenous knowledge into nutritional understandings. The contributors to this volume point the way toward further needed research about how sociocultural dimensions of food use can inform policy decisions. They suggest that nutritionists need to pay attention to sociocultural factors surrounding food, such as how food and drink are defined and categorized in the local language, before collecting data on food intake. Knowledge of cultural factors allows for the development of culturally appropriate messages regarding optimal nutrition and suggested dietary changes. The media and policymakers can benefit from the approaches presented here to communicate with the public about food and hunger issues. Contributors to this volume encourage policymakers and all of us to think critically about our most taken-for-granted assumptions surrounding food and lead us to deeper understandings of ourselves and others, and the relations between "us" and "them."

NOTES

We thank Marlie Wasserman, formerly Anthropology Editor at Routledge, now director of Rutgers University Press, for suggesting that we compile this reader. We are grateful to our students Karen Lindenberg (Millersville University) and Afsaneh Hojabri (York University) for their editorial and proffreading assistance. We thank, Jim Taggart and John Van Esterik, for their support in so many.

1. See Fisher 1954, 1983.
2. See also Spang's (1988) article about anthropologists' work on food during World War II.
3. Bourdieu 1984, Bryant et al. 1985, Douglas 1984, Farb and Armelagos 1980, Goody 1982, Harris 1986, Jerome 1980, Quandt and Ritenbaugh 1986.
4. Adams 1990, Bell 1985, Brumberg 1988, Bynum 1986, Curtin and Heldke 1992, Hinz 1991, Mennell 1985, Visser 1988, Weinberger 1987.
5. In addition to Fisher's spectacular corpus of memoirs and essays centered on food, there are a plethora of novels and short stories with food at their heart, most notably Esquivel's recent *Like Water For Chocolate* (1989).
6. Fisher (1954:350), from "How to Cook a Wolf."
7. See Young 1971, 1986 and Kahn 1986.
8. See De Garine and Pollock 1995 for a fascinating compendium of articles about fat across cultures and Beller 1977 for a holistic discussion of weight. See Sobal and Stunkard 1989 for a recent review of the obesity literature.
9. Archaeologists are increasingly turning specifically to a focus on foodways to interpret the past (Coe 1994, Cohen 1977, Hastorf and Johannessen 1993). A number of recent volumes illustrate the wide range of symbolic, materialist and ecological perspectives used in the field of nutritional anthropology: Bryant et al. 1985; Farb and Armelagos 1980; Harris 1986; Jerome 1980; Pelto et al. 1989; Quandt and Ritenbaugh 1986.
10. See Arnold 1988, Leghorn and Roodkowsky 1977, and Vaughan 1987.

11. Writings on women's use of food to attain power in a variety of cultural settings include Adams, 1990, Behar 1989, Bell 1985, Brumberg 1988, Bynum 1987, Charles and Kerr 1988, Chernin 1981, Counihan 1988, 1992, De Vault 1991, Friedlander 1978, Kahn 1986, Lawrence 1984, Massara 1989, Meigs 1984, Millman 1980, Styles 1980, Weismantel 1988.

WORKS CITED

Adams, Carol. 1990. *The Sexual Politics of Meat: A Feminist-Vegetarian Critical Theory*. New York: Continuum.

Arnold, David. 1988. *Famine: Social Crisis and Historical Change*. New York: Basil Blackwell.

Behar, Ruth. 1989. "Sexual Witchcraft, Colonialism, and Women's Powers: Views from the Mexican Inquisition." In *Sexuality and Marriage in Colonial Latin America*. Ed. Asuncíon Lavrin. Lincoln: University of Nebraska Press, pp. 178–206.

Bell, Rudolph M. 1985. *Holy Anorexia*. Chicago: University of Chicago Press.

Beller, Anne Scott (1977). *Fat and Thin: A Natural History of Obesity*. New York: Farrar, Straus and Giroux.

Bourdieu, Pierre. 1984. *Distinction*. Cambridge: Harvard University Press.

Bruch, Hilde. 1973. *Eating Disorders: Obesity, Anorexia Nervosa, and the Person Within*. New York: Basic Books.

———. 1978. *The Golden Cage: The Enigma of Anorexia Nervosa*. New York: Vintage.

Brumberg, Joan Jacobs. 1988. *Fasting Girls: The Emergence of Anorexia Nervosa as a Modern Disease*. Cambridge: Harvard University Press.

Bryant, Carol et al. 1985. *The Cultural Feast*. St. Paul: West Publishing Co.

Bynum, Caroline Walker. 1987. *Holy Feast and Holy Fast: The Religious Significance of Food to Medieval Women*. Berkeley: University of California Press.

Charles, Nickie, and Marion Kerr. 1988. *Women, Food and Families*. Manchester: Manchester University Press.

Chernin, Kim. 1981. *The Obsession: Reflections on the Tyranny of Slenderness*. New York: Harper and Row.

Coe, Sophie. 1994. *America's First Cuisines*. Austin: University of Texas Press.

Cohen, Mark. 1977. *Food in Prehistory*. New Haven: Yale University Press.

Counihan, Carole M. 1988. "Female Identity, Food and Power in Contemporary Florence." In *Anthropological Quarterly* 61, 2:51–62.

———. 1992. "Food Rules in the United States: Individualism, Control, and Hierarchy." In *Anthropological Quarterly* 65, 2:55–66.

Curtin, Deane W., and Lisa M. Heldke, eds. 1992. *Cooking, Eating, Thinking: Transformative Philosophies of Food*. Bloomington: Indiana University Press.

De Garine, Igor, and Nancy Pollock, eds. 1995. *The Social Aspects of Obesity*. New York: Gordon and Breach.

Dettwyler, Katherine. 1994. *Dancing Skeletons: Life and Death in West Africa*. Prospect Heights, IL: Waveland.

De Vault, Marjorie L. 1991. *Feeding the Family: The Social Organization of Caring as Gendered Work*. Chicago: University of Chicago Press.

Douglas, Mary. 1984. *Food in the Social Order: Studies of Food and Festivities in Three American Communities*. New York: Russell Sage Foundation.

Du Bois, Cora. 1941. "Attitudes toward Food and Hunger in Alor." In *Language, Culture, and Personality*. Eds. Leslie Spier, et al. Menasha, WI: Sapir Memorial Publication Fund, pp. 272–281.

Esquivel, Laura. 1989. *Like Water for Chocolate*. New York: Doubleday.

Farb, Peter, and George Armelagos. 1980. *Consuming Passions: The Anthropology of Eating*. Boston: Houghton-Mifflin.

Firth, Raymond. 1934. "The Sociological Study of Native Diet." In *Africa* 7, 4:401–414.

Fisher, M. F. K. 1954. *The Art of Eating*. Cleveland: World Publishing.

———. 1961. *A Cordial Water: A Garland of Odd and Old Receipts to Assuage the Ills of Man and Beast*. New York: North Point Press.

———. 1983. *As They Were*. New York: Vintage.

Fitchen, Janet M. 1988. "Hunger, Malnutrition and Poverty in the Contemporary United States: Some Observations on Their Social and Cultural Context." In *Food and Foodways* 2, 3: 309–333.

Fortes, Myer, and S. L. Fortes. 1936. "Food in the Domestic Economy of the Tallensi." In *Africa* 9, 2:237–276.

Friedlander, Judith. 1978. "Aesthetics of Oppression: Traditional Arts of Women in Mexico." In *Heresies* 1, 4:3–9.
Goody, Jack. 1982. *Cooking, Cuisine and Class: A Study in Comparative Sociology*. New York: Cambridge University Press.
Harris, Marvin. 1985. *Good to Eat: Riddles of Food and Culture*. New York: Simon Schuster.
———, and Eric Ross. 1987. *Food and Evolution*. Philadelphia: Temple University Press.
Hastorf, Christine, and Sissel Johannessen. 1993. "Pre-Hispanic Political Change and the Role of Maize in the Central Andes of Peru." In *American Anthropologist*, 95:115–138.
Hinz, Evelyn, ed. 1991. *Diet and Discourse: Eating, Drinking and Literature*. Special Issue of *Mosaic*. Winnipeg: University of Manitoba.
Jerome, Norge. 1980. *Nutritional Anthropology: Contemporary Approaches to Diet and Culture*. Pleasantville, NY: Redgrave Publishers.
Kahn, Miriam. 1986. *Always Hungry, Never Greedy: Food and the Expression of Gender in a Melanesian Society*. New York: Cambridge University Press.
Lappé, Frances Moore, and Joseph Collins. 1986. *World Hunger: Twelve Myths*. New York: Grove Press.
Lawrence, Marilyn. 1984. *The Anorexic Experience*. London: Women's Press.
Leghorn, Lisa, and Mary Roodkowsky. 1977. *Who Really Starves? Women and World Hunger*. New York: Friendship Press.
Lévi-Strauss, Claude. 1978. *The Origin of Table Manners*. London: Jonathan Cape.
Massara, Emily Bradley. 1989. *Que Gordita: A Study of Weight Among Women in a Puerto Rican Community*. New York: AMS Press.
Mauss, Marcel. 1967 (orig. 1925). *The Gift: Forms and Functions of Exchange in Archaic Societies*. New York: Norton.
Meigs, Anna S. 1984. *Food, Sex and Pollution: A New Guinea Religion*. New Brunswick, NJ: Rutgers University Press.
Mennell, Stephen. 1985. *All Manners of Food: Eating and Taste in England and France from the Middle Ages to the Present*. New York: Basil Blackwell.
Millman, Marcia. 1980. *Such a Pretty Face: Being Fat in America*. New York: Norton.
Mintz, Sidney W. 1985. *Sweetness and Power: The Place of Sugar in Modern History*. New York: Penguin.
Ohnuki-Tierney, Emiko. 1993. *Rice as Self: Japanese Identities through Time*. New Brunswick, NJ: Princeton University Press.
Pelto, Gretel et al. 1989. *Research Methods in Nutritional Anthropology*. New York: United Nations University Press.
Pollock, Nancy J. 1992. *These Roots Remain: Food Habits in Islands of the Central and Eastern Pacific since Western Contact*. Honolulu: Institute for Polynesian Studies.
Quandt, Sara, and Cheryl Ritenbaugh, eds. 1986. *Training Methods in Nutritional Anthropology*. Washington, D.C.: American Anthropological Association.
Richards, Audrey I. 1932. *Hunger and Work in a Savage Tribe*. London: Routledge.
———. 1939. *Land, Labour and Diet in Northern Rhodesia: An Economic Study of the Bemba Tribe*. Oxford: Oxford University Press.
Sahlins, Marshall. 1972. *Stone Age Economics*. Hawthorne, NY: Aldine.
Sobal, Jeffery, and Albert J. Stunkard. 1989. "Socioeconomic Status and Obesity: A Review of the Literature." *Psychological Bulletin* 105, 2:260–275.
Spang, Rebecca L. 1988. "The Cultural Habits of a Food Committee." In *Food and Foodways* 2, 4: 359–391.
Vaughan, Megan. 1987. *The Story of an African Famine: Gender and Famine in Twentieth Century Malawi*. New York: Cambridge University Press.
Visser, Margaret. 1988. *Much Depends on Dinner*. New York: Collier.
Weinberger, David. 1987. "Philosophical Vegetarianism." In *Food and Foodways* 2, 1: 81–92.
Weismantel, M. J. 1988. *Food, Gender and Poverty in the Ecuadorian Andes*. Philadelphia: University of Pennsylvania Press.
Young, Michael W. 1971. *Fighting with Food: Leadership, Values and Social Control in a Massim Society*. Cambridge: Cambridge University Press.
———. 1986. "'The Worst Disease': The Cultural Definition of Hunger in Kalauna." In *Shared Wealth and Symbol: Food, Culture and Society in Oceania and Southeast Asia*, ed. Lenore Manderson. New York: Cambridge University Press.

Food, Meaning, and Voice

We begin this section with Margaret Mead's essay, "The Changing Significance of Food." This pithy piece reveals Mead's famous ability to get to the heart of a matter and inspire others to further research that was the hallmark of her career. Mead, following Audrey Richards' pioneering work, was one of the earliest anthropologists to articulate the centrality of foodways to human culture and thus to social science.

We have included the classic articles by Barthes and Lévi-Strauss on food's ability to convey meaning. While many have critiqued the specifics of Lévi-Strauss's "culinary triangle," and he himself later revised his formulations, this piece remains a classic structuralist statement. Barthes's ruminations on "the psychosociology of contemporary food consumption" also constitute a seminal account of the semiotic and symbolic power of foodways. Both Mary Douglas and Jean Soler build on Barthes and Lévi-Strauss to create divergent and fascinating explanations of Jewish dietary law, a much-debated topic in the field of food and culture. Marvin Harris rejects semiotic explanations of the abomination of pigs and offers a cultural-materialist explanation based on economic and ecological utility. The question of why Jews do not eat pork and other foods, even when these are available and "edible," is a test case for exploring the apparently whimsical gustatory selectivity of all human groups. It also provides a wonderful example of how the same cultural phenomenon can be explained from three different theoretical viewpoints, all of which are in some ways satisfactory but in some ways incomplete. Presenting three diverse accounts allows readers to evaluate whether symbolic, semiotic, or materialist explanations best account for Jewish and other dietary choices. Anderson closes this section with a discussion of the beliefs and practices surrounding the relationship between food, health, and illness in Chinese culture.

1

The Changing Significance of Food

MARGARET MEAD

We live in a world today where the state of nutrition in each country is relevant and important to each other country, and where the state of nutrition in the wealthy industrialized countries like the United States has profound significance for the role that such countries can play in eliminating famine and providing for adequate nutrition throughout the world. In a world in which each half knows what the other half does, we cannot live with hunger and malnutrition in one part of the world while people in another part are not only well nourished, but over-nourished. Any talk of one world, of brotherhood, rings hollow to those who have come face to face on the television screen with the emaciation of starving children and to the people whose children are staring as they pore over month-old issues of glossy American and European magazines, where full color prints show people glowing with health, their plates piled high with food that glistens to match the shining textures of their clothes. Peoples who have resolutely tightened their belts and put up with going to bed hungry, peoples who have seen their children die because they did not have the strength to resist disease, and called it fate or the will of God, can no longer do so, in the vivid visual realization of the amount and quality of food eaten—and wasted—by others.

Through human history there have been many stringent taboos on watching other people eat, or on eating in the presence of others. There have been attempts to explain this as a relationship between those who are involved and those who are not simultaneously involved in the satisfaction of a bodily need, and the inappropriateness of the already satiated watching others who appear—to the satisfied—to be shamelessly gorging. There is undoubtedly such an element in the taboos, but it seems more likely that they go back to the days when food was so scarce and the onlookers so hungry that not to offer them half of the little food one had was unthinkable, and every glance was a plea for at least a bite.

In the rural schools of America when my grandmother was a child, the better-off children took apples to school and, before they began to eat them, promised the poor children who had no apples that they might have the cores. The spectacle of the poor in rags at the rich man's gate and of hungry children pressing their noses against the glass window of the rich man's restaurant have long been invoked to arouse human compassion. But until the advent of the mass media and travel, the sensitive and sympathetic could protect themselves by shutting themselves away from the sight of the starving, by gifts of food to the poor on religious holidays, or perpetual bequests for the distribution of a piece of meat "the size of a child's head" annually. The starving in India and China saw only a few feasting foreigners and could not know how well or ill the poor were in countries from which they came. The proud poor hid their hunger behind a facade that often included insistent hospitality to the occasional visitor; the beggars flaunted their hunger and so, to a degree, discredited the hunger of their respectable compatriots.

But today the articulate cries of the hungry fill the air channels and there is no escape from the knowledge of the hundreds of millions who are seriously malnourished, of the periodic famines that beset whole populations, or of the looming danger of famine in many other parts of the world. The age-old divisions between one part of the world and another, between one class and another, between the rich and the poor everywhere, have been broken down, and the tolerances and insensitivities of the past are no longer possible.

But it is not only the media of communication which can take a man sitting at an overloaded breakfast table straight into a household where some of the children are too weak to stand. Something else, something even more significant, has happened. Today, for the first time in the history of mankind, we have the productive capacity to feed everyone in the world, and the technical knowledge to see that their stomachs are not only filled but that their bodies are properly nourished with the essential ingredients for growth and health. The progress of agriculture—in all its complexities of improved seed, methods of cultivation, fertilizers and pesticides, methods of storage, preservation, and transportation—now make it possible for the food that is needed for the whole world to be produced by fewer and fewer farmers, with greater and greater certainty. Drought and flood still threaten, but we have the means to prepare for and deal with even mammoth shortages—if we will. The progress of nutritional science has matched the progress of agriculture; we have finer and finer-grained knowledge of just which substances—vitamins, minerals, proteins—are essential, especially to growth and full development, and increasing ability to synthesize many of them on a massive scale.

These new twentieth-century potentialities have altered the ethical position of the rich all over the world. In the past, there were so few who lived well, and so many who lived on the edge of starvation, that the well-to-do had a rationale and indeed almost a necessity to harden their hearts and turn their eyes away. The jewels of the richest rajah could not have purchased enough food to feed his hungry subjects for more than a few days; the food did not exist, and the knowledge of how to use it was missing also. At the same time, however real the inability of a war-torn and submarine-ringed Britain to respond to the famine in Bengal, this inability was made bearable in Britain only by the extent to which the British were learning how to share what food they had among all the citizens, old and young. "You do not know," the American consul, who had come to Manchester from Spain, said to me: "you do not know what it means to live in a country where no child has to cry itself to sleep with hunger." But this was only achieved in Britain in the early 1940s. Before the well-fed turned away their eyes, in the feeling that they were powerless to alleviate the perennial poverty and hunger of most of their own people and the peoples in their far-flung commonwealth. And such turning away the eyes, in Britain and in the United States and elsewhere, was accompanied by the rationalizations, not only of the inability of the well-to-do—had they given all their wealth—to feed the poor, but of the undeservingness of the poor, who had they only been industrious and saving would have had enough, although of course of a lower quality, to keep "body and soul together."

When differences in race and in cultural levels complicated the situation, it was only too easy to insist that lesser breeds somehow, in some divinely correct scheme, would necessarily be less well fed, their alleged idleness and lack of frugality combining with such matters as sacred cows roaming over the landscapes—in India—or nights spent in the pub or the saloon—at home in Britain or America—while fathers drank up their meager pay checks and their children starved. So righteous was the assumed association between industriousness and food that, during the Irish famine, soup kitchens were set up out of town so that the starving could have the moral advantage of a long walk to receive the ration that stood between them and death. (The modern version of such ethical acrobatics can be found in the United States, in the mid-1960s, where food stamps were so expensive, since they had to be bought in large amounts, that only those who have been extraordinary frugal, saving, and lucky could afford to buy them and obtain the benefits they were designed to give.)

The particular ways in which the well-to-do of different great civilizations have rationalized the contrast between rich and poor have differed dramatically, but ever since the agricultural revolution, we have been running a race between our capacity to produce enough food to make it possible to assemble great urban centers, outfit huge armies and armadas, and build and elaborate the institutions of civilization and our ability to feed and care for the burgeoning population which has always kept a little, often a great deal, ahead of the food supply.

In this, those societies which practiced agriculture contrasted with the earlier simpler societies in which the entire population was engaged in subsistence activities. Primitive peoples may be well or poorly fed, feasting seldom, or blessed with ample supplies of fish or fruit, but the relations between the haves and the have-nots were in many ways simpler. Methods by which men could obtain permanent supplies of food and withhold them from their fellows hardly existed. The sour, barely edible breadfruit mash which was stored in breadfruit pits against the ravages of hurricanes and famines in Polynesia was not a diet for the table of chiefs but a stern measure against the needs of entire communities. The chief might have a right to the first fruits, or to half the crop, but after he had claimed it, it was redistributed to his people. The germs of the kinds of inequities that later entered the world were present: there was occasional conspicuous destruction of food, piled up for prestige, oil poured on the flames of self-glorifying feasts, food left to rot after it was offered to the gods. People with very meager food resources might use phrases that made it seem that each man was the recipient of great generosity on the part of his fellow, or on the other hand always to be giving away a whole animal, and always receiving only small bits.

The fear of cannibalism that hovered over northern peoples might be elaborated into cults of fear, or simply add to the concern that each member of a group had for all, against the terrible background that extremity might become so great that one of the group might in the end be sacrificed. But cannibalism could also be elaborated into a rite of vengeance or the celebration of victories in war, or even be used to provision an army in the field. Man's capacity to elaborate man's inhumanity to man existed before the beginning of civilization, which was made possible by the application of an increasingly productive technology to the production of food.

With the rise of civilizations, we also witness the growth of the great religions that made the brotherhood of all men part of their doctrine and the gift of alms or the life of voluntary poverty accepted religious practices. But the alms were never enough, and the life of individual poverty and abstinence was more efficacious for the individual's salvation than for the well-being of the poor and hungry, although both kept alive an ethic, as yet impossible of fulfillment, that it was right that all should be fed. The vision preceded the capability.

But today we have the capability. Whether that capability will be used or not becomes not a technical but an ethical question. It depends, in enormous measure, on the way in which the rich, industrialized countries handle the problems of distribution, or malnutrition and hunger, within their own borders. Failure to feed their own, with such high capabilities and such fully enunciated statements of responsibility and brotherhood, means that feeding the people of other countries is almost ruled out, except for sporadic escapist pieces of behavior where people who close their eyes to hunger in Mississippi can work hard to send food to a "Biafra." The development of the international instruments to meet food emergencies and to steadily improve the nutrition of the poorer countries will fail, unless there is greater consistency between ideal and practice at home.

And so, our present parlous plight in the United States, with the many pockets of rural unemployment, city ghettos, ethnic enclaves, where Americans are starving and an estimated tenth of the population malnourished, must be viewed not only in its consequences for ourselves, as a viable political community, but also in its consequences for the world. We need to examine not only the conditions that make this possible, to have starving people in the richest country in the world, but also the repercussions of American conditions on the world scene.

Why, when twenty-five years ago we were well on the way to remedying the state of the American people who had been described by presidential announcement as "one third ill-housed, ill-clothed, and ill-fed," when the vitamin deficiency diseases had all but vanished, and a variety of instruments for better nutrition had been developed, did we find, just two short years ago, due to the urgent pleading of a few crusaders, that we have fallen so grievously behind? The situation is complex, closely related to a series of struggles for regional and racial justice, to the spread of automation and resulting unemployment, to changes in crop economies, as well as to population growth and the inadequacy of many of our institutions to deal with it. But I wish to single out here two conditions which have, I believe, seriously contributed to our blindness to what was happening: the increase in the diseases of affluence and the growth of commercial agriculture.

In a country pronounced only twenty years before to be one third ill-fed, we suddenly began to have pronouncements from nutritional specialists that the major nutritional disease of the American people was overnutrition. If this had simply meant overeating, the old puritan ethics against greed and gluttony might have been more easily invoked, but it was overnutrition that was at stake. And this in a country where our idea of nutrition had been dominated by a dichotomy which distinguished food that was "good for you, but not good" from food that was "good, but not good for you." This split in man's needs, into our cultural conception of the need for nourishment and the search for pleasure, originally symbolized in the rewards for eating spinach or finishing what was on one's plate if one wanted to have a dessert, lay back of the movement to produce, commercially, nonnourishing foods. Beverages and snacks came in particularly for this demand, as it was the addition of between-meal eating to the three square, nutritionally adequate meals a day that was responsible for much of the trouble.

We began manufacturing, on a terrifying scale, foods and beverages that were guaranteed not to nourish. The resources and the ingenuity of industry were diverted from the preparation of foods necessary for life and growth to foods nonexpensive to prepare, expensive to buy. And every label reassuring the buyer that the product was not nourishing increased our sense that the trouble with Americans was that they were too well nourished. The diseases of affluence, represented by new forms of death in middle-age, had appeared before we had, in the words of Jean Mayer, who has done so much to define the needs of the country and of the world, conquered the diseases of poverty—the ill-fed pregnant women and lactating women, sometimes resulting in irreversible damage to the ill-weaned children, the school children so poorly fed that they could not learn.

It was hard for the average American to believe that while he struggled, and paid, so as not to be overnourished, other people, several millions, right in this country, were hungry and near starvation. The gross contradiction was too great. Furthermore, those who think of their country as parental and caring find it hard to admit that this parental figure is starving their brothers and sisters. During the great depression of the 1930s, when thousands of children came to school desperately hungry, it was very difficult to wring from children the admission that their parents had no food to give them. "Or what man is there of you, whom, if his son ask bread, will he give a stone?"

So today we have in the United States a situation not unlike the situation in Germany under Hitler, when a large proportion of the decent and law-abiding simply refuse to believe that what is happening can be happening. "Look at the taxes we pay," they say, or they point to the millions spent on welfare; surely with such quantities assigned to the poor, people can't be really hungry, or if they are, it is because they spend their money on TV sets and drink. How can the country be overnourished and undernourished at the same time?

A second major shift, in the United States and in the world, is the increasing magnitude of commercial agriculture, in which food is seen not as food which nourishes men, women, and children, but as a staple crop on which the prosperity of a country or region and the economic prosperity—as opposed to the simple livelihood—of the individual farmer depend. This is pointed up on a world scale in the report of the Food and Agriculture Organization of the United Nations for 1969, which states that there are two major problems in the world: food deficits in the poor countries, which mean starvation, hunger, and malnutrition on an increasing scale, and food surpluses in the industrialized part of the world, serious food surpluses.

On the face of it, this sounds as foolish as the production of foods guaranteed not to nourish, and the two are not unrelated. Surpluses, in a world where people are hungry! Too much food, in a world where children are starving! Yet we lump together all *agricultural* surpluses, such as cotton and tobacco, along with food, and we see these surpluses as threatening the commercial prosperity of many countries, and farmers in many countries. And in a world politically organized on a vanishing agrarian basis, this represents a political threat to those in power. However much the original destruction of food, killing little pigs, may have been phrased as relieving the desperate situation of little farmers or poor countries dependent upon single crop exports, such situations could not exist if food as something which man needs to provide growth and maintenance had not been separated from food as a cash crop, a commercial as opposed to a basic maintenance enterprise. When it becomes the task of government to foster the economic prosperity of an increasingly small, but politically influential, sector of the electorate at the expense of the well-being of its own and other nations' citizens, we have reached an ethically dangerous position.

And this situation, in the United States, is in part responsible for the previous state of our poor and hungry and for the paralysis that still prevents adequate political action. During the great depression, agriculture in this country was still a viable way of life for millions. The Department of Agriculture had responsibility, not only for food production and marketing, but also for the well-being from the cradle to the grave, in the simplest, most human sense, of every family who lived in communities under 2,500. Where the needs of urban man were parceled out among a number of agencies—Office of Education, Children's Bureau, Labor Department—there was still a considerable amount of integration possible in the Department of Agriculture, where theory and practices of farm wives, the education of children and youth, the questions of small loans for small landowners all could be considered together. It was in the Department of Agriculture that concerned persons found, during the depression, the kind of understanding of basic human needs which they sought.

There were indeed always conflicts between the needs of farmers to sell crops and the needs of children to be fed. School lunch schemes were tied to the disposal of surplus commodities. But the recognition of the wholeness of human needs was still there, firmly related to the breadth of the responsibilities of the different agencies within the Department of Agriculture. Today this is no longer so. Agriculture is big business in the United States. The subsidies used to persuade farmers to withdraw their impoverished land from production, like the terrible measures involving the slaughter of little pigs, are no longer ways of helping the small farmer on a family farm. The subsidies go to the rich commercial farmers, many of them the inheritors of old exploitive plantation traditions, wasteful of manpower and land resources, often in the very countries where the farm workers, displaced by machinery, are penniless, too poor to move away, starving. These subsidies exceed the budget of the antipoverty administration.

So today, many of the reforms which are suggested, in the distribution of food or distribution of income from which food can be bought, center on removing food relief programs from the Department of Agriculture and placing them under the Department of Health, Education, and Welfare. In Britain, during World War II, it was necessary to have a Ministry of Food, concerned primarily in matching the limited food supplies with basic needs.

At first sight, this proposal is sound enough. Let us remove from an agency devoted to making a profit out of crops that are treated like any other manufactured product the responsibility for seeing that food actually feeds people. After all, we do not ask clothing manufacturers to take the responsibility for clothing people, or the house-building industry for housing them. To the extent that we recognize them at all, these are the responsibilities of agencies of government which provide the funds to supplement the activities of private industry. Why not also in food? The Department of Health, Education, and Welfare is concerned with human beings; they have no food to sell on a domestic or world market and no constituents to appease. And from this step it is simply a second step to demand that the whole system of distribution be re-oriented, that a basic guaranteed annual income be provided each citizen, on the one hand, and that the government police standards, on behalf of the consumer, on the other.

But neither of these changes, shifting food relief programs from Agriculture to Health, Education, and Welfare, or shifting the whole welfare program into a guaranteed income, really meet the particular difficulties that arise because we are putting food into two compartments with disastrous effects; we are separating food that nourishes people from food out of which some people, and some countries, derive their incomes. It does not deal with the immediacy of the experience of food by the well-fed, or with the irreparability of food deprivation during prenatal and postnatal growth, deprivation that can never be made up. Human beings have maintained their dignity in incredibly bad conditions of housing and clothing, emerged triumphant from huts and log cabins, gone from ill-shod childhood to Wall Street or the Kremlin. Poor housing and poor clothing are demeaning to the human spirit when they contrast sharply with the visible standards of the way others live.

But food affects not only man's dignity but the capacity of children to reach their full potential, and the capacity of adults to act from day to day. You can't eat either nutrition or part of a not yet realized guaranteed annual income, or political promises. You can't eat hope. We know that hope and faith have enormous effects in preventing illness and enabling people to put forth the last ounce of energy they have. But energy is ultimately dependent upon food. No amount of rearrangement of priorities in the future can provide food in the present. It is true that the starving adult, his efficiency enormously impaired by lack of food, may usually be brought back again to his previous state of efficiency. But this is not true of children. What they lose is lost for good.

What we do about food is therefore far more crucial, both for the quality of the next generation, our own American children, and children everywhere, and also for the quality of our responsible action in every field. It is intimately concerned with the whole problem of the pollution and exhaustion of our environment, with the danger that man may make this planet uninhabitable within a short century or so. If food is grown in strict relationship to the needs of those who will eat it, if every effort is made to reduce the costs of transportation, to improve storage, to conserve the land, and there, where it is needed, by recycling wastes and water, we will go a long way toward solving many of our environmental problems also. It is as a responsible gardener on a small, limited plot, aware of the community about him with whom he will face adequate food or famine, that man has developed what conserving agricultural techniques we have.

Divorced from its primary function of feeding people, treated simply as a commercial commodity, food loses this primary significance; the land is mined instead of replenished and conserved. The Food and Agriculture Organization, intent on food production, lays great stress on the increase in the use of artificial fertilizers, yet the use of such fertilizers with their diffuse runoffs may be a greater danger to our total ecology than the industrial wastes from other forms of manufacturing. The same thing is true of pesticides. With the marvels of miracle rice and miracle wheat, which have brought the resources of international effort and scientific resources together, go at present prescriptions for artificial fertilizer and pesticides. The innovative industrialized countries are exporting, with improved agricultural methods, new dangers to the environment of the importing countries. Only by treating food, unitarily, as a substance necessary to feed people, subject first to the needs of people and only second to the needs of commercial prosperity—whether they be the needs of private enterprise or of a developing socialist country short of foreign capital—can we hope to meet the ethical demands that our present situation makes on us. For the first time since the beginning of civilization, we can feed everyone, now. Those who are not fed will die or, in the case of children, be permanently damaged.

We are just beginning to develop a world conscience. Our present dilemma is due to previous humanitarian moves with unanticipated effects. Without the spread of public health measures, we would not have had the fall in infant death rates which has resulted in the population explosion. Without the spread of agricultural techniques, there would not have been the food to feed the children who survived. The old constraints upon population growth—famine, plague, and war—are no longer acceptable to a world whose conscience is just barely stirring on behalf of all mankind. As we are groping our way back to a new version of the full fellow-feeling and respect for the natural world which the primitive Eskimo felt when food was scarce, so we are trembling on the edge of a new version of the sacrifice to cannibalism of the weak, just as we have the technical means to implement visions of responsibility that were very recently only visions.

The temptation is to turn aside, to deny what is happening to the environment, to trust to the "green revolution" and boast of how much rice previously hungry countries will export, to argue about legalities while people starve and infants and children are irreparably damaged, to refuse to deal with the paradoxes of hunger in plenty, and the coincidences of starvation and overnutrition. The basic problem is an ethical one; the solution of ethical problems can be solved only with a full recognition of reality. The children of the agricultural workers of the rural South, displaced by the machine, are hungry; so are the children in the Northern cities to which black and white poor have fled in search of food. On our American Indian reservations, among the Chicanos of California and the Southwest, among the seasonally employed, there is hunger now. If this hunger is not met now, we disqualify ourselves, we cripple ourselves, to deal with world problems.

We must balance our population so that every child that is born can be well fed. We must cherish our land, instead of mining it, so that food produced is first related to those who need it; and we must not despoil the earth, contaminate, and pollute it in the interests of immediate gain. Behind us, just a few decades ago, lies the vision of André Mayer and John Orr, the concepts of a world food bank, the founding of the United Nations Food and Agriculture Organization; behind us lie imaginative vision and deep concern. In the present we have new and various tools to make that vision into concrete actuality. But we must resolve the complications of present practice and present conceptions if the very precision and efficiency of our new knowledge is not to provide a stumbling block to the exercise of fuller humanity.

NOTES

The Proctor Prize given by the Scientific Research Society of America (RESA) was designed in 1969 to fit in with the emphasis on hunger and malnutrition in the program of the annual meetings in Boston of the American Association for the Advancement of Science. The subject, "The Changing Significance of Food," was selected before the White House Conference on "Food, Nutrition, and Health" was announced, and therefore the Association-wide symposia, of which the symposium based on the RESA lecture was one, came as a follow-up of the White House Conference rather than as a prelude. The chairman of the session was Dr. Jean Mayer, whose vision had piloted the White House Conference through very troubled waters, and the panel consisted of Mr. Robert Choate, crusader for a recognition of hunger in America; Nick Kotz, whose book *Let Them Eat Promises* was to appear in two weeks; Dr. Effie Ellis, Director of Maternal and Child Health, Ohio State Department of Health; and Mrs. L. C. Dorsey of the North Bolivar County Farmers' Cooperative Inc., of Mound Bayou, Mississippi, to which Dr. Mead had asked that the RESA prize money be given in recognition of an effort which promised increasing strength and experience for black people in the rural South and which was also directly related to providing food now.

During World War II, Margaret Mead was executive secretary of the Committee on Food Habits of the National Research Council, where she developed a theoretical and practical background for the relationship between the behavioral sciences and the nutritional sciences. She has twice brought this material up to date: in *Cultural Patterns and Technical Change*, which she edited, published by UNESCO, in 1955, and in *Food Habits Research: Problems of the 1950s, Special Publication 1225 of the National Research Council,* 1964, Address: American Museum of Natural History, New York, NY 10024.

2

Toward a Psychosociology of Contemporary Food Consumption

ROLAND BARTHES

The inhabitants of the United States consume almost twice as much sugar as the French.[1] Such a fact is usually a concern of economics and politics. But this is by no means all. One needs only to take the step from sugar as merchandise, an abstract item in accounts, to sugar as food, a concrete item that is "eaten" rather than "consumed," to get an inkling of the (probably unexplored) depth of the phenomenon. For the Americans must do something with all that sugar. And as a matter of fact, anyone who has spent time in the United States knows that sugar permeates a considerable part of American cooking; that it saturates ordinarily sweet foods, such as pastries; makes for a great variety of sweets served, such as ice creams, jellies, syrups; and is used in many dishes that French people do not sweeten, such as meats, fish, salads, and relishes. This is something that would be of interest to scholars in fields other than economics, to the psychosociologist, for example, who will have something to say about the presumably invariable relation between standard of living and sugar consumption. (But is this relation really invariable today? And if so, why?)[2] It could be of interest to the historian also, who might find it worthwhile to study the ways in which the use of sugar evolved as part of American culture (the influence of Dutch and German immigrants who were used to "sweet-salty" cooking?). Nor is this all. Sugar is not just a foodstuff, even when it is used in conjunction with other foods; it is, if you will, an "attitude," bound to certain usages, certain "protocols," that have to do with more than food. Serving a sweet relish or drinking a Coca-Cola with a meal are things that are confined to eating habits proper; but to go regularly to a dairy bar, where the absence of alcohol coincides with a great abundance of sweet beverages, means more than to consume sugar; through the sugar, it also means to experience the day, periods of rest, traveling, and leisure in a specific fashion that is certain to have its impact on the American. For who would claim that in France wine is only wine? Sugar or wine, these two superabundant substances are also institutions. And these institutions necessarily imply a set of images, dreams, tastes, choices, and values. I remember an American hit song: *Sugar Time*. Sugar is a time, a category of the world.[3]

I have started out with the example of the American use of sugar because it permits us to get outside of what we, as Frenchmen, consider "obvious." For we do not see our own food or, worse, we assume that it is insignificant. Even—or perhaps especially—to the scholar, the subject of food connotes triviality or guilt.[4] This may explain in part why the psychosociology of French eating habits is still approached only indirectly and in passing when more weighty subjects, such as life-styles, budgets, and advertising, are under discussion. But at least the sociologists, the historians of the present—since we are talking only about contemporary eating habits here—and the economists are already aware that there is such a thing.

Thus P. H. Chombart de Lauve has made an excellent study of the behavior of French working-class families with respect to food. He was able to define areas of frustration and to outline some of the mechanisms by which needs are transformed into values, necessities into alibis.[5] In her book *Le mode de vie des familles bourgeoises de 1873 à 1953,* M. Perrot came to the conclusion that economic factors played a less important role in the changes that have taken place in middle-class food habits in the last hundred years than changing tastes; and this really means ideas, especially about nutrition.[6] Finally, the development of advertising has enabled the economists to become quite conscious of the ideal nature of consumer goods; by now everyone knows that the product as bought—that is, experienced—by the consumer is by no means the real product; between the former and the latter there is a considerable production of false perceptions and values. By being faithful to a certain brand and by justifying this loyalty with a set of "natural" reasons, the consumer gives diversity to products that are technically so identical that frequently even the manufacturer cannot find any differences. This is notably the case with most cooking oils.[7]

It is obvious that such deformations or reconstructions are not only the manifestation of individual, anomic prejudices, but also elements of a veritable collective imagination showing the outlines of a certain mental framework. All of this, we might say, points to the (necessary) widening of the very notion of food. For what is food? It is not only a collection of products that can be used for statistical or nutritional studies. It is also, and at the same time, a system of communication, a body of images, a protocol of usages, situations, and behavior. Information about food must be gathered wherever it can be found: by direct observation in the economy, in techniques, usages, and advertising; and by indirect observation in the mental life of a given society.[8] And once these data are assembled, they should no doubt be subjected to an internal analysis that should try to establish what is significant about the way in which they have been assembled before any economic or even ideological determinism is brought into play. I should like to give a brief outline of what such an analysis might be.

When he buys an item of food, consumes it, or serves it, modern man does not manipulate a simple object in a purely transitive fashion; this item of food sums up and transmits a situation; it constitutes an information; it signifies. That is to say that it is not just an indicator of a set of more or less conscious motivations, but that it is a real sign, perhaps the functional unit of a system of communication. By this I mean not only the elements of *display* in food, such as foods involved in rites of hospitality,[9] for all food serves as a sign among the members of a given society. As soon as a need is satisfied by standardized production and consumption, in short, as soon as it takes on the characteristics of an institution, its function can no longer be dissociated from the sign of that function. This is true for clothing;[10] it is also true for food. No doubt, food is, anthropologically speaking (though very much in the abstract), the first need; but ever since man has ceased living off wild berries, this need has been highly structured. Substances, techniques of preparation, habits, all become part of a system of differences in signification; and as soon as this happens, we have communication by way of food. For the fact that there is communication is proven, not by the more or less vague consciousness that its users may have of it, but by the ease with which all the facts concerning food form a structure analogous to other systems of communication.[11] People may very well continue to believe that food is an immediate reality (necessity or pleasure), but this does not prevent it from carrying a system of communication; it would not be the first thing that people continue to experience as a simple function at the very moment when they constitute it into a sign.

If food is a system, what might be its constituent units? In order to find out, it would obviously be necessary to start out with a complete inventory of all we know of the food in a given society (products, techniques, habits), and then to subject these facts to what the linguists call transformational analysis, that is, to observe whether the passage from one fact to another produces a difference in signification. Here is an example: the changeover from ordinary bread to *pain de mie* involves a difference in what is signified: the former signifies day-to-day life, the latter a party. Similarly, in contemporary terms, the changeover from white to brown bread corresponds to a change in what is signified in social terms, because, paradoxically, brown bread has become a sign of refinement. We are therefore justified in considering the varieties of bread as units of signification—at least these varieties—for the same test can also show that there are insignificant varieties as well, whose use has nothing to do with a collective institution, but simply with individual taste. In this manner, one could, proceeding step by step, make a compendium of the differences in signification regulating the system of our food. In other words, it would be a matter of separating the significant from the insignificant and then of reconstructing the differential system of signification by constructing, if I may be permitted to use such a metaphor, a veritable grammar of foods.

It must be added that the units of our system would probably coincide only rarely with the products in current use in the economy. Within French society, for example, bread as such does not constitute a signifying unit: in order to find these we must go further and look for certain of its varieties. In other words, these signifying units are more subtle than the commercial units and, above all, they have to do with subdivisions with which production is not concerned, so that the sense of the subdivision can differentiate a single product. Thus it is not at the level of its cost that the sense of a food item is elaborated, but at the level of its preparation and use. There is perhaps no brute item of food that signifies anything in itself, except for a few deluxe items such as salmon, caviar, truffles, and so on, whose preparation is less important than their absolute cost.

If the units of our system of food are not the *products* of our economy, can we at least have some preliminary idea of what they might be? In the absence of a systematic inventory, we may risk a few hypotheses. A study by P. F. Lazarsfeld[12] (it is old, concerned with particulars, and I cite it only as an example) has shown that certain sensorial "tastes" can vary according to the income level of the social groups interviewed: lower-income persons like sweet chocolates, smooth materials, strong perfumes; the upper classes, on the other hand, prefer bitter substances, irregular materials, and light perfumes. To remain within the area of food, we can see that signification (which, itself, refers to a twofold social phenomenon: upper classes/lower classes) does not involve kinds of products, but flavors: *sweet* and *bitter* make up the opposition in signification, so that we must place certain units of the system of food on that level. We can imagine other classes of units, for example, opposite substances such as dry, creamy, watery ones, which immediately show their great psychoanalytical potential (and it is obvious that if the subject of food had not been so trivialized and invested with guilt, it could easily be subjected to the kind of "poetic" analysis that G. Bachelard applied to language). As for what is considered tasty, C. Lévi-Strauss has already shown that this might very well constitute a class of oppositions that refers to national characters (French versus English cuisine, French versus Chinese or German cuisine, and so on).[13]

Finally, one can imagine opposites that are even more encompassing, but also more subtle. Why not speak, if the facts are sufficiently numerous and sufficiently clear, of a certain "spirit" of food, if I may be permitted to use this romantic term? By this I mean that a coherent set of food traits and habits can constitute a complex but homogeneous dominant feature useful for defining a general system of tastes and habits. This "spirit" brings together different units (such as flavor or substance), forming a composite unit with a single signification, somewhat analogous to the suprasegmental prosodic units of language. I should like to suggest here two very different examples. The ancient Greeks unified in a single (euphoric) notion the ideas of succulence, brightness, and moistness, and they called it *yávos*. Honey had *yávos*, and wine was the *yávos* of the vineyard.[14] Now this would certainly be a signifying unit if we were to establish the system of food of the Greeks, even though it does not refer to any particular item. And here is another example, modern this time. In the United States, the Americans seem to oppose the category of sweet (and we have already seen to how many different varieties of foods this applies) with an equally general category that is not, however, that of salty—understandably so, since their food is salty and sweet to begin with—but that of *crisp* or *crispy. Crisp* designates everything that crunches, crackles, grates, sparkles, from potato chips to certain brands of beer; *crisp*—and this shows that the unit of food can overthrow logical categories—*crisp* may be applied to a product just because it is ice cold, to another because it is sour, to a third because it is brittle. Quite obviously, such a notion goes beyond the purely physical nature of the product; *crispness* in a food designates an almost magical quality, a certain briskness and sharpness, as opposed to the soft, soothing character of sweet foods.

Now then, how will we use the units established in this manner? We will use them to reconstruct systems, syntaxes ("menus"), and styles ("diets")[15] no longer in an empirical but in a semantic way—in a way, that is, that will enable us to compare them to each other. We now must show, not that which is, but that which signifies. Why? Because we are interested in human communication and because communication always implies a system of signification, that is, a body of discrete signs standing out from a mass of indifferent materials. For this reason, sociology must, as soon as it deals with cultural "objects" such as clothing, food, and—not quite as clearly—housing, structure these objects before trying to find out what society does with them. For what society does with them is precisely to structure them in order to make use of them.

To what, then, can these significations of food refer? As I have already pointed out, they refer not only to display,[16] but to a much larger set of themes and situations. One could say that an entire "world" (social environment) is present in and signified by food. Today we have a tool with which to isolate these themes and situations, namely, advertising. There is no question that advertising provides only a projected image of reality; but the sociology of mass communication has become increasingly inclined to think that large-scale advertising, even though technically the work of a particular group, reflects the collective psychology much more than it shapes it. Furthermore, studies of motivation are now so advanced that it is possible to analyze cases in which the response of the public is negative. (I already mentioned the feelings of guilt fostered by an advertising for sugar which emphasized pure enjoyment. It was bad advertising, but the response of the public was nonetheless psychologically most interesting.)

A rapid glance at food advertising permits us rather easily, I think, to identify three groups of themes. The first of these assigns to food a function that is, in some sense, commemorative: food permits a person (and I am here speaking of French themes) to partake each day of the national past. In this case, this historical quality is obviously linked to food techniques (preparation and cooking). These have long roots, reaching back to the depth of the French past. They are, we are told, the repository of a whole experience, of the accumulated wisdom of our ancestors. French food is never supposed to be innovative, except when it rediscovers long-forgotten secrets. The historical theme, which was so often sounded in our advertising, mobilizes two different values: on the one hand, it implies an aristocratic tradition (dynasties of manufacturers, *moutarde du Roy,* the Brandy of Napoleon); on the other hand, food frequently carries notions of representing the flavorful survival of an old, rural society that is itself highly idealized.[17] In this manner, food brings the memory of the soil into our very contemporary life; hence the paradoxical association of gastronomy and industrialization in the form of canned "gourmet dishes." No doubt the myth of French cooking abroad (or as expressed to foreigners) strengthens this "nostalgic" value of food considerably; but since the French themselves actively participate in this myth (especially when traveling), it is fair to say that through his food the Frenchman experiences a certain national continuity. By way of a thousand detours, food permits him to insert himself daily into his own past and to believe in a certain culinary "being" of France.[18]

A second group of values concerns what we might call the anthropological situation of the French consumer. Motivation studies have shown that feelings of inferiority were attached to certain foods and that people therefore abstained from them.[19] For example, there are supposed to be masculine and feminine kinds of food. Furthermore, visual advertising makes it possible to associate certain kinds of foods with images connoting a sublimated sexuality. In a certain sense, advertising eroticizes food and thereby transforms our consciousness of it, bringing it into a new sphere of situations by means of a pseudocausal relationship.

Finally, a third area of consciousness is constituted by a whole set of ambiguous values of a somatic as well as psychic nature, clustering around the concept of *health*. In a mythical way, health is indeed a simple relay midway between the body and the mind; it is the alibi food gives to itself in order to signify materially a pattern of immaterial realities. Health is thus experienced through food only in the form of "conditioning," which implies that the body is able to cope with a certain number of day-to-day situations. Conditioning originates with the body but goes beyond it. It produces *energy* (sugar, the "powerhouse of foods," at least in France, maintains an "uninterrupted flow of energy"; margarine "builds solid muscles"; coffee "dissolves fatigue"); *alertness* ("Be alert with Lustucru"), and *relaxation* (coffee, mineral water, fruit juices, Coca-Cola, and so on).

In this manner, food does indeed retain its physiological function by giving strength to the organism, but this strength is immediately sublimated and placed into a specific situation (I shall come back to this in a moment). This situation may be one of conquest (alertness, aggressiveness) or a response to the stress of modern life (relaxation). No doubt, the existence of such themes is related to the spectacular development of the science of nutrition, to which, as we have seen, one historian unequivocally attributes the evolution of food budgets over the last fifty years. It seems, then, that the acceptance of this new value by the masses has brought about a new phenomenon, which must be the first item of study in any psychosociology of food: it is what might be called nutritional consciousness. In the developed countries, food is henceforth *thought out*, not by specialists, but by the entire public, even if this thinking is done within a framework of highly mythical notions. Nor is this all. This nutritional rationalizing is aimed in a specific direction. Modern nutritional science (at least according to what can be observed in France) is not bound to any moral values, such as asceticism, wisdom, or purity,[20] but on the contrary, to values of *power*. The energy furnished by a consciously worked out diet is mythically directed, it seems, toward an adaptation of man to the modern world. In the final analysis, therefore, a representation of contemporary existence is implied in the consciousness we have of the function of our food.[21]

For, as we said before, food serves as a sign not only for themes, but also for situations; and this, all told, means for way of life that is emphasized, much more than expressed, by it. To eat is a behavior that develops beyond its own ends, replacing, summing up, and signalizing other behaviors, and it is precisely for these reasons that it is a sign. What are these other behaviors? Today, we might say all of them: activity, work, sports, effort, leisure, celebration—every one of these situations is expressed through food. We might almost say that this "polysemia" of food characterizes modernity; in the past, only festive occasions were signalized by food in any positive and organized manner. But today, work also has its own kind of food (on the level of a sign, that is): energy-giving and light food is experienced as the very sign of, rather than only a help toward, participation in modern life. The snack bar not only responds to a new need, it also gives a certain dramatic expression to this need and shows those who frequent it to be modern men, managers who exercise power and control over the extreme rapidity of modern life. Let us say that there is an element of "Napoleonism" in this ritually condensed, light, and rapid kind of eating. On the level of institutions, there is also the business lunch, a very different kind of thing, which has become commercialized in the form of special menus: here, on the contrary, the emphasis is placed on comfort and long discussions; there even remains a trace of the mythical conciliatory power of conviviality. Hence, the business lunch emphasizes the gastronomic, and under certain circumstances traditional, value of the dishes served and uses this value to stimulate the euphoria needed to facilitate the transaction of business. Snack bar and business lunch are two very closely related work situations, yet the food connected with them signalizes their differences in a perfectly readable manner. We can imagine many others that should be catalogued.

This much can be said already: today, at least in France, we are witnessing an extraordinary expansion of the areas associated with food: food is becoming incorporated into an ever-lengthening list of situations. This adaptation is usually made in the name of hygiene and better living, but in reality, to stress this fact once more, food is also charged with signifying the situation in which it is used. It has a twofold value, being nutrition as well as protocol, and its value as protocol becomes increasingly more important as soon as the basic needs are satisfied, as they are in France. In other words, we might say that in contemporary French society *food has a constant tendency to transform itself into situation.*

There is no better illustration for this trend than the advertising mythology about coffee. For centuries, coffee was considered a stimulant to the nervous system (recall that Michelet claimed that it led to the Revolution), but contemporary advertising, while not expressly denying this traditional function, paradoxically associates it more and more with images of "breaks," rest, and even relaxation. What is the reason for this shift? It is that coffee is felt to be not so much a substance[22] as a circumstance. It is the recognized occasion for interrupting work and using this respite in a precise protocol of taking sustenance. It stands to reason that if this transferral of the food substance to its use becomes really all-encompassing, the power signification of food will be vastly increased. Food, in short, will lose in substance and gain in function; this function will be general and point to activity (such as the business lunch) or to times of rest (such as coffee); but since there is a very marked opposition between work and relaxation, the traditionally festive function of food is apt to disappear gradually, and society will arrange the signifying system of its food around two major focal points: on the one hand, activity (and no longer work), and on the other hand, leisure (no longer celebration). All of this goes to show, if indeed it needs to be shown, to what extent food is an organic system, organically integrated into its specific type of civilization.

NOTES

This article originally appeared in "Vers une psycho-sociologie de l'alimentation moderne" by Roland Barthes, in *Annales: Économies, Sociétés, Civilisations* no. 5 (September–October 1961), pp. 977–986. Reprinted by permission of *Annales.*

1. Annual sugar consumption in the United States is 43 kg. per person; in France 25 kg. per person.
2. F. Charny, *Le sucre,* Collection "Que sais-je?" (Paris: P. U. F., 1950), p. 8.
3. I do not wish to deal here with the problem of sugar "metaphors" or paradoxes, such as the "sweet" rock singers or the sweet milk beverages of certain "toughs."
4. Motivation studies have shown that food advertisements openly based on enjoyment are apt to fail, since they make the reader feel guilty (J. Marcus-Steiff, *Les études de motivation* [Paris: Hermann, 1961], pp. 44–45).
5. P. H. Chombart de Lauve, *La vie quotidienne des familles ouvières* (Paris: C.N.R.S., 1956).
6. Marguerite Perrot, *Le mode de vie des familles bourgeoises,* 1873–1953 (Paris: Colin, 1961). "Since the end of the nineteenth century, there has been a very marked evolution in the dietary habits of the middle-class families we have investigated in this study. This evolution seems related, not to a change in the standard of living, but rather to a transformation of individual tastes under the influence of a greater awareness of the rules of nutrition" (p. 292).
7. J. Marcus-Steiff, *Les études de motivation,* p. 28.
8. On the latest techniques of investigation, see again J. Marcus-Steiff, *Les études de motivation.*
9. Yet on this point alone, there are many known facts that should be assembled and systematized: cocktail parties, formal dinners, degrees and kinds of display by way of food according to the different social groups.
10. R. Barthes, "Le bleu est à la mode cette année: Note sur la recherche des unités signifiantes dans le vêtement de mode," *Revue française de sociologie* 1 (1960): 147–162.
11. I am using the word *structure* in the sense that it has in linguistics: "an autonomous entity of internal dependencies" (L. Hjelnislev, *Essais linguistiques* [Copenhagen, 1959], p. 1).
12. P. F. Lazarsfeld, "The Psychological Aspect of Market Research," *Harvard Business Review* 13 (1934): 54–71.

13. C. Lévi-Strauss, *Anthropologie structurale* (Paris: Plon, 1958), p. 99.
14. H. Jeanmaire, *Dionysos* (Paris: Payot), p. 510.
15. In a semantic analysis, vegetarianism, for example (at least at the level of specialized restaurants), would appear as an attempt to copy the appearance of meat dishes by means of a series of artifices that are somewhat similar to "costume jewelry" in clothing, at least the jewelry that is meant to be seen as such.
16. The idea of social *display* must not be associated purely and simply with vanity; the analysis of motivation, when conducted by indirect questioning, reveals that worry about appearances is part of an extremely subtle reaction and that social strictures are very strong, even with respect to food.
17. The expression *cuisine bourgeoise,* used at first in a literal, then in a metaphoric way, seems to be gradually disappearing while the "peasant stew" is periodically featured in the photographic pages of the major ladies' magazines.
18. The exotic nature of food can, of course, be a value, but in the French public at large, it seems limited to coffee (tropical) and pasta (Italian).
19. This would be the place to ask just what is meant by "strong" food. Obviously, there is no psychic quality inherent in the thing itself. A food becomes "masculine" as soon as women, children, and old people, for nutritional (and thus fairly historical) reasons, do not consume it.
20. We need only to compare the development of vegetarianism in England and France.
21. Right now, in France, there is a conflict between traditional (gastronomic) and modern (nutritional) values.
22. It seems that this stimulating, re-energizing power is now assigned to sugar, at least in France.

3

The Culinary Triangle

CLAUDE LÉVI-STRAUSS

Linguistics has familiarized us with concepts like "minimum vocalism" and "minimum consonantism" which refer to systems of oppositions between phonemes of so elementary a nature that every known or unknown language supposes them; they are in fact also the first oppositions to appear in the child's language, and the last to disappear in the speech of people affected by certain forms of aphasia.

The two concepts are moreover not really distinct, since, according to linguists, for every language the fundamental opposition is that between consonant and vowel. The subsequent distinctions among vowels and among consonants result from the application to these derived areas of such contrasts as compact and diffuse, open and closed, acute and grave.

Hence, in all the languages of the world, complex systems of oppositions among phonemes do nothing but elaborate in multiple directions a simpler system common to them all: the contrast between consonant and vowel which, by the workings of a double opposition between compact and diffuse, acute and grave, produces on the one hand what has been called the "vowel triangle":[1]

a
u   i

and on the other hand the "consonant triangle":

k
p   t

It would seem that the methodological principle which inspires such distinctions is transposable to other domains, notably that of cooking which, it has never been sufficiently emphasized, is with language a truly universal form of human activity: if there is no society without a language, nor is there any which does not cook in some manner at least some of its food.

We will start from the hypothesis that this activity supposes a system which is located—according to very difficult modalities in function of the particular cultures one wants to consider—within a triangular semantic field whose three points correspond respectively to the categories of the raw, the cooked and the rotted. It is clear that in respect to cooking the raw constitutes the unmarked pole, while the other two poles are strongly marked, but in different directions: indeed, the cooked is a cultural transformation of the raw, whereas the rotted is a natural transformation. Underlying our original triangle, there is hence a double opposition between *elaborated/unelaborated* on the one hand, and *culture/nature* on the other.

No doubt these notions constitute empty forms: they teach us nothing about the cooking of any specific society, since only observation can tell us what each one means by "raw," "cooked" and "rotted," and we can suppose that it will not be the same for all. Italian cuisine has recently taught us to eat *crudités* rawer than any in traditional French cooking, thereby determining an enlargement of the category of the raw. And we know from some incidents that followed the Allied landings in 1944 that American soldiers conceived the category of the rotted in more extended fashion than we, since the odor given off by Norman cheese dairies seemed to them the smell of corpses, and occasionally prompted them to destroy the dairies.

Consequently, the culinary triangle delimits a semantic field, but from the outside. This is moreover true of the linguistic triangles as well, since there are no phonemes *a, i, u* (or *k, p, t)* in general, and these ideal positions must be occupied, in each language, by the particular phonemes whose distinctive natures are closest to those for which we first gave a symbolic representation: thus we have a sort of concrete triangle inscribed within the abstract triangle. In any cuisine, nothing is simply cooked, but must be cooked in one fashion or another. Nor is there any condition of pure rawness: only certain foods can really be eaten raw, and then only if they have been selected, washed, pared or cut, or even seasoned. Rotting, too, is only allowed to take place in certain specific ways, either spontaneous or controlled.

Let us now consider, for those cuisines whose categories are relatively well-known, the different modes of cooking. There are certainly two principal modes, attested in innumerable societies by myths and rites which emphasize their contrast: the roasted and the boiled. In what does their difference consist? Roasted food is directly exposed to the fire; with the fire it realizes an unmediated conjunction, whereas boiled food is doubly mediated, by the water in which it is immersed, and by the receptacle that holds both water and food.

On two grounds, then, one can say that the roasted is on the side of nature, the boiled on the side of culture: literally, because boiling requires the use of a receptacle, a cultural object; symbolically, in as much as culture is a mediation of the relations between man and the world, and boiling demands a mediation (by water) of the relation between food and fire which is absent in roasting.

The natives of New Caledonia feel this contrast with particular vividness: "Formerly," relates M. J. Barrau, "they only grilled and roasted, they only 'burned' as the natives now say . . . The use of a pot and the consumption of boiled tubers are looked upon with pride . . . as a proof of . . . civilization."

A text of Aristotle, cited by Salomon Reinach (*Cultes, Mythes, Religions*, V, p. 63), indicates that the Greeks also thought that "in ancient times, men roasted everything."

Behind the opposition between roasted and boiled, then, we do in fact find, as we postulated at the outset, the opposition between nature and culture. It remains to discover the other fundamental opposition which we put forth: that between elaborated and unelaborated.

In this respect, observation establishes a double affinity: the roasted with the raw, that is to say the unelaborated, and the boiled with the rotted, which is one of the two modes of the elaborated. The affinity of the roasted with the raw comes from the fact that it is never uniformly cooked, whether this be on all sides, or on the outside and the inside. A myth of the Wyandot Indians well evokes what might be called the paradox of the roasted: the Creator struck fire, and ordered the first man to skewer a piece of meat on a stick and roast it. But man was so ignorant that he left the meat on the fire until it was black on one side, and still raw on the other. . . . Similarly, the Poconachi of Mexico interpret the roasted as a compromise between the raw and the burned. After the universal fire, they relate, that which had not been burned became white, that which had been burned turned black, and what had only been singed turned red. This explanation accounts for the various colors of corn and beans. In British Guiana, the Waiwai sorcerer must respect two taboos, one directed at roast meat, the other red paint, and this again puts the roasted on the side of blood and the raw.

If boiling is superior to roasting, notes Aristotle, it is because it takes away the rawness of meat, "roast meats being rawer and drier than boiled meats" (quoted by Reinach, *loc. cit.*).

As for the boiled, its affinity with the rotted is attested in numerous European languages by such locutions as *pot pourri, olla podrida,* denoting different sorts of meat seasoned and cooked together with vegetables; and in German, *zu Brei zerkochetes Fleisch,* "meat rotted from cooking." American Indian languages emphasize the same affinity, and it is significant that this should be so especially in those tribes that show a strong taste for gamey meat, to the point of preferring, for example, the flesh of a dead animal whose carcass has been washed down by the stream to that of a freshly-killed buffalo. In the Dakota language, the same stem connotes putrefaction and the fact of boiling pieces of meat together with some additive.

These distinctions are far from exhausting the richness and complexity of the contrast between roasted and boiled. The boiled is cooked within a receptacle, while the roasted is cooked from without: the former thus evokes the concave, the latter the convex. Also the boiled can most often be ascribed to what might be called an "endo-cuisine," prepared for domestic use, destined to a small closed group, while the roasted belongs to "exo-cuisine," that which one offers to guests. Formerly in France, boiled chicken was for the family meal, while roasted meat was for the banquet (and marked its culminating point, served as it was after the boiled meats and vegetables of the first course, and accompanied by "extraordinary fruits" such as melons, oranges, olives and capers).

The same opposition is found, differently formulated, in exotic societies. The extremely primitive Guayaki of Paraguay roast all their game, except when they prepare the meat destined for the rites which determine the name of a new child: this meat must be boiled. The Caingang of Brazil prohibit boiled meat for the widow and widower, and also for anyone who has murdered an enemy. In all these cases, prescription of the boiled accompanies a tightening, prescription of the roasted a loosening of familial or social ties.

Following this line of argument, one could infer that cannibalism (which by definition is an endo-cuisine in respect to the human race) ordinarily employs boiling rather than roasting, and that the cases where bodies are roasted—cases vouched for by ethnographic literature—must be more frequent in exo-cannibalism (eating the body of an enemy) than in endo-cannibalism (eating a relative). It would be interesting to carry out statistical research on this point.

Sometimes, too, as is often the case in America, and doubtless elsewhere, the roasted and the boiled will have respective affinities with life in the bush (outside the village community) and sedentary life (inside the village). From this comes a subsidiary association of the roasted with men, the boiled with women. This is notably the case with the Trumai, the Yagua and the Jivaro of South America, and with the Ingalik of Alaska. Or else the relation is reversed: the Assiniboin, on the northern plains of North America, reserve the preparation of boiled food for men engaged in a war expedition, while the women in the villages never use receptacles, and only roast their meat. There are some indications that in certain Eastern European countries one can find the same inversion of affinities between roasted and boiled and feminine and masculine.

The existence of these inverted systems naturally poses a problem, and leads one to think that the axes of opposition are still more numerous than one suspected, and that the peoples where these inversions exist refer to axes different from those we at first singled out. For example, boiling conserves entirely the meat and its juices, whereas roasting is accompanied by destruction and loss. One connotes economy, the other prodigality; the former is plebeian, the latter aristocratic. This aspect takes on primary importance in societies which prescribe differences of status among individuals or groups. In the ancient Maori, says Prytz-Johansen, a noble could himself roast his food, but he avoided all contact with the steaming oven, which was left to the slaves and women of low birth. Thus, when pots and pans were introduced by the whites, they seemed infected utensils; a striking inversion of the attitude which we remarked in the New Caledonians.

These differences in appraisal of the boiled and the roasted, dependent on the democratic or aristocratic perspective of the group, can also be found in the Western tradition. The democratic Encyclopedia of Diderot and d'Alembert goes in for a veritable apology of the boiled: "Boiled meat is one of the most succulent and nourishing foods known to man. . . .One could say that boiled meat is to other dishes as bread is to other kinds of nourishment" (Article "Bouilli"). A half-century later, the dandy Brillat-Savarin will take precisely the opposite view: "We professors never eat boiled meat out of respect for principle, and because we have pronounced *ex cathedra* this incontestable truth: boiled meat is flesh without its juice. . . . This truth is beginning to become accepted, and boiled meat has disappeared in truly elegant dinners; it has been replaced by a roast filet, a turbot, or a matelote" (*Physiologie du goût,* VI, ¶2).

Therefore if the Czechs see in boiled meat a man's nourishment, it is perhaps because their traditional society was of a much more democratic character than that of their Slavonic and Polish neighbors. One could interpret in the same manner distinctions made—respectively by the Greeks, and the Romans and the Hebrews—on the basis of attitudes toward roasted and boiled, distinctions which have been noted by M. Piganiol in a recent article ("Le rôti et le bouilli," *A Pedro Bosch-Gimpera,* Mexico City, 1963).

Other societies make use of the same opposition in a completely different direction. Because boiling takes place without loss of substance, and within a complete enclosure, it is eminently apt to symbolize cosmic totality. In Guiana as well as in the Great Lakes region, it is thought that if the pot where game is boiling were to overflow even a little bit, all the animals of the species being cooked would migrate, and the hunter would catch nothing more. The boiled is life, the roasted death. Does not world folklore offer innumerable examples of the cauldron of immortality? But there has never been a spit of immortality. A Cree Indian rite admirably expresses this character of cosmic totality ascribed to boiled food. According to them, the first man was commanded by the Creator to boil the first berries gathered each season. The cup containing the berries was first presented to the sun, that it might fulfill its office and ripen the berries; then the cup was lifted to the thunder, whence rain is expected; finally the cup was lowered toward the earth, in prayer that it bring forth its fruits.

Hence we rejoin the symbolism of the most distant Indo-European past, as it has been reconstructed by Georges Dumézil: "To Mitra belongs that which breaks of itself, that which is cooked in steam, that which is well sacrificed, milk . . . and to Varuna that which is cut with the axe, that which is snatched from the fire, that which is ill-sacrificed, the intoxicating soma" (*Les dieux des Germains,* p. 60). It is not a little surprising—but highly significant—to find intact in genial mid-nineteenth-century philosophers of cuisine a consciousness of the same contrast between knowledge and inspiration, serenity and violence, measure and lack of measure, still symbolized by the opposition of the boiled and the roasted: "One becomes a cook but one is born a roaster" (Brillat-Savarin); "Roasting is at the same time nothing, and an immensity" (Marquis de Cussy).

Within the basic culinary triangle formed by the categories of raw, cooked and rotted, we have, then, inscribed two terms which are situated: one, the roasted, in the vicinity of the raw; the other, the boiled, near the rotted. We are lacking a third term, illustrating the concrete form of cooking showing the greatest affinity to the abstract category of the cooked. This form seems to us to be smoking, which like roasting implies an unmediated operation (without receptacle and without water) but differs from roasting in that it is, like boiling, a slow form of cooking, both uniform and penetrating in depth.

Let us try to determine the place of this new term in our system of opposition. In the technique of smoking, as in that of roasting, nothing is interposed between meat and fire except air. But the difference between the two techniques comes from the fact that in one the layer of air is reduced to a minimum, whereas in the other it is brought to a maximum. To smoke game, the American Indians (in whose culinary system smoking occupies a particularly important place) construct a wooden frame (a buccan) about five feet high, on top of which they place the meat, while underneath they light a very small fire which is kept burning for forty-eight hours or more. Hence for one constant—the presence of a layer of air—we note two differentials which are expressed by the opposition *close/distant* and *rapid/slow.* A third differential is created by the absence of a utensil in the case of roasting (any stick doing the work of a spit), since the buccan is a constructed framework, that is, a cultural object.

In this last respect, smoking is related to boiling, which also requires a cultural means, the receptacle. But between these two utensils a remarkable difference appears, or more accurately, is instituted by the culture precisely in order, it seems, to create the opposition, which without such a difference might have remained too ill-defined to take on meaning. Pots and pans are carefully cared for and preserved utensils, which one cleans and puts away after use in order to make them serve their purpose as many times as possible; but the buccan *must be destroyed immediately after use,* otherwise the animal will avenge itself, and come in turn to smoke the huntsman. Such, at least, is the belief of those same natives of Guiana whose other symmetrical belief we have already noted:

that a poorly conducted boiling, during which the cauldron overflowed, would bring the inverse punishment, flight of the quarry, which the huntsman would no longer succeed in overtaking. On the other hand, as we have already indicated, it is clear that the boiled is opposed both to the smoked and the roasted in respect to the presence or absence of water.

But let us come back for a moment to the opposition between a perishable and a durable utensil which we found in Guiana in connection with smoking and boiling. It will allow us to resolve an apparent difficulty in our system, one which doubtless has not escaped the reader. At the start we characterized one of the oppositions between the roasted and the boiled as reflecting that between nature and culture. Later, however, we proposed an affinity between the boiled and the rotted, the latter defined as the elaboration of the raw by natural means. Is it not contradictory that a cultural method should lead to a natural result? To put it in other terms, what, philosophically, will be the value of the invention of pottery (and hence of culture) if the native's system associates boiling and putrefaction, which is the condition that raw food cannot help but reach spontaneously in the state of nature?

The same type of paradox is implied by the problematics of smoking as formulated by the natives of Guiana. On the one hand, smoking, of all the modes of cooking, comes closest to the abstract category of the cooked; and—since the opposition between raw and cooked is homologous to that between nature and culture—it represents the most "cultural" form of cooking (and also that most esteemed among the natives). And yet, on the other hand, its cultural means, the buccan, is to be immediately destroyed. There is striking parallel to boiling, a method whose cultural means (the receptacles) are preserved, but which is itself assimilated to a sort of process of auto-annihilation, since its definitive result is at least verbally equivalent to that putrefaction which cooking should prevent or retard.

What is the profound sense of this parallelism? In so-called primitive societies, cooking by water and smoking have this in common: one as to its means, the other as to its results, is marked by duration. Cooking by water operates by means of receptacles made of pottery (or of wood with peoples who do not know about pottery, but boil water by immersing hot stones in it): in all cases these receptacles are cared for and repaired, sometimes passed on from generation to generation, and they number among the most durable cultural objects. As for smoking, it gives food that resists spoiling incomparably longer than that cooked by any other method. Everything transpires as if the lasting possession of a cultural acquisition entailed, sometimes in the ritual realm, sometimes in the mythic, a concession made in return to nature: when the result is durable, the means must be precarious, and vice-versa.

This ambiguity, which marks similarly, but in different directions, both the smoked and the boiled, is that same ambiguity which we already know to be inherent to the roasted. Burned on one side and raw on the other, or grilled outside, raw within, the roasted incarnates the ambiguity of the raw and the cooked, of nature and culture, which the smoked and the boiled must illustrate in their turn for the structure to be coherent. But what forces them into this pattern is not purely a reason of form: hence the system demonstrates that the art of cooking is not located entirely on the side of culture. Adapting itself to the exigencies of the body, and determined in its modes by the way man's insertion in nature operates in different parts of the world, placed then between nature and culture, cooking rather represents their necessary articulation. It partakes of both domains, and projects this duality on each of its manifestations.

But it cannot always do so in the same manner. The ambiguity of the roasted is intrinsic, that of the smoked and the boiled extrinsic, since it does not derive from things themselves, but from the way one speaks about them or behaves toward them. For here again a distinction becomes necessary: the quality of naturalness which language confers upon boiled food is purely metaphorical: the "boiled" is not the "spoiled"; it simply resembles it. Inversely, the transfiguration of the smoked into a natural entity does not result from the nonexistence of the buccan, the cultural instrument, but from its voluntary destruction. This transfiguration is thus on the order of metonymy, since it consists in acting as if the effect were really the cause. Consequently, even when the structure is added to or transformed to overcome a disequilibrium, it is only at the price of a new disequilibrium which manifests itself in another domain. To this ineluctable dissymmetry the structure owes its ability to engender myth, which is nothing other than an effort to correct or hide its inherent dissymmetry.

To conclude, let us return to our culinary triangle. Within it we traced another triangle representing recipes, at least the most elementary ones: roasting, boiling and smoking. The smoked and the boiled are opposed as to the nature of the intermediate element between fire and food, which is either air or water. The smoked and the roasted are opposed by the smaller or larger place given to the element air; and the roasted and the boiled by the presence of absence of water. The boundary between nature and culture, which one can imagine as parallel to either the axis of air or the axis of water, puts the roasted and the smoked on the side of nature, the boiled on the side of culture as to means; or, as to results, the smoked on the side of culture, the roasted and the boiled on the side of nature:

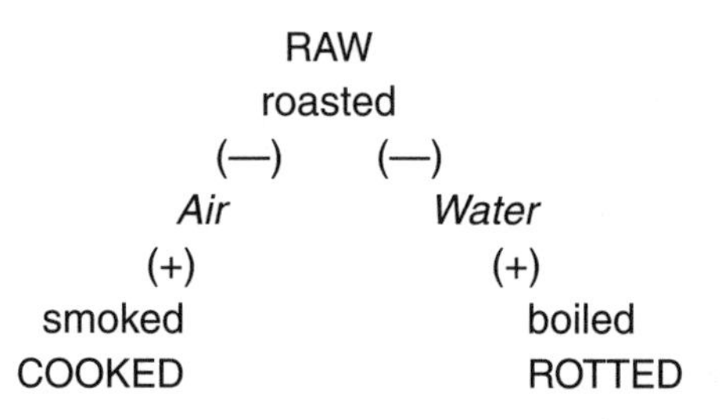

The operational value of our diagram would be very restricted did it not lend itself to all the transformations necessary to admit other categories of cooking. In a culinary system where the category of the roasted is divided into roasted and grilled, it is the latter term (connoting the lesser distance of meat from fire) which will be situated at the apex of the recipe triangle, the roasted then being placed, still on the air-axis, halfway between the grilled and the smoked. We may proceed in similar fashion if the culinary system in question makes a distinction between cooking with water and cooking with steam; the latter, where the water is at a distance from the food, will be located halfway between the boiled and the smoked.

A more complex transformation will be necessary to introduce the category of the fried. A tetrahedron will replace the recipe triangle, making it possible to raise a third axis, that of oil, in addition to those of air and water. The grilled will remain at the apex, but in the middle of the edge joining smoked and fried one can place roasted-in-the-oven (with the addition of fat), which is opposed to roasted-on-the-spit (without this addition). Similarly, on the edge running from fried to boiled will be braising (in a base of water and fat), opposed to steaming (without fat, and at a distance from the water). The plan can be still further developed, if necessary, by addition of the opposition between animal and vegetable foodstuffs (if they entail differentiating methods of cooking), and by the distinction of vegetable foods into cereals and legumes, since unlike the former (which one can simply grill), the latter cannot be cooked without water or fat, or both (unless one were to let the cereals ferment, which requires water but excluded fire during the process of transformation). Finally, seasonings will take their place in the system according to the combinations permitted or excluded with a given type of food.

After elaborating our diagram so as to integrate all the characteristics of a given culinary system (and no doubt there are other factors of a diachronic rather than a synchronic nature; those concerning the order, the presentation and the gestures of the meal), it will be necessary to seek the most economical manner of orienting it as a grille, so that it can be superposed on other contrasts of a sociological, economic, esthetic or religious nature: men and women, family and society, village and bush, economy and prodigality, nobility and commonality, sacred and profane, etc. Thus we can hope to discover for each specific case how the cooking of a society is a language in which it unconsciously translates its structure—or else resigns itself, still unconsciously, to revealing its contradictions.

NOTES

Translated from the French by Peter Brooks.

1. On these concepts, see Roman Jakobson, *Essais de linguistique générale*. Editions de Minuit, Paris, 1963.

4

Deciphering a Meal

MARY DOUGLAS

If language is a code, where is the precoded message? The question is phrased to expect the answer: nowhere. In these words a linguist is questioning a popular analogy.[1] But try it this way: if food is a code, where is the precoded message? Here, on the anthropologist's home ground, we are able to improve the posing of the question. A code affords a general set of possibilities for sending particular messages. If food is treated as a code, the messages it encodes will be found in the pattern of social relations being expressed. The message is about different degrees of hierarchy, inclusion and exclusion, boundaries and transactions across the boundaries. Like sex, the taking of food has a social component, as well as a biological one.[2] Food categories therefore encode social events. To say this is to echo Roland Barthes[3] on the sartorial encoding of social events. His book, *Système de la mode,* is primarily about methodology, about code-breaking and code-making taken as a subject in itself. The next step for the development of this conceptual tool is to take up a particular series of social events and see how they are coded. This will involve a close understanding of a microscale social system. I shall therefore start the exercise by analyzing the main food categories used at a particular point in time in a particular social system, our home. The humble and trivial case will open the discussion of more exalted examples.

Sometimes at home, hoping to simplify the cooking, I ask, "Would you like to have just soup for supper tonight? I mean a good thick soup—instead of supper. It's late and you must be hungry. It won't take a minute to serve." Then an argument starts: "Let's have soup now, and supper when you are ready." "No no, to serve two meals would be more work. But if you like, why not start with the soup and fill up with pudding?" "Good heavens! What sort of a meal is that? A beginning and an end and no middle." "Oh, all right then, have the soup as it's there, and I'll do a Welsh rarebit as well." When they have eaten soup, Welsh rarebit, pudding, and cheese: "What a lot of plates. Why do you make such elaborate suppers?" They proceed to argue that by taking thought I could satisfy the full requirements of a meal with a single, copious dish. Several rounds of this conversation have given me a practical interest in the categories and meaning of food. I needed to know what defines the category of a meal in our home.

The first source for enlightenment will obviously be Claude Lévi-Strauss's *The Raw and the Cooked* and the other volumes of his *Mythologiques*[4] which discuss food categories and table manners. But this is only a beginning. He fails us in two major respects. First, he takes leave of the small-scale social relations which generate the codification and are sustained by it. Here and there his feet touch solid ground, but mostly he is orbiting in rarefied space where he expects to find universal food meanings common to all mankind. He is looking for a precoded, panhuman message in the language of food, and thus exposing himself to the criticism implicit in the quoted linguist's question. Second, he relies entirely on the resources of binary analysis. Therefore he affords no techniques for assessing the relative value of the binary pairs that emerge in a local set of expressions. Worse than clumsy, his technical apparatus produces meanings which cannot be validated. Yea, or nay, he and Roman Jakobson may be right on the meanings in a sonnet of Baudelaire's.[5] But even if the poet himself had been able to judge between theirs and Riffaterre's alternative interpretation of the same work[6] and to say that one was closer to his thought than the other, he would be more likely to agree that all these meanings are there. This is fair for literary criticism, but when we are talking of grammar, coding, and the "science of the concrete,"[7] it is not enough.

For analyzing the food categories used in a particular family the analysis must start with why those particular categories and not others are employed. We will discover the social boundaries which the food meanings encode by an approach which values the binary pairs according to their position in a series. Between breakfast and the last nightcap, the food of the day comes in an ordered pattern. Between Monday and Sunday, the food of the week is patterned again. Then there is the sequence of holidays and fast days through the year, to say nothing of life cycle feasts, birthdays, and weddings. In other words, the binary or other contrasts must be seen in their syntagmatic relations. The chain which links them together gives each element some of its meaning. Lévi-Strauss discusses the syntagmatic relation in his earlier book, *The Savage Mind,* but uses it only for the static analysis of classification systems (particularly of proper names). It is capable of a much more dynamic application to food categories, as Michael Halliday has shown. On the two axes of syntagm and paradigm, chain and choice, sequence and set, call it what you will, he has shown how food elements can be ranged until they are all accounted for either in grammatical terms, or down to the last lexical item.[8]

> Eating, like talking, is patterned activity, and the daily menu may be made to yield an analogy with linguistic form. Being an analogy, it is limited in relevance; its purpose is to throw light on, and suggest problems of the categories of grammar by relating these to an activity which is familiar and for much of which a terminology is ready to hand.

The presentation of a framework of categories for the description of eating might proceed as follows:

Units: Daily menu
Meal
Course
Helping
Mouthful

Unit: Daily Menu	→
Elements of primary structure	E, M, L, S ("early," "main," "light," "snack")
Primary structures	EML EMLS (conflated as EML (S))
Exponents of these elements (primary classes of unit "meal")	E: I (breakfast) M: 2 (dinner) L: 3 } (no names available; see S: 4 } classes)
Secondary structures	ELaSaM ELaM EMLbSb EMSaLa
Exponents of secondary elements (systems of secondary classes of unit "meal")	La: 3.1 (lunch) Lb: 3.2 (high tea) La: 3.3 (supper) Sa: 4.1 (afternoon tea) Sb: 4.2 (nightcap)
System of sub-classes of unit "meal"	E: 1.1 (English breakfast) 1.2 (continental breakfast)

Passing to the rank of the "meal," we will follow through the class "dinner":

Unit: Meal, Class: dinner	→
Elements of primary structure	F, S, M, W, Z ("first," "second," "main," "sweet," "savoury")
Primary structures	MW MWZ MZW FMW FMWZ FMZW FSMW FSMWZ FSMZW (conflated as (F(S)MW(Z))
Exponents of these elements (primary classes of unit "course")	F: 1 (antipasta) S: 2 (fish) M: 3 (entrée) W: 4 (dessert) Z: 5 (cheese*)
Secondary structures	(various, involving secondary elements Fa.d, Ma.b, Wa.c)
Exponents of secondary elements (systems of secondary classes of unit "course")	Fa:1.1 (soup) Fb:1.2 (hors d'oeuvres) Fe:1.3 (fruit) Fd:1.4 (fruit juice) Ma: 3.1 (meat dish) Mb: 3.2 (poultry dish) Wa: 4.1 (fruit*) Wb: 4.2 (pudding) We: 4.3 (ice cream*)

Systems of sub-classes of unit "course"	Fa:1.11 (clear soup*) 1.12 (thick soup*) S: 2.01 (grilled fish*) 2.02 (fried fish*) 2.03 (poached fish*) Wb: 4.21 (steamed pudding*) 4.22 (milk pudding*)
Exponential systems operating in meal structure	Fa: grapefruit juice/pineapple juice/tomato juice Ma: beef/mutton/pork Mb: chicken/turkey/duck/goose

At the rank of the "course," the primary class "entrée" has secondary classes "meat dish" and "poultry dish." Each of these two secondary classes carries a grammatical system whose terms are formal items. But this system accounts only for simple structures of the class "entrée," those made up of only one member of the unit "helping." The class "entrée," also displays compound structures, whose additional elements have as exponents the (various secondary classes of the) classes "cereal" and "vegetable." We will glance briefly at these:

Unit: Course, Class: entrée

Elements of primary structure	J, T, A ('joint," "staple," "adjunct")
Primary structures	J JT JA JTA (conflated as J ((T)(A))
Exponents of these elements (primary classes of unit "helping")	J: 1 (flesh) T: 2 (cereal) A: 3 (vegetable)
Secondary structures	(various, involving—among others—secondary elements Ja,b, Ta,b, Aa,b)
Exponents of secondary elements (systems of secondary classes of unit "helping")	Ja: 1.1 (meat) } systems as at M in meal structure Jb: 1.2 (pountry) } Ta: 2.1 (potato) Tb: 2.2 (rice) Aa: 3.1 (green vegetable*) Ab: 3.2 (root vegetable*)

And so on, until everything is accounted for either in grammatical systems or in classes made up of lexical items (marked *). The presentation has proceeded down the rank scale, but shunting is presupposed throughout: there is mutual determination among all units, down to the gastronomic morpheme, the "mouthful."

This advances considerably the analysis of our family eating patterns. First, it shows how long and tedious the exhaustive analysis would be, even to read. It would be more taxing to observe and record. Our model of ethnographic thoroughness for a microscopic example should not be less exact than that practiced by anthropologists working in exotic lands. In India social distinctions are invariably accompanied by distinction in commensality and categories of edible and inedible foods. Louis Dumont's important work on Indian culture, *Homo Hierarchicus,* discusses the purity of food as an index of hierarchy. He gives praise to Adrian Mayer's detailed study of the relation between food categories and social categories in a village in Central India.[9] Here twenty-three castes group themselves according to the use of the same pipe, the provision of ordinary food for common meals, and the provision of food for feasts. Higher castes share the pipe with almost all castes except four. Between twelve and sixteen castes smoke together, though in some cases a different cloth must be placed between the pipe and the lips of the smoker. When it comes to their food, a subtler analysis is required. Castes which enjoy power in the village are not fussy about what they eat or from whom they receive it. Middle range castes are extraordinarily restrictive, both as to whom they will accept food from and what they will eat. Invited to family ceremonies by the more powerful and more ritually relaxed castes they puritanically insist on being given their share of the food raw and retire to cook it themselves in their own homes.[10] If I were to follow this example and to include all transmission of food from our home my task would be greater. For certainly we too know situations in which drink is given to be consumed in the homes of the recipient. There are some kinds of service for which it seems that the only possible recognition is half or even a whole bottle of whiskey. With the high standards of the Indian research in mind, I try now to identify the relevant categories of food in our home.

The two major contrasted food categories are meals versus drinks. Both are social events. Outside these categories, of course, food can be taken for private nourishment. Then we speak only of the lexical item itself: "Have an apple. Get a glass of milk. Are there any sweets?" If likely to interfere with the next meal, such eating is disapproved. But no negative attitude condemns eating before drinks. This and other indices suggest that meals rank higher.

Meals contrast with drinks in the relation between solids and liquids. Meals are a mixture of solid foods accompanied by liquids. With drinks the reverse holds. A complex series of syntagmatic associations governs the elements in a meal, and connects the meals through the day. One can say: "It can't be lunchtime. I haven't had breakfast yet," and at breakfast itself cereals come before bacon and eggs. Meals in their sequence tend to be named. Drinks sometimes have named categories: "Come for cocktails, come for coffee, come for tea," but many are not named events: "What about a drink? What shall we have?" There is no structuring of drinks into early, main, light. They are not invested with any necessity in their ordering. Nor is the event called drinks internally structured into first, second, main, sweet. On the contrary, it is approved to stick with the same kind of drink, and to count drinks at all is impolite. The judgment "It is too early for alcohol" would be both rare and likely to be contested. The same lack of structure is found in the solid foods accompanying drinks. They are usually cold, served in discrete units which can be eaten tidily with fingers. No order governs the choice of solids. When the children were small and tea was a meal, bread and butter preceded scones, scones preceded cake and biscuits. But now that the adult-child contrast no longer dominates in this family, tea has been demoted from a necessary place in the daily sequence of meals to an irregular appearance among weekend drinks and no rules govern the accompanying solids.

Meals properly require the use of at least one mouth-entering utensil per head, whereas drinks are limited to mouth-touching ones. A spoon on a saucer is for stirring, not sucking. Meals require a table, a seating order, restriction on movement and on alternative occupations. There is no question of knitting during a meal. Even at Sunday breakfast, reaching for the newspapers is a signal that the meal is over. The meal puts its frame on the gathering. The rules which hedge off and order one kind of social interaction are reflected in the rules which control the internal ordering of the meal itself. Drinks and their solids may all be sweet. But a meal is not a meal if it is all in the bland-sweet-sour dimensions. A meal incorporates a number of contrasts, hot and cold, bland and spiced, liquid and semi-liquid, and various textures. It also incorporates cereals, vegetables, and animal proteins. Criticism easily fastens on the ordering of these elements in a given case.

Obviously the meanings in our food system should be elucidated by much closer observation. I cut it short by drawing conclusions intuitively from the social categories which emerge. Drinks are for strangers, acquaintances, workmen, and family. Meals are for family, close friends, honored guests. The grand operator of the system is the line between intimacy and distance. Those we know at meals we also know at drinks. The meal expresses close friendship. Those we only know at drinks we know less intimately. So long as this boundary matters to us (and there is no reason to suppose it will always matter) the boundary between drinks and meals has meaning. There are smaller thresholds and halfway points. The entirely cold meal (since it omits a major contrast within a meal) would seem to be such a modifier. So those friends who have never had a hot meal in our home have presumably another threshold of intimacy to cross. The recent popularity of the barbecue and of more elaborately structured cocktail events which act as bridges between intimacy and distance suggests that our model of feeding categories is a common one. It can be drawn as in Figure 1. Thus far we can go on the basis of binary oppositions and the number of classes and subclasses. But we are left with the general question which must be raised whenever a correspondence is found between a given social structure and the structure of symbols by which it is expressed, that is, the question of consciousness. Those who vehemently reject the possibility of a meal's being constituted by soup and pudding, or cake and fruit, are certainly not conscious that they are thereby sustaining a boundary between share-drinks and share-meals-too. They would be shocked at the very idea. It would be simplistic to trace the food categories direct to the social categories they embrace and leave it at Figure 1. Evidently the external boundaries are only a small part of the meaning of the meal. Somewhere else in the family system some other cognitive activity is generating the internal structuring.

We can go much further toward discovering the intensity of meanings and their anchorage in social life by attending to the sequence of meals. For the week's menu has its climax at Sunday lunch. By contrasting the structure of Sunday lunch with weekday lunches a new principle emerges. Weekday lunches tend to have a tripartite structure, one element stressed accompanied by two or more unstressed elements, for example a main course and cold supporting dishes. But Sunday lunch has two main courses, each

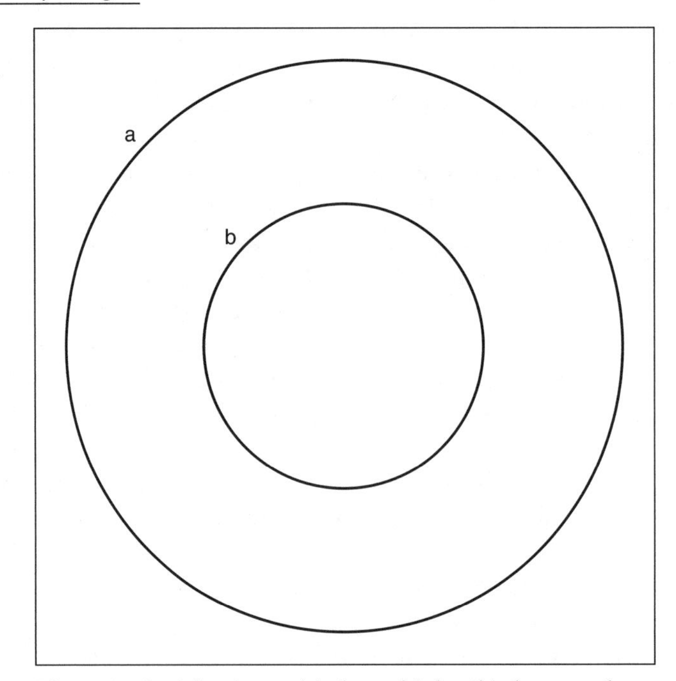

Figure 1 Social universe (a) share-drinks; (b) share-meals-too

of which is patterned like the weekday lunch—say, first course, fish or meat (stressed) and two vegetables (unstressed), second course, pudding (stressed), cream and biscuits (unstressed). Christmas lunch has three courses, each on the same tripartite model. Here we stop and realize that the analogy may be read in the reverse sense. Meals are ordered in scale of importance and grandeur through the week and the year. The smallest, meanest meal metonymically figures the structure of the grandest, and each unit of the grand meal figures again the whole meal—or the meanest meal. The perspective created by those repetitive analogies invests the individual meal with additional meaning. Here we have the principle we were seeking, the intensifier of meaning, the selection principle. A meal stays in the category of meal only in so far as it carries this structure which allows the part to recall the whole. Hence the outcry against allowing the sequence of soup and pudding to be called a meal.

As to the social dimension, admission to even the simplest meal incorporates our guest unwittingly into the pattern of solid Sunday dinners, Christmases, and the gamut of life cycle celebrations. Whereas the sharing of drinks (note the fluidity of the central item, the lack of structuring, the small, unsticky accompanying solids) expresses by contrast only too clearly the detachment and impermanence of simpler and less intimate social bonds.

Summing up, syntagmatic relations between meals reveal a restrictive patterning by which the meal is identified as such, graded as a minor or major event of its class, and then judged as a good or bad specimen of its kind. A system of repeated analogies upholds the process of recognition and grading. Thus we can broach the question of interpretation which binary analysis by itself leaves untouched. The features which a single copious dish would need to display before qualifying as a meal in our home would be something like those of the famous chicken Marengo of Napoleon after his defeat of the Austrians.[11]

> Bonaparte, who, on the day of a battle, ate nothing until after it was over, had gone forward with his general staff and was a long way from his supply wagon. Seeing his enemies put to flight, he asked Dunand to prepare dinner for him. The masterchef at once sent men of the quartermaster's staff and ordnance in search of provisions. All they could find were three eggs, four tomatoes, six crayfish, a small hen, a little garlic, some oil and a saucepan . . . the dish was served on a tin plate, the chicken surrounded by the fried eggs and crayfish, with the sauce poured over it.

There must have been many more excellent meals following similar scavenging after the many victories of those campaigns. But only this one has become famous. In my opinion the reason is that it combines the traditional soup, fish, egg, and meat courses of a French celebratory feast all in a *plat unique*.

If I wish to serve anything worthy of the name of supper in one dish it must preserve the minimum structure of a meal. Vegetable soup so long as it had noodles and grated cheese would do, or poached eggs on toast with parsley. Now I know the formula. A proper meal is A (when A is the stressed main course) plus 2B (when B is an unstressed course). Both A and B contain each the same structure, in small, a + 2b, when a is the stressed item and b the unstressed item in a course. A weekday lunch is A; Sunday lunch is 2A; Christmas, Easter, and birthdays are A + 2B. Drinks by contrast are unstructured.

To understand the categories we have placed ourselves at the hub of a small world, a home and its neighborhood. The precoded message of the food categories is the boundary system of a series of social events. Our example made only oblique references to costs in time and work to indicate the concerns involved. But unless the symbolic structure fits squarely to some demonstrable social consideration, the analysis has only begun. For the fit between the medium's symbolic boundaries and the boundaries between categories of people is its only possible validation. The fit may be at different levels, but without being able to show some such matching, the analysis of symbols remains arbitrary and subjective.

The question that now arises is the degree to which a family uses symbolic structures which are available from the wider social system. Obviously this example reeks of the culture of a certain segment of the middle classes of London. The family's idea of what a meal should be is influenced by the Steak House and by the French *cuisine bourgeoise*. Yet herein is implied a synthesis of different traditions. The French version of the grand meal is dominated by the sequence of wines. The cheese platter is the divide between a mounting crescendo of individual savory dishes and a descending scale of sweet ones ending with coffee. Individual dishes in the French sequence can stand alone. Compare the melon course in a London restaurant and a Bordeaux restaurant. In the first, the half slice is expected to be dusted with powdered ginger and castor sugar (a + 2b) or decorated with a wedge of orange and crystallized cherry (a + 2b). In the second, half a melon is served with no embellishment but its own perfume and juices. A + 2B is obviously not a formula that our family invented, but one that is current in our social environment. It governs even the structure of the cocktail canapé. The latter, with its cereal base, its meat or cheese middle section, its sauce or pickle topping, and its mixture of colors, suggests a mock meal, a minute metonym of English middle-class meals in general. Whereas the French pattern is more like: $C^1 + B^1 + A^1/A^2 + B^2 + C^2$, when the cheese course divides A^1 (the main savory dish) from A^2 (the main sweet). It would be completely against the spirit of this essay to hazard a meaning for either structure in its quasi environmental form. French families reaching out to the meal structure of their cultural environment develop it and interact with it according to their intentions. English families reach out and find another which they adapt to their own social purposes. Americans, Chinese, and others do likewise. Since these cultural environments afford an ambient stream of symbols, capable of differentiating and intensifying, but not anchored to a stable social base, we cannot proceed further to interpret them. At this point the analysis stops. But the problems which cannot be answered here, where the cultural universe is unbounded, can usefully be referred to a more closed environment.

To sum up, the meaning of a meal is found in a system of repeated analogies. Each meal carries something of the meaning of the other meals; each meal is a structured social event which structures others in its own image. The upper limit of its meaning is set by the range incorporated in the most important member of its series. The recognition which allows each member to be classed and graded with the others depends upon the structure common to them all. The cognitive energy which demands that a meal look like a meal and not like a drink is performing in the culinary medium the same exercise that it performs in language. First, it distinguishes order, bounds it, and separates it from disorder. Second, it uses economy in the means of expression by allowing only a limited number of structures. Third, it imposes a rank scale upon the repetition of structures. Fourth, the repeated formal analogies multiply the meanings that are carried down any one of them by the power of the most weighty. By these four methods the meanings are enriched. There is no single point in the rank scale, high or low, which provides the basic meaning or real meaning. Each exemplar has the meaning of its structure realized in the examples at other levels.

From coding we are led to a more appropriate comparison for the interpretation of a meal, that is, versification. To treat the meal as a poem requires a more serious example than I have used hitherto. I turn to the Jewish meal, governed by the Mosaic dietary rules. For Lu Chi, a third-century Chinese poet, poetry traffics in some way between the world and mankind. The poet is one who "traps Heaven and Earth in a cage of form."[12] On these terms the common meal of the Israelites was a kind of classical poem. Of the Israelites, table, too, it could be said that it enclosed boundless space. To quote Lu Chi again:[13]

> We enclose boundless space in a square-foot of paper;
> We pour out deluge from the inch-space of the heart.

But the analogy slows down at Lu Chi's last line. For at first glance it is not certain that the meal can be a traffic medium. The meal is a kind of poem, but by a very limited analogy. The cook may not be able to express the powerful things a poet can say.

In *Purity and Danger* [14] I suggested a rational pattern for the Mosaic rejection of certain animal kinds. Ralph Bulmer has very justly reproached me for offering an animal taxonomy for the explanation of the Hebrew dietary laws. The principles I claimed to discern must remain, he argued, at a subjective and arbitrary level, unless they could take account of the multiple dimensions of thought and activity of the Hebrews concerned.[15] S. J. Tambiah has made similarly effective criticisms of the same shortcoming in my approach.[16] Both have provided from their own fieldwork distinguished examples of how the task should be conducted. In another publication I hope to pay tribute to the importance of their research. But for the present purpose, I am happy to admit the force of their reproach. It was even against the whole spirit of my book to offer an account of an ordered system of thought which did not show the context of social relations in which the categories had meaning. Ralph Bulmer let me down gently by supposing that the ethnographic evidence concerning the ancient Hebrews was too meager. However, reflection on this new research and methodology has led me to reject that suggestion out of hand. We know plenty about the ancient Hebrews. The problem is how to recognize and relate what we know.

New Guinea and Thailand are far apart, in geography, in history, and in civilization. Their local fauna are entirely different. Surprisingly, these two analyses of animal classification have one thing in common. Each society projects on to the animal kingdom categories and values which correspond to their categories of marriageable persons. The social categories of descent and affinity dominate their natural categories. The good Thailand son-in-law knows his place and keeps to it: disordered, displaced sex is reprobated and the odium transferred to the domestic dog, symbol of dirt and promiscuity. From the dog to the otter, the transfer of odium is doubled in strength. This amphibian they class as wild, counterpart-dog. But instead of keeping to the wild domain it is apt to leave its sphere at flood time and to paddle about in their watery fields. The ideas they attach to incest are carried forward from the dog to the otter, the image of the utterly wrong son-in-law. For the Karam the social focus is upon the strained relations between affines and cousins. A wide range of manmade rules sustains the categories of a natural world which mirrors those anxieties. In the Thailand and Karam studies, a strong analogy between bed and board lies unmistakably beneath the system of classifying animals. The patterns of rules which categorize animals correspond in form to the patterns of rules governing human relations. Sexual and gastronomic consummation are made equivalents of one another by reasons of analogous restrictions applied to each. Looking back from these examples to the classifications of Leviticus we seek in vain a statement, however oblique, of a similar association between eating and sex. Only a very strong analogy between table and altar stares us in the face. On reflection, why should the Israelites have had a similar concern to associate sex with food? Unlike the other two examples, they had no rule requiring them to exchange their womenfolk. On the contrary, they were allowed to marry their parallel first cousins. E. R. Leach has reminded us how strongly exogamy was disapproved at the top political level,[17] and within each tribe of Israel endogamy was even enjoined (Numbers 36). We must seek elsewhere for their dominant preoccupations. At this point I turn to the rules governing the common meal as prescribed in the Jewish religion. It is particularly interesting that these rules have remained the same over centuries. Therefore, if these categories express a relevance to social concerns we must expect those concerns to have remained in some form alive. The three rules about meat are: (1) the rejection of certain animal kinds as unfit for the table (Leviticus 11; Deuteronomy 14), (2) of those admitted as edible, the separation of the meat from blood before cooking (Leviticus 17: 10; Deuteronomy 12: 23–7), (3) the total separation of milk from meat, which involved the minute specialization of utensils (Exodus 23: 19; 34: 26; Deuteronomy 14: 21).

I start with the classification of animals whose rationality I claim to have discerned. Diagrams will help to summarize the argument (first outlined in *Purity and Danger*, 1966).

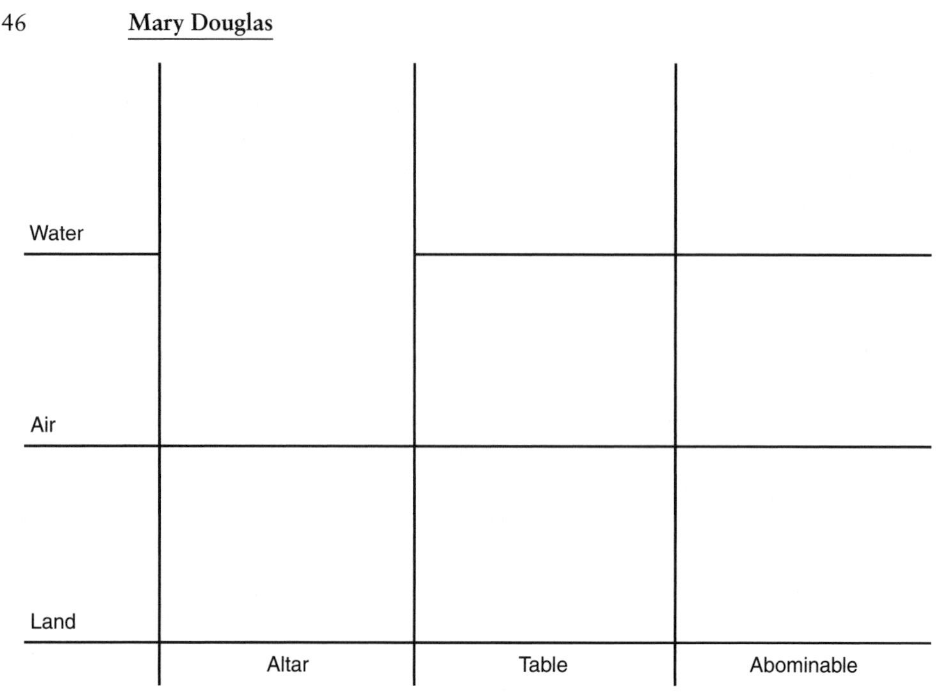

Figure 2 Degrees of holiness

First, animals are classified according to degrees of holiness (see Figure 2). At the bottom end of the scale some animals are abominable, not to be touched or eaten. Others are fit for the table, but not for the altar. None that are fit for the altar are not edible and vice versa, none that are not edible are sacrificeable. The criteria for this grading are coordinated for the three spheres of land, air, and water. Starting with the simplest, we find the sets as in Figure 3.

Water creatures, to be fit for the table, must have fins and scales (Leviticus 13: 9–12; Deuteronomy 14: 19). Creeping swarming worms and snakes, if they go in the water or on the land, are not fit for the table (Deuteronomy 14: 19; Leviticus 11: 41–3). "The term swarming creatures (*shéreç)* denotes living things which appear in swarms and is applied both to those which teem in the waters (Genesis 1: 20; Leviticus 11: 10) and to those which swarm on the ground, including the smaller land animals, reptiles and creeping insects."[18] Nothing from this sphere is fit for the altar. The Hebrews only sanctified domesticated animals and these did not include fish. "When any one of you brings an offering to Jehovah, it shall be a domestic animal, taken either from the herd or from the flock" (Leviticus 1: 2). But, Assyrians and others sacrificed wild beasts, as S. R. Driver and H. A. White point out.

Air creatures (see Figure 4) are divided into more complex sets: set (a), those which fly and hop on the earth (Leviticus 11: 12), having wings and two legs, contains two subsets, one of which contains the named birds, abominable and not fit for the table, and the rest of the birds (b), fit for the table. From this latter subset a sub-subset (c) is drawn, which is suitable for the altar—turtledove and pigeon (Leviticus 14; 5: 7–8) and the sparrow (Leviticus 14: 49–53). Two separate sets of denizens of the air are abominable, untouchable creatures; (f), which have the wrong number of limbs for their habitat,

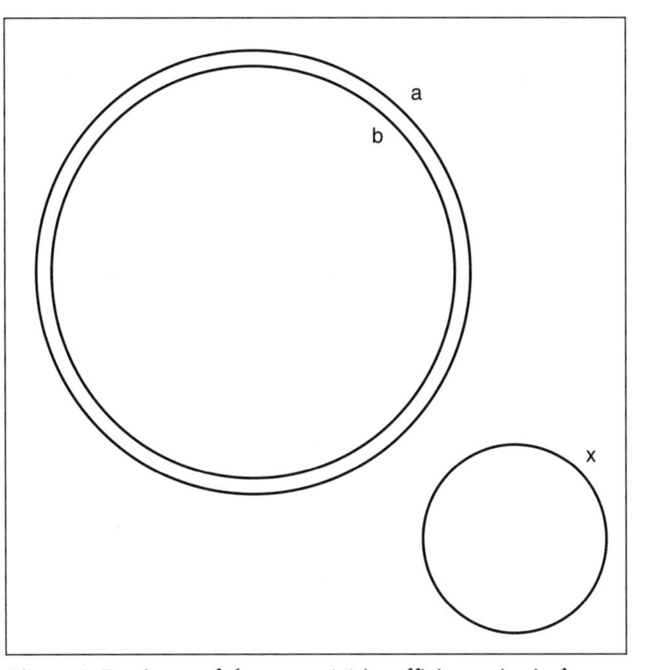

Figure 3 Denizens of the water (a) insufficient criteria for (b); (b) fit for table; (x) abominable: swarming

four legs instead of two (Leviticus 9: 20), and (x), the swarming insects we have already noted in the water (Deuteronomy 14: 19).

The largest class of land creatures (a) (see Figure 5) walk or hop on the land with four legs. From this set of quadrupeds, those with parted hoofs and which chew the cud (b) are distinguished as fit for the table (Leviticus 11: 3; Deuteronomy 14: 4–6) and of this set a subset consists of the domesticated herds and flocks (c). Of these the first born (d) are to be offered to the priests (Deuteronomy 24: 33). Outside the set (b) which part

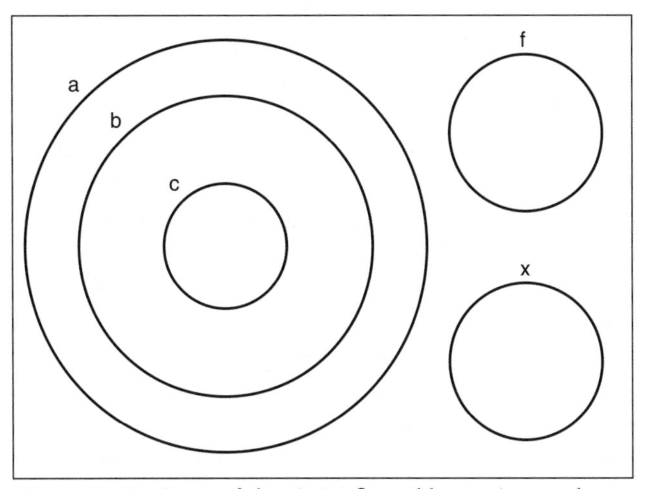

Figure 4 Denizens of the air (a) fly and hop: wings and two legs; (b) fit for table; (c) fit for altar; (f) abominable: insufficient criteria for (a); (x) abominable: swarming

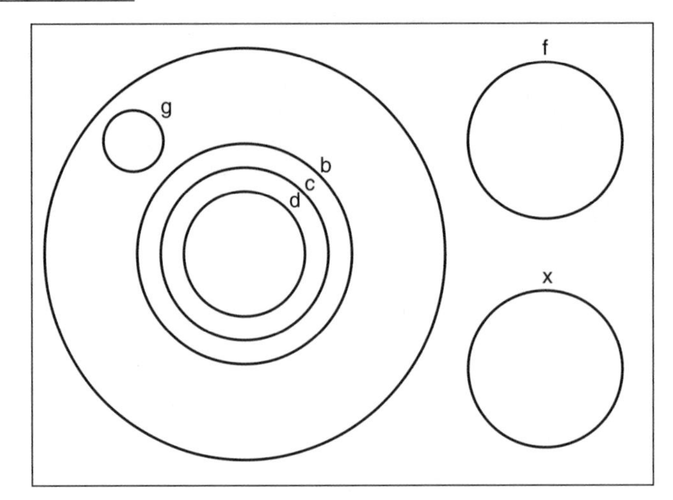

Figure 5 Denizens of the land (a) walk or hop with four legs; (b) fit for table; (c) domestic herds and flocks; (d) fit for altar; (f) abominable: insufficient criteria for (a); (g) abominable: insufficient criteria for (b); (x) abominable: swarming

the hoof and chew the cud are three sets of abominable beasts: (g) those which have either the one or the other but not both of the required physical features; (f) those with the wrong number of limbs, two hands instead of four legs (Leviticus 11: 27 and 29: 31; and see Proverbs 30: 28); (x) those which crawl upon their bellies (Leviticus 11: 41–4).

The isomorphism which thus appears between the different categories of animal classed as abominable helps us to interpret the meaning of abomination. Those creatures which inhabit a given range, water, air, or land, but do not show all the criteria for (a) or (b) in that range are abominable. The creeping, crawling, teeming creatures do not show criteria for allocation to any class, but cut across them all.

Here we have a very rigid classification. It assigns living creatures to one of three spheres, on a behavioral basis, and selects certain morphological criteria that are found most commonly in the animals inhabiting each sphere. It rejects creatures which are anomalous, whether in living between two spheres, or having defining features of members of another sphere, or lacking defining features. Any living being which falls outside this classification is not to be touched or eaten. To touch it is to be defiled and defilement forbids entry to the temple. Thus it can be summed up fairly by saying that anomalous creatures are unfit for altar and table. This is a peculiarity of the Mosaic code. In other societies anomaly is not always so treated. Indeed, in some, the anomalous creature is treated as the source of blessing and is specially fit for the altar (as the Lele pangolin), or as a noble beast, to be treated as an honorable adversary, as the Karam treat the cassowary. Since in the Mosaic code every degree of holiness in animals has implications one way or the other for edibility, we must follow further the other rules classifying humans and animals. Again I summarize a long argument with diagrams. First, note that a category which divides some humans from others also divides their animals from others. Israelites descended from Abraham and bound to God by the Covenant between God and Abraham are distinguished from all other peoples and similarly the rules which Israelites obey as part of the Covenant apply to their animals (see Figure 6). The rule that the womb opener or first born is consecrated to divine service applies to firstlings

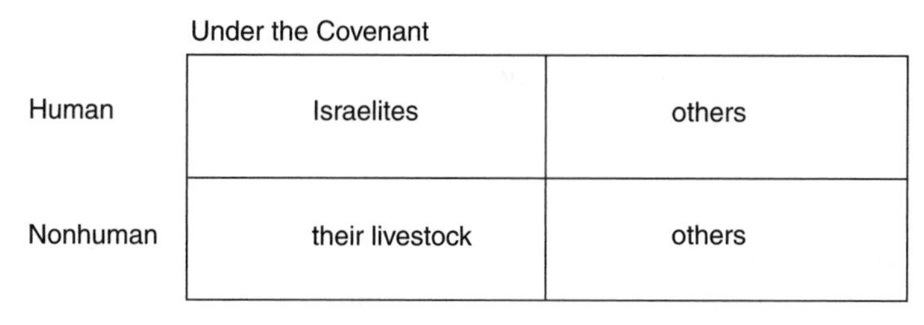

Figure 6 Analogy between humans and nonhumans

of the flocks and herds (Exodus 22: 29–30; Deuteronomy 24: 23) and the rule of Sabbath observance is extended to work animals (Exodus 20: 10). As human and animal firstlings are to God, so a man's own first born is unalterably his heir (Deuteronomy 21: 15–17). The analogy by which Israelites are to other humans as their livestock are to other quadrupeds develops by indefinite stages the analogy between altar and table.

Since Levites who are consecrated to the temple service represent the first born of all Israel (Numbers 3: 12 and 40) there is an analogy between the animal and human firstlings. Among the Israelites, all of whom prosper through the Covenant and observance of the Law, some are necessarily unclean at any given time. No man or woman with issue of seed or blood, or with forbidden contact with an animal classed as unclean, or who has shed blood or been involved in the unsacralized killing of an animal (Leviticus 18), or who has sinned morally (Leviticus 20), can enter the temple. Nor can one with a blemish (Deuteronomy 23) enter the temple or eat the flesh of sacrifice or peace offerings (Leviticus 8: 20). The Levites are selected by pure descent from all the Israelites. They represent the first born of Israel. They judge the cleanness and purify the uncleanness of Israelites (Leviticus 13, 14, 10: 10; Deuteronomy 21: 5). Only Levites who are without blemish (Leviticus 21: 17–23) and without contact with death can enter the Holy of Holies. Thus we can present these rules as sets in Figures 7 and 8. The analogy between

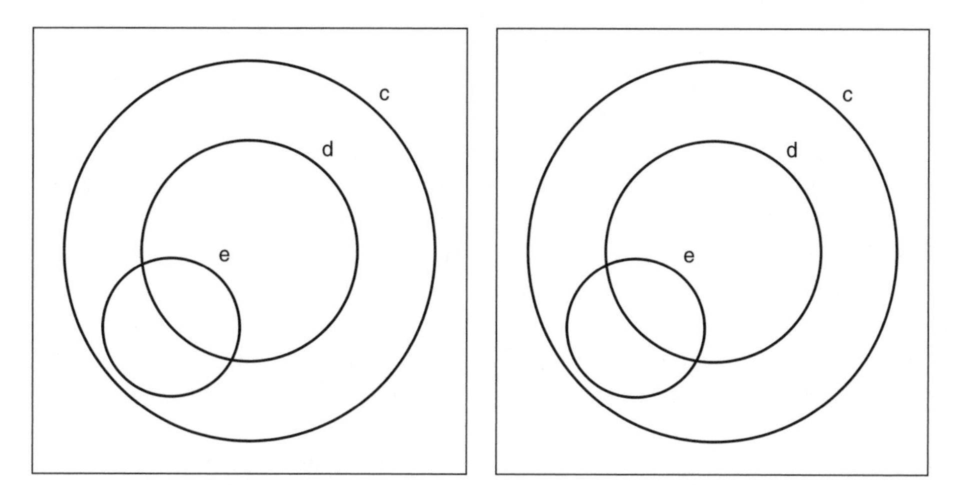

Figure 7 The Israelites (c) under the Covenant; (d) fit for temple sacrifice: no blemish; (e) consecrated to temple service, first born

Figure 8 Their livestock (c) under the Covenant; (d) fit for temple sacrifice: no blemish; (e) consecrated to temple service, first born

humans and animals is very clear. So is the analogy created by these rules between the temple and the living body. Further analogies appear between the classification of animals according to holiness (Figure 2) and the rules which set up the analogy of the holy temple with its holier and holier inner sanctuaries, and on the other hand between the temple's holiness and the body's purity and the capability of each to be defiled by the self-same forms of impurity. This analogy is a living part of the Judeo-Christian tradition which has been unfaltering in its interpretation of New Testament allusions. The words of the Last Supper have their meaning from looking backward over the centuries in which the analogy had held good and forward to the future celebrations of that meal. "This is my body . . . this is my blood" (Luke 22: 19–20; Mark 14: 22–4; Matthew 26: 26–8). Here the meal and the sacrificial victim, the table and the altar are made explicitly to stand for one another.

Lay these rules and their patternings in a straight perspective, each one looking forward and backward to all the others, and we get the same repetition of metonyms that we found to be the key to the full meaning of the categories of food in the home. By itself the body and its rules can carry the whole load of meanings that the temple can carry by itself with its rules. The overlap and repetitions are entirely consistent. What then are these meanings? Between the temple and the body we are in a maze of religious thought. What is its social counterpart? Turning back to my original analysis (in 1966) of the forbidden meats we are now in a much better position to assess intensity and social relevance. For the metonymical patternings are too obvious to ignore. At every moment they are in chorus with a message about the value of purity and the rejection of impurity. At the level of a general taxonomy of living beings the purity in question is the purity of the categories. Creeping, swarming, teeming creatures abominably destroy the taxonomic boundaries. At the level of the individual living being impurity is the imperfect, broken, bleeding specimen. The sanctity of cognitive boundaries is made known by valuing the integrity of the physical forms. The perfect physical specimens point to the perfectly bounded temple, altar, and sanctuary. And these in their turn point to the hard-won and hard-to-defend territorial boundaries of the Promised Land. This is not reductionism. We are not here reducing the dietary rules to any political concern. But we are showing how they are consistently celebrating a theme that has been celebrated in the temple cult and in the whole history of Israel since the first Covenant with Abraham and the first sacrifice of Noah.

Edmund Leach, in his analysis of the genealogy of Solomon, has reminded us of the political problems besetting a people who claim by pure descent and pure religion to own a territory that others held and others continually encroached upon.[19] Israel is the boundary that all the other boundaries celebrate and that gives them their historic load of meaning. Remembering this, the orthodox meal is not difficult to interpret as a poem. The first rule, the rejection of certain animal kinds, we have mostly dealt with. But the identity of the list of named abominable birds is still a question. In the Mishnah it is written: "The characteristics of birds are not stated, but the Sages have said, every bird that seizes its prey (to tread or attack with claws) is unclean."[20] The idea that the unclean birds were predators, unclean because they were an image of human predation and homicide, so easily fits the later Hellenicizing interpretations that it has been suspect. According to the late Professor S. Hooke (in a personal communication), Professor R. S. Driver once tried out the idea that the Hebrew names were onomatopoeic of the screeches and calls of the birds. He diverted an assembly of learned divines with ingenious vocal exercises combining ornithology and Hebrew scholarship. I have not traced the record of this meeting. But following the method of analysis I have been using, it seems very likely that the traditional predatory idea is sufficient, considering its compatibility with the second rule governing the common meal.

According to the second rule, meat for the table must be drained of its blood. No man eats flesh with blood in it. Blood belongs to God alone, for life is in the blood.[21] This rule relates the meal systematically to all the rules which exclude from the temple on grounds of contact with or responsibility for bloodshed. Since the animal kinds which defy the perfect classification of nature are defiling both as food and for entry to the temple, it is a structural repetition of the general analogy between body and temple to rule that the eating of blood defiles. Thus the birds and beasts which eat carrion (undrained of blood) are likely by the same reasoning to be defiling. In my analysis, the Mishnah's identifying the unclean birds as predators is convincing.

Here we come to a watershed between two kinds of defilement. When the classifications of any metaphysical scheme are imposed on nature, there are several points where it does not fit. So long as the classifications remain in pure metaphysics and are not expected to bite into daily life in the form of rules of behavior, no problem arises. But if the unity of Godhead is to be related to the unity of Israel and made into a rule of life, the difficulties start. First, there are the creatures whose behavior defies the rigid classification. It is relatively easy to deal with them by rejection and avoidance. Second, there are the difficulties that arise from our biological condition. It is all very well to worship the holiness of God in the perfection of his creation. But the Israelites must be nourished and must reproduce. It is impossible for a pastoral people to eat their flocks and herds without damaging the bodily completeness they respect. It is impossible to renew Israel without emission of blood and sexual fluids. These problems are met sometimes by avoidance and sometimes by consecration to the temple. The draining of blood from meat is a ritual act which figures the bloody sacrifice at the altar. Meat is thus transformed from a living creature into a food item.

As to the third rule, the separation of meat and milk, it honors the procreative functions. The analogy between human and animal parturition is always implied, as the Mishnah shows in its comment on the edibility of the afterbirth found in the slaughtered dam: if the afterbirth had emerged in part, it is forbidden as food; "it is a token of young in a woman and a token of young in a beast."[22] Likewise this third rule honors the Hebrew mother and her initial unity with her offspring.

In conclusion I return to the researches of Tambiah and Bulmer. In each case a concern with sexual relations, approved or disapproved, is reflected on to the Thailand and Karam animal classifications. In the case of Israel the dominant concern would seem to be with the integrity of territorial boundaries. But Edmund Leach has pointed out how over and over again they were concerned with the threat to Israel's holy calling from marriages with outsiders. Foreign husbands and foreign wives led to false gods and political defections. So sex is not omitted from the meanings in the common meal. But the question is different. In the other cases the problems arose from rules about exchanging women. In this case the concern is to insist on not exchanging women.

Perhaps I can now suggest an answer to Ralph Bulmer's question about the abhorrence of the pig.

> Dr. Douglas tells us that the pig was an unclean beast to the Hebrew quite simply because it was a taxonomic anomaly, literally as the Old Testament says, because like the normal domestic animals it has a cloven hoof, whereas *un*like other cloven-footed beasts, it does not chew the cud. And she pours a certain amount of scorn on the commentators of the last 2,000 years who have taken alternative views and drawn attention to the creature's feeding habits, etc.

Dr. Bulmer would be tempted to reverse the argument and to say that the other animals are prohibited as part of an elaborate exercise for rationalizing[23]

> the prohibition of a beast for which there were probably multiple reasons for avoiding. It would seem equally fair, on the limited evidence available, to argue that the pig was accorded anomalous taxonomic status because it was unclean as to argue that it was unclean because of its anomalous taxonomic status.

On more mature reflection, and with the help of his own research, I can now see that the pig to the Israelites could have had a special taxonomic status equivalent to that of the otter in Thailand. It carries the odium of multiple pollution. First, it pollutes because it defies the classification of ungulates. Second, it pollutes because it eats carrion. Third, it pollutes because it is reared as food (and presumably as prime pork) by non-Israelites. An Israelite who betrothed a foreigner might have been liable to be offered a feast of pork. By these stages it comes plausibly to represent the utterly disapproved form of sexual mating and to carry all the odium that this implies. We now can trace a general analogy between the food rules and the other rules against mixtures: "Thou shalt not make the cattle to gender with beasts of any other kind" (Leviticus 19: 19). "Thou shalt not copulate with any beast" (Leviticus 18: 23). The common meal, decoded, as much as any poem, summarizes a stern, tragic religion.

We are left the question of why, when so much else had been forgotten[24] about the rules of purification and their meaning, the three rules governing the Jewish meal have persisted. What meanings do they still encode, unmoored as they partly are from their original social context? It would seem that whenever a people are aware of encroachment and danger, dietary rules controlling what goes into the body would serve as a vivid analogy of the corpus of their cultural categories at risk. But here I am, contrary to my own strictures, suggesting a universal meaning, free of particular social context, one which is likely to make sense whenever the same situation is perceived. We have come full-circle to Figure 1, with its two concentric circles. The outside boundary is weak, the inner one strong. Right through the diagrams summarizing the Mosaic dietary rules the focus was upon the integrity of the boundary at (b). Abominations of the water are those finless and scaleless creatures which lie outside that boundary. Abominations of the air appear less clearly in this light because the unidentified forbidden birds had to be shown as the widest circle from which the edible selection is drawn. If it be granted that they are predators, then they can be shown as a small subset in the unlisted set, that is as denizens of the air not fit for table because they eat blood. They would then be seen to threaten the boundary at (b) in the same explicit way as among the denizens of the land the circle (g) threatens it. We should therefore not conclude this essay without saying something more positive about what this boundary encloses. In the one case it divides edible from inedible. But it is more than a negative barrier of exclusion. In all the cases we have seen, it bounds the area of structured relations. Within that area rules apply. Outside it, anything goes. Following the argument we have established by which each level of meaning realizes the others which share a common structure, we can fairly say that the ordered system which is a meal represents all the ordered systems associated with it. Hence the strong arousal power of a threat to weaken or confuse that category. To take our analysis of the culinary medium further we should study what the poets say about the disciplines that they adopt. A passage from Roy Fuller's lectures helps to explain the flash of recognition and confidence which welcomes an ordered pattern. He is quoting Allen Tate, who said: "Formal versification is the primary structure of poetic order, the assurance to the reader and to the poet himself that the poet is in control of the disorder both outside him and within his own mind."[25]

The rules of the menu are not in themselves more or less trivial than the rules of verse to which a poet submits.

NOTES

I am grateful to Professor Basil Bernstein and to Professor M. A. K. Halliday for valuable suggestions and for criticisms, some of which I have not been able to meet. My thanks are due to my son James for working out the Venn diagrams used in this article.

1. Michael A. K. Halliday (1963), "Categories of the Theory of Grammar," *World, Journal of the Linguistic Circle of New York,* 17, 241–91.
2. The continuing discussion between anthropologists on the relation between biological and social facts in the understanding of kinship categories is fully relevant to the understanding of food categories.
3. Roland Barthes (1967), *Système de la mode,* Paris Editions Seuil.
4. Claude Lévi-Strauss (1970), *The Raw and the Cooked: Introduction to a Science of Mythology,* I, London, Jonathan Cape. The whole series in French is *Mythologiques:* I. *Le Cru et le cuit,* II. *Du Miel aux cendres,* III. *L'Origine des manières de table* (Paris, Plon, 1964–8).
5. Roman Jakobson and Claude Lévi-Strauss (1962), "Les Chats de Charles Baudelaire," *L'Homme,* 3, 5–21.
6. Michael Riffaterre (1967), "Describing Poetic Structures: Two Approaches to Baudelaire's *Les Chats,*" *Structuralism,* Yale French Studies 36 and 37.
7. Claude Lévi-Strauss, *The Savage Mind* (London, Weidenfeld & Nicolson, 1966; University of Chicago Press, 1962, 1966).
8. Halliday, "Categories of the Theory of Grammar," pp. 277–9.
9. Adrian C. Mayer (1960), *Caste and Kinship in Central India: A Village and its Region,* London, Routledge.
10. Louis Dumont, *Homo Hierarchicus: The Caste System and its Implications,* trans. M. Sainsbury (London, Weidenfeld & Nicolson, 1970; French ed., Gallimard, 1966), pp. 86–9.
11. See under "Marengo," *Larousse Gastronomique* (Hamlyn, 1961).
12. A. MacLeish (1960), *Poetry and Experience,* London, Bodley Head, p. 4.
13. *Ibid.*
14. Mary Douglas (1966), *Purity and Danger: An Analysis of Concepts of Pollution and Taboo,* London, Routledge.
15. Ralph Bulmer (1967), "Why Is the Cassowary not a Bird? A Problem of Zoological Taxonomy among the Karam of the New Guinea Highlands," *Man, new sex,* 2, 5–25.
16. S. J. Tambiah (1969), "Animals Are Good to Think and Good to Prohibit," *Ethnology,* 7, 423–59.
17. E. R. Leach (1969), "The Legitimacy of Solomon," *Genesis as Myth and Other Essays,* London, Jonathan Cape.
18. S. R. Driver and H. A. White, *The Polychrome Bible, Leviticus,* v.1. fn. 13.

19. Leach, "Legitimacy of Solomon."
20. H. Danby, trans. (1933), *The Mishnah,* London, Oxford University Press, p. 324.
21. See Jacob Milgrom (1971), "A Prolegomena to Leviticus 17: 17," *Journal of Biblical Literature,* 90, II: 149–56. This contains a textual analysis of the rules forbidding eating flesh with the blood in it which is compatible with the position herein advocated.
22. *Ibid,* p. 520.
23. Bulmer, "Why Is the Cassowary not a Bird?" p. 21.
24. Moses Maimonides (1904), *Guide for the Perplexed,* trans. M. Friedlander, London, Routledge, first ed., 1881.
25. Roy Fuller (1971), *Owls and Artificers: Oxford Lectures on Poetry,* London, Deutsch, p. 64.

5

The Semiotics of Food in the Bible

JEAN SOLER

How can we explain the dietary prohibitions of the Hebrews? To this day these rules—with variations, but always guided by the Mosaic laws—are followed by many orthodox Jews. Once a number of false leads, such as the explanation that they were hygienic measures, have been dismissed, the structural approach appears to be enlightening.

Lévi-Strauss has shown the importance of cooking, which is peculiar to man in the same manner as language. Better yet, cooking is a language through which a society expresses itself. For man knows that the food he ingests in order to live will become assimilated into his being, will become himself. There must be, therefore, a relationship between the idea he has formed of specific items of food and the image he has of himself and his place in the universe. There is a link between a people's dietary habits and its perception of the world.

Moreover, language and dietary habits also show an analogy of form. For just as the phonetic system of a language retains only a few of the sounds a human being is capable of producing, so a community adopts a dietary regime by making a choice among all the possible foods. By no means does any given individual eat everything; the mere fact that a thing is edible does not mean that it will be eaten. By bringing to light the logic that informs these choices and the interrelation among its constituent parts—in this case the various foods—we can outline the specific characteristics of a society, just as we can define those of a language.

The study of my topic is made easier by the existence of a *corpus* whose boundaries cannot be considered arbitrary; the dietary laws of the Hebrews have been laid down in a book, the Book, and more precisely in the first five books of the Bible, which are known as the Torah to the Jews and the Pentateuch to the Christians. This set of writings is composed of texts from various eras over a wide span of time. But to the extent that they have been sewn together, have coexisted and still do coexist in the consciousness of a people, it is advisable to study them together. I shall therefore leave aside the historical dimension in order to search for the rules that give cohesion to the different laws constituting the Law.

It is true, of course, that these five books tell a story, running from the creation of the world to the death of Moses, the man to whom these laws, and even this set of writings, are attributed. Attention will therefore have to be given to the order of the narrative; but whether and when the events mentioned in it actually occurred, whether or not the persons mentioned actually existed, and if so, when, has no bearing whatsoever on my analysis, any more than does the existence or nonexistence of God.

Man's food is mentioned in the very first chapter of the first book. It has its place in the plan of the Creation: "Behold, I have given you every plant yielding seed which is upon the face of all the earth, and every tree with seed in its fruit; you shall have them for your food" (Gen. 1: 29),[1] says Elohim. Paradise is vegetarian.

In order to understand why meat eating is implicitly but unequivocally excluded, it must be shown how both God and man are defined in the myth by their relationship to each other. Man has been made "in the image" of God (Gen. 1: 26–27), but he is not, nor can he be, God. This concept is illustrated by the dietary tabu concerning the fruit of two trees. After Adam and Eve have broken this prohibition by eating the fruit of one of these trees, Elohim says: "Behold, the man has become like one of us, knowing good and evil; and now, lest he put forth his hand and take also of the tree of life, and eat, and live forever" (Gen. 3: 22). This clearly marked distance between man and God, this fundamental difference, is implicitly understood in a threefold manner.

First, the immortality of the soul is unthinkable. All life belongs to God, and to him alone. God is Life, and man temporarily holds only a small part of it. We know that the notion of the immortality of the soul did not appear in Judaism until the second century B.C. and that it was not an indigenous notion.

Secondly, killing is the major prohibition of the Bible. Only the God who gives life can take it away. If man freely uses it for his own ends, he encroaches upon God's domain and oversteps his limits. From this it follows that meat eating is impossible. For in order to eat an animal, one must first kill it. But animals, like man, belong to the category of beings that have within them "a living soul." To consume a living being, moreover, would be tantamount to absorbing the principle that would make man God's equal.

The fundamental difference between man and God is thus expressed by the difference in their foods. God's are the living beings, which in the form of sacrifices (either human victims, of which Abraham's sacrifice represents a relic, or sacrificial animals) serve as his "nourishment" according to the Bible; man's are edible plants (for plants are not included among the "living things"). Given these fundamental assumptions, the origins of meat eating constitute a problem. Did men, then, at one point find a way to kill animals and eat them without prompting a cataclysm?

This cataclysm did indeed take place, and the Bible does speak of it. It was the Flood, which marks a breaking point in human history. God decided at first to do away with his Creation, and then he spared one family, Noah's, and one pair of each species of animal. A new era thus began after the Flood, a new Creation, which coincided with the appearance of a new dietary regime. "Every moving thing that lives shall be food for you; as I gave you the green plants, I give you everything" (Gen. 9: 3).

Thus, it is not man who has taken it upon himself to eat meat; it is God who has given him the right to do so. And the cataclysm does not come after, but before the change, an inversion that is frequently found in myths. Nevertheless, it must be understood that meat eating is not presented as a reward granted to Noah. If God has wanted "to destroy all flesh in which is the breath of life from under heaven" (Gen. 6: 17), it is because man has "corrupted" the entire earth: "and the earth was filled with violence" (Gen. 6:17), in other words, with murder. And while it is true that he spares Noah because Noah is "just" and even "perfect" (Gen. 6: 9), the human race that will come from him will not escape the evil that had characterized the human race from which he issued. The Lord says, after the Flood: "I will never again curse the ground because of man, for the imagination of man's heart is evil from his youth; neither will I ever again destroy every living creature as I have done" (Gen. 8: 21). In short, God takes note of the evil that is in man. A few verses later, he gives Noah permission to eat animals. Meat eating is given a negative connotation.

Yet even so, it is possible only at the price of a new distinction; for God adds the injunction: "Only you shall not eat flesh with its life, that is, its blood" (Gen. 9: 4). Blood becomes the signifier of the vital principle, so that it becomes possible to maintain the distance between man and God by expressing it in a different way with respect to food. Instead of the initial opposition between the eating of meat and the eating of plants, a distinction is henceforth made between flesh and blood. Once the blood (which is God's) is set apart, meat becomes desacralized—and permissible. The structure remains the same, only the signifying elements have changed.

At this stage the distinction between clean and unclean animals is not yet present, even though three verses in the account of the Flood refer to it. Nothing is said that would permit Noah to recognize these two categories of animals, and this distinction is out of place here, since the power to eat animals he is given includes all of them: "Every moving thing that lives shall be food for you."

It is not until Moses appears that a third dietary regime comes into being, one that is based on the prohibition of certain animals. Here we find a second breaking point in human history. For the covenant God had concluded with Noah included all men to be born from the sole survivor of the Flood (the absence of differentiation among men corresponded to the absence of differentiation among the animals they could consume), and the sign of that covenant was a cosmic and hence universal sign, the rainbow (Gen. 9: 12–17). The covenant concluded with Moses, however, concerns only one people, the Hebrews; to the new distinction between men corresponds the distinction of the animals they may eat: "I am the Lord your God, who have separated you from the peoples. You shall therefore make a distinction between the clean beasts and the unclean; and between the unclean bird and the clean; you shall not make yourselves abominable by beast or by bird or by anything with which the ground teems, which I have set apart for you to hold unclean" (Lev. 20: 24–25). The signs of this new covenant can only be individual, since they will have to become the distinctive traits of the Hebrew people. In this manner the Mosaic dietary code fulfills the same function as circumcision or the institution of the Sabbath. These three signs all involve a cut (a cut on the male sex organ: a partial castration analogous to an offering, which in return will bring God's blessing upon the organ that ensures the transmission of life and thereby the survival of the Hebrew people; a cut in the regular course of the days: one day of every seven is set apart, so that the sacrificed day will desacralize the others and bring God's blessing on their work; a cut in the *continuum* of the created animals—added to the already accomplished cut, applying to every animal, between flesh and blood, and later to be strengthened by an additional cut within each species decreed to be clean between the first-born, which are God's, and the others, which are thereby made more licit). The cut is at the origin of differentiation, and differentiation is the prerequisite of signification.

Dietary prohibitions are indeed a means of cutting a people off from others, as the Hebrews learned at their own expense. When Joseph's brothers journeyed to Egypt in order to buy wheat, he had a meal containing meat served to them: "They served him by himself, and them by themselves, for the Egyptians might not eat bread with the Hebrews, for that is an abomination to the Egyptians" (Gen. 43: 32). It is likely that the nomadic Hebrews already had dietary prohibitions but, according to Biblical history, they began to include their dietary habits among the defining characteristics of their people only after the exodus, as if they were taking their model from the Egyptian civilization.

Dietary habits, in order to play their role, must be different; but different from what? From those, unquestionably, of the peoples with whom the Hebrews were in contact. Proof of this is the famous injunction: "You shall not boil a kid in its mother's milk," for here a custom practiced among the people of that region was forbidden. Yet the dietary regime of the Hebrews was not contrary to the regimes of other peoples in every point; had this been the case they would have had very few things to eat! Why, then, did they strictly condemn some food items and not others? The answer must not be sought in the nature of the food item, any more than the sense of a word can be sought in the word itself. (It is contained in the dictionary, which defines that word by other words, which refer to yet other words, with all of these operations taking place within the dictionary). A social sign—in this case a dietary prohibition—cannot be understood in isolation. It must be placed into the context of the signs in the same area of life, together with which it constitutes a system; and this system in turn must be seen in relation to the systems in other areas, for the interaction of all these systems constitutes the socio-cultural system of a people. The constant features of this system should yield the fundamental structures of the Hebrew civilization or—and this may be the same thing—the underlying thought patterns of the Hebrew people.

One first constant feature naturally comes to mind in the notion of "cleanness," which is used to characterize the permissible foods. In order to shed light on this notion, it must first of all be seen as a conscious harking back to the Origins. To the extent that the exodus from Egypt and the revelation of Sinai represent a new departure in the history of the World, it can be assumed that Moses—or the authors of the system that bears his name—felt very strongly that this third Creation, lest it too fall into degradation, would have be to patterned after the myth of Genesis (whether that account was elaborated or only appropriated by Moses). Man's food would therefore be purest of all if it were patterned as closely as possible upon the Creator's intentions. Now the myth tells us that the food originally given to man was purely vegetarian. Has there been, historically, an attempt to impose a vegetarian regime on the Hebrews? There is no evidence to support this hypothesis, but the Bible does contain traces of such an attempt or, at any rate, of such an ideal. One prime trace is the fact that manna, the only daily nourishment of the Hebrews during the exodus, is shown as a vegetal substance: "It was like coriander seed, white, and the taste of it was like wafer made with honey" (Exod. 16: 31). Moreover, the Hebrews had large flocks, which they did not touch. Twice, however, the men rebelled against Moses because they wanted to eat meat. The first time, this happened in the wilderness of Sin: "Would that we have died by the hand of the Lord in the land of Egypt, when we sat by the flesh-pots" (Exod. 16: 3). God thereupon granted them the miracle of the quails. The second rebellion is reported in Numbers (11: 4): "O that we had meat to eat," wail the Hebrews. God agrees to repeat the miracle of the quails, but does so only unwillingly and even in great wrath: "You shall not eat one day, or two days, or five days, or ten days, or twenty days, but a whole month, until it comes out at your nostrils and becomes loathsome to you" (Num. 11: 19–20). And a great number of the Hebrews who fall upon the quails and gorge themselves die on the spot. Here, as in the myth of the Flood, meat is given a negative connotation. It is a concession God makes to man's imperfection.

Meat eating, then, will be tolerated by Moses, but with two restrictions. The tabu against blood will be reinforced, and certain animals will be forbidden. The setting apart of the blood henceforth becomes the occasion of a ritual. Before the meat can be eaten, the animal must be presented to the priest, who will perform the "peace offering," in which he pours the blood upon the altar. This is not only a matter of separating God's share from man's share; it also means that the murder of the animal that is to be eaten is redeemed by an offering. Under the elementary logic of retribution, any murder requires in compensation the murder of the murderer; only thus can the balance be restored. Since animals, like men, are "living souls," the man who kills an animal should himself be killed. Under this basic assumption, meat eating is altogether impossible. The solution lies in performing a ritual in which the blood of the sacrificial animal takes the place of the man who makes the offering.[2] "For the life of the flesh is in the blood, and I have given it for you upon the altar to make atonement for your souls; for it is the blood that makes atonement, by reason of the life" (Lev. 17: 11). But if a man kills an animal himself in order to eat it, "bloodguilt shall be imputed to that man; he has shed blood; and that man shall be cut off from among his people" (Lev. 17: 4); that is, he shall be killed. The importance of the blood tabu thus becomes very clear. It is not simply one prohibition among others; it is the *conditio sine qua non* that makes eating meat possible.

It should be noted that this ritual is attenuated in Deuteronomy. With the institution of a single sanctuary in Jerusalem, it became difficult for the Hebrews who lived outside the city to go to Jerusalem every time they wanted to eat meat. In this case, they were permitted to perform the offering of animals themselves. The procedure was to be the same as for hunting, where ritual offerings obviously could not be performed: "You may slaughter and eat flesh within any of your towns . . . as of the gazelle and as of the hart. Only you shall not eat the blood; you shall pour it out upon the earth like water" (Deut. 12: 15–16). This is a tangible example of how the variations of a system must adapt to the given infrastructure of geography.[3]

As for the prohibition of certain animals, we must now analyze two chapters (Lev. 11 and Deut. 14) devoted to the distinction between clean and unclean species. Neither of these texts, which are essentially identical, provides any explanation. The Bible only indicates the particular traits the clean animals must possess—though not always; for when dealing with the birds, it simply enumerates the unclean species.

The text first speaks of the animals living on land. They are "clean" if they have a "hoofed foot," a "cloven hoof," and if they "chew the cud." The first of these criteria is clearly meant to single out the herbivorous animals. The Hebrews had established a relationship between the foot of an animal and its feeding habits. They reasoned like Cuvier, who said, "All hoofed animals must be herbivorous, since they lack the means of seizing a prey."[4]

But why are herbivorous animals clean and carnivorous animals unclean? Once again, the key to the answer must be sought in Genesis, if indeed the Mosaic laws intended to conform as much as possible to the original intentions of the Creator. And in fact, Paradise was vegetarian for the animals as well. The verse dealing with human food, "I have given you every plant yielding seed which is upon the face of all the earth, and every tree with seed in its fruit; you shall have them for your food," is followed by a verse about the animals (and here, incidentally, we note a secondary differentiation, serving to mark the distance between humankind and the various species of animals): "And to every beast of the earth, and to every bird of the air, and to everything that has the breath of life, I have given every green plant for food" (Gen. 1: 29–30). Thus carnivorous animals are not included in the plan of the Creation. Man's problem with meat eating is compounded when it involved eating an animal that has itself consumed meat and killed other animals in order to do so. Carnivorous animals are unclean. If man were to eat them, he would be doubly unclean. The "hoofed foot" is thus the distinctive trait that contrasts with the claws of carnivorous animals—dog, cat, felines, etc.—for these claws permit them to seize their prey. Once this point is made, the prohibition against eating most of the birds that are cited as unclean becomes comprehensible: they are carnivorous, especially such birds of prey as "the eagle," which is cited at the head of the list.

But to return to the beasts of the earth. Why is the criterion "hoofed foot" complemented by two other criteria? The reason is that it is not sufficient to classify the true herbivores, since it omits pigs. Pigs and boars have hoofed feet, and while it is true that they are herbivores, they are also carnivorous.[5] In order to isolate the true herbivores it is therefore necessary to add a second criterion, "chewing the cud." One can be sure that ruminants eat grass; in fact, they eat it twice. In theory, this characteristic should be sufficient to distinguish true herbivores. But in practice, it is difficult to ascertain, especially in wild animals, which can properly be studied only after they are dead. Proof of this is the fact that the hare is considered to be a ruminant by the Bible (Lev. 11: 6 and Deut. 14: 7), which is false; but the error arose from mistaking the mastication of the rodents for rumination. This physiological characteristic therefore had to be reinforced by an anatomical criterion, the hoof, which in turn was strengthened by using as a model the hoof of the ruminants known to everyone: cows and sheep. (In the myth of Creation, livestock constitutes a separate category, distinct from the category of wild animals. There is no trace of the domestication of animals; livestock was created tame). This is why clean wild animals must conform to the domestic animals that may be consumed; as it happens, cows and sheep tread the ground on two toes, each encased in a layer of horn. This explains the third criterion listed in the Bible: the "cloven hoof."

One important point must be made here: The presence of the criterion "cloven hoof" eliminates a certain number of animals, even though they are purely herbivorous (the horse, the ass, and especially the three animals expressly cited in the Bible as "unclean": the camel, the hare, and the rock badger). A purely herbivorous animal is therefore not automatically clean. This is a necessary, though not a sufficient condition. In addition, it must also have a foot analogous to the foot that sets the norm: that of domestic animals. Any foot shape deviating from this model is conceived as a blemish, and the animal is unclean.

This notion of the "blemish" and the value attributed to it is elucidated in several passages of the Bible. Leviticus prohibits the sacrificing of animals, even of a clean species, if the individual animal exhibits any anomaly in relation to the normal type of the species: "And when any one offers a sacrifice of peace offerings to the Lord, to fulfill a vow or as a freewill offering, from the herd or from the flock, to be accepted it must be perfect; there shall be no blemish in it. Animals blind or disabled or mutilated or having a discharge or an itch or scabs, you shall not offer to the Lord or make of them an offering by fire upon the altar of the Lord" (Lev. 22: 21). This prohibition is repeated in Deuteronomy (17: 1): "You shall not sacrifice to the Lord your God an ox or a sheep in which is a blemish, any defect whatever; for that is an abomination to the Lord your God." The equation is stated explicitly: the blemish is an evil. A fundamental trait of the Hebrews' mental structures is uncovered here. There are societies in which impaired creatures are considered divine.

What is true for the animal is also true for man. The priest must be a wholesome man and must not have any physical defects. The Lord says to Aaron (Lev. 21: 17–18): "None of your descendants throughout their generations who has a blemish may approach to offer the bread of his God. For no one who has a blemish shall draw near, a man blind or lame, or one who has a mutilated face or a limb too long, or one who has an injured foot or an injured hand, or a hunchback, or a dwarf, or a man with a defect in his sight or an itching disease or scabs or crushed testicles; no man of the descendants of Aaron the priest who has a blemish shall come near to offer the Lord's offerings by fire." The men who participate in cultic acts must be true men: "He whose testicles are crushed or whose male member is cut off shall not enter the assembly of the Lord" (Deut. 23: 1). To be whole is one of the components of "cleanness"; eunuchs and castrated animals are unclean.

To the blemish must be added alteration, which is a temporary blemish. Periodic losses of substance are unclean, whether they be a man's emission of semen or a woman's menstruation (Lev. 15). The most unclean thing of all will therefore be death, which is the definitive loss of the breath of life and the irreversible alteration of the organism. And indeed, death is the major uncleanness for the Hebrews. It is so strong that a high priest (Lev. 21: 11) or a Nazirite (Num. 6: 6–7) may not go near a dead body, even if it is that of his father or his mother, notwithstanding the fact that the Ten Commandments order him to "honor" them.

The logical scheme that ties cleanness to the absence of blemish or alteration applies to things as well as to men or animals. It allows us to understand the status of ferments and fermented substances. I shall begin with the prohibition of leavened bread during the Passover. The explanation given in the Bible does not hold; it says that it is a matter of commemorating the exodus from Egypt when the Hebrews, in their haste, did not have time to let the dough rise (Exod. 12: 34). If this were the reason, they would have been obliged to eat poorly leavened or half-baked bread; but why bread without leavening? In reality, even if the Passover is a celebration whose meaning may have changed in the course of the ages—and this is the case with other institutions, notably the Sabbath—it functions as a commemoration of the Origins, a celebration not only of the exodus from Egypt and the birth of a nation but also of the beginning of the religious year at the first full moon after the vernal equinox. The Passover feast is a sacrifice of renewal, in which the participants consume the food of the Origins.[6] This ritual meal must include "bitter herbs," "roasted meat," and "unleavened bread" (Exod. 12: 8). The bitter herbs must be understood, it would seem, as the opposite of vegetables, which are produced by agriculture. Roast meat is the opposite of boiled meat, which explicitly proscribed in the text (Exod. 12: 9): the boiling of meat, which implies the use of receptacles obtained by an industry, albeit a rudimentary form of it, is a late stage in the preparation of food. As for the unleavened bread, it is the bread of the Patriarchs. Abraham served cakes made of fine meal to the three messengers of God on their way to Sodom (Gen. 18: 6). These cakes were undoubtedly identical to those that Lot prepared shortly thereafter for the same messengers: "and he made them a feast, and baked unleavened bread, and they ate" (Gen. 19: 3). But unleavened bread is clean not only because it is the bread of the Origins. It is clean also and above all because the flour of which it is made is not changed by the ferment of the leavening; it is true to its natural state. This interpretation allows us to understand why fermented foods cannot be used as offerings by fire: "No cereal offering which you shall bring to the Lord shall be made with leaven; for you shall burn no leaven nor any honey as an offering by fire to the Lord" (Lev. 2: 11). A fermented substance is an altered substance, one that has become other. Fermentation is the equivalent of a blemish. Proof *a contrario* is the fact that just as fermentation is forbidden, so salt is mandatory in all offerings (Lev. 2: 13). Thus, there is a clear-cut opposition between fermentation, which alters a substance's being, and salt, which preserves it in its natural state. Leavened bread, honey,[7] and wine all have the status of secondary food items; only the primary foods that have come from the hands of the Creator in their present form can be used in the sacred cuisine of the offering. It is true, of course, that wine is used in cultic libations. But the priest does not consume it; indeed he must abstain from all fermented liquids before officiating in order to "distinguish between the holy and the common, and between the clean and the unclean" (Lev. 10: 10). Fermented liquids alter man's judgment because they are themselves altered substances. The libation of wine must be seen as the parallel of the libation of blood, which it accompanies in burnt offerings. Wine is poured upon the altar exactly like blood, for it is its equivalent in the plant; wine is the "blood of the grapes" (Gen. 49: 11, etc.).

To return to my argument, then, the clean animals of the earth must conform to the plan of the Creation, that is, be vegetarian; they must also conform to their ideal models, that is, be without blemish. In order to explain the distinction between clean and unclean fish, we must once again refer to the first chapter of Genesis. In the beginning God created the three elements, the firmament, the water, and the earth; then he created three kinds of animals out of each of these elements: "Let the waters bring forth swarms of living creatures, and let birds fly above the earth across the firmament of the heavens" (Gen. 1: 20); "Let the earth bring forth living creatures according to their kinds, cattle and creeping things and beasts of the earth according to their kinds" (Gen. 1: 24). Each animal is thus tied to one element, and one only.[8] It has issued from that element and must live there. Chapter 11 of Leviticus and chapter 14 of Deuteronomy reiterate this classification into three groups: creatures of the earth, the water, and the air. Concerning the animals of the water, the two texts only say: "Everything in the waters that has fins and scales . . . you may eat." All other creatures are unclean. It must be understood that the fin is the proper organ of locomotion for animals living in the water. It is the equivalent of the leg in the animal living on land and of the wing in the animal that lives in the air. Recall also that locomotion distinguishes animals from ants, which in the Bible are not included in the category of "living" things. In this manner, the animals of the earth must walk, fish must swim, and birds must fly. Those creatures of the sea that lack fins and do not move about (mollusks) are unclean. So are those that have legs and can walk (shellfish), for they live in the water yet have the organs of a beast of the earth and are thus at home in two elements.

In the same manner, scales are contrasted with the skin of the beasts of the land and with the feathers of the birds. As far as the latter are concerned, the Biblical expression "birds of the air" must be taken quite literally; it is not a poetic image but a definition. In the formulation "the likeness of any winged bird that flies in the air" (Deut. 4: 17), the three distinctive traits of the clean bird are brought together: "winged," "which flies," and "in the air." If a bird has wings but does not fly, (the ostrich, for instance, that is cited in the text), it is unclean. If it has wings and can fly but spends most of its time in the water instead of living in the air, it is unclean (and the Bible mentions the swan, the pelican, the heron, and all the stilted birds). Insects pose a problem. "All winged animals that go upon all fours are an abomination to you," says Leviticus (11: 20). This is not a discussion of four-legged insects, for the simple reason that all insects have six. The key expression is "go upon" (walk). The insects that are meant here are those that "go upon all fours," like the normal beasts of the earth, the quadrupeds. Their uncleanness comes from the fact that they walk rather than fly, even though they are "winged." The exception mentioned in Leviticus (11: 21) only confirms the rule; no uncleanness is imputed to insects that have "legs above their feet, with which to leap on the earth." Leaping is a mode of locomotion midway between walking and flying. Leviticus feels that it is closer to flying and therefore absolves these winged grasshoppers. Deuteronomy, however, is not convinced and prohibits all winged insects (14: 19).

Leviticus also mentions, toward the end, some unclean species that cannot be fitted into the classification of three groups, and it is for this reason, no doubt, that Deuteronomy does not deal with them. The first of these are the reptiles. They belong to the earth, or so it seems, but have no legs to walk on. "Upon your belly you shall go," God had said to the serpent (Gen. 3: 14). This is a curse. Everything that creeps and goes on its belly is condemned. These animals live more under the earth than on it. They were not really "brought forth" by the earth, according to the expression of Genesis 1: 24. They are not altogether created. And like the serpent, the centipede is condemned (Lev. 11: 30) in the expression "whatever has many feet" (Lev. 11: 42). Having too many feet or none at all falls within the same category; the clean beast of the earth has four feet, and not just any kind of feet either, as we have seen.

All these unclean animals are marked with a blemish; they show an anomaly in their relation to the element that has "brought them forth" or to the organs characteristic of life, and especially locomotion, in that element. If they do not fit into any class, or if they fit into two classes at once, they are unclean. They are unclean because they are unthinkable. At this point, instead of stating once again that they do not fit into the plan of the Creation, I should like to advance the hypothesis that the dietary regime of the Hebrews, as well as their myth of the Creation, is based upon a taxonomy in which man, God, the animals, and the plants are strictly defined through their relationships with one another in a series of opposites. The Hebrews conceived of the order of the world as the order underlying the creation of the world. Uncleanness, then, is simply disorder, wherever it may occur.

Concerning the raising of livestock and agriculture, Leviticus 19: 19 mentions the following prohibition: "You shall not let your cattle breed with a different kind." A variant is found in Deuteronomy 22: 10: "You shall not plow with an ox and an ass together." The reason is that the animals have been created (or classified) "each according to its kind," an expression that is a very leitmotif of the Bible. Just as a clean animal must not belong to two different species (be a hybrid), so man is not allowed to unite two animals of different species. He must not mix that which God (or man) has separated, whether the union takes place in a sexual act or only under the yoke. Consider what is said about cultivated plants: "You shall not sow your field with two kinds of seeds" (Lev. 19: 19), an injunction that appears in Deuteronomy as: "You shall not sow your vineyard with two kinds of seed." The same prohibition applies to things: "nor shall there come upon you a garment of cloth made of two kinds of stuff" (Lev. 19: 19). In Deuteronomy 22: 11, this becomes: "You shall not wear a mingled stuff, wool and linen together." Here the part plant, part animal origin of the material further reinforces the distinction. In human terms, the same schema is found in the prohibition of mixed marriages—between Hebrews and foreigners—(Deut. 7: 3), and also in the fact that a man of mixed blood (offspring of a mixed marriage) or, according to a different interpretation, a bastard (offspring of adultery) may not enter the assembly of the Lord (Deut. 23: 3). This would seem to make it very understandable that the Hebrews did not accept the divine nature of Jesus. A God-man, or a God become man, was bound to offend their logic more than anything else.[9] Christ is the absolute hybrid.

A man is a man, or he is God. He cannot be both at the same time. In the same manner, a human being is either a man or a woman, not both: homosexuality is outlawed (Lev. 18: 22). The prohibition is extended even to clothes: "A woman shall not wear anything that pertains to a man, nor shall a man put on a woman's garment" (Deut. 22: 5). Bestiality is also condemned (Lev. 18: 20) and, above all, incest (Lev. 18: 6 ff.): "She is your mother, you shall not uncover her nakedness." This tautological formulation shows the principle involved here: once a woman is defined as "mother" in relation to a boy, she cannot also be something else to him. The incest prohibition is a logical one. It thus becomes evident that the sexual and the dietary prohibitions of the Bible are coordinated. This no doubt explains the Bible's most mysterious prohibition: "You shall not boil a kid in its mother's milk" (Exod. 23: 19 and 34: 26; Deut. 14: 21). These words must be taken quite literally. They concern a mother and her young. They can be translated as: you shall not put a mother and her son into the same pot, any more than into the same bed.[10] Here as elsewhere, it is a matter of upholding the separation between two classes or two types of relationships. To abolish distinction by means of a sexual or culinary act is to subvert the order of the world. Everyone belongs to one species only, one people, one sex, one category. And in the same manner, everyone has only one God: "See now that I, even I, am he, and there is no God beside me" (Deut. 32: 39). The keystone of this order is the principle of identity, instituted as the law of every being.

The Mosaic logic is remarkable for its rigor, indeed its rigidity. It is a "stiff-necked" logic, to use the expression applied by Yahveh to his people. It is self-evident that the very inflexibility of this order was a powerful factor for unification and conservation in a people that wanted to "dwell alone."[11] On the other hand, however, the Mosaic religion, inseparable as it is from the sociocultural system of the Hebrews, could only lose in power of diffusion what it gained in power of concentration. Christianity could only be born by breaking with the structures that separated the Hebrews from the other peoples. It is not surprising that one of the decisive ruptures concerned the dietary prescriptions. Matthew quotes Jesus as saying: "Not what goes into the mouth defiles a man, but what comes out of the mouth, this defiles a man" (15: 11). Similar words are reported by Mark, who comments: "Thus he declared all foods clean" (7: 19). The meaning of this rejection becomes strikingly clear in the episode of Peter's vision at Jaffa (Acts 10): a great sheet descends from heaven with all kinds of clean and unclean animals in it, and God's voice speaks: "Rise, Peter; kill and eat." Peter resists the order twice, asserting that he is a good Jew and has never eaten anything unclean. But God repeats his order a third time. Peter's perplexity is dispelled by the arrival of three men sent by the Roman centurion Cornelius, who is garrisoned in Caesarea. Cornelius wants to hear Peter expound the new doctrine he is propagating. And Peter, who had hitherto been persuaded that Jesus' reform was meant only for the Jews, now understands that it is valid for the Gentiles as well. He goes to Caesarea, shares the meal of a non-Jew, speaks to Cornelius, and baptizes him. Cornelius becomes the first non-Jew to be converted to Christianity. The vision in which the distinction between clean and unclean foods was abolished had thus implied the abolition of the distinction between Jews and non-Jews.

From this starting point, Christianity could begin its expansion, grafting itself onto the Greco-Roman civilization, which, unlike the Hebrew civilization, was ready to welcome all blends, and most notably a God-man. A new system was to come into being, based on new structures. This is why the materials it took from the older system assumed a different value. Blood, for instance, is consumed by the priest in the sacrifice of the Mass in the form of its signifier: "the blood of the grape." This is because the fusion between man and God is henceforth possible, thanks to the intermediate term, which is Christ. Blood, which had acted as an isolator between two poles, now becomes a conductor. In this manner, everything that Christianity has borrowed from Judaism, every citation of the Biblical text in the text of Western civilization (in French literature, for example), must in some way be "tinkered with," to use Lévi-Strauss's comparison.[12]

By contrast, whatever variations the Mosaic system may have undergone in the course of history, they do not seem to have shaken its fundamental structures. This logic, which sets up its terms in contrasting pairs and lives by the rule of refusing all that is hybrid, mixed, or arrived at by synthesis and compromise, can be seen in action to this day in Israel, and not only in its cuisine.

NOTES

1. *The Oxford Annotated Bible with the Apocrypha,* rev. standard ed., ed. Herbert G. May and Bruce M. Metzger (New York and Oxford, 1965).
2. That the life of an animal can atone for/save the life of the men who have sacrificed it can be seen in the episode of Exod. 12, where, during the night preceding the exodus from Egypt, the Hebrews sacrifice a lamb (the Passover lamb) and paint the doors of their houses with its blood. During that night, God strikes all the first-born of Egypt, except those who live in the houses marked with the blood. In Abraham's sacrifice also, the life of an animal and the life of a child can be made to stand for each other.
3. In keeping with the principle of the arbitrary nature of the sign, life can have other signi-

fiers than blood. In certain societies, for instance, it is the head, the heart, or the womb. In Leviticus itself, the fat that covers the entrails is forbidden to man and set apart for God (3: 16–17). The metaphoric use of the word also seems to indicate that fat is conceived as the vital substance of the solid parts of the body: "and I will give you the best of the land of Egypt, and you shall eat the fat of the land" (Gen. 45: 18, etc.). The sciatic nerve, which also may not be eaten, is perhaps interpreted as the element *par excellence* of locomotion, a privilege that belongs to living beings only. As Jacob wrestled with the angel, he was paralyzed when this nerve was touched (Gen. 32: 26–33). Fat and the sciatic nerve may well be secondary variants of blood in a different context.

4. Cited in the *Dictionnaire Robert,* s.v. "*sabot.*" See also F. Jacob, *La Logique du vivant* (Paris, 1970), p. 119.
5. The boar with its tusks, which are hyperdeveloped canine teeth, was naturally included among the wild beasts of which it is said: "And I shall loose the wild beasts among you, which shall rob you of your children, and destroy your cattle" (Lev. 26: 22).
6. Cf. Mircea Eliade, *Aspects du mythe* (Paris, 1963), p. 59: "To take nourishment is not simply a physiological act, but also a 'religious' act: one eats the creations of the Supernatural Beings, and one eats them as the mythical ancestors ate them for the first time at the beginning of the world."
7. See C. Lévi-Strauss, *Du miel aux cendres* (Paris, 1964), p. 253: Honey is an already "prepared" item; "it can be consumed fresh or fermented"; and it "pours forth ambiguity from each one of its facets."
8. In *Purity and Danger* (London, 1966), a work that came to my attention after I had finished writing the present study, Mary Douglas adopts a similar approach, and the similarity of our conclusions on this particular point is striking indeed.
9. Cf. the Gospel according to John (10: 31–33): "The Jews took up stones again to stone him. Jesus answered them, 'I have shown you many good works from the Father; for which of these do you stone me?' The Jews answered him: 'It is not for a good work that we stone you but for blasphemy; because you being a man, make yourself God.'"
10. Cf. the prohibition against taking the mother, the young ones, or the eggs from a bird's nest. Here the eggs are sufficient to represent the young ones, just as the milk represents the kid's mother (Deut. 22: 6–7). See also the prohibition against sacrificing on the same day a cow or a ewe and her young (Lev. 22: 28). Both of these acts might lead to culinary incest.
11. "Lo, a people dwelling alone and not reckoning itself among the nations!" (Num. 23: 9).
12. C. Lévi-Strauss, *La Pensée sauvage* (Paris, 1962), pp. 26 ff. English translation. *The Savage Mind* (Chicago, 1966).

6

The Abominable Pig

MARVIN HARRIS

An aversion to pork seems at the outset even more irrational than an aversion to beef. Of all domesticated mammals, pigs possess the greatest potential for swiftly and efficiently changing plants into flesh. Over its lifetime a pig can convert 35 percent of the energy in its feed to meat compared with 13 percent for sheep and a mere 6.5 percent for cattle. A piglet can gain a pound for every three to five pounds it eats while a calf needs to eat ten pounds to gain one. A cow needs nine months to drop a single calf, and under modern conditions the calf needs another four months to reach four hundred pounds. But less than four months after insemination, a single sow can give birth to eight or more piglets, each of which after another six months can weigh over four hundred pounds. Clearly, the whole essence of pig is the production of meat for human nourishment and delectation. Why then did the Lord of the ancient Israelites forbid his people to savor pork or even to touch a pig alive or dead?

> Of their flesh you shall not eat, and their carcasses you shall not touch; they are unclean to you (Lev. 11: 1) . . . everyone who touches them shall be unclean. (Lev. 11:24)

Unlike the Old Testament, which is a treasure trove of forbidden flesh, the Koran is virtually free of meat taboos. Why is it the pig alone who suffers Allah's disapproval?

> These things only has He forbidden you: carrion, blood, and the flesh of swine.
> (Holy Koran 2, 168)

For many observant Jews, the Old Testament's characterization of swine as "unclean" renders the explanation of the taboo self-evident: "Anyone who has seen the filthy habits of the swine will not ask why it is prohibited," says a modern rabbinical authority. The grounding of the fear and loathing of pigs in self-evident piggishness goes back at least to the time of Rabbi Moses Maimonides, court physician to the Islamic emperor Saladin during the twelfth century in Egypt. Maimonides shared with his Islamic hosts a lively disgust for pigs and pig eaters, especially Christian pigs and pig eaters: "The principal reason why the law forbids swine-flesh is to be found in the circumstance that its habits and food are very filthy and loathsome." If the law allowed Egyptians and Jews to raise pigs, Cairo's streets and houses would become as filthy as those of Europe, for "the mouth of a swine is as dirty as dung itself." Maimonides could only tell one side of the story. He had never seen a clean pig. The pig's penchant for excrement is not a defect of its nature but of the husbandry of its human masters. Pigs prefer and thrive best on roots, nuts, and grains; they eat excrement only when nothing better presents itself. In fact, let them get hungry enough, and they'll even eat each other, a trait which they share with other omnivores, but most notably with their own masters. Nor is wallowing in filth a natural characteristic of swine. Pigs wallow to keep themselves cool; and they much prefer a fresh, clean mudhole to one that has been soiled by urine and feces.

In condemning the pig as the dirtiest of animals, Jews and Moslems left unexplained their more tolerant attitude toward other dung-eating domesticated species. Chickens and goats, for example, given motivation and opportunity, also readily dine on dung. The dog is another domesticated creature which easily develops an appetite for human feces. And this was especially true in the Middle East, where dung-eating dogs filled the scavenging niche left vacant by the ban on pigs. Jahweh prohibited their flesh, yet dogs were not abominated, bad to touch, or even bad to look at, as were pigs.

Maimonides could not be entirely consistent in his efforts to attribute the abstention from pork to the pig's penchant for feces. The Book of Leviticus prohibits the flesh of many other creatures, such as cats and camels, which are not notably inclined to eat excrement. And with the exception of the pig, had not Allah said all the others were good to eat? The fact that Maimonides's Moslem emperor could eat every kind of meat except pork would have made it impolitic if not dangerous to identify the biblical sense of cleanliness exclusively with freedom from the taint of feces. So instead of adopting a cleaner-than-thou attitude, Maimonides offered a proper court physician's theory of the entire set of biblical aversions: the prohibited items were not good to eat because not only was one of them—the pig—filthy from eating excrement but all of them were not good for you. "I maintain," he said, "that food forbidden by the Law is unwholesome." But in what ways were the forbidden foods unwholesome? The great rabbi was quite specific in the case of pork: it "contained more moisture than necessary and too much superfluous matter." As for the other forbidden foods, their "injurious character" was too self-evident to merit further discussion.

Maimonides's public health theory of pork avoidance had to wait seven hundred years before it acquired what seemed to be a scientific justification. In 1859 the first clinical association between trichinosis and undercooked pork was established, and from then on it became the most popular explanation of the Jewish and Islamic pork taboo. Just as Maimonides said, pork was unwholesome. Eager to reconcile the Bible with the findings of medical science, theologians began to embroider a whole series of additional public health explanations for the other biblical food taboos: wild animals and beasts of burden were prohibited because the flesh gets too tough to be digested properly; shellfish were to be avoided because they serve as vectors of typhoid fever; blood is not good to eat because the bloodstream is a perfect medium for microbes. In the case of pork this line of rationalization had a paradoxical outcome. Reformist Jews began to argue that since they now understood the scientific and medical basis of the taboos, pork avoidance was no longer necessary; all they had to do was to see to it that the meat was thoroughly cooked. Predictably, this provoked a reaction among Orthodox Jews, who were appalled at the idea that the book of God's law was being relegated to the "class of a minor medical text." They insisted that God's purpose in Leviticus could never be fully comprehended; nonetheless the dietary laws had to be obeyed as a sign of submission to divine will.

Eventually the trichinosis theory of pork avoidance fell out of favor largely on the grounds that a medical discovery made in the nineteenth century could not have been known thousands of years ago. But that is not the part of the theory that bothers me. People do not have to possess a scientific understanding of the ill effects of certain foods in order to put such foods on their bad-to-eat list. If the consequences of eating pork had been exceptionally bad for their health, it would not have been necessary for the Israelites to know about trichinosis in order to ban its consumption. Does one have to understand the molecular chemistry of toxins in order to know that some mushrooms are dangerous? It is essential for my own explanation of the pig taboo that the trichinosis theory be laid to rest on entirely different grounds. My contention is that there is absolutely nothing exceptional about pork as a source of human disease. All domestic animals are potentially hazardous to human health. Undercooked beef, for example, is a prolific source of tapeworms, which can grow to a length of sixteen to twenty feet inside the human gut, induce a severe case of anemia, and lower the body's resistance to other diseases. Cattle, goat, and sheep transmit the bacterial disease known as brucellosis, whose symptoms include fever, aches, pains, and lassitude. The most dangerous disease transmitted by cattle, sheep, and goats is anthrax, a fairly common disease of both animals and humans in Europe and Asia until the introduction of Louis Pasteur's anthrax vaccine in 1881. Unlike trichinosis, which does not produce symptoms in the majority of infected individuals and rarely has a fatal outcome, anthrax runs a swift course that begins with an outbreak of boils and ends in death.

If the taboo on pork was a divinely inspired health ordinance, it is the oldest recorded case of medical malpractice. The way to safeguard against trichinosis was not to taboo pork but to taboo undercooked pork. A simple advisory against undercooking pork would have sufficed: "Flesh of swine thou shalt not eat until the pink has been cooked from it." And come to think of it, the same advisory should have been issued for cattle, sheep, and goats. But the charge of medical malpractice against Jahweh will not stick.

The Old Testament contains a rather precise formula for distinguishing good-to-eat flesh from forbidden flesh. This formula says nothing about dirty habits or unhealthy meat. Instead it directs attention to certain anatomical and physiological features of animals that are good to eat. Here is what Leviticus 11: 1 says:

> Whatever parts the hoof and is cloven footed and chews the cud among animals, you may eat.

Any serious attempt to explain why the pig was not good to eat must begin with this formula and not with excrement or wholesomeness, about which not a word is said. Leviticus goes on to state explicitly of the pig that it only satisfies one part of the formula. "It divideth the hoof." But the pig does not satisfy the other part of the formula: "It cheweth not the cud."

To their credit, champions of the good-to-eat school have stressed the importance of the cud-chewing, split-hoof formula as the key to understanding Jahweh's abomination of the pig. But they do not view the formula as an outcome of the way the Israelites used domestic animals. Instead they view the way the Israelites used domestic animals as an outcome of the formula. According to anthropologist Mary Douglas, for example, the cud-chewing, split-hoof formula makes the split-hoof but non-cud-chewing pig a thing that's "out of place." Things that are "out of place" are dirty, she argues, for the essence of dirt is "matter out of place." The pig, however, is more than out of place; it is neither here nor there. Such things are both dirty and dangerous. Therefore the pig is abominated as well as not good to eat. But doesn't the force of this argument lie entirely in its circularity? To observe that the pig is out of place taxonomically is merely to observe that Leviticus classifies good-to-eat animals in such a way as to make the pig bad to eat. This avoids the question of why the taxonomy is what it is.

Let me attend first to the reason why Jahweh wanted edible animals to be cud-chewers. Among animals raised by the ancient Israelites, there were three cud-chewers: cattle, sheep, and goats. These three animals were the most important food-producing species in the ancient Middle East not because the ancients happened capriciously to think that cud-chewing animals were good to eat (and good to milk) but because cattle, sheep, and goats are ruminants, the kind of herbivores which thrive best on diets consisting of plants that have a high cellulose content. Of all domesticated animals, those which are ruminants possess the most efficient system for digesting tough fibrous materials such as grasses and straw. Their stomachs have four compartments which are like big fermentation "vats" in which bacteria break down and soften these materials. While cropping their food, ruminants do little chewing. The food passes directly to the rumen, the first of the compartments, where it soon begins to ferment. From time to time the contents of the rumen are regurgitated into the mouth as a softened bolus—the "cud"—which is then chewed thoroughly and sent on to the other "vats" to undergo further fermentation.

The ruminant's extraordinary ability to digest cellulose was crucial to the relationship between humans and domesticated animals in the Middle East. By raising animals that could "chew the cud," the Israelites and their neighbors were able to obtain meat and milk without having to share with their livestock the crops destined for human consumption. Cattle, sheep, and goats thrive on items like grass, straw, hay, stubble, bushes, and leaves—feeds whose high cellulose content renders them unfit for human consumption even after vigorous boiling. Rather than compete with humans for food, the ruminants further enhanced agricultural productivity by providing dung for fertilizer and traction for pulling plows. And they were also a source of fiber and felt for clothing, and of leather for shoes and harnesses.

I began this puzzle by saying that pigs are the most efficient mammalian converters of plant foods into animal flesh, but I neglected to say what kinds of plant foods. Feed them on wheat, maize, potatoes, soybeans, or anything low in cellulose, and pigs will perform veritable miracles of transubstantiation; feed them on grass, stubble, leaves, or anything high in cellulose, and they will lose weight.

Pigs are omnivores, but they are not ruminants. In fact, in digestive apparatus and nutrient requirements pigs resemble humans in more ways than any mammal except monkeys and apes, which is why pigs are much in demand for medical research concerned with atherosclerosis, calorie-protein malnutrition, nutrient absorption, and metabolism.

But there was more to the ban on pork than the pig's inability to thrive on grass and other high-cellulose plants. Pigs carry the additional onus of not being well adapted to the climate and ecology of the Middle East. Unlike the ancestors of cattle, sheep, or goats, which lived in hot, semiarid, sunny grasslands, the pig's ancestors were denizens of well-watered, shady forest glens and riverbanks. Everything about the pig's body heat-regulating system is ill suited for life in the hot, sun-parched habitats which were the homelands of the children of Abraham. Tropical breeds of cattle, sheep, and goats can go for long periods without water, and can either rid their bodies of excess heat through perspiration or are protected from the sun's rays by light-colored, short fleecy coats (heat-trapping heavy wool is a characteristic of cold-climate breeds). Although a perspiring human is said to "sweat like a pig," the expression lacks an anatomical basis. Pigs can't sweat—they have no functional sweat glands. (Humans are actually the sweatiest of all animals.) And the pig's sparse coat offers little protection against the sun's rays. Just how does the pig keep cool? It does a lot of panting, but mostly it depends on wetting itself down with moisture derived from external sources. Here, then, is the explanation for the pig's love of wallowing in mud. By wallowing, it dissipates heat both by evaporation from its skin and by conduction through the cool ground. Experiments show that the cooling effect of mud is superior to that of water. Pigs whose flanks are thoroughly smeared with mud continue to show peak heat-dissipating evaporation for more than twice as long as pigs whose flanks are merely soaked with water, and here also is the explanation for some of the pig's dirty habits. As temperatures rise above thirty degrees celsius (eighty-six degrees Fahrenheit), a pig deprived of clean mudholes will become desperate and begin to wallow in its own feces and urine in order to avoid heat stroke. Incidentally, the larger a pig gets, the more intolerant it becomes of high ambient temperatures.

Raising pigs in the Middle East therefore was and still is a lot costlier than raising ruminants, because pigs must be provided with artificial shade and extra water for wallowing, and their diet must be supplemented with grains and other plant foods that humans themselves can eat.

To offset all these liabilities pigs have less to offer by way of benefits than ruminants. They can't pull plows, their hair is unsuited for fiber and cloth, and they are not suited for milking. Uniquely among large domesticated animals, meat is their most important produce (guinea pigs and rabbits are smaller equivalents; but fowl produce eggs as well as meat).

For a pastoral nomadic people like the Israelites during their years of wandering in search of lands suitable for agriculture, swineherding was out of the question. No arid-land pastoralists herd pigs for the simple reason that it is hard to protect them from exposure to heat, sun, and lack of water while moving from camp to camp over long distances. During their formative years as a nation, therefore, the ancient Israelites could not have consumed significant quantities of pork even had they desired it. This historical experience undoubtedly contributed to the development of a traditional aversion to pig meat as an unknown and alien food. But why was this tradition preserved and strengthened by being written down as God's law long after the Israelites had become settled farmers? The answer as I see it is not that the tradition born of pastoralism continued to prevail by mere inertia and ingrown habit, but that it was preserved because pig raising remained too costly.

Critics have opposed the theory that the ancient Israelite pork taboo was essentially a cost/benefit choice by pointing to evidence of pigs being raised quite successfully in many parts of the Middle East including the Israelite's promised land. The facts are not in dispute. Pigs have indeed been raised for ten thousand years in various parts of the Middle East—as long as sheep and goats, and even longer than cattle. Some of the oldest Neolithic villages excavated by archaeologists—Jericho in Jordan, Jarmo in Iraq, and Argissa-Magulla in Greece—contain pig bones with features indicative of the transition from wild to domesticated varieties. Several Middle Eastern pre-Bronze Age villages (4000 B.C. to 2000 B.C.) contain concentrated masses of pig remains in association with what archaeologists interpret as altars and cultic centers, suggestive of ritual pig slaughter and pig feasting. We know that some pigs were still being raised in the lands of the Bible at the beginning of the Christian era. The New Testament (Luke) tells us that in the country of the Garadines near Lake Galilee Jesus cast out devils from a man named Legion into a herd of swine feeding on the mountain. The swine rushed down into the lake and drowned themselves, and Legion was cured. Even modern-day Israelites continue to raise thousands of swine in parts of northern Galilee. But from the very beginning, fewer pigs were raised than cattle, sheep, or goats. And more importantly, as time went on, pig husbandry declined throughout the region.

Carlton Coon, an anthropologist with many years of experience in North America and the Levant, was the first scholar to offer a cogent explanation of why this general decline in pig husbandry had occurred. Coon attributed the fall of the Middle Eastern pig to deforestation and human population increase. At the beginning of the Neolithic period, pigs were able to root in oak and beech forests which provided ample shade and wallows as well as acorns, beechnuts, truffles, and other forest floor products. With an increase in human population density, farm acreage increased and the oak and beech forests were destroyed to make room for planted crops, especially for olive trees, thereby eliminating the pig's ecological niche.

To update Coon's ecological scenario, I would add that as forests were being destroyed, so were marginal farmlands and grazing lands, the general succession being from forest to cropland to grazing land to desert, with each step along the way yielding a greater premium for raising ruminants and a greater penalty for raising swine. Robert Orr Whyte, former director general of the United Nations Food and Agricultural Organization, estimated that in Anatolia the forests shrank from 70 percent to 13 percent of the total land area between 5000 B.C. and the recent past. Only a fourth of the Caspian shorefront forest survived the process of population increase and agricultural intensification; half of the Caspian mountainous humid forest; a fifth to a sixth of the oak and juniper forests of the Zagros Mountains; and only a twentieth of the juniper forests of the Elburz and Khorassan ranges.

If I am right about the subversion of the practical basis of pig production through ecological succession, one does not need to invoke Mary Douglas's "taxonomic anomaly" to understand the peculiarly low status of the pig in the Middle East. The danger it posed to husbandry was very tangible and accounts quite well for its low status. The pig had been domesticated for one purpose only, namely to supply meat. As ecological conditions became unfavorable for pig raising, there was no alternative function which could redeem its existence. The creature became not only useless, but worse than useless—harmful, a curse to touch or merely to see—a pariah animal. This transformation contrasts understandably with that of cattle in India. Subject to a similar series of ecological depletions—deforestation, erosion, and desertification—cattle also became bad to eat. But in other respects, especially for traction power and milk, they became more useful than ever—a blessing to look at or to touch—animal godheads.

In this perspective, the fact that pig raising remained possible for the Israelites at low cost in certain remnant hillside forests or swampy habitats, or at extra expense where shade and water were scarce, does not contradict the ecological basis of the taboo. If there had not been some minimum possibility of raising pigs, there would have been no reason to taboo the practice. As the history of Hindu cow protection shows, religions gain strength when they help people make decisions which are in accord with preexisting useful practices, but which are not so completely self-evident as to preclude doubts and temptations. To judge from the Eight-fold Way or the Ten Commandments, God does not usually waste time prohibiting the impossible or condemning the unthinkable.

Leviticus consistently bans all vertebrate land animals that do not chew the cud. It bans, for example, in addition to swine, equines, felines, canines, rodents, and reptiles, none of which are cud-chewers. But Leviticus contains a maddening complication. It prohibits the consumption of three land-dwelling vertebrates which it specifically identifies as cud-chewers: the camel, the hare, and a third creature whose name in Hebrew is *shāphān*. The reason given for why these three alleged cud-chewers are not good to eat is that they do not "part the hoof":

> Nevertheless, these shall ye not eat of them that chew the cud . . . the camel because he . . . divideth not the hoof. And the *shāphān* because he . . . divideth not the hoof . . . And the hare, because he. . . divideth not the hoof. (Lev. 11: 4–6)

Although strictly speaking camels are not ruminants, because their cellulose-digesting chambers are anatomically distinct from those of the ruminants, they do ferment, regurgitate, and chew the cud much like cattle, sheep, and goats. But the classification of the hare as a cud-chewer immediately casts a pall over the zoological expertise of the Levite priests. Hares can digest grass but only by eating their own feces—which is a very uncudlike solution to the problem of how to send undigested cellulose through the gut for repeated processing (the technical term for this practice is coprophagy). Now as to the identify of the *shāphān*. As the following stack of Bibles shows, *shāphān* is either the "rock badger," "cherogrillus," or "cony":

Bibles Translating Shāphān as "Rock Badger"

The Holy Bible. Berkeley: University of California Press.

The Bible. Chicago: University of Chicago Press, 1931.

The New Schofield Reference Library Holy Bible (Authorized King James Version). New York: Oxford University Press, 1967.

The Holy Bible. London: Catholic Truth Society, 1966.

The Holy Bible. (Revised Standard Version). New York: Thomas Nelson and Sons, 1952.

The American Standard Bible. (Reference Edition). La Habra, CA: Collins World, 1973.

The New World Translation of the Holy Scriptures. Brooklyn, NY: Watchtower Bible and Tract Society of Pennsylvania, 1961.

Bibles Translating Shāphān as "Cony"

The Pentateuch: The Five Books of Moses. Edited by William Tyndale. Carbondale: Southern Illinois University Press, 1967.
The Interpreter's Bible: The Holy Scriptures. 12 vols. New York: Abingdon Press, 1953.
The Holy Bible. King James Version (Revised Standard Version). Nashville: Thomas Nelson and Sons, 1971.
Holy Bible. Authorized version. New York: Harpers.
Holy Bible. Revised. New York: American Bible Society, 1873.
Modern Readers Bible. Edited by Richard Moulton. New York: Macmillan, 1935.

Bibles Translating Shāphān as "Cherogrillus"

Holy Bible. (Duay, translated from Vulgate.) Boston: John Murphy and Co., 1914.
The Holy Bible. (Translated from the Vulgate by John Wycliffe and his followers.) Edited by Rev. Josiah Forshall and Sir Frederick Madden. Oxford: Oxford University Press, 1850.

All three terms refer to a similar kind of small, furtive, hoofed herbivore about the size of a squirrel that lives in colonies on rocky cliffs or among boulders on hilltops. It has two other popular aliases: "dassie" and "damon." It could have been any of these closely related species: *Hyrax capensia, Hyrax syriacus,* or *Procavia capensis*. Whichever it was, it had no rumen and it did not chew the cud.

This leaves the camel as the only bona fide cud-chewer that the Israelites couldn't eat. Every vertebrate land animal that is not a ruminant was forbidden flesh. And only one vertebrate land animal that is a ruminant, the camel, was forbidden. Let me see if I can explain this exception as well as the peculiar mixup about hares and shāphān.

My point of departure is that the food laws in Leviticus were mostly codifications of preexisting traditional food prejudices and avoidances. (The Book of Leviticus was not written until 450 B.C.—very late in Israelite history.) I envision the Levite authorities as undertaking the task of finding some simple feature which good-to-eat vertebrate land species shared in common. Had the Levites possessed a better knowledge of zoology, they could have used the criterion of cud-chewing alone and simply added the proviso, "except for the camel." For, as I have just said, with the exception of the camel, all land animals implicitly or explicitly forbidden in Leviticus—all the equines, felines, canines, rodents, rabbits, reptiles, and so forth—are nonruminants. But given their shaky knowledge of zoology, the codifiers could not be sure that the camel was the only undesirable species which was a cud-chewer. So they added the criterion of split hooves—a feature which camels lacked but which the other familiar cud-chewers possessed (the camel has two large flexible toes on each foot instead of hooves).

But why was the camel not a desirable species? Why spurn camel meat? I think the separation of the camel from the other cud-chewers reflects its highly specialized adaptation to desert habitats. With their remarkable capacity to store water, withstand heat, and carry heavy burdens over great distances, and with their long eyelashes and nostrils that shut tight for protection against sandstorms, camels were the most important possession of the Middle Eastern desert nomads. (The camel's hump concentrates fat—not water. It acts as an energy reserve. By concentrating the fat in the hump, the rest of the skin needs only a thin layer of fat, and this facilitates removal of body heat.) But as village farmers, the Israelites had little use for camels. Except under desert conditions, sheep and goats and cattle are more efficient converters of cellulose into meat and milk. In addition, camels reproduce very slowly. The females are not ready to bear offspring and the males are not ready to copulate until six years of age. To slow things down further, the males have a once-a-year rutting season (during which they emit an offensive odor), and gestation takes twelve months. Neither camel meat nor camel milk could ever have constituted a significant portion of the ancient Israelites' food supply. Those few Israelites such as Abraham and Joseph who owned camels would have used them strictly as a means of transport for crossing the desert.

This interpretation gains strength from the Moslem acceptance of camel meat. In the Koran, pork is specifically prohibited while camel flesh is specifically allowed. The whole way of life of Mohammed's desert-dwelling, pastoral Bedouin followers was based on the camel. The camel was their main source of transport and their main source of animal food, primarily in the form of camel milk. While camel meat was not daily fare, the Bedouin were often forced to slaughter pack animals during their desert journeys as emergency rations when their regular supplies of food were depleted. An Islam that banned camel flesh would never have become a great world religion. It would have been unable to conquer the Arabian heartlands, to launch its attack against the Byzantine and Persian empires, and to cross the Sahara to the Sahel and West Africa.

If the Levite priests were trying to rationalize and codify dietary laws, most of which had a basis in preexisting popular belief and practice, they needed a taxonomic principle which connected the existing patterns of preference and avoidance into a comprehensive cognitive and theological system. The preexisting ban on camel meat made it impossible to use cud-chewing as the sole taxonomic principle for identifying land vertebrates that were good to eat. They needed another criterion to exclude camels. And this was how "split hooves" got into the picture. Camels have conspicuously different feet from cattle, sheep, or goats. They have split toes instead of split hooves. So the priests of Leviticus added "parts the hoof" to "chews the cud" to make camels bad to eat. The misclassification of the hare and *shāphān* suggest that these animals were not well known to the codifiers. The authors of Leviticus were right about the feet—hares have paws and *Hyrax* (and *Procavia*) have tiny hooves, three on the front leg and five on the rear leg. But they were wrong about the cud-chewing—perhaps because hares and shāphān have their mouths in constant motion.

Once the principle of using feet to distinguish between edible and inedible flesh was established, the pig could not be banned simply by pointing to its nonruminant nature. Both its cud-chewing status and the anatomy of its feet had to be considered, even though the pig's failure to chew the cud was its decisive defect.

This, then, is my theory of why the formula for forbidden vertebrate land animals was elaborated beyond the mere absence of cud-chewing. It is a difficult theory to prove because no one knows who the authors of Leviticus were or what was really going on inside their heads. But regardless of whether or not the good-to-eat formula originated in the way I have described, the fact remains that the application of the expanded formula to hare and *shāphān* (as well as to pig and camel) did not result in any dietary restrictions that adversely affected the balance of nutritional or ecological costs and benefits. Hare and *shāphān* are wild species; it would have been a waste of time to hunt them instead of concentrating on raising far more productive ruminants.

To recall momentarily the case of the Brahman protectors of the cow, I do not doubt the ability of a literate priesthood to codify, build onto, and reshape popular foodways. But I doubt whether such "top-down" codifications generally result in adverse nutritional or ecological consequences or are made with blithe disregard of such consequences. More important than all the zoological errors and flights of taxonomic fancy is that Leviticus correctly identifies the classic domesticated ruminants as the most efficient source of milk and meats for the ancient Israelites. To the extent that abstract theological principles result in flamboyant lists of interdicted species, the results are trivial if not beneficial from a nutritional and ecological viewpoint. Among birds, for example, Leviticus bans the flesh of the eagle, ossifrage, osprey, ostrich, kite, falcon, raven, nighthawk, sea gull, hawk, cormorant, ibis, waterhen, pelican, vulture, stork, hoopoe, and bat (not a bird of course). I suspect but again cannot prove that this list was primarily the result of a priestly attempt to enlarge on a smaller set of prohibited flying creatures. Many of the "birds," especially the sea birds like pelicans and cormorants, would rarely be seen inland. Also, the list seems to be based on a taxonomic principle that has been somewhat overextended: most of the creatures on it are carnivores and "birds of prey." Perhaps the list was generated from this principle applied first to common local "birds" and then extended to the exotic sea birds as a validation of the codifiers' claim to special knowledge of the natural and supernatural worlds. But in any event, the list renders no disservice. Unless they were close to starvation and nothing else was available, the Israelites were well advised not to waste their time trying to catch eagles, ospreys, sea gulls, and the like, supposing they were inclined to dine on creatures that consist of little more than skin, feathers, and well-nigh indestructible gizzards in the first place. Similar remarks are appropriate vis-à-vis the prohibition of such unlikely sources of food for the inland-dwelling Israelites as clams and oysters. And if Jonah is an example of what happened when they took to the sea, the Israelites were well advised not to try to satisfy their meat hunger by hunting whales.

But let me return to the pig. If the Israelites had been alone in their interdictions of pork, I would find it more difficult to choose among alternative explanations of the pig taboo. The recurrence of pig aversions in several different Middle Eastern cultures strongly supports the view that the Israelite ban was a response to recurrent practical conditions rather than to a set of beliefs peculiar to one religion's notions about clean and unclean animals. At least three other important Middle Eastern civilizations—the Phoenicians, Egyptians, and Babylonians—were as disturbed by pigs as were the Israelites. Incidentally, this disposes of the notion that the Israelites banned the pig to "set themselves off from their neighbors," especially their unfriendly neighbors. (Of course, after the Jews dispersed throughout pork-eating Christendom, their abomination of the pig became an ethnic "marker." There was no compelling reason for them to give up their ancient contempt for pork. Prevented from owning land, the basis for their livelihood in Europe had to be crafts and commerce rather than agriculture. Hence there were no ecological or economic penalties associated with their rejection of pork while there were plenty of other sources of animal foods.)

In each of the additional cases, pork had been freely consumed during an earlier epoch. In Egypt, for example, tomb paintings and inscriptions indicate that pigs were the object of increasingly severe opprobrium and religious interdiction during the New Kingdom (1567–1085 B.C.). Toward the end of late dynastic times (1088–332 B.C.) Herodotus visited Egypt and reported that "the pig is regarded among them as an unclean animal so much so that if a man in passing accidentally touches a pig, he instantly hurries to the river and plunges in with all his clothes on." As in Roman Palestine when Jesus drove the Garadine swine into Lake Galilee, some Egyptians continued to raise pigs. Herodotus described these swineherds as an in-marrying pariah caste who were forbidden to set foot in any of the temples.

One interpretation of the Egyptian pig taboo is that it reflects the conquest of the northern pork-eating followers of the god Seth by the southern pork-abstaining followers of the god Osiris and the imposition of southern Egyptian food preferences on the northerners. The trouble with this explanation is that if such a conquest occurred at all, it took place at the very beginning of the dynastic era and therefore does not account for the evidence that the pig taboo got stronger in late dynastic times.

My own interpretation of the Egyptian pig taboo is that it reflected a basic conflict between the dense human population crowded into the treeless Nile Valley and the demands made by the pig for the plant foods that humans could consume. A text from the Old Kingdom clearly shows how during hard times humans and swine competed for subsistence: " . . . food is robbed from the mouth of the swine, without it being said, as before 'this is better for thee than for me,' for men are so hungry." What kinds of foods were robbed from the swine's mouth? Another text from the Second Intermediate period, boasting of a king's power over the lands, suggests it was grains fit for human consumption: "The finest of their fields are ploughed for us, our oxen are in the Delta, wheat is sent for our swine." And the Roman historian, Pliny, mentions the use of dates as a food used to fatten Egyptian pigs. The kind of preferential treatment needed to raise pigs in Egypt must have engendered strong feelings of antagonism between poor peasants who could not afford pork and the swineherds who catered to the tastes of rich and powerful nobles.

In Mesopotamia, as in Egypt, the pig fell from grace after a long period of popularity. Archaeologists have found clay models of domesticated pigs in the earliest settlements along the lower Tigris and Euphrates rivers. About 30 percent of the animal bones excavated from Tell Asmar (2800–2700 B.C.) came from pigs. Pork was eaten at Ur in predynastic times, and in the earliest Sumerian dynasties there were swineherds and butchers who specialized in pig slaughter. The pig seems to have fallen from favor when the Sumerians' irrigated fields became contaminated with salt, and barley, a salt-tolerant but relatively low-yielding plant, had to be substituted for wheat. These agricultural problems are implicated in the collapse of the Sumerian Empire and the shift after 2000 B.C. of the center of power upstream to Babylon. While pigs continued to be raised during Hammurabi's reign (about 1900 B.C.), they virtually disappear from Mesopotamia's archaeological and historical record thereafter.

The most important recurrence of the pig taboo is that of Islam. To repeat, pork is Allah's only explicitly forbidden flesh. Mohammed's Bedouin followers shared an aversion to pig found everywhere among arid-land nomadic pastoralists. As Islam spread westward from the Arabian Peninsula to the Atlantic, it found its greatest strength among North African peoples for whom pig raising was also a minor or entirely absent component of agriculture and for whom the Koranic ban on pork did not represent a significant dietary or economic deprivation. To the east, Islam again found its greatest strength in the belt of the semiarid lands that stretch from the Mediterranean Sea through Iran,

Afghanistan, and Pakistan to India. I don't mean to say that none of the people who adopted Islam had previously relished pork. But for the great mass of early converts, becoming a Moslem did not involve any great upending of dietary or subsistence practices because from Morocco to India people had come to depend primarily on cattle, sheep, and goats for their animal products long before the Koran was written. Where local ecological conditions here and there strongly favored pig raising within the Islamic heartland, pork continued to be produced. Carlton Coon described one such pork-tolerant enclave—a village of Berbers in the oak forests of the Atlas Mountains in Morocco. Although nominally Moslems, the villagers kept pigs which they let loose in the forest during the day and brought home at night. The villagers denied that they raised pigs, never took them to market, and hid them from visitors. These and other examples of pig-tolerant Moslems suggest that one should not overestimate the ability of Islam to stamp out pig eating by religious precept alone if conditions are favorable for pig husbandry.

Wherever Islam has penetrated to regions in which pig raising was a mainstay of the traditional farming systems, it has failed to win over substantial portions of the population. Regions such as Malaysia, Indonesia, the Philippines, and Africa south of the Sahara, parts of which are ecologically well suited for pig raising, constitute the outer limits of the active spread of Islam. All along this frontier the resistance of pig-eating Christians has prevented Islam from becoming the dominant religion. In China, one of the world centers of pig production, Islam has made small inroads and is confined largely to the arid and semiarid western provinces. Islam, in other words, to this very day has a geographical limit which coincides with the ecological zones of transition between forested regions well suited for pig husbandry and regions where too much sun and dry heat make pig husbandry a risky and expensive practice.

While I contend that ecological factors underlie religious definitions of clean and unclean foods, I also hold that the effects do not all flow in a single direction. Religiously sanctioned foodways that have become established as the mark of conversion and as a measure of piety can also exert a force of their own back upon the ecological and economic conditions which gave rise to them. In the case of the Islamic pork taboos, the feedback between religious belief and the practical exigencies of animal husbandry has led to a kind of undeclared ecological war between Christians and Moslems in several parts of the Mediterranean shores of southern Europe. In rejecting the pig, Moslem farmers automatically downgrade the importance of preserving woodlands suitable for pig production. Their secret weapon is the goat, a great devourer of forests, which readily climbs trees to get at a meal of leaves and twigs. By giving the goat free reign, Islam to some degree spread the conditions of its own success. It enlarged the ecological zones ill suited to pig husbandry and removed one of the chief obstacles to the acceptance of the words of the Prophet. Deforestation is particularly noticeable in the Islamic regions of the Mediterranean. Albania, for example, is divided between distinct Christian pig-keeping and Moslem pig-abominating zones, and as one passes from the Moslem to the Christian sectors, the amount of woodland immediately increases.

It would be wrong to conclude that the Islamic taboo on the pig caused the deforestation wrought by the goat. After all, a preference for cattle, sheep, and goats and the rejection of pigs in the Middle East long antedated the birth of Islam. This preference was based on the cost/benefit advantages of ruminants over other domestic animals as sources of milk, meat, traction, and other services and products in hot, arid climates. It represents an unassailably "correct" ecological and economic decision embodying thousands of years of collective wisdom and practical experience. But as I have already pointed out in relation to the sacred cow, no system is perfect. Just as the combination of population growth and political exploitation led to a deterioration of agriculture in India, so too population growth and political exploitation took their toll in Islamic lands. If the response to demographic and political pressures had been to raise more pigs rather than goats, the adverse effects on living standards would have been even more severe and would have occurred at a much lower level of population density.

All of this is not to say that a proseletyzing religion such as Islam is incapable of getting people to change their foodways purely out of obedience to divine commandments. Priests, monks, and saints do often refuse delectable and nutritious foods out of piety rather than practical necessity. But I have yet to encounter a flourishing religion whose food taboos make it more difficult for ordinary people to be well nourished. On the contrary, in solving the riddle of the sacred cow and abominable pig, I have already shown that the most important food aversions and preferences of four major religions—Hinduism, Buddhism, Judaism, and Islam—are on balance favorable to the nutritional and ecological welfare of their followers.

REFERENCES

Coon, Carleton, 1951. *Caravan.* New York: Henry Holt.

Douglas, Mary. 1966. *Purity and Danger: An Analysis of Concepts of Pollution and Taboo.* New York: Praeger.

7

Traditional Medical Values of Food

E. N. ANDERSON

"It is hard to find a dish in the Middle Kingdom that is not based upon the recipe of some sage who lived centuries ago and who had an hygienic principle in mind when he designed it." So wrote E. H. Nichols in 1902, with pardonable exaggeration. The truth is, of course, less extreme, but the point is well taken: the Chinese have a complex and very ancient science of nutrition.

In the Chou Dynasty, the *Chou Li* (Rituals of Chou) prescribed that nutritionists be attached to the court as part of the highest class of medical personnel. The imperial household had a large number of specialized cooks. The high position of nutritional medicine and of culinary art, in and out of the imperial court, continued to be characteristic of Chinese civilization throughout historic times.

Chinese tradition categorizes food in several different ways. Foodstuffs are classified according to biological relationship. All these categories cross-cut; a given item can be classed under many heads, depending on context or purpose. This chapter concerns the traditional classification of the foodstuffs themselves—their traditional and folk biology. I begin with some comparisons of Chinese and English names for foods.

The earliest record of plant names from China is provided by the Book of Songs, supposedly compiled by Confucius. Hsuan Keng found and identified seventy-five plant names therein. Almost all of them, and all the food plant names, are simple, basic terms. By the time of the first agricultural manuals known, in the Han Dynasty, several compound terms were in general use. Today, most common plant names are binomial compounds; scientists can readily give any plant in the world such a name. Some of the more recent coinages are complicated: kohlrabi is "ball-stalked sweet vegetable." Some are delightful: a citron that looks like a clenched fist is "Buddha's hand fruit," and so the Mexican chayote, which looks like the citron, is "Buddha's hand gourd." Some are borrowed words: fenugreek is *hu lu pa* from Arabic *hulba*. Some are translations: grapefruit is *p'u t'ao yu* (grape pomelo). Some are descriptive: fig is "flowerless fruit" (fig flowers are tiny and hidden inside the "fruit," which is actually a swollen twig).

Meanwhile, in English, someone has recently turned the luffa or silk gourd into "Chinese okra," though it is neither related nor similar to okra. Apricot kernels are used in China as the Western world uses almonds, so the former are often confused with the latter. (The apricot kernels are cooked, which eliminates the poisonous hydrocyanates and makes the kernels into a good throat-soothing foods. They are usually powdered.) "Chinese artichokes" are the roots of a mint (*Stachys*) and "Chinese olives" are not related to olives. Such problems are inevitable when different languages meet. But it is a shame when they go uncorrected—especially if you are trying to make an authentic Chinese dish and the recipe translates the ingredients wrong.

Today, Chinese traditional medical and nutritional beliefs persist, and they are in no danger of disappearing. Arthur Kleinman (1980), studying a large sample of Chinese on Taiwan, found that in 93 percent of sickness episodes, diet was altered—usually the first thing done, initiated by the patient or family. My sketchier figures from Hong Kong are even higher. Certainly the vast majority of Chinese react almost immediately to physical distress of any kind by changing what they eat. Diet therapy grades into herbal medicine with no sharp separation: ginseng, white fungus, birds' nests, stewed wild birds, and the like are foods but are considered to be of almost purely medical use.

The Chinese traditional science of nutrition is based on the commonsense observation that foods provide energy for the body. Different amounts of energy are contained in different foods, and the energy takes different forms. Some foods are extremely strengthening; others are weakening, if eaten to excess. (For a very full account of Chinese nutrition and food therapy, see Lu 1986.)

The traditional word for energy is *ch'i*, which literally means "breath." Like the Latin *spiritus*, it was generalized to mean "spirit"—not *a* spirit but spiritual or invisible energy. Air and gas are ch'i as well (carbonated water is ch'i water). A "ch'i vehicle" is one powered by an internal combustion engine. Ch'i in reference to the human body, or any other natural object, usually means "energy" unless the context makes it obvious that breath or spiritual nature is meant. When Chinese talk of food providing ch'i, however, they do not mean energy in the limited Western sense. The forms or qualities of bodily ch'i are different from anything known to Western science.

The most basic division of the cosmos, in traditional Chinese thought, is between *yang* and *yin*. Originally, yang meant the sunny side of a hill—the southwest face—and yin meant the shady side. The character for yang includes a small abstract picture of a hill and the character for "sun" written over what might be a slope. Yang is thus the bright, dry, warm aspect of the cosmos; yin is the dark, moist, cool one. Note that these are aspects of a single hill (or person, or universe), not really things in themselves. Males have more yang quality, and the penis is politely known as the "yang organ"; females are more yin. However, each sex has some of the other's quality; indeed, all things have both aspects.

Another key division of the cosmos is into the Five Phases (see Liu and Liu 1980; Porkert 1974; Unschuld 1985), earth, metal, fire, wood, and water. These have been called "elements" in English, likening them to the Greek elements, but the Chinese concept is fundamentally different. The Five Phases deal with phases of the cosmos and everything in it rather than with things. (Ch'i is not a phase; it pervades everything.) The full cosmology of the phases was elaborated by Ts'ou Yen in the Warring States period and became the basis of science and cosmology, including nutrition and medicine, in the early Han Dynasty. The thinkers of those ages classified everything in the universe by fives. Preoccupation with fives has lasted to this day in China. The compass directions—including the center—are the most classic and universally known set of five, and everything else probably stems from this basic perception; even the phases may have been set at five to fit them to the all-important directions. Of particular importance for food are the Five Smells (rancid, scorched, fragrant, rotten, and putrid) and the Five Flavors (sour, bitter, sweet, pungent [piquant, "hot"], and salt). These equate with the compass directions: east, south, center, west, north. The tastes are apparently those the Han scholars found to be characteristic of the regional cuisines of those times. At least the *Yellow Emperor's Classic of Internal Medicine* (Veith 1966)—the great Han medical text—says that people in the respective regions eat foods flavored accordingly. The alternative idea—that the coding was arbitrary and the scholars merely imagined, post hoc, the regional cuisines—seems too forced even for the highly scholastic Han academies.

The Five Flavors remained to classify foods, but nutritional medicine was soon to be transformed. Sometime between the *Yellow Emperor's Classic* and the great fifth-century herbal and agricultural texts, Western medicine reached China. The nutritional medicine of the Western world at that time was based on the humoral system and was shared by the Hippocratic-Galenic, Vedic, and Near Eastern medical traditions. No one knows where or how it started; Hippocrates in the fifth century B.C. speaks of it as old. Greece, the Near East, and India all take credit for it. The Greeks have "prior publication" on their side, and the system could well have spread with Alexander's world conquests. It may have reached China from many sources, but there is little doubt that the main impetus for its adoption was Buddhism (Sivin 1980). Independent origins of similar beliefs in several places may also be involved.

The humoral theory, in its most general form, holds that the human body is affected by heat, cold, wetness, and dryness. These "qualities" or "valences" must remain in balance if the body is to remain healthy. Most illness is caused (or exacerbated) by imbalance. The model is of a person working in the hot sun and suffering heatstroke or falling into cold water and suffering from a chill. The ancient Greeks noted that illness varied with season and climate and naturally assumed that the weather had a direct effect—which was true up to a point. Lacking microscopes, they had no way of knowing how typhoid (commoner in summer) differed from heatstroke or winter pneumonia from frostbite and exposure. Another observation was that certain foods increased body heat, others seemed to make the body colder. For a long time, modern scientists thought this was all nonsense—purely arbitrary and irrational—but we now know that the ancients were really attending to something. High-calorie foods were quite correctly seen as more heating; they raise body heat in a malnourished people in winter. Perhaps the ancients saw that such foods burn with much heat when dry. Low-calorie foods don't maintain body heat in a malnourished person in cold weather (unless unrealistic amounts of them are eaten), thus such things as lettuce and cabbage were considered cooling. Salty water was seen to prevent heatstroke, thus it is classed as cooling. (The truth is that heatstroke can be caused by salt depletion through sweating.) Water itself chills the body if one falls into cold water, so it was obviously cooling (see Anderson 1980, 1982, 1984; Gould-Martin 1978).

Once these simple observations had been incorporated into a simple system, everything seemed to fit. Some foods have an effect on the skin that is similar to a burn; thus ginger, pepper, and (much later) chilis were obviously heating. The effects of alcohol make it obviously heating. A "neutral" category arose for foods that are everyday staples—bread in the Western world, rice and fish in the Eastern—these mainstays were (surely) the perfectly balanced foods. Since bread is much higher in calories than rice, bread is considered heating in the Orient; the stage is set for argument. Meat, even when not very high in calories, was seen as strengthening and body-building; so it was coded as heating, but much less so than fat or sugar. Sour foods seemed cooling (think of lemonade); bitter ones often heated. Finally, foods of hot colors—red, orange, brilliant yellow—were often coded as hot, while foods of cool colors—icy white, green—were cool. Foods of a pale brown or dull chalk-white were neutral. Disagreement over such items led to locally different lists and eventually to frequent modern dismissal of the whole system as sheer superstition. We can now see that, while the system is not perfect, it was a plausible extension of real home truths. Nothing succeeds like a simple extrapolation from everyday reality. This is not to say that the full system is simple. Much remains to be learned about how and why it produces effective therapy. There are no doubt many values still to be discovered in the old hot/cold coding system (see Lu 1986), and Chinese medical research continues on it.

Wetness and dryness—obviously relevant in climate—were also seen in humans. Weeping rashes, bloating, and edema were due to excess of wetness. Dry throat, feverish wasting, and a scratchy, rough feeling could be due to excess of dryness. Foods that promoted these must be wetting and drying. In China, several foods that often cause allergic rashes—wet and succulent—are wetting; shellfish are a common example. Foods that are thought to produce a dry, scratchy feeling in the throat, including coffee and dry-roasted peanuts, are drying. This dimension was never as important as heating/cooling, however.

In recent decades, many Chinese have abandoned the heating/cooling dichotomy, although it and the related concept of "rising fire" (*shang huo*) widely persist in China and throughout East Asia.

There is a category of "cold" (*han*) foods that are quite separate from cooling foods. Several foods are both cooling *and* cold. "Coldness" is not very salient, and my informants do not have a very clear picture of the quality; such foods are thought to give one a cold feeling in the stomach or to make the body actually feel icy. They are the opposite not of heating foods in general, but of those specific heating foods that are standardly used in winter to make one feel warm: dog meat, snake meat, guava, and the like.

Remember that the actual temperature of the foods is not relevant here; their effect on the body is what counts. Cooling foods may thus be used to treat fever, rash, sores, red places, and other overhot or burnlike conditions, as well as constipation and other binding symptoms. Heating foods are used to treat low temperature (as from shock or chronic tuberculosis), pallor, frequent chills, wasting, weakness, and diarrhea. Observation often bore out the value of these cures. When people in old China suffered from chronic sores, dry skin, and redness, the problem was very frequently due to (or exacerbated by) vitamin deficiency, especially vitamins A and C. The cooling foods are usually vegetables high in one or both of these vitamins: Chinese cabbage, watercress, carrots, green radishes, and so on. Similarly, pallor and weakness usually involved anemia. Warming and strengthening foods were typically chicken stew, pigs' blood or internal organs, Chinese wolfthorn berries, and other excellent sources of iron and other minerals. On the other hand, the system sometimes had disastrous results. In particular, diarrhea is considered a cold symptom; water, vegetables, fruits, and other foods would be withheld from the sufferer, a practice often fatal. Children with diarrhea might go without vegetables and fruits and suffer from malnutrition (especially loss of vitamin C). However, on balance, the system fitted observation and cured many more people than it killed. The few deaths would likely have been among children, who in old China were almost expected to die; at least among the poor, infant mortality frequently exceeded 50 percent. The system's failures thus attracted too little notice. (Moreover, the belief in withholding vegetables and fruits was far from universal.)

By the nineteenth century, humoral nutritional medicine was believed and practiced in China, India, the Near East, most of Europe, and most of Latin America. It was widespread in the Philippines (where Spanish influences met Chinese), northern Africa, Japan (but identified there as a Chinese import), and Southeast Asia. Most of these areas' medical systems included concerns with heat and cold even before identifiable Greco-Arab-Indian influence reached them. Today, the humoral system remains the basis of folk medicine in all the less developed parts of this vast realm, and it is an important scientific field in China, Japan, and the Indian subcontinent. Modern nutritional science was not advanced enough to challenge it in the English-speaking world until the late nineteenth or early twentieth century. Indeed, a few remnants of humoral medicine are still with us—not just such metaphors as "cool as a cucumber" or the use of "hot" to mean "spicy" but also beliefs in such things as the curative value of chicken soup and the weakening effects of getting one's feet wet or standing in a draft. No medical belief system in all human history has influenced more people or lasted so long in the popular mind. Directly or indirectly, humoral nutrition affects the diet of literally every Chinese who still eats any traditional food. Indeed, few people in the world have not been influenced a bit by the system's teachings. The wide use of many vegetables, chicken soup, and several cooking herbs is dependent on it.

Heating and cooling caught on as an idea in China not only because the system worked but also because it fit so beautifully with the age-old yang and yin even before the Greco-Indian ideas entered. By 550 A.D., Greco-Indian codings dominated Chinese ones. But the whole logic of the system is beautifully Chinese; it stressed balance, order, and harmony, the greatest of all virtues in the Confucian worldview.

Once the system became popular, little was added to it. New foods were added; disagreements arose about the codings of some items. But the whole simply based system never changed. Significantly, the Greek concept of actual humors—sanguine, choleric, melancholic, and phlegmatic—was never accepted in Eastern Asia. Hot, cold, wet, and dry energies were enough to explain what needed to be explained. Actual bodily secretions were thought to be just secretions, and none received the special pride of place that blood, phlegm, and bile obtained in Europe.

In its final form, as seen today (and for many centuries past), the system classifies some foods as dangerously heating, to be avoided except by those in good health. These include fried and long-baked foods, strong alcoholic drinks, and hot spices. Milder heating foods are strengthening and restorative, good for those with too much cool energy: most meats, red beans, ginger, ginseng (some kinds are cooling, though), a few vegetables like chrysanthemum greens (they are spicy tasting), and so on. Neutral foods are the great mainstays, starch staples and ordinary white-fleshed fish. Cooling foods used routinely as medicine or dietary aid include Chinese cabbages, green beans (fresh or dry), radishes (green ones are cooler), watercress, and many other vegetables.

But not all diet therapy is based on the humoral dimensions. Almost as important is the concept of *pu*: "strengthening, supplementing, patching up." Such foods are initially those that promote tissue repair, cure anemia, or show general tonic action. Analysis of supplementing foods shows that they usually do have some such action, but also that they are striking in appearance. Often sympathetic magic is at work: walnut meats have a reputation for strengthening the brain because they look like a brain; red jujubes and port wine are thought to strengthen the blood mainly because of their red color. However, usually foods regarded as pu are not only appropriate but also effective. The vast majority of pu foods are easily digestible, high-quality protein. Fowl—especially wild—are probably most used. Much stronger are sea cucumbers, birds' nests, raccoon-dogs, deer antlers, shark fins, pangolins, and many other wild animals and animal products. Many of these are famous worldwide as examples of the bizarre things human beings will eat and pay high prices for.

One of the most expensive is ginseng. The plant is on the border between food and medicine, categories that merge in Chinese. It is called medicine (*yüch*) but is eaten in quantity by those who can afford it, either cooked with foods or drunk in powder or tincture form. Ginseng's actual effect appears to be tonic. Drugs within it, including panaquin and panaquilone, have a mildly stimulating effect; the taker feels energized but not "wired" (as with caffeine). Many other drugs of tonic effect are regarded as pu, and frequently are placed—as ginseng is—in the highest class of medicines: the "heaven" or "ruler" class, which strengthens the body or increases its energy, rather than treating or helping a particular condition.

It is the combination of real effect and apparent oddness or weirdness that gives some foods their special reputation as pu. Sympathetic magic enters in ways other than appearance; for instance, the male genitalia of deer are believed to be especially strengthening to the human equivalent, doubtless because one male deer can service approximately seventy does during rut season. This is the nearest thing to an aphrodisiac in Chinese medicine; the dozens of items so listed in salacious books are all pu rather than actually aphrodisiac (i.e., directly irritating or stimulating to the sex organs), and few are specifically pu to the genitalia. Of course, all of them can work as aphrodisiacs, since nothing is more responsive to placebo effects than sexual functions. The general tonic and stimulant effects of such medicines as ginseng are obviously useful in such cases, too. But the main reason for the sexual effects of most pu foods is (I feel sure) that malnutrition rapidly and drastically weakens sexual performance and interest. Foods rich in minerals and protein were just what was needed in the bad old days. They are described as *chuang yang* (helping the yang organ), *pu shen* (strengthening the testicles), and so on.

In short, pu is a system generated by—and explicable only by—the interaction of empirical truth and psychological construction. On a solid base of observed fact, people erected a structure of extrapolation and inference. Psychosomatic effects appear to validate much of this structure; traditional Chinese has never seen any reason to separate the power of suggestion from other medical powers.

Rare, exotic, and unusual foods are considered pu not just because of cost and strangeness, though these are certainly factors, and conspicuous consumption is a very major part of their use. More basic is the concept of ch'i. In the traditional Chinese worldview, bodily energy, spiritual energy, and the flow of energy in the natural world are all part of one great system. This is true of modern physics too, but the Chinese belief is more extreme, claiming that people can draw on natural energy flow by eating creatures that have a great deal of energy or even by positioning themselves in places that are appropriately located to take advantage of the flow of ch'i. The striking appearance of such creatures as pangolins and raccoon-dogs is thought to indicate great energy or unusual energy patterns. Powerful creatures like eagles—to say nothing of the sexually hyperpotent deer—are also obvious sources of energy. Unfortunately, desire for such pu items as antlers, bear gall, snake meat, and rhinoceros horn is leading to vast worldwide poaching and the extermination of many species of wildlife. Only large-scale farming or ranching offers hope for the survival and continued use of these animals. The Chinese are moving (but perhaps too slowly) in this direction.

A further powerful factor in directing Chinese attention to the vital importance of balance and harmony (*ho*, "harmony," is the term most often used) is the social importance of this value.

Other foods impart pu to other bodily systems. The Doctrine of Similarities is important here. Stewed lungs of animals improve the lungs; steamed pig or chicken or duck blood supplements the blood (which is perfectly correct if one looks at the assimilable iron value). Blood is strengthened not only by animal blood, but by port wine and many other blood-resembling items.

Almost all animal foods are pu to some degree if prepared correctly—usually by steaming or simmering slowly, especially with herbs. Essentially, all pu things are heating, but gently so; they are at the low-calorie, low-fat, low-irritant end of the hotness scale. The slow simmering is intended to reduce their heating ch'i still more. The idea is to provide a gentle warming rather than a sudden shock of heat. Prepared this way, pu foods are almost always easy to digest, by both Chinese and modern scientific criteria. They are also usually rich in protein and often in mineral nutrients. Such items as chicken are often stewed with enough vinegar to leach some of the calcium from the bones and otherwise pick up mineral nutrients. Many of the herbal items, such as ginseng, also actually have some tonic or nutrient effect. The similarity of the ginseng root to the human body is also relevant. In short, it is not enough for an item to look like an organ. The Chinese do not take the simplistic attitude once found in Europe, that any liver-shaped leaf is good for the liver, or any yellow plant is good for jaundice. They will accept an item as pu only if it does have some discernible nutrient, drug, or medical effect—though in China as in nineteenth-century America it is sometimes to be strongly suspected that the only effect of some items is produced by the strong alcohol content. (Not only port wine, but a vast variety of native wines and tinctures, are pu.)

White tree-fungus, abalone, and other anomalous creatures and plants are also pu. Such routine creatures as chickens are still less pu. It is almost safe to say that the more bizarre and striking an item is, the more pu it will be. This is an obvious instance of Mary Douglas' famous generalizations about anomalous animals, and all her comments about the pangolin in Africa are apposite—I think—in China too (Douglas 1966, 1975). My fisherman informants in Hong Kong told me that the giant grouper (which may reach five hundred pounds) often has a tiny crustacean parasite in its gills; if the grouper is caught and dies, all its ch'i goes into the crustacean, which is thus the richest possible source and the most powerful of all tonics. This is definitely a folk explanation. Although Chinese doctors educated in the elite tradition often explain pu action in terms of actual tonic chemicals alleged to exist in the foods, the folk explanation is probably the older.

Another key term in Chinese medical nutrition is *tu*. This literally means "poison," but it is used in two different senses, and almost all informants note (often spontaneously) that they are really quite distinct. One, identical to the English word, refers to things that are directly toxic if eaten, like puffer-fish liver. The other is used in reference to foods that are not poisonous in themselves but bring out or potentiate any poisons in the body of the eater. The classic foods in this category are uncastrated male poultry. In a study of cancer epidemiology, I found that cancer victims and often their families rigorously abstained from all poultry they did not actually see killed and cleaned, for fear of getting even the tiniest bit of an uncastrated male; they believed cancer would be stimulated by such foods. Beef is often considered tu, lamb and mutton sometimes. Several fish are poisonous in this sense, as are some nuts, seeds, and vegetables, although lists differ widely from informant to informant and in the various classical Chinese medical herbals. As Carol Laderman (1981) points out, allergic reactions—specifically hives and rashes—are often at the root of such ascriptions, especially in regard to seafoods. Since rashes are often seen as internal poisons breaking out at or through the surface of the body, responding to a food with a rash is often taken as a sign that the food is poisonous. Alternatively, though, it may indicate the food is heating and wetting, for this humoral combination brings out or stimulates certain poisons, notably those of venereal disease.

Due to the lack of agreement about what foods are poisonous, generalization is risky, but one things stands out: the foods usually considered poisonous and/or hot and wet are either similar to or specific forms of those that are pu. Most pu foods are nonpoisonous, but the poisonous foods tend to be pu to some degree. Many herbal remedies—those of the lower herbal classes—are poisonous in their action (sometimes they act by "using poison to drive out poison," as the Chinese used to "use barbarians to control barbarians," another social-medical analogy.) The ideal pu foods and medicines are nonpoisonous, but it is clear that there is some association. Perhaps uncastrated male poultry and the like are seen to strengthen the internal poisons, nourish the cancer and give it power, for example. The tremendous amount of yang energy in a rooster or drake converts it from a gentle nourisher and cherisher of the body to an uncontrolled, dangerous nourisher of both the body and the body's enemies. I am thus tentatively persuaded—pending a much fuller study of ascription of foods to the tu category—that poison-potentiation is a logical extension of pu, or perhaps of a more general category of pharmacologically broadly effective things. Poison-potentiators are effective but hard-to-control drugs. They are, of course, conceptually very close to drugs that actually have toxic side effects.

Harder to explain are the many poisonous combinations. Here the belief is not merely that these combinations are poison-potentiating; certain foods, eaten together, are supposed to react to produce actual, virulent poisons. Gould-Martin lists "in Taiwan, crab and pumpkin, port and liquorice, mackerel and plums and, in Hong Kong, garlic and honey, crab and persimmon, dog meat and green beans" (1978:43). Very long lists of these can easily be compiled by anyone with access to informants or traditional medical books. At present I am completely at a loss to explain them. Informants tell me the combinations were arrived at empirically rather than through theory or logic; yet none of them is empirically demonstrable to be harmful in the slightest degree. No one dares actually experiment (except modern Chinese outside the traditional framework), so the belief goes untested. A delightful article by Libin Cheng (1936) recounts his daring experiments with allegedly poisonous combinations. He survived unhurt, as did his experimental animals, and he gives a good overview and summary of the whole matter. Cheng suspects the complex may be traceable to experiences with allergy, bacterially contaminated foods, adulterated foods, and the like. But why these particular combinations were chosen seems impossible to determine.

I pass briefly over such minor problems as foods said to bloat or cause flatulence—here people describe reality. One other key concept underlies the concepts of *ch'ing* (cleaning) and *hsiao* (dispelling, clearing away). In both cases, the idea is to get rid of undesirable matter or essences in the body. Ch'ing gets rid of waste products and any poisons built up in the system. One clears away (hsiao) excess wetness, "wind," and other pathogenic natural forces that have entered the body. Curing inflammation, edema, and the like involves clearing away the accumulated ill humors. Some foods, licorice and honey, for example, free one from poisons (*chieh tu*). "Dirt" (not the same thing as tu) can also be dispelled. Foods particularly good at cleaning are honey, brown sugar, and sugar cane juice (Gould-Martin 1978:40), some vegetables, a number of herbs. A very common herbal mixture, sold in all Chinese drug and general stores, is the *ch'ing pu liang* (cleaning, strengthening, and cooling) herbal tea or soup mix. *Ch'ing liang* foods relieve heat; *lin nao* foods are diuretic (often because of potassium content in the context of a high-salt diet).

The Chinese have lived with famine and malnutrition for a very long time and have accumulated countless observations connected therewith. From these they have constructed a folk nutritional science—rather, both a folk and an elite nutritional science—that subsumes the observations under a set of simple principles or concepts. Some of these broad concepts stand the test of modern science. Others merely illustrate the truth of the remark attributed to H. L. Mencken (among others) that "for every problem there is a solution that is simple, plausible, and wrong."

The study of medicinal classification has implications for the study of Chinese thought. More important are its implications for the study of human thought in general. In actual working taxonomy (so to speak), people appear to go up the taxonomic tree and then back down. They classify things by seeing some particularly important general, shared qualities. They then overgeneralize and overextend these qualities to produce a simple, grand, overarching, high-level system. They then use deduction to classify new or unusual items: if a new item has quality X, it is classified under the appropriate heading. Often the new item should not be so classified, in terms of its actual effects, but the assignment of it to a particular category is thoroughly system-driven: logic takes precedence over mere fact. On the other hand, feedback from experience guarantees that any widespread system has some truth or value. Nutritional medicine, in particular, must be grounded in experiential reality.

"Chunking" enters in that people do not generalize along a smooth gradient. They recognize the natural "chunking" of the world—for example, into biological species—and oversharpen this distinction by treating members of a chunk as if they were pretty much identical but very different from members of any other chunk. All pangolins are about equally pu. This apparently simple matter—oversharpening of some distinctions and blurring of the distinctions not so oversharpened—is at the root of many human mistakes and misperceptions (Nisbett and Ross 1980).

Last is analogical thinking. It was once said in philosophy classes that the Chinese are analogical rather than logical in their thought processes. During the critical formative period of Chinese philosophic thought, syllogistic logic vied for place with argument based on analogy. The latter won, but not without being affected by the former. Chinese thinkers recognized that there were analogies and analogies—even if you do answer Lewis Carroll's question, "why is a raven like a writing desk?" you haven't learned anything very exciting. Philosophers argued by analogy, but the one whose analogy actually included a homology (or something like one) was the winner. The idea was that if two things share a common quality, they may share a common substrate. There are qualities that are real and shared but superficial and trivial (whiteness) and ones that are real, shared, and basic (energy, in moving systems, for instance). Chinese philosophy, as it took form, focused on pragmatic, existential reality and on process. Thus, what was shared was, most importantly, certain types of effect, of energy, of transforming ability and transforming power. The Western tradition of idealism (focusing on essential reality and on unchanging, ultimate Form) was unacceptable to the Chinese, even though it was often introduced, for example with some schools of Buddhist thought, from westward.

There is little "essential" difference between logic and analogic. One can set up analogies as syllogisms:

Things that are strikingly unlike other natural things have a particular
and powerful ch'i.
The pangolin is strikingly distinctive.
Therefore, it has a lot of powerful ch'i.

One can set up syllogisms as analogies:

All the people I know of who reached a great age died.
Socrates is like these other people I know of—not in everything, but in what
I think are key respects.
Thus, we can expect Socrates to die too.

Either way, one carries out similar inductive and deductive processes.

On balance, Chinese traditional beliefs worked very well to keep people healthy and to keep the food production system diverse. Many plants and animals that would not otherwise have been domesticated, or kept in domestication, were grown because of their alleged medical values. While few of these were as medicinally effective as traditional doctors thought, they did provide a richer and more varied resource base for agriculture. Thus more ecological riches were used; nutrients and land were employed more efficiently, since each cultigen had its special requirements and habitat that would often have lain unused if people had wished to grow only the cheap grains. Deer and raccoon-dog farms, for instance, provide a valuable economic resource in areas otherwise too cold and too far from markets to produce much. Only the high price commanded by these animals justifies cropping the areas where they occur. In the central lands, such crops as watercress and Chinese wolfthorn make efficient use of marginal bits of land. (What besides watercress would grow profitably in shallow, cold water?) Such crops also provide insurance; an agriculture that specializes in the two or three most productive crops dooms its users to famine when the crops fail. Chinese agriculture was so diverse that the people were relatively buffered against famine—or, more accurately, more people could be supported when famine struck. A knowledge of wild edible plants, gained through use thereof as medicines, also stood the peasants in good stead at such times.

But, also, the system is based on much empirical observation. The Chinese explained these observations as best they could; lacking modern laboratories and having a rather primitive, although extensive, analytic chemistry, they could not possibly have discovered those compounds and analyzed them. They thus came up with reasonable, plausible, logical inferences, which we now know to be often incorrect. But they were often very close to the truth—as in the similarity of the heating/cooling dimension to our concept of calories. (The calorie is a measure of heat.) Therefore, they worked reasonably well. To an old man who had never had much protein-rich and mineral-rich food, or for that matter to one who had been rich enough to indulge in the fatty, greasy, salty diets of banqueting luxuriants, a diet of birds' nests and sea cucumbers would be nothing but helpful.

We still have much to learn from Chinese traditional medicine and nutrition. Recent discoveries of hormones in deer velvet, stimulants in ginseng, and literally thousands of valuable drugs in Chinese herbal remedies should drive us back to the laboratories and clinical trials to see if other traditional foods have values that we do not yet know about. Mineral availability, enzyme systems, undiscovered animal medicines, and synergistic effects of various foods seem particularly promising avenues for research. I do not believe that we know all the reasons why pu foods strengthen, why cooling foods seem to heal sores, why honey seems so soothing, or why licorice seems almost magically effective at harmonizing medicines in mixed doses and preventing bad side effects. I can personally testify to such benefits as relief of cold symptoms and sore throats by loquat syrup and pear syrup. The whole concept of a medical therapy based on gentle, inexpensive, everyday means of strengthening the body and soothing its aches has much to contribute to our modern system with its powerful and dangerous remedies that all too often create iatrogenic pathologies of their own.

WORKS CITED

Anderson, E. N., Jr. 1980. Heating and Cooling Foods in Hong Kong and Taiwan. *Social Science Information* 19:237–68.

———. 1982. Ecology and Ideology in Chinese Folk Nutritional Therapy. Paper, American Anthropological Association Annual Meeting.

———. 1984. Heating and Cooling Foods Re-Examined. *Social Science Information* 4/5:755–73.

Cheng, Libin. 1936. Are the So-Called Poisonous Food-Combinations Really Poisonous? *Contributions, Biological Laboratory, Science Society of China, Zoological Services* II, 9:307–16.

Douglas, Mary. 1966. *Purity and Danger*. London: Barrie and Rockliff.

———. 1975. *Implicit Meanings*. London: Routledge and Kegan Paul.

Gould-Martin, Katherine. 1978. Hold Cold Clean Poison Dirt: Chinese Folk Medical Categories. *Social Science Information* 12:39–46.

Kleinman, Arthur. 1980. *Patients and Healers in the Context of Culture*, vol. 1. Berkeley: University of California Press.

Laderman, Carol. 1981. Symbolic and Empirical Reality: A New Approach to the Analysis of Food Avoidances. *American Ethnologist* 3:468–93.

Liu, Frank, and Liu Yan Mau. 1980. *Chinese Medical Terminology*. Hong Kong: Commercial Press.

Lu, Henry. 1986. *Chinese System of Food Cures*. New York: Sterling.
Nichols, E. H. 1902. *Through Hidden Shensi*. New York: Charles Scribner's Sons.
Nisbett, Richard and Lee Ross. 1980. *Human Inference*. Englewood Cliffs, N.J.: Prentice-Hall.
Porkert, Manfred. 1974. *Theoretical Foundations of Chinese Medicine*. Cambridge: MIT Press.
Sivin, Nathan. 1980. Science in China's Past. In *Science in Contemporary China*, ed. Leo Orleans, pp. 1–29. Stanford: Stanford University Press.
Unschuld, Paul. 1985. *Medicine in China: A History of Ideas*. Berkeley: University of California Press.
Veith, Ilza (ed. and trans.). 1966. *The Yellow Emperor's Classic of Internal Medicine*. Berkeley: University of California Press.

Commensality and Fasting

Giving, Receiving, and Refusing Food

Food-sharing is so prevalent that it has been taken for granted in many studies of foodways. The exception is anthropology, where food-sharing has been studied as the social cement holding groups together. Anna Meigs defines food as a "cultural construction" among the Hua of Papua New Guinea and shows how its ability to augment or deplete *nu* or "vital essence" among men and women as they exchange food creates gender balance and malleability in their society. Her exploration of gender through food among the Hua throws into relief the rigid and dualistic Western conceptions of gender, which are explored in other articles. Anna Freud offers a brilliant psychoanalytic perspective on the mother-child relationship as enacted through infant feeding. In two related articles, Dorothy Shack and William Shack explore the institutionalization of food deprivation among the Gurage of Ethiopia and show its effects on personality development and social pathology. Caroline Bynum examines "the religious significance of food to medieval women." She shows that women's use of food for personal religious expression—through fasting, exuding miraculous foods from the body, and distributing food to the poor—was one of few channels open to women in the Middle Ages for personal and cultural power. Joan Brumberg explores the meanings of food consumption and fasting to Victorian girls. Marjorie De Vault closes this section with a sociological exploration of women's role in feeding their families in the United States. She shows that this role often reproduces women's subordination while simultaneously offering some avenues for self-expression and influence.

8

Food as a Cultural Construction

ANNA MEIGS

Eating for the Hua people of the Eastern Highlands of Papua New Guinea is one of several ways in which physical properties and "vital essence"[1] are transferred between objects
ɔod and eating as part of a central process that

ɔr whom food is so important? Prominent among
e of Mary Douglas, who has consistently argued
posed to the nutritive and physiological) aspects
focused on indigenous classifications of food and
ssifications to those of social groups. Her much-
ɔollution is a quality attributed to things that do
nal, interstitial, or anomalous (see Douglas 1966).
undaries of the underlying system of categories.
food prohibitions as motivated by an underly-
ell established and highly applicable to the Hua.
emphasizing not how the categorization of food
s and classes but rather how exchanges of food
l solidarity (cf. Robertson Smith 1907, and Mauss
d and eating (and the rules associated with both)
ɔarently separate and diverse objects and organ-
lly, in a single life. This eating-induced unity has
derstanding of self.
) slash-and-burn horticulturists resident on the
near the Lufa District Office, about forty miles
uinea. Although sweet potato is the year-round
anas, sugarcane, and a variety of leafy greens are
European contact since the 1950s has led to the
cucumbers, beans, tomatoes, and cabbage. The
raised by women, and consume large quantities
ut pork is minor in terms of overall dietary intake.
ence in Hua culture (see Meigs 1984). Food and
oduction, preparation, and consumption are fre-

quent topics of conversation. Foods are heavily featured in local folk tales and are one of the most common topics of remembered dreams. Indeed, the Hua word for "everything" is *do 'ado' na,* literally "that which can be eaten and that which cannot."[2]

THE CULTURAL CONSTRUCTION OF FOOD

Among the Hua all persons must observe food rules, and all edibles are prescribed or proscribed to some category of persons. Much of the Hua construction of food is implicit in these rules. To explore that construction, I analyze a representative sample of Hua food and associated avoidance rules.

All Hua rules are either absolute or relative. Absolute rules define a relation between the consumer and a certain kind of food: prototypically, "X should/should not eat Y." The relationship between X and Y is determined largely by properties intrinsic to the food, properties that may help or harm the consumer. Relative rules define a relationship between a consumer, a food, and a source: prototypically, "X should/should not eat Y from Z." Here the relationship between X and Y is determined principally by the social relationship that exists between X and Z and is relatively independent of the nature of the food itself.

Absolute Rules and Contagion

Absolute rules are about contagion. The law of contagion, described most clearly by Frazer (1978), states that things which have once been in contact retain a permanent trace of this contact. Thus many properties that we understand as inherent or innate the Hua construct as added to or subtracted from an object or organism by the process of contagion. Typical examples of the many absolute rules among the Hua are:

1. Young male initiates may not eat food cooked in ashes unless the ashes have been completely removed.

The properties of dryness and ashiness of skin will, it is feared, be transferred from the food to the initiate's skin. The Hua prize an oily skin as indicative of an abundance of internal fluids and vitality.

2. A girl at menarche should not eat sugarcane, *kito'* (leafy vegetables), *pitpit* (a vegetable), pig fat, cucumber, or *hevia'* (Acanthaceae *Rungia blossii,* a leafy green vegetable) but should eat *zau* (a small pinelike bush that grows on the higher slopes of Mount Michael), *hakrua'* (Symplocaceae *Symplocos,* a hardwooded tree), *rka'* (Fagaceae *Nothofagus,* a hardwood tree), and *finmu* (Melastomataceae *Memecylon,* a tree used in fence building because of its exceptionally hard wood).

The girl at menarche is trying to reduce her menstrual flow, first, by avoiding foods with a high fluid content (sugarcane, leafy vegetables, cucumber, etc.); second, by eating foods that are exceptionally dry and hard. Quantity of menstrual flow is determined, the Hua assumes, not by innate processes but by external relationships, that is, through receipt of the contagious properties of plants.

3. A pregnant woman may not eat cat, dog, and possum.

According to the Hua these animals are capable only of *dri dri ke,* "babble, confused stuttering," and if the pregnant woman eats any of them her child may never be capable of more than babble. Inability to speak correctly is in the Hua construction something due not to malfunction of an internal organ or process but rather to the transmission of the quality of inarticulateness from an animal to a human.

4. Men prior to battle should eat *kaso' namo,* "sharp scratchy things."

The Hua hope that the sharp, wounding, scratchy, blood-drawing qualities of such plants will be transmitted to their eaters, making them more fierce and effective fighters. Again the Hua construct what we deem an internally constituted personality trait as the consequence of relationships of exchange with the surrounding material environment.

5. A woman in the last months of pregnancy should not eat any of the *hakeri'a*, "dry and hard," varieties of taro, yam, and banana, or food from either end of a bamboo cooking tube (some informants say just from the lower end). On the other hand, the woman should eat pig fat, frog, *kva kito'* (a vegetable green), and *maita'* (Amarantaceae *Amaranthus*, a particularly juicy green vegetable).

The *hakeri'a* varieties of yam, taro, and banana may cause the womb and birth canal to dry out, resulting in a difficult delivery. Food from either end of a bamboo cooking tube may cause a difficult birth in that such food is packed tightly and tends to stick. Juicy foods are required eating because they are *hrupo'di na*, "slippery thing," and consumption of them will facilitate delivery. In Hua thinking the quality of a woman's labor is determined not by internal organic processes and body states but by what the woman eats, more generally by the connections she makes with the contagious properties of the surrounding world.

These and several hundred other absolute rules operate from the perspective of a fluid reality. In this reality physical properties are not anchored immutably in the bodies of their owners, placed there irrevocably by natural processes. Rather, they can be traded around, can flow and percolate from one object or organism to another through the process of contagion. Physical characteristics such as ashy skin, inarticulate speech, fierceness in battle, light menstrual flow, and ease in childbirth are not understood as the products of the unique organic life of a single autonomous, and separate individual. Such characteristics, rather, are understood to flow between humans, other organisms, and objects, moving along the lines established by the web of their interconnections. The absolute rules describe a world in which physical properties and traits are in flux, a world in which the boundaries between individuals become blurred.

Relative Rules and Nurture

Relative rules define a relationship between a consumer, a food, and a source. They deal with *nu*, a central concept in Hua culture and, in particular, in the Hua theory of nurture. (Whereas absolute rules embody the Hua theory of contagion, relative rules embody their theory of nurture.)

In Hua thinking *nu* is associated quintessentially with body fluids (blood, sweat, sexual substances, etc.), but it also includes any body substance (hair, fingernail, flesh itself) and is extended to any product of the body (feces, footprint, shadow) or its labor. In this last category appear such items as bows, arrows, axes, string bags, flutes, and (most important in this horticultural society) food.[3]

All foods with the exception of wild species contain the *nu* of their producer. Foods nourish by virtue of the *nu* that is in them. *Nu* can be growth- and health-producing, leading its consumer to *kosi'*, "grow, increase in strength and health," or it can lead the consumer to be *keva ro*, "stunted of growth, lacking in strength and health." The positive or negative charge attributed to *nu* in a specific instance is determined by the relationship of the consumer to the producer of the *nu*. If that relationship is positive—is one of warmth and friendship—then the effect of eating the *nu* in the food will be health- and growth-enhancing. If, on the other hand, the *nu* comes from a person with whom the consumer has a relationship of hostility, or even of emotional ambivalence, then its consumption will lead to losses in growth and strength. Food, in other words, does not nourish by virtue of its innate properties. The same sweet potato that nourishes one person will cause debility and sickness in the next. The food is merely the avenue in an interactional system for the transmission of *nu* from one node to another.

The fetus is created prenatally out of the semen *nu* of the father and menstrual blood *nu* of the mother and then nourished postnatally by *nu* in all its many forms: breast milk, blood (let from veins and fed to growing children), sweat (rubbed from real and classificatory parents onto the skin of growing children), and, of course, food (understood to promote health and growth by virtue of the *nu* it contains). A child grows from *nu*, and parents age from the same process.

Aging

Aging is the consequence of the expenditure of *nu* by the senior generation on the junior generation, and so aging is a process of "nurturing oneself out." A small set of relative rules are relevant:

1. A person must participate in the eating of the corpse of the same sex parent.
2. A person may not eat the corpse of any person one generation junior to him or her.
3. A person may not eat the blood of any person one generation junior to him or her.
4. A person may not eat pig raised by, an animal shot by, or the largest and best garden produce from the garden of, a person one generation junior to him or her.
5. A person may not eat pig he or she has raised, an animal he or she has shot, or the largest and best produce of his or her own garden.

Nu is the source of nourishment and growth; it is *the* substance of nurture. The most powerful forms of *nu* are the most direct: blood (rule 3) and flesh (rules 1 and 2). When a child's growth is stunted or an adult is sick, he or she will be fed blood let from the vein of a same sex consanguine of the same or one senior generation. *Nu* in foods appears to be less powerful; nevertheless, what is significant in a food is its *nu*. The pigs people raise (rules 4 and 5) contain the people's *nu*; so also with the animals that people have shot. The same is true of garden produce, particularly the largest and best. According to Hua informants, these vegetables and fruits are the ones they have most labored over and which, therefore, contain a larger portion of their *nu*.

Whereas rules 1–3 imply that body substances are food and rules 4 and 5 that foods are a kind of body substance (all body substances being *nu*), all five rules attempt to regulate the transfer of body substances. Only transfers in one direction are deemed appropriate: from the senior to the junior generation. For parents to eat their real or classificatory children's pigs, prize garden produce, or blood or flesh would be, the Hua say, like a dog licking its genitals, a physically disgusting and morally repugnant act.[4]

It would also be dangerous. The Hua say that if one eats the body substances of one's child, one is eating one's own *nu*. One's own body substances, on being reintroduced into one's own body, will *bi aina pto,* "put the existing *nu* down": the effect is losses of strength and health. The principle that people may not consume their own body substances, no matter how indirectly, is more explicitly spelled out in rule 5.

Mature individuals must donate their blood, breast milk, semen, the foods (particularly the largest and best) they have labored to produce and which thus contain a portion of their own *nu*, to others but in particular to their real and classificatory children. Only through the receipt of these substances can children grow. They feed off the *nu* of their parents, real and classificatory; and only by doing so can they mature.

Nu, however, is in short supply. Individuals contain only finite amounts, the amounts that they received from their real and classificatory parents when they were children and that they supplement through *nu* intake as adults. Each donation of *nu* to a child, whether it be direct as in blood letting or indirect as in giving food, involves an irreparable loss for the donor. The cumulative effect of these donations causes aging. Aging, in other words, is the outcome of being a parent, of spending one's *nu* on one's child. The Hua think, as do other Melanesian peoples (Kelly 1976, Counts and Counts 1985, Scaletta 1985), that the child's gain is the parent's loss.

In the Hua conception, therefore, aging is not produced by internal body processes, by the machine wearing down. Rather, aging occurs because the senior generation is giving its *nu* to the junior. What produces the phenomenon of aging is not the body itself as an autonomous agent but the social relationships of nurture. (Theoretically, people would not age if they did not nurture and feed children, but no such adults exist among the Hua. Those who are unable to have children adopt them.)

The Motility of Nu

In Hua thinking *nu* is a highly volatile and nonstatic substance that is always in motion. *Nu* is passed directly from person to person not only through intentional acts (for example, blood letting and cannibalism) but also through unintentional acts (transfers of body oils, sweat, and so forth in casual contacts). Furthermore, all food transactions involve transfers and exchanges of *nu*. The most indirect transfers are those in which the *nu* moves from its producer through several intermediate substances to its consumer. Several relative rules protect the individual from intake of multiply transmitted (and negatively charged) *nu*.

1. A young initiated male may not eat any food cooked over a fire into which a menstruating woman has blown (in order to liven up the coals).

Informants maintain that the *nu* of the woman's breath may be transmitted by the fire to the food cooked in it and from there to its eater (the woman ⟶ her breath ⟶ fire ⟶ food ⟶ eater).

2. Young male initiates may not eat foods associated with the ground.

In Hua thinking the ground is the repository of many of the *nu* substances of the people who inhabit it—their urine, feces, menstrual blood, sweat, body oils, sexual substances. The ground not only receives the substances of the people who live on it but also transmits these substances to others. The ground thus serves as a transmitter of *nu*. It may transmit directly (for example, young initiated males do not touch the ground with their hands in women's houses for fear of direct contact with female menstrual blood) or indirectly. Plants or animals especially associated with the ground are thought of as containing its substances most strongly (in the case of plants, ground-hugging plants such as the short sugarcanes and ground-growing mushrooms; in the case of animals, mouse, dog, pig, and those species of bird and possum which feed on the ground). The chain of connection is human ⟶ his or her *nu* in whatever form ⟶ ground ⟶ plants or animals ⟶ human eaters.

3. Ego may not wear the clothes or carry the string bags of certain alters.

This rule provides protection against the sweat *nu* and body oil *nu* of contaminating alters, in which category generally fall affines, firstborn children, age-mates, cowives, and real or classificatory wives in the earlier years of marriage. The pattern of transmission is alter ⟶ sweat and body oils ⟶ string bags or clothes ⟶ ego.

These relative rules, together with those described in the section on aging, regulate the flow of *nu* in a world understood in a holistic manner to be highly porous and interconnected.

The processes of nurture (the *nu* transfers regulated by relative rules) and contagion (the transfer of physical qualities regulated by absolute rules) emphasize that the individual is caught up in patterns of participation. Each individual is commingled with and is consubstantial with others. The traits that make up individual identity come from a multiplicity of sources, and the patchwork quilt that is the individual (made up of the speed of growth of the banana, the fierceness of the white pandanus, the hardness and dryness of the black palm) is always being changed by new contacts, new relationships. Food is instrumental in these patterns of flow.

Where food is understood to be alive not only with its own contagious properties but also with the vital essence of its producers and preparers, it plays an important role in the intellectual construction of the self. Anthropologically, it is of considerable significance.

THE CULTURAL CONSTRUCTION OF FOOD RULES

Mary Douglas has for many years been calling for a recognition of the social significance of food and eating (1966, 1972, 1982, 1984). She has focused on categories relevant to food and eating systems. Food systems are, like myth or ritual systems, Douglas says, codes wherein the patterns by which a culture "sees" are embedded. Through analysis of food and eating systems one can gain information about how a culture understands some of the basic categories of its world. The first step in such an analysis is the discovery of the constituent units or categories of the eating system; the second, the discovery of how this system of classification for foods relates to the wider system of social classification (see, for example, Douglas 1975a).

Purity and Danger, first published in 1966, is among other things an attempt to understand rules of pollution—an important feature of many eating systems. What makes one food polluting and another food not? In part the answer, according to Douglas, has to do with ideas of form and formlessness. Formlessness is analogous to the visual chaos that the once-blind person encounters. Formlessness is the property of those parts of reality which have not been neatly classified or which do not occupy a secure, established place within the system of already established categories. Pollution is, in other words, a negative power attributed to areas of confusion, ambiguity, and disorder within the set of forms or system of classifications.

> To plot a map of the powers and dangers in a primitive universe, we need to underline the interplay of ideas of form and formlessness. So many ideas about power are based on an idea of society as a series of forms contrasted with surrounding non-form. There is power in the forms and other power in the inarticulate area, margins, confused lines, and beyond the external boundaries. (1966:98)

That which is prohibited (and labeled as polluting) is that which is formless (or anomalous, and/or is viewed as threatening to the structure of categories) (1966:98–104). Douglas phrases this same insight in terms of uncleanness: "In general the underlying principle of cleanness in animals is that they shall conform fully to their class" (1966:55).

That which is unclean or polluted does not conform. "Dirt," according to Douglas (1975:48), "is the by-product of a systematic ordering and classification of matter," and it is as matter out of place that it is proscribed. In other words, Douglas (1966) understands ritual prohibitions as protecting established categories. Pollution is triggered when a category boundary or distinction is violated (1975:113). Ritual prescriptions and proscriptions are understood as protecting and defending the correct definition, the relevant discriminations, and the proper order of social categories.[5]

This approach, with its emphasis on difference and opposition, is appropriate for many Hua food rules. Listed below are some Hua food rules in which the clear intent is to accentuate relations of opposition and difference (and also hierarchy) between categories of person.

> 1. No Hua person may eat food produced, prepared, or served by a person who is a stranger (that is, a person with whom she or he does not share *nu* through coresidence or common birth).

This rule clearly works (in the manner Douglas describes) to enhance a perception of distinctiveness, to protect a border or a category—that between kin and strangers. If a Hua eats a food produced, prepared, or served by a stranger, then he or she is eating *siro na,* literally "dirty thing," and the consequence will be losses in health and strength if not outright sickness. The Hua notion of pollution serves here to preserve social and cognitive order by keeping apart those categories which are, in terms of Hua thought, meant to be separate.

Rules enjoined upon the young male initiate are designed to strengthen the opposition between male and female.

2. Young male initiates may not eat *kito'* because it is like women and because women wear it as decoration on their buttocks.
3. Young male initiates may not eat any of the mushrooms that grow during *a'di kuna,* "women's time" (the new moon, when all Hua women allegedly menstruate).
4. Young male initiates may not eat the *dguripa* mushroom because its shape resembles a young woman's breasts before she has children.

If young male initiates were to eat any of these foods, they would be eating *siro na* and their growth would be stunted. (The red *kito',* the specified mushrooms, the *dguripa* mushroom, and the many other foods that resemble aspects of female reproductive anatomy are thought by the Hua male initiates to be not only dangerous (because polluting) but also disgusting.) The concept of pollution here works to preserve and enhance the males' perception of their difference from and superiority to females.

Alternative Approaches

Douglas's work is an effort to overcome a bias toward materialistic interpretations of food and food rules, interpretations that focus exclusively on the nutritive and physiological aspects of food and eating. Douglas has consistently argued for approaches that explore social meanings and functions. Within anthropology a second sociological perspective on food exists, however, one that differs in angle of vision. Whereas Douglas focuses on how food and eating systems reflect distinctions of social category, the work of this second group of anthropologists centers on how food is used to develop social relationships of exchange and alliance between the various individuals and larger social units of the system.

Marcel Mauss's *The Gift*, first published in 1925, is important to this line of thought. *The Gift* is a reflection on economic systems organized not around sale but around the principle of reciprocal gift giving. Mauss focuses on how reciprocal gift exchanges, often of food, bind members of a society together in relations of mutual participation and unity. Mauss quotes a New Caledonian: "Our feasts are the movement of the needle which sews together the parts of our reed roofs, making of them a single roof, one single world" (1967:19).

Mauss stresses that the obligation to repay the original gift derives from the fact that each gift contains some of the self of the giver: "To give something is to give a part of oneself. . . . In this system of ideas one gives away what is in reality a part of one's nature and substance, while to receive something is to receive a part of someone's spiritual essence" (1967:10).

The Gift is permeated with an idea analogous to the Hua *nu*: that some of the "self" is inherent in objects that a person gives away. Thus to give a gift is to give some of oneself, and to receive is to take in some of the self of another person. A gift economy, according to Mauss, is not only a system that promotes the circulation of material goods (as in a market economy) but also a system that promotes the circulation of "selves" and thus creates a mystical economy of participation between individuals, organisms, and objects.

Much anthropological labor has been invested in showing how food exchanges develop and express bonds of solidarity and alliance, how exchanges of food are parallel to exchanges of sociality, and how commensality corresponds to social communality (see, for example, Lévi-Strauss 1969, Strathern 1971, Young 1971, Khare 1976, Schieffelin 1976, Weiner 1977, Manderson 1986). Authors in this tradition generally focus on food as an instrument in the creation of social alliance, food as a political tool in the deliberate construction of social solidarity. I stress here a subtheme in their work: namely, how food and eating are intimately connected with cultural conceptions of self. Food as object and eating as act resonate with attitudes and emotions related to the individual's understandings and feelings about self and other and the relationship between. This emphasis is consistent with the tenets of the interpretive or hermeneutical school of anthropology (see Geertz 1983, Clifford and Marcus 1986, Turner and Bruner 1986, Marcus and Fischer 1987).

Food in Hua thinking contains the self and the feelings of its producer. To eat a sweet potato produced by a person with whom one enjoys a relationship of friendship is to feel confident and to be nourished; to eat the same sweet potato produced by a personal enemy is to feel fear and to court physical degeneration, illness, possibly death. To eat a food produced by another person is to experience that person, both physiologically and emotionally.

Mauss continually alluded to such notions ("to give something is to give a part of oneself. . . . In this system of ideas one gives away what is in reality a part of one's nature and substance, while to receive something is to receive a part of someone's spiritual essence" [1967:10]). It was, however, Robertson Smith, a Scottish theologian, who first discussed this experience of physical communality and mystical sharing. *Lectures on the Religion of the Semites,* first published in 1889, is a refutation of prevailing views of sacrifice as a means of making obeisance to the deity. According to Robertson Smith, the purpose of sacrifice is to share a common meal with the deity and in so doing to establish a bond of common life. Eating, the sharing of food, is a means by which to establish physical commingling, interdependence, and oneness. Sacrifice is a ritual by means of which tribesmen "cement and seal their mystic unity with one another and with their god" (1907:313). The "sacred cement" produced in the sacrifice ritual "creates or keeps alive a living bond of union between the worshippers and their god" (1907:313).

Eating is understood and analyzed by Robertson Smith as a way of establishing a physical unity not only with the deity but also with one's fellow humans. The basis of the idea of kinship is a notion of common life.

> The members of one kindred looked on themselves as one living whole, a single animated mass of blood, flesh, and bones, of which no member could be touched without all the members suffering. . . . Now, if kinship means participating in a common mass of flesh, blood, and bones, it is natural that it should be regarded as dependent, not merely on the fact that a man was born of his mother's body, and so was from his birth a part of her flesh, but also on the not less significant fact that he was nourished by her milk. And so we find that among the Arabs there is a tie of milk, as well as of blood, which unites the foster-child to his foster-mother and her kin. Again, after the child is weaned, his flesh and blood continue to be nourished and renewed by the food which he shares with his commensals, so that commensality can be thought of (1) as confirming or even (2) as constituting kinship in a very real sense. (1907:274)

Robertson Smith is interested in how ancient Semitic communities understood themselves to be mystical and physical unities, as sharing a common life, as participating in an intense communality with one another and with their deity. Food and eating he understands as one of the means by which that communality is established and enhanced.

CONCLUSION

Food, according to our dictionaries, is something material, a thing made up of different kinds of similarly impersonal and material things. Most persons in the industrialized West do not grow food or see or know those who do. We encounter food in tins, in boxes, under plastic. We know food as inert matter dissociated from its human producers and natural context. In fact, children in societies such as ours must be taught that the impersonal lifeless packages we call "food" were originally living animals or plants.

The Hua never lose sight of that perception. Foods are not inert objects, "things" to be bought and sold.[6] Rather, they possess the vitality and dynamism of living beings. They are alive; alive not only with their own contagious qualities (their rates of growth, textures, smells, and so forth) but also with the transmittable vitality, essence, *nu* of their human producers. Last but not least, foods are alive with the feelings, the emotional intents, of their producers (and to a lesser extent their preparers). People fear to eat the foods of those who bear them ill will and, on becoming sick, speculate as to which people poisoned them—not with a specific identifiable poison but rather with the vague but virulent hostility of their *nu*. In this connection, the concept of pollution (Hua *siro na*) reflects not only the violations of social boundaries, as suggested by Douglas, but also the experience and emotions of fear. Pollution is attributed by the Hua to foods and body substances produced by people with whom they do not enjoy supportive and friendly relationships. Pollution for the Hua is not motivated exclusively by things sociological, having to do with maintenance of category distinctions, but also by things personal and emotional, matters of individual feeling and experience.

To eat a food, in the Hua conception, is not only an economic, social, and nutritive event but also an emotional and mystical one. In the act of eating one is connecting oneself with the world, opening one's body and one's self to the dynamic influence of properties, vital essence, and emotions of other organisms. Similarly, to refuse to eat a food is an emotional and mystical as well as an economic, social, and nutritive event: it is to refuse a certain kind of participation between self and other.

Underlying the Hua understanding of food and food rules is a world view that emphasizes relatedness. All organisms are linked in chains of mutual influence; borders between bodies are permeable; there is a grand cycle of flow by which *nu* circulates through all living beings. Implicit in this understanding is a notion of self and other as involved in a continuous and dynamic process of participation. The self is blended in and through the surrounding world and, conversely, that world is blended in and through the self. Through his or her continual acts of food exchange, both as producer and as consumer, the individual is constituted as part of a physically commingled and communal whole. At the heart of the Hua understanding of food and eating is a statement of understanding about identity and the boundaries of self.

Food has a distinctive feature, one that sets it off from the rest of material culture: it is ingested, it is eaten, it goes inside.[7] In a small-scale society, moreover, it is and is understood to be the product of the labor of known individuals, the output of their blood, their sweat, their tears. As output of one person and as input into another, food is a particularly apt vehicle for symbolizing and expressing ideas about the relationship of self and other.

NOTES

I am grateful for the comments and criticisms of John Haiman, John Caughey, and Mac Marshall. Many of the ideas for this article were developed when I was working with the support of a Bush Grant from Macalester College.

1. There is a problem of translation here. The Hua word in question is *nu*, described below. I translate it "vital essence" because the phrase conveys the Hua sense of *nu* as a substance or essence that is vital to the life of its owner. This translation connects the Hua notion of *nu* with a well-known concept in the anthropological literature to which it has a close resemblance: vital essence, soul substance (see, for example, Lévy-Bruhl 1966, Frazer 1978, and Rivers 1920).
2. It was not my original intention to study Hua food beliefs and rules, but Hua enthusiasm for this topic is intense. I conducted fieldwork on food in the villages of Kemerake and Sara intermittently between 1970 and 1975. Most of the information and ideas presented here were developed in conversations with two men, Aza Kevao and Busa Hgaiteme. Their thinking was corroborated, elaborated, and expanded upon (and contradicted) in interviews with eleven other Hua, most of them male (the food rule system being one that applies primarily to males). I began a second study focusing on female conceptions of marriage in 1987.
3. *Nu* thus conforms closely to the classic definition of "soul substance" or "vital essence" promulgated by an earlier generation of anthropologists.
4. In classificatory kinship systems, collateral kin are terminologically equated with lineal kin: father's brother with father, mother's sister with mother, and so on. In a Hua village a child calls all people of the same generation and sex as his or her mother *ita,* "mother," and as his or her father *davu,* "father."
5. Douglas (1975) modifies her earlier position, acknowledging that the extreme concern with form, with clearly defined categories and highly marked boundaries, is not a universal characteristic of cultures but rather a specific feature of cultures such as the ancient Hebrew, which for social and political reasons was highly concerned with preservation of identity.
6. Babcock (1986:319), in a discussion of the general anthropological treatment of the pottery of Pueblo cultures, makes a similar point. She argues that Pueblo potteries are seldom described or thought of by anthropologists as other than mere things or lifeless artifacts. They are, she says, rarely conceived as the symbolic forms they are, forms in and through which Pueblo conceptions of the person, the social order, and the cosmology are articulated and displayed. Weltfish (1960) argues that anthropology's most profound dilemma is to figure out the significance of material objects and to include these understandings in its museum collections, its archaeological accounts, and its ethnographies.
7. Food can be considered as part of material culture rather than as a natural object in that, like pottery, it is the product of individual labor and reflects cultural conception and design.

REFERENCES

Babcock, Barbara. 1986. "Modeled Selves: Helen Cordero's Little People." In V. Turner and E. Bruner, eds. *The Anthropology of Experience.* Champaign: University of Illinois Press.

Clifford, James, and George Marcus. 1986. *Writing Culture.* Berkeley: University of California Press.

Counts, Dorothy, and David Counts. 1985. "I'm Not Dead Yet! Aging and Death: Process and Experience in Kaliai." In Counts and Counts, eds. *Aging and Its Transformations.* ASAO Monograph no. 10. Lanham, Md.: University Press of America.

Douglas, Mary. 1966. *Purity and Danger.* London: Routledge & Kegan Paul.

——. 1975. *Implicit Meanings.* London: Routledge & Kegan Paul.

——. 1982. *In the Active Voice.* London: Routledge & Kegan Paul.

——. 1984. *Food in the Social Order.* New York: Russell Sage Foundation.

Frazer, James. 1978. *The Golden Bough* (1922). London: Macmillan.

Geertz, Clifford. 1983. *Local Knowledge.* New York: Basic Books.
Kelly, Raymond. 1976. "Witchcraft and Sexual Relations: An Exploration in the Social and Semantic Implications of the Structure of Belief." In Paula Brown and Georgina Buchbinder, eds., *Man and Woman in the New Guinea Highlands.* Special Publications of the American Anthropological Association, vol. 8. Washington, D.C.
Khare, R. S. 1976. *The Hindu Hearth and Home.* New Delhi: Vikas.
Lévi-Strauss, Claude. 1969. *The Elementary Structures of Kinship* (1949). Boston: Beacon.
Lévy-Bruhl, Lucien. 1966. *The "Soul" of the Primitive.* Chicago: Regnery.
Manderson, Lenore. 1986. "Introduction." In Manderson, ed., *Shared Wealth and Symbol.* Cambridge: Cambridge University Press.
Marcus, George, and Michael Fischer. 1987. *Anthropology as Cultural Critique.* Chicago: University of Chicago Press.
Mauss, Marcel. 1967. *The Gift* (1925). New York: Norton.
Meigs, Anna. 1984. *Food, Sex, and Pollution: A New Guinea Religion.* New Brunswick: Rutgers University Press.
Rivers, W. H. R. 1920. "The Concept of 'Soul-Substance' in New Guinea and Melanesia." *Folklore* 31: 48–69.
Scalleta, Naomi. 1985. "Death by Sorcery: The Social Dynamics of Dying in Bariai, West New Britain." In Dorothy Counts and David Counts, eds., *Aging and Its Transformations.* ASAO Monograph no. 10. Lanham, Md.: University Press of America.
Schieffelin, Edward. 1976. *The Sorrow of the Lonely and the Burning of the Dancers.* New York: St. Martin's Press.
Smith, W. Robertson. 1907. *Lectures on the Religion of the Semites.* (1889). London: Adam & Charles Black.
Strathern, Andrew. 1971. *The Rope of Moka.* Cambridge: Cambridge University Press.
Turner, Victor, and E. Bruner, eds. 1986. *The Anthropology of Experience.* Champaign: University of Illinois Press.
Weiner, Annette. 1976. *Women of Value and Men of Renown.* Austin: University of Texas Press.
Weltfish, Gene. 1960. "The Anthropologist and the Question of the Fifth Dimension." In Stanley Diamond, ed., *Culture in History: Essays in Honor of Paul Radin.* New York: Columbia University Press.
Young, Michael. 1971. *Fighting with Food.* Cambridge: Cambridge University Press.

9

The Psychoanalytic Study of Infantile Feeding Disturbances

ANNA FREUD

In the psychoanalytic study of children interest frequently has been concentrated on one or the other of the feeding problems of infancy and childhood. The first disorders of this kind to attract the attention of analytic authors were the upsets of feeding after weaning (Freud, Abraham (1), Bernfeld (3)). These were investigated in the beginning indirectly through the after-effects for the individual's emotional life as they showed up under analytic treatment in adulthood; later on directly, during the observation and treatment of young children. Other feeding problems gradually came into the field of analytic vision. Endre Petö (14) devoted a paper to the emotional attitude of the mother, as an important factor for the success of breastfeeding. Merrel P. Middlemore (12) made a systematic study of the "suckling situation" between mother and newborn infant, and interpreted some of her findings in the light of Melanie Klein's theories of the conflicts of the oral phase. Edith Jackson (10) and G. J. Mohr (13) stressed the importance of emotional factors in nutrition work with infants and children. Editha Sterba (16) drew attention to the interrelations between habit training and feeding disorders; Otto Fenichel (7), James Strachey (17), Melitta Schmideberg (15) and others, to the connection between feeding inhibitions and inhibitions of intellectual activities. Emmy Sylvester (18) in a recently published case of psychogenic anorexia traced the influence of the mother-child relationship on the origin and course of the disturbance. Refusal of food owing to the repression of oral sadism and oral introjection has played a large part in the psychoanalytic theory of depressive states and melancholia (see Freud (8), Abraham (2), Melanie Klein (11)).

Psychoanalytic studies of this kind have been instrumental in shedding light on the origin and meaning of specific feeding disorders, especially of the graver types and those which occur as single symptoms within the framework of a neurotic illness. Less notice was taken of the common feeding difficulties which occur in the everyday life of otherwise normal children. Nor have the findings of the various authors been correlated and systematically applied to the wide field of feeding problems which ranges from manifestations like simple fluctuations of appetite, and transitory food fads, to severe disorders which endanger the child's health, and sometimes its life.

The function of eating serves primarily the biological bodily need for nourishment and operates in agreement with the id forces and ego forces which are jointly directed towards the self-preservation of the individual. The function of eating, as such, lies therefore outside the sphere of psychological conflict (see Hartmann's "konfliktfreie Sphäre"). Eating may, on the other hand, become invested with sexual and aggressive meaning and thereby, secondarily, become the symbolic representative of id forces which are opposed by the ego.

The need for nourishment announces itself to the child's awareness by the sensations of hunger. The painful tension which is created by the sensations of hunger urges the child to take appropriate action (announcing its hunger to the environment by crying, later on asking for food, or helping itself to food). The appeasement of hunger through the intake of nourishment is felt as satisfactory and is accompanied by pleasure. Since the infant's behavior is dominated by the urge to avoid pain and discomfort and to gain pleasure, the urge for self-preservation through feeding is reinforced by the urge to gain pleasure through feeding.

There are, according to these conditions, three main ways in which the function of eating is open to disturbance:

(i) through changes in the organisms which directly or indirectly affect the organism's drive to survive, or the need for nourishment. (*Organic feeding disturbances.)*

(ii) through changes in the pleasurable character of the function. (*Non-organic disturbance of the instinctive process itself.)*

(iii) through sexualisation or aggressive use of the function, which involves the activity of feeding in conflicts with the ego forces and leads to states of neurotic anxiety, inhibition and symptom formation. (*Neurotic feeding disturbances.)*

(I) ORGANIC FEEDING DISTURBANCES

The organic feeding disorders lie outside the field of analytic interest except in those cases where they become the basis of, or otherwise combine with, non-organic disorders. In states of severe physical illness, weakness, exhaustion, strain, and in certain states of convalescence, the organism is forced to a lower level of adjustment (9, Gesell and Ilg, p. 112), and the need for food, with the accompanying sensations of hunger, is decreased. Some children show a steady sustained appetite even during illness or malaise (Gesell and Ilg, l. c.); but with the majority the appetite falls to a low level with a consequent reduction of intake. Such children are at these times "bad eaters" for physiological reasons.

Where it is necessary, on medical grounds, to urge the child to eat beyond the limits of its appetite, or where mothers, for their own reassurance, force the child to eat against its will, emotional factors may enter into an otherwise simple feeding situation. Eating then becomes symbolic of a struggle between mother and child, in which the child can find an outlet for its passive or active, sadistic or masochistic tendencies towards the mother. To win a victory in this battle may, for the child, become more important than to satisfy its returning appetite. In such cases the phase of "bad eating" outlasts by far the phases of illness and convalescence, or may become the starting point for permanent non-organic feeding troubles. Where the child can be permitted in illness to adjust its feeding to the level of its appetite, it will return to its former normal feeding standards as soon as the need for food returns to normal level.

(II) DISTURBANCES OF THE INSTINCTIVE PROCESS

The satisfaction of hunger constitutes the first experience of instinctual gratification in the child's life. An infant who feeds successfully is a contented and "happy" infant. So far as the mother, by providing nourishment, guarantees this satisfaction and thereby provides for a pleasurable experience, the child, its instinctive need, and the environment, are all in perfect harmony.

On the other hand no mother gives food to the child without imposing on it at the same time a feeding regime which constitutes on the part of the environment the first serious interference with an instinctive desire of the child.

The current feeding schedules for the first year of life are based on detailed physiological knowledge of the infant's bodily functioning. The number of mothers who do not follow the advice either of their own pediatricians or of the Medical Officers and nurses of the Welfare Clinics diminishes in England and America from year to year. The mixed diet of the toddlers is, in the middle classes, chosen according to similar advice; in the poorer parts of the population it is left almost entirely in the hands of the mothers, and determined by the food habits and circumstances of the other members of the family. With older children scientific planning of a balanced diet may be introduced again where the child shares in nursery- or school-meals.

Some feeding regimes are, thus, based on medical-hygienic knowledge; others are the result of preconceived ideas, sometimes sensible, sometimes outdated and often superstitious, handed down to mothers by the former generation. In each case they are the embodiment of what a specific environment believes to be wholesome, advantageous and suitable for the various ages. It is a common feature of all the current feeding regimes in our civilization that they take maximum account of the bodily requirements for physiological health, growth and development, and little or no account of the pleasure which should be an invariable accompaniment and inducement to the feeding process.

The amount of pleasure which an individual child gains from eating depends only partly on the adequate fulfilment of bodily requirements; for an equally large part it is dependent on the manner in which the food is given. The child finds feeding most pleasurable when it can eat what it likes, how much or how little it likes and in whatever way it likes. The average feeding regime which regulates the child's meals according to quality, quantity, frequency and procedure, therefore inevitably interferes with the element of pleasure in all these respects.

In the last decades the feeding schedules for infants have inclined towards strict and fixed hygienic regulation and have left little room for individual fluctuations. Recently many authors show a growing tendency to stress the importance of individualization in the supervision of infant feeding (see Gesell and Ilg, l. c.), and to allow for a certain amount of adjustment of the schedule to the individual child. But even with a more flexible feeding regime of this nature the interference with the natural process will remain considerable. It is inevitable under conventional feeding conditions that infants and toddlers are made to cease feeding while they still feel unsatisfied; or that they are urged to continue when they feel that they have had enough; or that they are given foods which are considered necessary for the diet but which they dislike; or that they are offered sweets at moments when they prefer savouries, or savouries when they prefer sweets; that the temperature of liquids or the consistency of solids is not according to preference; that they are passively fed at an age when they desire to handle the food actively;

or forced to use implements when they prefer using their fingers, etc. On all these occasions the child will feel displeased, frustrated, uncomfortable, and will connect these painful sensations with the feeding process.

Further discrepancies between an imposed schedule and the child's wishes arise about the incidence of feedings. The former provides for meals at set times, while the wishes turn to food, or the satisfaction connected with it, for ulterior motives too, to allay anxiety, loneliness, longing, boredom, tiredness or any other emotional upset. This means that for the child food acts as an important general comforter, a function which is disregarded in most feeding schemes. Where the child's wish for food remains unfulfilled at such moments, it feels deeply dissatisfied.

A similar discrepancy exists between child and environment about the question of timing meals. The adult's conception of the length of waiting times after sensations of hunger have appeared is different from the infant's. In infantile life instinctive needs are of overwhelming urgency. There is no organized ego, able to postpone wish-fulfillment with the help of the thought processes or other inhibitory functions. Nothing therefore will diminish the painful and distressing tension of the need except immediate satisfaction. Where hungry infants or toddlers are made to wait for their meals, even for minutes, they suffer acute distress to a degree which may prevent them from enjoying the meal when it finally arrives.

On all occasions enumerated the child experiences sensations of a disagreeable and painful nature instead of pleasure. Where the disappointments, dissatisfactions and frustrations connected with the feeding experiences become too frequent they may, in time, outweigh the pleasures and ultimately spoil the child's attitude to the whole feeding process.

Urged by the physiological need of the body and by pressure from the environment, children will continue to eat even when the process of feeding has become dissociated from the powerful urge for pleasure which originally characterizes every instinctive drive. But meal times will then have lost their former attraction and instead be tiresome tasks; forced labor rather than an occasion for wish-fulfillments. The children will then eat slowly instead of greedily; be easily distracted from their meals, or demand to be entertained while eating; object to more types of food and be more distrustful of new foods than they otherwise would; need considerably more urging to take in sufficient nourishment; and cannot, as their mothers express it, "be bothered to eat," or will be "too deep in play" to come for their meals. They have become "bad eaters" owing to loss of pleasure in the function.[1]

If doctors, mothers and nurses are right in taking the child's desire for food as a blind instinctive force, without the discrimination and self-regulation which are essential factors in the parallel feeding situations of young animals, then the imposing of feeding schedules from without, with the loss of pleasure and the eating disturbances consequent on it, are inevitable. On the other hand recent studies on these lines, made under completely changed feeding conditions, do not confirm this distrust in the self-regulating powers of the child's appetite. Clara M. Davis (4, 5, 6) has shown in her experiments with nursing infants, infants of weaning age, and older children in a hospital setting that infants and children can, under carefully regulated conditions, be trusted to make their own choice according to quantity and quality among selected foods rich in the nutritional elements essential to growth, and that under these conditions they will have better appetites, eat more and have happier mealtimes than children fed in the usual manner. The pleasure which they obtain from the gratification of their appetite will be fully maintained and form an essential element in the smooth functioning of the feeding process.[2]

(III) NEUROTIC FEEDING DISTURBANCES

Eating, more than any other bodily function, is drawn into the circle of the child's emotional life and used as an outlet for libidinal and aggressive tendencies.

a) *The relationship between eating and the stages of object love:* The newborn infant is self-centered and self-sufficient as a being when it is not in a state of tension. When it is under the pressure of urgent bodily needs, as for instance hunger, it periodically establishes connections with the environment which are withdrawn again after the needs have been satisfied and the tension is relieved. These occasions are the child's first introduction to experiences of wish-fulfillment and pleasure. They establish centres of interest to which libidinal energy becomes attached. An infant who feeds successfully "loves" the experience of feeding. (Narcissistic love.)

When the child's awareness develops sufficiently to discern other qualities besides those of pain and pleasure, the libido cathexis progresses from the pleasurable experience of feeding to the food which is the source of pleasure. The infant in this second stage "loves" the milk, breast, or bottle. (Since, on this level of development, no certain distinctions are made between the child's self and the environment, this libido attachment forms a transitional stage between narcissism and object love.) When its powers of perception permit the child to form a conception of the person through whose agency it is fed, its "love" is transferred to the provider of food, that is to the mother or mother-substitute. (Object love.)

It is not difficult to pursue the line of development which leads from these crude beginnings of object attachment to the later forms of love. The infant's first love for the mother is directed towards material satisfaction. (Stomach love, cupboard love, egoistic love; "to be fed.") In a next stage object love is still egoistic but directed toward non-material satisfactions, i.e., to receive love, affection, approval from the mother; "to be loved." As the child progresses from the oral and anal to the phallic level, object attachment loses its egoistic character; the qualities of the object increase in libidinal importance while the immediate benefit from the relationship becomes less important. The next and highest stage of development is the ability to love the object regardless of benefit (altruistic love).

Where infants are breast-fed, and the milk and breast are in fact part of the mother and not merely, as with bottle-fed babies, symbolic of her, the transition from narcissism to object love is easier and smoother. The image of food and the mother-image remain merged into one until the child is weaned from the breast.

Psychoanalytic authors have been repeatedly accused of exaggerating the upsets to the feeding situation caused by weaning (see for instance Gesell and Ilg, l. c.). Pediatricians and psychologists stress the fact that where weaning takes place in slow stages with the gradual introduction of other foods and other means of feeding (spoon, cup), no shock is felt by the child. In the author's opinion this is only reliably so where weaning takes place in the first period of narcissistic enjoyment of feeding, when the child's unfavorable reactions are to alterations of the feeding condition which has proved satisfactory and where upsets can be avoided by the avoidance of sudden changes. At the latter stages of developing object-relationship weaning signifies, besides the changes in food and the means of feeding, an entry into a new phase of the mother-relationship to which certain children are unable to adjust themselves smoothly. The difficulties on the emotional side may find their outlet as difficulties of adjusting to the new food.

Though food and the mother become separated for the conscious mind of all children from the second year onwards, the identity between the two images remains so far as the child's unconscious is concerned. Much of the child's conflicting behavior towards food does not originate from loss of appetite or a lessened need to eat, etc., but from conflicting emotions towards the mother which are transferred on to the food which is a symbol for her. Ambivalence towards the mother may express itself as fluctuations between over-eating and refusal of food; guilty feelings towards the mother and a consequent inability to enjoy her affection as an inability to enjoy food; obstinacy and hostility towards the mother as a struggle against being fed. Jealously of the mother's love for the other children of the family may find its outlet in greediness and insatiableness. At the stage of repression of the oedipus complex refusal of food may accompany, or be substituted for, the inner rejection of the phallic sexual strivings towards the mother.

Eating disturbances of this type normally disappear in adolescence when repressions of the infantile relationship to the parents are revised and solutions for them have to be found on a different level. Where the feeding disorders arising from the child-mother relationship have been especially severe, they may return in adult life in the form of psychosomatic disorders of the stomach or digestive tract.

Mothers, though they do not produce these feeding difficulties in their children, nevertheless may behave in a manner which aggravates the pathogenic elements in the situation. Under the influence of their own unconscious fantasies they often continue much longer than necessary to act as the connecting link between the child and the food, and on their side to treat the food which they offer as if it were a part of themselves; they are pleased and affectionate when the child accepts the food, and as offended when food is rejected as if their love for the child had suffered a rebuff; they beg a badly eating child to eat "for their sake," etc. This attitude of the mother coincides with the unconscious attitude of the child and thereby strengthens the unconscious emotional tendencies which are a threat to feeding. Mothers cannot alter the unconscious fantasies of their children. But they can, by their actions, strengthen the healthy, conscious moves towards the next stage of development. That is, they can give the child direct access to food as early as possible, trust the self-regulating powers of its appetite within sensible limits and thereby increasingly withdraw from the feeding situation, in the measure in which the child learns to handle food independently.

With children in the pre-oedipal phase the feeding disorders of this origin normally disappear when mother and child are separated (away from home, in nursery-school, in hospital). In all later phases, the conflicts arising from the mother-relationship persist as inner conflicts regardless of the mother's presence or behavior. The feeding disorders which are dependent on them then transfer themselves automatically to every available mother-substitute.

b) *The relationship between feeding and the oral pleasures:* The connection between feeding and the oral component of infantile sexuality is so close and its pathogenic influence so obvious, that analysts often make the mistake of diagnosing all infantile feeding disturbances automatically as "oral disturbances."

From the beginning of suckling the infant derives from the milk flow two different kinds of satisfaction: one from the appeasement of hunger, the other from the stimulation of the mucous membranes of the mouth. The latter, oral, satisfaction remains from then onwards through life a constant accompaniment of every feeding situation. When a mixed diet begins, oral satisfaction is gained from the taste and consistency of the various food substances and plays an important part in the formation of the various individual likes and dislikes for food. (Preference for sweets or salty foods, for hot or cold liquids, etc.) This winning of libidinal pleasure from an otherwise non-libidinal body function is an added stimulus for the child's feeding, the significance of which cannot be too strongly emphasized. An infant's or toddler's diet in which this element is disregarded (i.e., a drab, dull diet, or one in which too many items are "distasteful" to the child) defeats its own ends by lowering the total gain in satisfaction from the intake of food with a consequent decrease of the child's appetite.

In the oral phase of libido development, oral pleasure, though originally discovered in conjunction with feeding, is sought and reproduced independently of the food situation in the numerous forms of thumb-sucking, as an auto-erotic oral activity. As such it may be pursued by the infant as a substitute for feeding (while waiting for food or when feeding has to be interrupted before the child is fully satisfied). It may enter into competition with feeding (when infants are unwilling to remove the thumb from the mouth to take the teat of the bottle or the spoon filled with food). It further plays an important part, completely independent of feeding, as a general comforter (like food) before sleep, when the child is lonely, dull, etc.

Through the intimate connection between feeding pleasure, oral pleasure, and the roots of object-relationship, oral tendencies become the first carriers of the libidinal attachment to the mother. Oral attitudes are consequently as decisive for shaping the child-mother relationship as the latter is decisive in determining the child's attitude to food.

The oral pleasures are an asset to feeding while the child can enjoy them without interference from the environment or from its own ego forces. When they are separated, or otherwise rejected by the ego, serious upsets for feeding result. It is impossible for the child to give up or ward off the element of oral enjoyment without losing simultaneously its enjoyment of food and its wish to eat.

c) *The relationship between eating and the aggressive instinct:* Since Abraham's study of the oral-sadistic phase of libido development, the aggressive significance of eating has received constant attention in psychoanalytic literature. According to Abraham oral sadism is at its height after teething, when eating symbolizes an aggressive action against the food which is in this way attacked and consumed, or against the love object that is represented by the food. Melanie Klein and her followers emphasize the significance of oral aggressive phantasies for early childhood and their after-effects for later normal and abnormal development. (According to Melanie Klein aggressive meaning is attached even to the child's first feeding experiences, independent of the possession of teeth.)

Oral-sadistic (cannibalistic) fantasies are under no circumstances tolerated in consciousness, not even when the ego is immature. They are rejected with the help of all the defence mechanisms available to the child in this early period of life.

The consequences for feeding are inhibitions of eating, refusal to bite, to chew, or to swallow the food. These feeding disorders may reach their height at the toddler stage, though at that time the child still freely uses its teeth for aggressive biting, as a weapon in fights with other children, or to express anger and resentment against the mother.

Where the repression or other defence mechanisms used against cannibalistic wishes are not completely successful the child remains anxious about its oral sadism, not only in the oral phase but all through childhood, with serious consequences for the pleasure felt in eating. Children of this type feel guilty when they enjoy food and eat only under pressure of need or under compulsion from the environment, and with no freedom or abandon. They eat slowly and sometimes keep unchewed food in their mouths for long times. They show certain well known dislikes and minor or greater food fads which originate in the fear of hurting or destroying a living creature. They anxiously watch their eggs for possible signs of a partly hatched chicken; they are simultaneously attracted and revolted by the idea of eating sweets, biscuits, cakes, etc., which in their shape imitate a human body, or an animal, or any recognizable part of it; they are not able to eat chickens, rabbits, pigs, etc., which they have known alive. Sometimes a compulsive vegetarianism becomes the last safeguard before a far-reaching eating inhibition is established.

In extreme instances, the defence against oral sadism leads to neurotic self-starvation. In this case the mechanism used is that of turning the aggression away from objects to the individual's own body which is thereby seriously threatened or even destroyed.

d) *The relationship between eating and the anal pleasures:* While infants are still passively fed they accompany the process with certain movements of their hands or fingers which indicate an impulse towards action. When they are encouraged to help themselves to food or to handle the spoon, it becomes evident that their intention is not directed towards feeding themselves but towards handling the food, playing with it, smearing it over the table or over themselves, etc. It is an error to ascribe this messing of the young child to lack of skill. The child's actions in this respect are deliberate and intentional. They are motivated by the pleasure of smearing, an anal-erotic activity transferred from the excrements to the foodstuffs which are similar to the former in consistency, color, temperature, etc. Behavior of this kind begins approximately at the age of eleven months and reaches its peak at the height of the anal phase of libido development. Where it is tolerated by the environment the anal pleasure gained from the actions contributes appreciably to the pleasure of feeding. The handling of food which is in the beginning stages merely messy and possessive merges gradually into purposeful actions of self-feeding. Food which is at first only held tightly or squeezed in the fist, finds its way into the child's mouth, etc. Children who are permitted to develop their self-feeding methods on the basis of this pleasurable anal attitude towards food become skillful in feeding themselves with their hands and fingers earlier than others and shortly afterwards make an easy transition to the use of the spoon.

Where messing with food is interfered with too strictly by the environment and the child is prevented from adding anal pleasure to the other feeding pleasures, appetite suffers. To keep a child in the anal phase from smearing food necessitates its being kept passively fed for a longer time which has, secondarily, an adverse influence on eventual self-selection of food, on the child's relationship to the mother, etc.

At the period when anal tendencies become repressed and relegated to the unconscious, a whole series of feeding difficulties make their appearance, especially in those cases where habit training has been too sudden or too severe. The reaction-formation of disgust which is set up in consciousness to prevent the return from the unconscious of former wishes to play with excrement, take it into the mouth, etc., is transferred to all those foodstuffs which by their look, touch or smell remind the child of the now forbidden dirty matter. As a result, many children form violent dislikes of squashy and smeary foods, of green or brown substances, of sausages, occasionally of all sauces and creams, regardless of taste. Where children are forced to eat the foods which disgust them, they react according to the strength of the anal repressions, with being sick, with loss of appetite, with a widening refusal of foods, etc. Where their disgust is understood as an inevitable outcome of the defence against anal urges, and where the resultant food fads are tolerated, the disturbances remain limited to certain substances, and transitory in character. When the anal repressions are firmly established, anxiety with regard to the underlying unconscious tendencies diminishes, and the majority of temporarily "disgusting" foods are readmitted by the child into its diet.

In the interest of the child's appetite, anal pleasure should be permitted to enter into the combination of feeding pleasures, and, for the same and other similar reasons, habit training should be carried out gradually and leniently. This is in contrast to the conventional but psychologically unsound attitude that children should as early as possible be in control of their excretory functions and acquire good table manners.

The intimate connection between the child's behaviour in these two respects can be proved experimentally. Where children in the latency period are urged to make a sudden advance in their table manners, they almost invariably react by regression to smearing and messiness in the lavatory, or vice versa.

Other anal admixtures to the feeding situation arise from a connection, made during the transition from orality to anality, between the ideas of intake of food and output of excrement, and between the body openings which serve the two functions. The feeding disorders which originate from them are of the type described in Editha Sterba's paper quoted above. (Retention of food in the mouth as equivalent of retention of stool in the rectum.)

e) *The relationship between eating and certain typical fantasies of the phallic phase:* Certain fantasies of the pre-oedipal and oedipal phase have a specific bearing on the neurotic eating disturbances.

During the conflicts and struggles of the oedipus complex, many children escape anxiety by regressing from the phallic level to the earlier pregenital levels of libido development. This leads to the conception of parental intercourse by mouth, a fantasy which is frequently reinforced by actual observation of fellatio acts. The consequent sexualization of the mouth (by means of genital as well as oral libido) with the repressions following on it endangers the function of eating by producing hysterical symptoms, as for instance globus hystericus and hysterical vomiting.

In one of the typical infantile theories of sex it is asserted that babies are conceived through the mouth (oral conception) and born through the rectum (anal birth), the intestines being substituted for the womb. Anxiety and guilt which become attached to these birth phantasies lead to refusal of food (warding off the wish to be impregnated), to a horror of getting fat (in defence against the phantasy of being with child).

Where fixations to the oral level and the regression to them are especially powerful the wish to be impregnated takes the form of a fear of being poisoned which leads to severe eating inhibitions.

Guilt feelings arising from sexual competition with the parents and from death wishes against them lead to the masochistic desire not to grow up, which may express itself by means of refusal of food.

The penis envy of the little girl which may lead to the fantasy of biting off the male genital combines oral-sadistic with phallic elements. It frequently causes symptoms of hysterical vomiting without reference to specific foods.

CONCLUSIONS

The various types of eating disturbances, which have been separated off from each other in this paper for the purpose of theoretical evaluation, are invariably intermixed and interrelated when observed clinically. Organic feeding disorders become the basis for the non-organic types. Neurotic disturbances arise more easily where loss of pleasure in the function of eating has prepared the ground for them. Considerate handling of the child's feeding, with a reasonable amount of self-determination, to safeguard the child's appetite, makes the function of eating less vulnerable and less favorable ground for neurotic superstructures.

NOTES

1. In a discussion of these problems in 1933 Dr. Grete Bibring suggested that the very widespread reduction in the pleasure of eating in the European children of the twentieth century might well be a consequence of the growing rigidity of the then current schedules of infant feeding.
2. The author can confirm these findings from experience with toddlers in the Jackson Nursery in Vienna 1937–38 and the Hampstead Nurseries in London 1940–45.

BIBLIOGRAPHY

1. Abraham, K. "Pregenital Stage of the Libido," *Selected Papers*, 1916.
2. Abraham, K. "A Short Study of the Development of the Libido Viewed in the Light of Mental Disorders," *ibid*, 1924.
3. Bernfeld, S. *The Psychology of the Infant,* New York, 1929.
4. Davis, C. M. "Self-selection of Diet by Newly-Weaned Infants," *Amer. J. Diseases of Children,* 36, 1928.
5. Davis, C. M. "Self-selection of Food by Children," *Amer. J. Nursing,* 35, 1935.
6. Davis, C. M. "Choice of Formulas made by Three Infants Throughout the Nursing Period," *Amer. J. Diseases of Children,* 50, 1935.
7. Fenichel, O. *The Psychoanalytic Theory of Neurosis,* Norton, New York, 1945.
8. Freud, S. "Mourning and Melancholia," *Coll. Papers,* IV, 1925.
9. Gesell, A. and Ilg, G. *Feeding Behavior of Infants,* Lippincott, Philadelphia, 1937.
10. Jackson, E. "Prophylactic Considerations for the Neonatal Period," *Amer. J. Orthopsychiatry,* XV, 1945.
11. Klein, M. *The Psycho-Analysis of Children,* London, 1932.
12. Middlemore, M. P. *The Nursing Couple,* Hamish Hamilton Medical Books, London, 1941.
13. Mohr, G. J. "Emotional Factors in Nutrition Work with Children," *Ment. Hyg.* XII, 1928.
14. Petö, E. "Säugling und Mutter," *Zeit f. psa. Päd.,* XI, 1937, 3/4.
15. Schmideberg, M. "Intellektuelle Hemmung and Esstörung," *Zeit. J. psa. Päd.,* 3/4, 1934.
16. Sterba, E. "An Important Factor in Eating Disturbances of Childhood," *Psa. Quarterly,* X, 1941.
17. Strachey, J. "Some Unconscious Factors in Reading," *Int. J. Psa.,* XI, 1930.
18. Sylvester, E. "Analysis of Psychogenic Anorexia in a Four-year-old," this *Annual,* I.

10

Nutritional Processes and Personality Development among the Gurage of Ethiopia[1]

DOROTHY N. SHACK

For several decades anthropologists have shown concern for the relationship of culture to personality development. However, few students have centered on the role of cultural systems of nutrition. This paper, essentially exploratory in nature, is an attempt to advance the understanding of these processes by examining the way in which food comes to have meaning outside the nutritional sphere for the Gurage people of Southwest Ethiopia, and by suggesting that certain manifestations of Gurage personality are related to their nutritional system. In the analysis of these data I have leaned heavily for orientation upon the studies of Audrey Richards (1948)[2] and Cora Du Bois (1941). The significance of nutritional systems for cultural behavior was clearly underscored by Richards (1948: 9) when she wrote that "for men food acquires a series of values other than those which hunger provides." The expression and transmission of these values should be manifest in the way people are fed and the uses, other than nutritional, to which food is put.

The Gurage are sedentary agricultural people of patrilineal persuasion who speak a Semitic language and inhabit a sparsely fertile semi-mountainous region in Southwest Ethiopia.[3] Their tribal district is bounded by the Awash River on the north, the Ghibie (Omo) River on the southwest, and Lake Zway on the east. *Ensete edulis (äsät),* their staple food crop, commonly called the "false banana plant," is produced in abundance by each Gurage homestead.

Ensete is totally involved in every aspect of the daily social and ritual life of the Gurage, who, with several other tribes in Southwest Ethiopia, form what has been termed "the Ensete Culture Complex Area" (Shack 1963). From birth, when the umbilicus is tied off with a fiber drawn from an ensete frond, to death, when the corpse is wrapped and then covered over with ensete fronds, the life of the Gurage is enmeshed with various uses of ensete, not the least of which is nutritional.

Ensete is the primary source of food. The massive root *(wähta)* of the mature harvested plant, as well as the pseudostem, is decorticated to extract the edible food substance, which is then buried in deep pits and allowed to ferment. Ensete is prepared in several different ways, from the heavy daily bread *(wusa)* to the many more palatable and esthetically pleasing varieties made from *wähta.*

The utilitarian products of ensete are numerous. A fiber from the heavy stems of the plant's fronds is marketed for cash. Goods and personal items are wrapped in suitably sized sections of the pliable fronds, making a virtually water-tight bundle. On ceremonial occasions, food is served on "plates" shaped from small pieces of the fronds. When a fire breaks out, the thatched roofs of nearby huts are covered with the moist green fronds to provide protection from flying embers. These, though only a few of the uses to which ensete is put, are enough to indicate the range of utilitarian and food needs served by this staple crop.

In addition to maintaining a large number of ensete plants in each of the four growth phases (Shack 1966: 59), Gurage farmers also cultivate some minor cash crops such as coffee and *chat.*[4] Livestock are kept but mainly for the supply of milk and dung, the latter being essential in ensete horticulture. The normal diet of ensete is supplemented to a limited extent by green cabbage, cheese, butter, and roasted grains. Meat is eaten sparingly, usually on some ritual or ceremonial occasion when the sacrifice of a sheep or ox is required. The birth of a child is one such event.

One hour after birth, the first feeding of a Gurage child is ritually administered. The "godmother" places a small amount of soft rancid butter in the infant's mouth, where it slowly melts and is swallowed. Several hours later he is given the breast. From this time on the feeding rhythm is determined by the infant's crying, which Gurage mothers interpret as a demand for the breast.

For the first five days the child and mother lie on ensete fronds beside the open fire, partially shielded from visitors who have come to offer congratulations but who may be unsuspectingly harboring the "evil eye." At the end of this period, a small feast consisting of several specially prepared ensete dishes is held, signifying that the newborn has survived the first crisis stage of life, for Gurage children often die in infancy. Later the father constructs a crude bed, made either of logs or the stout stems of ensete fronds, and both mother and child are removed from the warmth of the fire and secluded behind a screen for two months.

At the end of this period, a village "coming-out" feast is held to herald the mother and child; it is at this time that the child is named. *Bra-brat,* a special kind of ensete food prepared from the plant's root, which is eaten when the placenta is buried and on other ritual occasions, is distributed among women of the village. This feast in part reflects the principles of reciprocity through the distribution of food, for it is the women of the village who relieve the mother from domestic responsibilities during her period of confinement. Liberated from these chores, the mother devotes herself almost exclusively to caring for the child. The infant is held except when asleep, and immediately upon awaking he is again cradled in his mother's arms and usually suckled. Until a child is weaned, some two to four years later, self-demand feeding is the norm, and the breast is frequently offered when his discomforts do not stem from hunger.

Intensive maternal care ends with the close of the two-month period of seclusion. In a simple ritual of reincorporation into the village community, the mother carries her child from homestead to homestead, pausing for a time at each to be served coffee spiced with butter,[5] roasted grains, or *wusa* bread. Afterwards she resumes her normal domestic duties, communal work obligations, house-to-house visiting, and weekly marketing. At this time much of the responsibility for the care of the child is shifted to a female surrogate, usually a young daughter or a father's brother's daughter, for the Gurage residential unit is a patrilineal extended family homestead. Among wealthy Gurage a servant is assigned the task. The infant is harnessed high on the back of the child caretaker in a *sama* (cotton shawl) and is carried most hours of the day.

Since child caretakers also have other domestic duties to perform, they often abandon their charges to the sleeping mat *(gipa)*. Here, without the security and warmth that being carried provides, young babies cry almost continuously from hunger and a lack of attention. When the mother is present, she habitually offers the child the breast either to satisfy his hunger or to serve as a pacifier,[6] but it is not unusual for a child to remain unfed for several hours. This inconsistent behavior in infant feeding would appear to establish a pattern of want and glut, contributing to the development of the personality characteristics discernible in adult Gurage.

Breast feeding affords a Gurage child neither physical nor emotional satisfaction. The amount of food taken from sources other than the breast is minimal, though occasionally a mother may feed a child of two or three months small amounts of cow's milk from the cupped palm of her hand. However, milk is mainly reserved to make butter and cheese, which are used in curative rituals.

In the common nursing practice, the mother lays the child across her knee with the nipple placed in his mouth, while she carries on with as little interruption as possible such other activities as churning milk or spinning cotton. If the child loses the nipple, she supports his head and holds her breast in a more functional nursing position. But nursing is usually halted from time to time as the mother moves about the hut to suit her own convenience.

The desire to resume chores abandoned because of the demands of child care, or to undertake outside activities which keep the mother away from home for several hours, takes priority over nursing. During a lengthy period of field observation, only one occasion was witnessed where a Gurage child was nursed for a period exceeding ten minutes, and then the mother frequently removed her breast from the child's mouth in hopeful expectation that nursing would stop peacefully. Gurage children are not given prime consideration by adults, at least overtly, at any time. No Gurage adult will suffer personal inconvenience for the sake of a child, even his own. As a Gurage child approaches the weaning period, the nursing situation appears to become increasingly less satisfactory. The amount of time allowed for breast feeding declines steadily, and each meal is characterized by the child's struggle to keep the breast in nursing position and thus to satisfy his hunger in the time allotted for feeding. If the child, in his eagerness to achieve satiety, bites the mother's breast, she delivers a sharp slap on his cheek.

Gurage mothers have no routine for introducing solid foods into the child's diet to supplement the decreasing intake of breast milk, but usually the mother, caretaker, or some other responsible adult begins randomly feeding the child bits of solid food as he enters the second year of life. These are merely casual handouts, mainly scraps of ensete bread, and they do not constitute a proper meal even by Gurage standards. Indeed, Gurage parents often say that children at this age refuse to accept any other source of food than the breast.[7] At the same time, Gurage constantly verbalize their concern over the inadequacy of the ensete diet for children. When pressed as to why they take no steps to round out this diet with more nourishing foodstuffs, most adults claim a lack of ready cash which would enable them to purchase them, even if they are readily available locally. Indeed some adult Gurage express a preference for the grain diet of teff *(Eragrostis abyssinica)* and barley of their linguistic kin, the Amhara of Northern Ethiopia.[8]

In any case, the dependence of children on solid foods increases as the first set of teeth is completed, for invariably weaning is encouraged at this time. Weaning is customarily enforced between the ages of two and four years, when, in an effort to repel the child, the mother applies a bitter substance to her nipples. There is evidence to suggest that, when weaning is difficult to enforce, the child may be left to the care of the older children in the household while the mother spends several days visiting with relatives in another village.[9] By the age of five the child shares fully the adult diet of ensete, even though he may be as yet unable to masticate raw meat. In general, coffee and other mild liquid stimulants—*tallä* beer, brewed from barley, and *təj,* a mead drink—are permitted at this time, but I have been informed that in some tribal districts girls abstain from drinking coffee until after marriage.

At daily mealtimes, when Gurage children eat with adults, they always are fed last, and then usually only food that adults leave behind. Exceptions occur at communal work feasts and at tribal ceremonials when the large quantities of food prepared are shared equitably by all who partake in the event. At daily morning and afternoon gatherings when coffee is passed around, it is the fortunate child who is given half a cup of coffee, often the remains of some elder's third serving, or a small portion of ensete bread, or a handful of roasted grains. Even so, children do not pester adults for food and children over six seldom beg. At mealtimes, they always keep to the back of the house, away from their parents, although not out of sight or earshot.

There is a marked change in eating behavior when a boy approaches the age for marriage, ideally at about eighteen. He is now treated somewhat as an adult, joining with his elder kinsmen in communal labor activities and sharing equally with them the bonus of food passed around as recompense for the services rendered. He sits alongside adults at the daily coffee breaks, and he receives his meals at home before his younger siblings are fed. In a word, marriage brings about significant changes in eating habits, at least for male Gurage. A man, in keeping with cultural norms, subsists on a minimal diet, but, as a food producer, he is able to determine within social and economic limits how much ensete he will grow,[10] when he will eat it, and how much of it he will consume.

There is evidence of a close correspondence between the status position of Gurage women within the family and the wider society, and child-feeding practices generally. This relationship takes on special significance for Gurage boys, who appear to develop greater anxiety over food than do girls. This may well relate to the fact that men are ultimately dependent upon women for the preparation of their food, since it would be abhorrent to a Gurage man to prepare his own meal under any circumstances. If it is therefore reasonable to assume that men develop greater and more enduring anxiety about food than do women, it can also be argued that Gurage women, lacking economic and social status and constantly faced with threats of domestic insecurity, develop anxieties which their control over the preparation of food enables them to release in covert aggressive behavior, especially toward their male dependents.

There are obvious sex differences in the feeding of children. Customarily brothers are always fed before their sisters and supposedly in greater quantity; sisters are required to serve their brothers at every meal and also to prepare food for them when necessary. On the frequent special occasions in the village when adults come together and young children are allowed to congregate in their midst, boys are almost always served food, girls almost never. A boy of six or seven, though he may receive food as a special treat, is not expected nor required to share it with girls, even his own sisters. The distribution of food between the sexes, I suggest, is one means by which Gurage reinforce notions of male superiority at an early stage.

Not only are children underfed, but Gurage cultural habits of feeding lead to the development of anxiety over nutrition which is manifested in institutionalized attitudes and behaviors toward food. The use of ensete in curing patients believed to be possessed by evil spirits is to be noted in this regard.[11] A few examples will serve to illustrate some variations of this theme.

When Gurage children are struck with an illness for whose cure their parents lack any effective household remedy, advice of the *sägwora* (wizard) is commonly sought to divine the cause. Once this is ascertained, the wizard prescribes a remedy, somewhat routinized, which usually entails the sacrifice of a sheep having certain peculiar markings and color, and a prayer formula is repeatedly recited until the cure is effected. The flesh of the sacrificial animal is eaten exclusively by the parents of the sick child and others who are present at the curing rite; no portion of the meat is consumed by the patient whose illness may well stem from an inadequate diet. Yet the Gurage appear to be aware of a more than chance or casual relationship between hunger and illness. This is attested in other contexts of ritual curing, where, for example, a sick adult or a parturient female is given a small amount of meat from an animal sacrificed on the patient's behalf. There is rarely a shortage of meat on these occasions, and the kinsmen gathered in the homestead to assist in the ritual commonly eat to the point of glut.

I have already touched on the Gurage practice of storing the edible substance extracted from ensete in deep pits in the earth between the rows of plants in the ensete field. Apart from the practical aspects of this custom in preserving food, the Gurage claim that burial in the earth for long periods of time—from not less than two months to as long as several years—increases the palatability of ensete, so that it is more highly favored than ensete eaten immediately after harvesting. In point of fact, a great deal of waste results from this practice, since the top layers of the stored food darken considerably in color, and Gurage women liken such blemished ensete to earth, meaning that it is "unclean," "contaminated," and therefore fit to be eaten only by slaves and members of outcaste occupational pariah groups. "Unclean" ensete is a common form of payment to slaves and pariahs for services rendered to Gurage of whatever status position. It may be noted in passing that the ensete horticulture assures the Gurage of a steady source of food supply, though its consumption at any given meal is not great. Large surpluses of ensete remain stored in pits, and are seldom removed for daily consumption. Starvation and famine are unknown in Gurage history.

In addition to serving the obvious nutritional needs of the Gurage, ensete is used as a recompense for labor and services rendered, e.g., to discharge obligations to neighboring kinsmen for sharing in communal labor activities. In this case, ensete is eaten by the laborers at the homestead of the host, and the rest of their families profit from the communal labor only through the obligation of the host to reciprocate by helping to cultivate the fields of those who have labored for him. A man does not share with his family the ensete he receives in compensation for his labor, even if they are in need, nor does he set aside a portion for future consumption.

Gurage are often forced to eat when they are not hungry, for the norms of food etiquette prevent a guest from refusing an offer of food, which is the first sign of hospitality, and, on the other hand, they often go hungry even though ample quantities of food are readily available. In ritual life, ensete in particular, and foodstuffs in general, are offered as tribute to dignitaries representing the major Gurage deities, and, as we have seen, ensete is also prescribed for use in sacrifices. Similarly, in life crises, food assumes symbolic importance. For example, neighbors who come to assist a woman in the ordeal of childbirth (mainly by repetitively reciting standard petitions invoking the deity responsible for good health) often become so engrossed in eating that they temporarily abandon the mother to her pain. In funeral rites, food for the bereaved family is prepared by the villagers and brought to them daily for at least one week.

Marriage, family, and the number of offspring, especially boys, rank high among Gurage values, and a Gurage woman who has given birth to eight children is ceremonially honored publicly. A large number of offspring is seen as a provision for old age, since children are expected to take care of their parents. But Gurage marriages are notably unstable. Until the *samər* feast, which is celebrated upon the birth of the eighth child, at which time the marriage is ritually blessed, a husband can, and frequently does, return his wife to her parents on the slightest provocation (Shack 1966: 125). Barrenness is considered shameful and ample ground for divorce, and infertility is probably the principal factor accounting for the relatively high instability of Gurage marriages.[12]

There exist no institutionalized patterns of parent-child behavior that foster the maximum physiological growth and development of the child, and, while no accurate figures are available, the mortality rate among Gurage children appears to be high. Parents show an awareness of the implications of their feeding practices and often attribute the noticeably small stature of their children and their slow rate of physical development to their meager diet of ensete. However, so tightly bound are the Gurage to their ensete culture that radical changes in food habits are deemed beyond their power to effect. Elderly Gurage suffer severely from physical and emotional deprivation, and people of all ages commonly express concern about whether there will be someone to take care of them in their old age.

The following general personality characteristics of the Gurage seem related to the nutritional system:

1. Selfishness. The Gurage carefully calculate the degree of reciprocity that can be expected from any output of energy. For example, a man who helps a neighbor build a house applies himself only to the point where he can be assured that the neighbor will have to reciprocate, and when the neighbor does repay in kind he attempts to extract from him much more labor than reciprocity would normally demand.
2. Emotional detachment. The emotional distance between parents and offspring is expressed by the fact that their relationships are mainly verbalized in terms of mutual obligations. An orphaned child, for example, says: "Now I have no one to feed me or buy my clothes." Similar emotional distance exists between married couples. Thus a married person without offspring or elder relatives verbalizes his condition with some such phrase as "I am all alone; there is no one to help me."

3. Unrelatedness. The behavior exhibited toward persons in need of social and economic assistance is essentially negative, though verbalized attitudes are sympathetic.
4. Passivity. Both verbal and nonverbal behavior suggests defenselessness in an environment that is envisaged as hostile. When asked what steps they will take to solve a seemingly simple problem, Gurage tends to say: "What can we do? We can do nothing."
5. Dependency. Even affluent Gurage expect to receive gifts from one who is considered slightly more fortunate, though he may have rendered no service to the latter.
6. Feelings of worthlessness. The Gurage demand constant verbal assurances that their land, their mode of life, their staple food, ensete,[13] and indeed their material culture in general are as good as those of any other Ethiopian people.

The attitudes and behaviors associated with nutrition among the Gurage have been examined in relationship to their influence on personality development. Much of the time and energy of adults and of children old enough to work is spent in the production and distribution of foodstuffs. No single homestead could successfully plant and harvest ensete without the cooperation of other homesteads in the village. Except for daily meals, every social and ritual event in Gurage society is prefaced or closed with the taking of food. It has been shown that the mother is seldom a dependable or satisfactory source of food for the child and it has been suggested that this leads to the development of anxiety where nutrition is concerned. With maturity this exhibits itself in the many values, other than nutritional, which the Gurage ascribe to foodstuffs, including the use of food as compensation for services and the role of edibles in ritual activities and life crises. Several traits of Gurage personality structure appear to be attributable to what I term "nutritional deprivation." Projective tests,[14] analyzed as yet only in terms of content, seem to confirm the significance of food and nutritional practices in the personality development of the Gurage.

NOTES

1. The research on which this paper is based was supported by National Institute of Health grants nos. MH 0714101–03 to William Shack, and was carried out between 1963–65 mainly among the Chaha Gurage, who, together with the Muher, Eza, Ennemor, Gyeto, Aklil, and Walani-Woriro, compose the larger tribal grouping of Western Gurage. No major cultural and social structural differences exist among the Western Gurage tribes. Thus, while the generalizations on Gurage personality and behavior advanced here refer specifically to the Chaha tribe, I am of the opinion that these assumptions hold equally true for other Western Gurage tribes as well.
2. I wish to express my thanks to Professor Audrey Richards, who was kind enough to read the original draft of this paper and to offer many useful criticisms.
3. For a detail account of the ethnography and social structure of the Western Gurage, see Shack (1966).
4. *Catha edulis,* a stimulant widely grown as a cash crop, especially by Muslim Gurage who habitually chew it throughout the day.
5. The Gurage use butter extensively in a variety of curing methods, both as a preventive and as an antidote. For example, it is taken internally for a mild laxative; for a headache, it is applied to the head and a small piece of ensete food is placed over the point of pain; relief is brought to rheumatic and arthritic discomforts by anointing the aching limb.
6. On this point, Jelliffe (1955: 153), referring to studies of several writers, comments: "the child appears to be breast fed at very frequent intervals, but it is quite possible that on many of these occasions the breast may act only as a comforter."
7. That a diet of breast milk is inadequate is corroborated by Jelliffe (1955: 133), who states: "At the same time, it must be emphasized that even an optimal flow of breast milk cannot

form the sole food of a growing infant after the age of six months, being inadequate in both calories and protein. A child fed on the breast alone after six months has been termed "breast starved."

8. A very low-grade, unrefined quality of teff is usually available in the large tribal markets held weekly, but even for those Gurage who eat grain bread with some regularity it remains only a supplement to ensete. For a description of the history and culture of the Northern Amhara see E. Ullendorff (1960).
9. It appears that this practice is resorted to especially if a mother becomes pregnant while still suckling a child.
10. On the social and economic determinants of land use in ensete cultivation see Shack (1966: 57ff.).
11. W. Shack deals with the problem of institutionalized hunger and nutritional deprivation, spirit possession, and ritual illness in a forthcoming paper.
12. For a further discussion of the several factors which make for marital instability among the Gurage see Shack (1966: 125ff.).
13. The Gurage are well aware that Northern Ethiopians generally, as well as other tribal groups, disdain ensete as a staple food, though the plant is often used by them as an emergency crop under conditions of famine. Gurage pottery and basketry, however, are highly regarded and are marketed throughout Ethiopia. The term "coolie," used in urban Ethiopia to refer to men engaged in degrading manual labor, was formerly applied mainly to Gurage, epitomizing the attitudes of other Ethiopians toward them.
14. A Thematic Aperception Test modified in terms of Gurage cultural themes and standard Rorschach cards were administered in the field to twenty Gurage boys and girls (ten male and ten female) between the ages of fourteen and nineteen. The statement advanced here is based on a partial analysis of these data.

BIBLIOGRAPHY

Barry, H., Child, I., and Bacon, M. K. 1959. Relation of Child Training to Subsistence Economy. *American Anthropologist* 61: 51–63.

Du Bois, C. 1941. Attitudes Toward Food and Hunger in *Alor. Language, Culture and Personality*, ed. L. Spier, A. I. Hallowell, and S. Newman, pp. 272–281. Menasha.

Jelliffe, D. B. 1955. *Infant Nutrition in the Subtropics and Tropics*. Geneva.

Richards, A. 1948. *Hunger and Work in a Savage Tribe*. Glencoe.

Shack, W. A. 1963. Some Aspects of Ecology and Social Structure in the Ensete Complex in South-West Ethiopia. *Journal of the Royal Anthropological Institute* 93: 72–79.

——1966. The Gurage: A People of the Ensete Culture. London.

Ullendorff, E. 1960. *The Ethiopians*. London.

11

Hunger, Anxiety, and Ritual

Deprivation and Spirit Possession among the Gurage of Ethiopia

WILLIAM A. SHACK

Several decades ago Audrey Richards (1948: 213) remarked that "the greater insecurity of . . . man's food supply throws into relief his nutritional institutions, and his whole scale of hopes, fears, and beliefs, as to his daily bread." The conceptual framework of Richards's study embraced the central question of how certain institutionalised attitudes and behaviour patterns of tribal peoples are related to their nutritional system. Elsewhere, in explaining how the Bemba of Zambia (formerly Northern Rhodesia), who practise shifting cultivation of the slash and burn type in rather poor soils, make their living, Richards (1939) took account of the ecological conditions of this inhospitable environment which give rise to the Bemba preoccupation with questions concerning food. Holmberg (1950) reached similar conclusions in his study of the Siriono of eastern Bolivia. These wandering hunting and collecting bands living along the banks of the Rio Blanco exist under conditions of severe food scarcity. The daily search for small game, wild roots and berries, which constitute the bulk of their diet, involves long, arduous expenditures of energy. Partly in consequence of uncertainty in the food quest, many Siriono magical practices appear to have as their principal function a reduction of the anxiety that centres round the satisfaction of hunger (Holmberg 1950:91). At a later stage in this article I shall refer to one of several studies (e.g. Hallowell 1941; Parker 1960) relating to the function of magico-religious apparatus as an anxiety-reducing device among some North American Indian tribes.

The Bemba, Siriono, Ojibwa and other tribal groups who make their living in similarly harsh ecological settings resort to utilitarian magic and spiritual practices frequently, seeking supernatural intervention to ensure success in the hunt or to increase horticultural yields. Ritual performances associated with the food quest serve, among other important functions, to reduce periodically such anxiety as centres round the satisfaction of hunger.[1] It seems reasonable to contend that hunger-frustration probably exists in those tribal societies which are inadequately equipped with cultural apparatus for dealing with their environment, and are thus unable to guarantee their members a constant food supply.

Social anthropological analysis and interpretation of a different order is demanded, however, when attention shifts towards tribal societies which have evolved relatively complicated social structures and elaborate food-producing techniques, and yet despite very favourable food-producing conditions, individual tribesmen conspicuously display anxiety over the prospect of unsated daily hunger. Such concern over procuring daily bread and satisfying hunger exists among the sedentary Gurage who inhabit the sparsely fertile plateaux of southwest Ethiopia, west of the Rift Valley lake chain. I have described elsewhere, at length, how Gurage tribesmen devote an inordinate amount of time and energy to the cultivation of their food staple, *Ensete edule* (Shack 1966). Throughout the year this hardy resilient cultigen provides a constant and abundant food supply. Severe drought and unseasonable heavy rains, which are not uncommon occurrences in Gurage-land and adjacent tribal districts, occasionally affect ensete growth. Crop damage has never been sufficient to cause famine and starvation—things unknown in Gurage history. However, as I explain later, dangers of destruction to fields and property owing to warfare formerly constituted very real threats to survival.

Even so hunger-frustration does exist. I shall attempt to show in this article that one form of anxiety aroused in individual Gurage stems from institutionalised food deprivation, and is expressed in spirit-afflicted illnesses, that is, spirit possession. Before suggesting how certain Gurage forms of spirit possession can be understood in terms of such frustration, a few preliminary remarks are necessary. We must first realise how Gurage cultural patterns of food production and consumption intensify their concern over food. Secondly we need to examine how the food practices of adult Gurage affect their infant feeding methods, and how differential patterns of food distribution among children of opposite sexes correlate closely with the sex pattern of spirit possession.

Of the cultivating techniques of those tribes which exhibit the "Ensete Culture Complex" (Smeds 1955; Shack 1966), that of the Gurage appears to involve an expenditure of labour which is grossly in excess of that necessary to ensure fruitful yields from ensete, the "false banana plant." Of the ensete cultivating peoples of this area, the Gurage alone employ a complex system of crop rotation and transplanting (Shack 1966: 59–63); exacting techniques which not only prolong over a period of eight years the growth cycle of ensete, but also delay the moment of achievement when the harvest gratifyingly rewards the labour spent in the field. These horticultural practices relate closely to many notions Gurage hold about daily food consumption. As I shall explain later, the storing of ensete in large earthpits several feet deep, as Gurage customarily do, symbolises a range of attitudes which they hold about the gratification of hunger, and anxiety over failure to satisfy it.

As regards patterns of land use, I need only mention that, even in those tribal districts where poor soil conditions have severely reduced the available arable land for ensete cultivation, most Gurage homesteads grow far more ensete than family requirements demand from one year to the next. Yearly accumulations of ensete-food, kept stored in the earth, provide each homestead with an almost inexhaustible food supply, let alone sufficient for ordinary daily meals, for barter, for ceremonial feastings, for rewarding services rendered, and for the occasional emergency when other crops have suffered damage from drought or heavy rains. Dispensing ensete unsparingly, to kinsmen and strangers alike, is a social act in which all Gurage who can afford to do so engage, for it connotes more than pragmatic generosity and hospitality. It is one of the few socially acceptable means by which Gurage men can openly display power and prestige. Men who command greatest respect and authority in all spheres of tribal life are in fact the best food providers.

If "food and beer are without doubt the most exciting and interesting topics of Bemba conversation" (Richards 1939: 44), the Gurage constantly discuss ensete. That is not all; a wide range of Gurage social relationships, which cut across kinship groupings and status differences, stem also from economic transactions involved in dispensing or receiving ensete. Bemba manifest in one way, it is said, their emotional attitude towards food by overeating, frequently suffering discomforts from digestive pains. The Gurage daily round of food taking is likewise typically characterised by individuals sparingly consuming the staple ensete and supplementary food stuffs. By western standards, none of the three daily meals of ensete served habitually in each Gurage household could be distinguished as being the "main meal"; neither the quantity of ensete prepared, nor the amount individuals consume varies markedly from one meal to the next. National religious festivals such as Mäsqal (Shack 1969), a ceremony performed to usher in the New Year, or commensal feasting at life crisis rituals, are the exceptional occasions where ritual license permits individuals to reverse the cultural norms of food etiquette. On such festive occasions, Gurage eat to the point of glut. In the daily round, however, the consumption of ensete is restricted to the minimal level required for subsistence, seldom varying from slight handfuls, to nibbles, to hardly any.

Food taboos prohibit the eating of the flesh of hunted animals and the Gurage share with their Cushitic neighbours a disdain for fish. Livestock provide the main source of protein intake in their diet. However, livestock are kept principally for their supply of milk and dung; the latter is essential in ensete horticulture. Because of the exceptionally poor breed of Gurage Zebu cattle, which normally produce at milking only a litre or less, the supply of milk is always scanty and is seldom drunk even by breast-starved children (Shack, D. 1969; this volume). Instead milk is churned into butter; and each household always keeps stored large quantities of spiced butter ageing in clay pots hung from the walls of the main hut, for some later therapeutic use in both the ritual and non-ritual treatment of illnesses. Gurage believe firmly that butter taken internally, or used as a body lotion or poultice, can effectively remedy a digestive upset or relieve uncomfortable aches and pains. A popular Gurage proverb expresses these notions about the special healing quality butter is believed to possess: "A sickness which has the upper-hand over butter is destined for death" (Leslau 1950: 134).

Sheep and oxen are hardly ever slaughtered by heads of households, except when the performance of a special ritual demands a sacrifice. To increase the efficacy of the rite, the sacrifice must often be repeated several times, thereby prolonging the feasting on the sacrificial meat. And those who partake in the ritual eat until they are replete.

Status differences are not reflected in Gurage norms of food etiquette. Individuals or groups of unequal status usually consume, in the daily round of meal taking, no more or no less ensete than others. Though I explain in greater detail below, in the context of Gurage spirit possession, I note here that the *mwakyar* monthly feasts provide for prosperous Gurage men an institutionalised and socially acceptable outlet for the gratification of hunger. As a normative value of food consumption, however, Gurage of whatever social status consider overeating to be an act of coarse vulgarity; an unrefined custom they associate with the landless pariah groups of low-caste artisans and hunters who live as their dependents (Shack 1964; 1966). It is also a behaviour pattern which Gurage hastily explain as being characteristic of their linguistic kin, the Amhara and Tigrinya peoples of northern Ethiopia. In particular, the Amhara practise what Gurage deem to be a most distasteful custom, that of stuffing into the mouth of a person a handful of food immediately after he has eaten his fill and is obviously satiated (Levine 1965: 225). Gurage values decree as in equally bad taste the practice of eating to the last all the ensete bread that a host sets before a gathering; there should always remain some ensete even after the most meagre serving has been passed round. These Gurage attitudes about food consumption would appear to have more than a casual relation to their concepts about male handsomeness and female beauty. Slightness of personal build and keenness of features rank uppermost in the Gurage scale of values and, among both men and women, obesity is rare.

Publicly dispensing and sharing ensete, a normal part of Gurage social intercourse, symbolise the values associated with kinship and neighbourliness. The restatement of these values takes place most frequently within the context of highly stylised food sharing occasions, when eating in the company of others enables those who partake of the meal to act out certain norms of etiquette. Most meals are scanty, however; they constitute little more than morning coffee or afternoon tea in western societies. Rarely does meal-taking seem to be an occasion primarily concerned with supplying nutritional substance.

Covertly, however, Gurage consume larger quantities of ensete and other foodstuffs within the closed confines of their homesteads; this is virtually an act of secrecy. After inter-homestead visiting in the village draws to a close for the day and families retire, hoarded scraps of meat or vegetables are eaten by the fireplace in the dimly lit huts, late at night. Here it might be suggested that the habitual practice of garnering food to be eaten later during the night stems from the reluctance to share food with others. For meal-taking during the day becomes a semi-public affair, inevitably attracting a crowd of onlookers—some kinsmen, others not. Weary travellers on their long journey to market, for instance, would not be at all hesitant about comforting themselves in the home of a complete stranger, clansman or not, unabashedly accepting his hospitality. Visitors never beg for morsels of food. But a host is reluctant not to make at least the gesture of sharing his food among the gathering for fear of offending them. In this context, it is worthwhile to recall Holmberg's observation (1950: 36) of Siriono food hoarding behaviour, for it seems to hold equally for the Gurage: by eating at odd hours during the night when everyone else has retired, each family member not only gets more food but also avoids the nuisance of having others around to share it. At the same time, hoarding affords a means of circumventing the norms of kinship and propinquity as regards dispensing food, without fear of public sanction through gossip and ridicule.

So far I have tried to show that Gurage methods of food production, and habits of distributing and eating food point to two interrelated factors which underlie the relationship between food-deprivation and anxiety behaviour. On the one hand, hunger and anxiety exist because of food insufficiencies which do not stem from environmentally created conditions of scarcity; rather, hunger and anxiety exist as a consequence of the values placed on hoarding and self-denial and which prevent the daily satiation of hunger. On the other hand, the network of kinship obligations and duties centering round reciprocity and commensality where food distribution is concerned, inevitably brings individuals into social and ritual situations which compel them to eat gluttonously when they are not hungry. Attitudes such as these on the part of adult Gurage towards food-taking undoubtedly affect their methods of infant feeding which I describe next. It is in the early training of children with respect to nutrition that anxieties about food are established, to be manifested at a later stage in adult life in the form of spirit possession.

I need outline here only those nutritional practices, and child rearing methods, which form a necessary background to the discussion of spirit possession which follows; a detailed, descriptive analysis of these processes is to be found elsewhere (Shack, D. 1969; this volume). Until a Gurage child is weaned at about the age of four, the amount of food taken from sources other than the breast is minimal, and, at the same time, breast-feeding appears to afford the child neither physical nor emotional satisfaction. The feeding situation is characterised by the child's constant struggle to keep the mother's nipple in its mouth while she continues to perform such domestic chores as churning milk or carding raw cotton. Domestic activities, which keep the mother away from home for lengthy periods, take priority over nursing, and it is not unusual for a child to remain unfed for several hours. Young babies cry almost continuously from hunger, or lack of attention. If weaning proves difficult to enforce, the child may be left to the care of the older children in the household, while the mother spends several days visiting with relatives in another village.

Attention has already been called to the contrasting attitudes of Gurage parents towards the feeding of male and female children. Customarily, brothers are always fed before their sisters and supposedly in greater quantity; sisters are required to serve brothers at every meal, and also to prepare food for them when necessary. Along with other cultural patterns of food distribution, these practices take on special significance in view of the greater anxiety Gurage boys appear to develop over food than do the girls.[2] In a related sense, these cultural habits of feeding children differently according to sex would seem to play some part in associating the dominant social values of food-getting and food-dispensing with male achievement. Since the gratification of their hunger is for Gurage infants a matter of chance, the mother is seldom seen as a dependable or satisfactory source of food for the child; she is indeed a major factor in dependency-frustration. Thus, inconsistent behaviour in infant-feeding establishes a pattern of want and glut; and anxiety, so aroused over failure to satisfy hunger in early life, increases with individual maturity. In adult life, this finds expression in a variety of ritual experiences which Gurage men undergo, and especially in those rituals which concern spirit afflicted illnesses. It is to the ritual performances of spirit exorcism which Gurage men employ to bring about temporary relief from periodic discomforts that I now turn.

Gurage make precise categorical distinctions in their interpretations of illness. Although one or more different species of ensete are customarily consumed as cures for any disorder, sickness and discomfort caused by what Gurage aetiologically refer to as "natural" forces (for which they lack effective therapeutic remedies) are plainly distinguished from illness attributed to malevolent forces, or evil spirits. Of the latter, subsequent distinctions are drawn between evil spirits which possess either males only, or females only, and other demons which take possession of unsuspecting victims irrespective of their sex. Young women who have not yet given birth to their first child, for example, are said to be especially vulnerable to possession by the *ənəas*-spirit. Gurage men acknowledge the existence of several kinds of "personal spirit," one being known as *abwädiga*. This spirit, for instance, is claimed to be inherited from a dead father who had once been cursed during his lifetime for having committed some serious infraction of social norms, and in atonement for which his sons make periodic sacrifices. Here I am concerned with that form of ritual illness which Gurage attribute to the spirit called *awre*, an affliction which only affects men.

Now, when a man complains of loss of appetite, nausea and intermittent attacks of severe stomach pains, immediately suspicion is awakened that any of several potentially harmful spirits has "touched" him. (Evil spirits are believed to dwell in the dense forests and alongside the streams which demarcate village settlements, forming spatial and social boundaries between areas of habitation.) Discomforts of this nature are culturally stereotyped "signs" of supernatural affliction. If such nagging symptoms of illness persist and his condition deteriorates, the patient usually sinks into a semi-stupor, and although he may occasionally become conscious, he is seldom able to take food or water. In this torpid or trance-like state the victim's breathing becomes difficult; the slow regular pace of breathing is interrupted momentarily by loud, hoarse wheezing, not unlike an asthmatic patient, and those attending him respond by reciting a short word of prayer. Violent seizures of body trembling often overcome the patient, compelling his attendants to use force in holding him down on the sleeping mat; and in severe cases, partial paralysis of the extremities may set in. If there are no signs of a "natural" recovery of health, as indicated by a recovery of appetite, a "wizard"[3] (*sägwära*) is summoned to divine the causes of the illness—that is, to ascertain the type of spirit believed to have taken temporary possession of the victim. After consulting oracles to reveal the spirit's name, in this case *awre*, the wizard prescribes a routine formula for exorcising the spirit, a conventional recipe which belongs to the common repertoire of adult Gurage knowledge about supernatural matters. Though not a permanent cure, spirit exorcism enables the possessed person to establish a viable relationship with the spirit concerned. Once afflicted, the victim is chronically liable to re-possession, the temporary recovery from which entails a repetition of the exorcist rite described below.

As in non-ritual contexts where food is offered, so also in preparation for the *awre* ritual, the victim's wife, mother, or some elder woman of the household, is responsible for preparing large quantities of the ensete-food known as *bra-brat* for the exorcist rite. The essential ingredient of this special dish is the finely minced root of the *gwariya* ensete plant, a species of ensete to which Gurage attribute miraculous medicinal qualities. Its therapeutic properties are believed to be strengthened when *bra-brat* is eaten to a ritual accompaniment. No additional medicinal herbs, or magical substances, are combined in the mixture; the only additives are freshly ground red pepper and liberal amounts of thoroughly aged butter. Once prepared, *bra-brat* is served to the patient in a *bitär*-dish, the largest Gurage clay utensil, one of which is included in every household's ritual paraphernalia.[4]

Before commencing the rite of exorcism, the house fire is partially extinguished, allowing the smouldering hot embers to emit a steady stream of dense smoke which clouds the hut into nearly total darkness. If need be, the patient is lifted from the sleeping mat and placed next to the fireplace so as to bring him into closer contact with the rising smoke. Satan and other evil beings dislike lingering in a smoke-filled atmosphere, Gurage say, and once exorcised from the possessed person, the dense smoke hastens their departure. If an unduly protracted possession has caused the patient to be severely weakened, he may require help to maintain himself in a sitting position. Otherwise no physical contact occurs during the exorcism between the patient and those attending the rite, such as body massaging, the laying on of hands, or the like. As kinsmen and neighbours arrive to participate in the ceremony, they seat themselves alongside the circular wall of the darkened hut, awaiting the signal to commence chanting to *awre*. When the *bitär*-filled brim high with *bra-brat* is placed on the floor in front of the seated victim, he lowers his head over the bowl, whereupon the women in charge of the rite drape over him a *sammä*, a long white cotton shawl, completely cloaking from view his person and the ritual food. Symbolically hidden beneath the *sammä*, the patient, using both hands, begins greedily and ravenously stuffing his mouth with *bra-brat*.[5] Simultaneously the exorcism commences; those assembled chant to *awre* a simple, monotonous, stylised prayer, setting a cadence by hand clapping, which rhythmic beat keeps pace with the patient's loud, hoarse wheezing. The prayer chant to *awre* translates literally somewhat as follows:

> ask the wizard *(šägwära)*
> to eat *bra-brat* says the wizard
> feed (the spirit) *awre*
> feed (the spirit) *awre*

Customarily, the curing rite begins at nightfall, continuing into the early morning. If the exorcism proves especially difficult to effect, the crowd of relatives and friends remain, and keep up their chanting throughout the next day, or longer, until the spiritual sign is revealed that *awre* has given up possession of the victim. This revelation is announced symbolically when the spirit, speaking through the possessed person, utters with a sigh, several times—"*täfwahum*"—"I am satisfied."

It is scarcely necessary to stress the obvious parallels here between the actual and symbolic, the ritual and non-ritual forms of Gurage food behaviour. It is sufficient to note that the act of secretly eating hoarded scraps of food late at night, as Gurage ordinarily do, is here reproduced in ritual facsimile: in the exorcism, convened after dark, the possessed victim, symbolically hidden from view of the audience behind a shawl, indulges himself gluttonously. What is more significant is the interpretation Gurage themselves give to the ritual exorcism. This they conceive as an act of "feeding the spirit," a means of restoring the victim's health; for it is *awre,* the spirit, Gurage explain, not the possessed person, which is symbolically satiated when the victim eats his fill.[6] However, the Gurage's own aetiology of *awre*-illness raises, in turn, at least two anthropological questions relating to the social structural correlates of spirit possession and hunger-produced anxiety. Why are some men possessed by *awre* and others not? Why do women appear to express a lower degree of hunger-produced anxiety than men?

On the first point, the few cases of *awre*-illness which I recorded all reveal that the possessed victims were "poor men," according to Gurage evaluations of wealth and status. These men owned small plots of land with few ensete plants, and they were encumbered by long-standing debts arising out of borrowing cattle and money. I lack sufficient evidence to substantiate local hearsay that these victims of *awre* possession had all recently undergone some personal crisis. Even so, in the strictest sense, such men of lowly means were not destitute, nor did they constitute a social category equivalent to the depressed occupational pariah groups and ritual outcasts found in Gurage society. But land, ensete, and cattle are the principal economic resources that determine the extent to which Gurage men are able to achieve recognisable status and prestige by generously dispensing food. At the same time, prestigious men also enjoy the added nutritional rewards of participating in the monthly feast exchanges organised by members of the *mwəkyar* association.

The mwəkyar associations, in their membership, cut across kinship, lineage, and territorial alignments; and, though Muslim Gurage form separate groups, Christians and those who still profess their traditional beliefs frequently belong to the same group. Every month the eight to twelve men who form the average *mwəkyar* group gather at sunset in the main house of the host whose turn it is to provide the feast. After reciting a sacrificial prayer, the host slaughters an extra-large bull which has been specially tended for several months in order to fatten it for the feast. The feasting on raw meat, ensete, and *talla* beer continues gluttonously throughout the night until the following mid-afternoon by which time the sacrificial bull is largely consumed. Apart from those public feasts which are of tribalwide religious significance or which mark life crisis rituals, the monthly *mwəkar* is the only institutional mechanism which Gurage men have to relieve their anxiety aroused over daily hunger. When questioned on the distribution of *awre*-spirit possession among poor and wealthy men, Gurage were unable to remember a case of such ritual illness among prestigious men who occupied positions of high status and authority.[7]

In contrast to men, Gurage girls are initiated after puberty into the *mwəyät*-possession cult and retain membership in it throughout their lives. Through *mwəyät* activities, Gurage girls and women become involved with considerable regularity in the performance of obligatory rituals which allow them to consume formidable amounts of food; no comparable obligations to perform rites of sacrifice are imposed upon Gurage men. For example, in each village, *mwəyät* girls undergo monthly rites in which they appeal to the female deity, *Dämwamwit*; these rites to the goddess are believed to bless the community with good health, and, especially to ward off attacks of the dreaded influenza *(gumfa)*. Other sacrificial rites which the *mweyät* girls periodically perform are intended to restore the well-being of cult members who have suffered some personal crisis: a miscarriage or stillbirth, an unexplained illness, or a temporary state of possession. To these events is added the annual initiation ceremony of novitiates into the *mwəyät cult*. All these ritual occasions for Gurage women are marked as much by the feasting permitted as by the sacred nature of the events. The frequency with which Gurage women participate generally in ritual activities, I suggest, enables them to satisfy their appetites with greater regularity and ease than is the case for Gurage men, and so women's anxiety over food is reduced. However, it can also be argued that, although Gurage women most probably do develop anxieties over food, their control over the preparation and dispensing of food not only permits them to obtain a great deal of perquisites on the side, but also enables them to release tensions in covert aggressive behaviour, especially towards their male dependents (Shack, D. 1969: 296).

The analysis I have so far advanced of the relationship between hunger, anxiety, and ritual among the Gurage, and the somewhat less well refined but related interpretation of food-deprivation in cultural and personality terms, seems to me to provide a much broader analytical framework for understanding spirit possession than by considering each of these factors as separate and independent sociocultural or "psychological" variables. Parker (1960) has utilised a similar method in his analysis of *wiitiko* spirit possession among the Ojibwa Indians of North America. The essential features there exhibit striking similarities with *awre*-possession among the Gurage.

Amongst the Ojibwa hunters and collectors, the *wiitiko* spirit is said to possess only males who have been unsuccessful in the food quest. The victim's initial symptoms of possession are feelings of morbid depression, nausea, distaste for ordinary foods, and sometimes periods of semi-stupor. Gradually the victim feels that he is possessed by the *wiitiko* monster, a fierce cannibalistic being to whose will he has become subjected.[8] Now the Ojibwa place a high value on food providing, and failure in the food quest provokes intense anxiety because it threatens starvation and lowers self-esteem. The point to be stressed here is that both the Ojibwa and the Gurage victim feels that he has been possessed by a spirit whose appetite must be served as his own. During the state of possession the victim satiates the spirit's appetite by consuming food ravenously. The prototype of the *wiitiko* monster is said to be the mother figure, who is a major agent of dependency frustration in early life; I have already suggested a similar explanation for hunger produced frustration in the early life of Gurage children. Likewise, the basis for the predisposition that later results in the fantasy of a persecuting monster, whether the *wiitiko* or the *awre* spirit, is established in the development of covert dependency cravings in the early socialisation process, particularly in the mother-son relationship. The fantasy is cultivated and given form by the cultural belief in such an evil spirit.

The awre and *wiitiko* types of therapeutic possession, and still other cases which abound in the anthropological literature, are all examples which serve to point up the close correspondence between forms of "relative deprivation" (Wallace 1956) and spirit possession. That the phenomena warrant considerably more attention than has been granted hitherto by social anthropologists is amply demonstrated by Lewis's (1966) comparative treatment of what he termed "deprivation cults." I am in substantial agreement with that attempt to advance an understanding of possession cult phenomena in terms of the structural position of deprived status groups. The possession of people in marginal social status positions does seem to relate to situations of stress and conflict, and not only in the tribal societies of Africa, Melanesia and aboriginal North America; medieval European peasants and racial minority groups in contemporary western societies have been known to behave ritually in very similar ways. However, I have attempted in this essay to advance further Lewis's line of argument by taking account also of the importance of "nutritional institutions," to use Richards's phrase, as defined in the summative sense, in seeking correlations between deprivation and spirit possession.

Anxiety and social tensions can, and most likely do, arise as a result of factors other than the built-in structural conflicts over prestige, authority, status, inheritance, co-wife jealousy, sibling rivalry, and the like. If this is so, it seems equally plausible to argue that ritual expressions of conflict in social relationships are in some though not all cases merely the sociocultural glosses over failure by certain categories of individuals and groups to satisfy some more basic human needs, nutrition being one of these. Structural correlates may be important in advancing social anthropological explanations of spirit possession among socially deprived persons. But plainly, the bread and butter issue is that an empty belly and gnawing hunger pains will stimulate anxiety behaviour among most normal men in any society. The overriding emphasis placed upon social structural factors, levels of technological development, and methods employed in the food quest often obscure more fundamental questions of the type I have attempted to raise in this article about the correspondence between deprivation and spirit possession. I have dwelt at length upon some of these questions; more especially I have tried to give answers to them in terms of how Gurage men behave ritually in consequence of their failure to satisfy their hunger, and what available social and ritual mechanisms enable tensions arising out of prolonged food-deprivation to find release in less disruptive ways. In the Gurage case, *awre*-spirit possession serves at one level to bring the deprived possessed person to the centre of societal attention. The honour extended to him by feeding during the state of ritual trance is an indirect means by which he can mildly redress status inequalities (cf. Lewis 1970: 299). At another level, the "license in ritual," which allows possessed men to behave in normally prohibited ways, such as eating gluttonously with both hands during the *awre*-rite, not only gives expression to cultural norms by reversal, it also emphasises certain deeper social alignments and conflicts involved in the social position of men of lowly means in Gurage society (cf Gluckman 1956: 109–36).

In a word, I have tried to show how Gurage conventional restraints about food consumption lead to self-maintained hunger. Men deprive themselves daily of the very same ensete food which they have laboriously produced in abundance. Symbolically, they gluttonously feed the earth, as it were, by storing ensete deep in the ground, while their cupboards remain perpetually bare. Power, prestige, acceptance, affection, worth, self-esteem, and like personal attributes all rank high as male values of achievement, the epitome of which being manifested, in one way, through dispensing food unstintingly in particular kinds of social situations. But not all Gurage men have command of sufficient economic resources to enable participation in the monthly ceremonial food exchanges which, apart from the social and ritual functions these semi-private feasting events serve, also allow for periodic satiation of hunger. There is, then, a built-in conflict in the structure of Gurage alimentary values. On the one hand, eating as a nutritional act, to get one's fill, is culturally discouraged; on the other hand, eating to the full is indulged as a social act through which to gain the satisfaction of personal prestige and status. This dilemma confronts wealthy and poor men alike in Gurage society, but it affects acutely men of lowly means who, as well, seem to be especially subject to *awre*-spirit possession. Thus the rite of *awre*-spirit exorcism is both an expression and resolution of the conflict: it enables the most status-deprived victims of institutionalised hunger to have their cake and eat it.

One logical question, essential to the argument I have advanced, remains to be considered: why does food-hoarding behaviour obtain when ensete is not wanting in Gurage society? I believe the answer is to be found in the historical and political, rather than in the environmental and technological, experiences of the Gurage, and particularly in that bygone era, when threats of physical annihilation and destruction of property and crop fields were real and immediate to Gurage tribesmen. All the following are significant here: fratricidal strife between lineage and clan segments; armed conflict spurred by powerful tribal sections; predatory expansion of hostile western Galla, Kambatta, and Sidamo tribes; and the political expansion of the powerful Amhara kingdoms from the seventeenth century onwards. These political conditions persisted with remarkable regularity, and they constituted very real threats to the continuing existence of the Gurage state for nearly four hundred years. I re-emphasise that these were major events of Gurage political history until the *pax Aethiopica* was established over the tribal district in 1889.[9] Though now a part of the memorable past, occasionally relived in oral tradition, these essential features of Gurage political history contribute directly to the Gurage's conceptual view of the world as being threatening, hostile, and fraught with insecurity. I have elsewhere remarked (Shack 1963), in explaining how Gurage express feelings of insecurity in interpersonal relationships through ritual action, that fear and suspicion are rife in Gurage society.

Space permits no more than this cursory reference to the core features of Gurage history which I have elsewhere (Shack 1967) described in the context of civil war and political process in Gurageland. Enough of these past events have been indicated to reveal their importance for the understanding of contemporary forms of Gurage sociocultural and ritual behaviour, especially in so far as certain of these actions and related ideas concern Gurage habits of hoarding ensete. The storage of large quantities of ensete in deep earth-pits, and the deliberate hiding of the exact locations of the storage pits (by covering over the openings with garden refuse), are cultural practices which have been exacerbated by the historical dangers of destruction.[10] For in the past, after the most devastating plundering and raiding of villages for slaves and cattle, the ensete-food buried deep in the earth eluded the pillage of hostile raiders, to be recovered later by survivors once the threat of danger had passed. It is, of course, no surprise that Gurage continue to exercise conventional restraint in the consumption and distribution of food despite the disappearance of the political and historical conditions which gave rise to such habits. Thus the conventions of eating in small amounts publicly, tucking into food privately, and stacking away large food supplies, arose out of rational fears about physical survival and these forms of learned behaviour persist now as institutional norms expressing the right thing to do. Cultural patterns start off in particular directions and their momentum carries them on.

I must stress that this article is essentially exploratory. The very nature of the subject of spirit possession has limitless, shifting boundaries which are often difficult to define for the purpose of social anthropological analysis and interpretation. Here I have attempted to define the problem in terms of three interrelated, yet distinguishable, dimensions of social behaviour—hunger, anxiety, and ritual.

NOTES

The observations of Gurage religious behaviour described here were mainly recorded during the course of field research carried out between 1963–65, principally among the Chaha Gurage. This research was generously supported by National Institute of Health grants nos. MH-071411–03.

1. I need to define at the outset the terms "hunger," anxiety," and "deprivation" which are used repeatedly throughout this article. "Anxiety" is to be understood in its sociocultural meaning, rather than in the biopsychological interpretation commonly advanced in Freudian psychoanalysis, although I recognise that the distinction is, at best, only utilitarian and analytical. Hence I follow Hallowell's (1941: 870) view of anxiety as: "... an affective reaction to danger situations, as *culturally defined*..." (my emphasis). The terms "hunger" and "deprivation," though to be interpreted literally, are here used in the context of Gurage societal values also. One of several constructive criticisms offered by Professor Raymond Firth on an earlier version of this article, which I acknowledge with gratitude, is that what I call "anxiety behaviour" might best be expressed as "conventional restraint"; i.e., learned behaviour which may have been the response to anxiety at an earlier period in Gurage society. While

this suggests a useful analytical distinction, it also implies that apprehension, insecurity, intense fear, etc., are not current traits of Gurage personality, a conclusion which the projective test data obtained in the field would not support (see note 2 below). However, I am prompted to employ both terms in this article, reserving "conventional restraint," as used in the context of alimentary processes, to refer to the cultural manifestations of anxiety.

It is to be noted, furthermore, that although ritual specialists of the pariah Fuga group are important intermediaries in Gurage female possession cults (see e.g. Shack 1966; Lewis 1970), they are neither involved as afflicted victims nor as ritual exorcists in male spirit possession of the type discussed here. So far as I know, this form of spirit possession occurs only among Gurage and is thus strongly associated with Gurage normative values concerning land ownership and ensete cultivation, from which Fuga and other social outcastes are ritually excluded. I obtained no data on spirit possession among pariah groups in Gurageland, if it occurs at all.

My data are insufficient to establish firmly the general distribution of the form of spirit possession described in this article among other Western Gurage tribes, but the few scattered examples obtained of cases in non-Chaha areas, together with statements from informants, suggest that the phenomenon is uniform throughout western Gurageland.

2. Shack, D. (1969: 298–99; this volume, pp. 112–13) has attributed several traits of Gurage personality to "nutritional deprivation." The tentative conclusions advanced about the relationship between nutritional processes and Gurage personality development have been based upon partially analysed psychological field data, consisting of standard Rorschach cards, and Thematic Aperception Tests modified in terms of Gurage cultural themes.
3. Following Middleton and Winter (1963: 3) I use the term "wizard" also to define ritual practitioners with divination being an additional art, as is the case of the Gurage *sägwära*.
4. Among numerous rites performed in the household in which the *bitär* is used, that of childbirth is to be mentioned. After delivery, the newborn is placed in the *bitär* whereupon the "god-mother" washes and anoints the child with butter. The umbilical cord is carried in the *bitär* to the rear of the house for burial.
5. Apart from the more obvious symbolic reversal of Gurage norms of food etiquette in the *awre*-ritual, I note that food is never eaten with the left hand in the daily round of meal taking; or at least not publicly.
6. C.f. Hamer & Hamer (1966: 397) on spirit possession among the Sidamo of southwest Ethiopia, where ritual exorcism is conceived of as an act of "feeding the spirit." Part of the dialogue with the spirit is recorded as follows:

   Spirit: I brought this illness to X because I wish to possess him; and he has refused to accept me; I am hungry and he has not fed me.

   Spokesman: Will you heal X if he feeds you?

   Spirit: Yes, I will make him well if he will feed me.

7. Some aspects of Gurage sociocultural history lend substantial support to this conclusion. In former times, Gurage men gained prestige and wealth principally through warfare and those old warriors still alive and their sons constitute principally the prestigious groups of *mwəkyər* associates today. In oral traditions, the military exploits of famed warriors are likened to their voraciousness and insatiable appetites which they would exhibit at *mwəkyər* gatherings. One text I recorded in the field, which relates aspects of the life of ɜmwar Farda, a hero of the Chaha tribe of Gurage, boasts of the old warrior as having the capacity to consume alone at a single meal, a sheep or young calf. Though perhaps not devoid of exaggeration, this tale accurately conveys Gurage attitudes about the correspondence between social status and institutionalised means of satiating hunger.
8. Parker writes that *wiitiko*-possession is a psychosis, " . . . *a bizarre form of mental disorder,*" (my emphasis), which is not only a reaction to severe environmental pressures, but ". . . a form of pathological adjustment of the modal Ojibwa personality to such environmental pressures" (1960: 604). Similar conclusions might be drawn about *awre*-possession but I do not encourage such conjecture. Throughout this article I have been deliberately naive about the shifting boundaries between social anthropological and psychological analysis and interpretation of possession cult behaviour.
9. Nearly fifty years of continuous warfare and slave raiding have devastated vast areas of Gurageland, the turmoil reaching its peak in the 1840s and lasting until the conquest of 1889. For

the early period see the accounts by Harris (1844) and Burton (1856); on the conquest period and after see Guebre Sellasie (1931).

10. In this regard, some developmental aspects of Gurage law are of interest since they reflect the politico-legal efforts of Gurage to redress old wrongs and maintain peace through the application of judicial authority backed by religious sanction. With the formation of a centralised judicial system for the politically acephalous Gurage tribal grouping, the destruction of property and ensete fields by arson was defined as a criminal offence and equated with "sin." (See Shack 1967.)

REFERENCES

Burton, R. F. 1856. *First footsteps in east Africa.* London: Longmans.

Gluckman, M. 1956. *Custom and conflict in Africa.* Oxford: Blackwell.

Guebre Sellassie. 1931. *Chronique du règne de Menelik II.* Paris: Librairie Orientale.

Hallowell, A. I. 1941. The social function of anxiety in a primitive society. *Am. sociol. Rev.* 6, 869–81.

Hamer, J. & I. Hamer 1966. Spirit-possession and its socio-psychological implications among the Sidamo of southwest Ethiopia. *Ethnology* 5, 392–408.

Harris, W. C. 1844. *The highlands of Æthiopia.* London: Longmans.

Holmberg, A. 1950. *Nomads of the long bow: the Siriono of eastern Bolivia* (Publ. Inst. social Anthrop. Smithson. Instn 10). Washington: Government Printing Office.

Leslau, W. 1950. *Ethiopic documents: Gurage* (Viking Fd Publ. Anthrop. 14). New York: Viking Fund Inc.

Levine, D. 1965. *Wax and gold: tradition and innovation in Ethiopian culture.* Chicago: Univ. of Chicago Press.

Lewis, I. M. 1966. Spirit possession and deprivation cults. *Man* (N.S.) I, 307–29.

———. 1970. A structural approach to witchcraft and spirit-possession. In *Witchcraft confessions and accusations.* (ed.) M. Douglas. London: Tavistock.

Middleton, J. & E. Winter (eds.). 1963. *Witchcraft and sorcery in east Africa.* London: Routledge & Kegan Paul.

Parker, S. 1960. The wiitiko psychosis in the context of Ojibwa personality and culture. *Am. Anthrop.* 62, 603–23.

Richards, A. I. 1939. *Land, labour and diet among the Bemba of Northern Rhodesia.* London: Oxford Univ. Press.

———. 1948. *Hunger and work in a savage tribe.* Glencoe: Free Press.

Shack, D. N. 1969. Nutritional processes and personality development among the Gurage. *Ethnology* 8, 292–300.

Shack, W. A. 1963. Religious ideas and social action in Gurage bond-friendship. *Africa* 33, 198–206.

Shack, W. A. 1964. Notes on occupational castes among the Gurage of south-west Ethiopia. *Man* 64, 50–2.

———. 1966. *The Gurage: a people of the ensete culture.* London: Oxford Univ. Press.

———. 1967. On Gurage judicial structure and African political theory. *J. Ethiop. Stud.* 5, 89–101.

———. 1969. The Mäsqal pole: religious conflict and social change in Gurageland. *Africa* 38, 457–68.

Smeds, E. 1955. The ensete planting cultures of eastern Sidamo, Ethiopia. *Acta geogr.* 13, 1–39.

Wallace, A. F. C. 1956. Revitalization movements. *Am. Anthrop.* 58, 264–81.

12

Fast, Feast, and Flesh

The Religious Significance of Food to Medieval Women

CAROLINE WALKER BYNUM

> In reading the lives of the [ancients] our lukewarm blood curdles at the thought of their austerities, but we remain strangely unimpressed by the essential point, namely, their determination to do God's will in all things, painful or pleasant.
>
> —Henry Suso,[1] German mystic of the fourteenth century

> Strange to say the ability to live on the eucharist and to resist starvation by diabolical power died out in the Middle Ages and was replaced by "fasting girls" who still continue to amuse us with their vagaries.
>
> —William Hammond,[2] nineteenth-century American physician and founder of the New York Neurological Society

Scholars have recently devoted much attention to the spirituality of the thirteenth, fourteenth, and fifteenth centuries. In studying late medieval spirituality they have concentrated on the ideals of chastity and poverty—that is, on the renunciation, for religious reasons, of sex and family, money and property. It may be, however, that modern scholarship has focused so tenaciously on sex and money because sex and money are such crucial symbols and sources of power in our own culture. Whatever the motives, modern scholars have ignored a religious symbol that had tremendous force in the lives of medieval Christians. They have ignored the religious significance of food. Yet, when we look at what medieval people themselves wrote, we find that they often spoke of gluttony as the major form of lust, of fasting as the most painful renunciation, and of eating as the most basic and literal way of encountering God. Theologians and spiritual directors from the early church to the sixteenth century reminded penitents that sin had entered the world when Eve ate the forbidden fruit and that salvation comes when Christians eat their God in the ritual of the communion table.[3]

In the Europe of the late thirteenth and fourteenth centuries, famine was on the increase again, after several centuries of agricultural growth and relative plenty. Vicious stories of food hoarding, of cannibalism, of infanticide, or of ill adolescents left to die when they could no longer do agricultural labor sometimes survive in the sources, suggesting a world in which hunger and even starvation were not uncommon experiences. The possibility of overeating and of giving away food to the unfortunate was a mark of privilege, of aristocratic or patrician status—a particularly visible form of what we call conspicuous consumption, what medieval people called magnanimity or largesse. Small wonder then that gorging and vomiting, luxuriating in food until food and body were almost synonymous, became in folk literature an image of unbridled sensual pleasure; that magic vessels which forever brim over with food and drink were staples of European folktales; that one of the most common charities enjoined on religious orders was to feed the poor and ill; or that sharing one's own meager food with a stranger (who might turn out to be an angel, a fairy, or Christ himself) was, in hagiography and folk story alike, a standard indication of heroic or saintly generosity. Small wonder too that voluntary starvation, deliberate and extreme renunciation of food and drink, seemed to medieval people the most basic asceticism, requiring the kind of courage and holy foolishness that marked the saints.[4]

Food was not only a fundamental material concern to medieval people; food practices—fasting and feasting—were at the very heart of the Christian tradition. A Christian in the thirteenth and fourteenth centuries was required by church law to fast on certain days and to receive communion at least once a year.[5] Thus, the behavior that defined a Christian was food-related behavior. This point is clearly illustrated in a twelfth-century story of a young man (of the house of Ardres) who returned from the crusades claiming that he had become a Saracen in the East; he was, however, accepted back by his family, and no one paid much attention to his claim until he insisted on eating meat on Friday. The full impact of his apostasy was then brought home, and his family kicked him out.[6]

Food was, moreover, a central metaphor and symbol in Christian poetry, devotional literature, and theology because a meal (the eucharist) was the central Christian ritual, the most direct way of encountering God. And we should note that this meal was a frugal repast, not a banquet but simply the two basic foodstuffs of the Mediterranean world: bread and wine. Although older Mediterranean traditions of religious feasting did come, in a peripheral way, into Christianity, indeed lasting right through the Middle Ages in various kinds of carnival, the central religious meal was reception of the two basic supports of human life. Indeed Christians believed it was human life. Already hundreds of years before transubstantiation was defined as doctrine, most Christians thought that they quite literally ate Christ's body and blood in the sacrament.[7] Medieval people themselves knew how strange this all might sound. A fourteenth-century preacher, Johann Tauler, wrote:

> St. Bernard compared this sacrament [the eucharist] with the human processes of eating when he used the similes of chewing, swallowing, assimilation and digestion. To some people this will seem crude, but let such refined persons beware of pride, which come from the devil: a humble spirit will not take offense at simple things.[8]

Thus food, as practice and as symbol, was crucial in medieval spirituality. But in the period from 1200 to 1500 it was more prominent in the piety of women than in that of men. Although it is difficult and risky to make any quantitative arguments about the Middle Ages, so much work has been done on saints' lives, miracle stories, and vision literature that certain conclusions are possible about the relative popularity of various practices and symbols. Recent work by André Vauchez, Richard Kieckhefer, Donald Weinstein, and Rudolph M. Bell demonstrates that, although women were only about 18 percent of those canonized or revered as saints between 1000 and 1700, they were 30 percent of those in whose lives extreme austerities were a central aspect of holiness and over 50 percent of those in whose lives illness (often brought on by fasting and other penitential practices) was the major factor in reputation for sanctity.[9] In addition, Vauchez has shown that most males who were revered for fasting fit into one model of sanctity—the hermit saint (usually a layman)—and this was hardly the most popular male model, whereas fasting characterized female saints generally. Between late antiquity and the fifteenth century there are at least thirty cases of women who were reputed to eat nothing at all except the eucharist,[10] but I have been able to find only one or possibly two male examples of such behavior before the well-publicized fifteenth-century case of the hermit Nicholas of Flüe.[11] Moreover, miracles in which food is miraculously multiplied are told at least as frequently of women as of men, and giving away food is so common a theme in the lives of holy women that it is very difficult to find a story in which this particular charitable activity does not occur.[12] The story of a woman's basket of bread for the poor turning into roses when her husband (or father) protests her almsgiving was attached by hagiographers to at least five different women saints.[13]

If we look specifically at practices connected with Christianity's holy meal, we find that eucharistic visions and miracles occurred far more frequently to women, particularly certain types of miracles in which the quality of the eucharist as food is underlined. It is far more common, for example, for the wafer to turn into honey or meat in the mouth of a woman. Miracles in which an unconsecrated host is vomited out or in which the recipient can tell by tasting the wafer that the Priest who consecrated it is immoral happen almost exclusively to women. Of fifty-five people from the later Middle Ages who supposedly received the holy food directly from Christ's hand in a vision, forty-five are women. In contrast, the only two types of eucharistic miracle that occur primarily to men are miracles that underline not the fact that the wafer is food but the power of the priest.[14] Moreover, when we study medieval miracles, we note that miraculous abstinence and extravagant eucharistic visions tend to occur together and are frequently accompanied by miraculous bodily changes. Such changes are found almost exclusively in women. Miraculous elongation of parts of the body, the appearance on the body of marks imitating the various wounds of Christ (called stigmata), and the exuding of wondrous fluids (which smell sweet and heal and sometimes are food—for example, manna or milk) are usually female miracles.[15]

If we consider a different kind of evidence—the *exempla* or moral tales that preachers used to educate their audiences, both monastic and lay—we find that, according to Frederic Tubach's index, only about 10 percent of such stories are about women. But when we look at those stories that treat specifically fasting, abstinence, and reception of the eucharist, 30 to 50 percent are about women.[16] The only type of religious literature in which food is more frequently associated with men is the genre of satires on monastic life, in which there is some suggestion that monks are more prone to greed.[17] But this pattern probably reflects the fact that monasteries for men were in general wealthier than women's houses and therefore more capable of mounting elaborate banquets and tempting palates with delicacies.[18]

Taken together, this evidence demonstrates two things. First, food practices were more central in women's piety than in men's. Second, both men and women associated food—especially fasting and the eucharist—with women. There are, however, a number of problems with this sort of evidence. In addition to the obvious problems of the paucity of material and of the nature of hagiographical accounts—problems to which scholars since the seventeenth century have devoted much sophisticated discussion—there is the problem inherent in quantifying data. In order to count phenomena the historian must divide them up, put them into categories. Yet the most telling argument for the prominence of food in women's spirituality is the way in which food motifs interweave in women's lives and writings until even phenomena not normally thought of as eating, feeding, or fasting seem to become food-related. In other words, food becomes such a pervasive concern that it provides both a literary and a psychological unity to the woman's way of seeing the world. And this cannot be demonstrated by statistics. Let me therefore tell in some detail one of the many stories from the later Middle Ages in which food becomes a leitmotif of stunning complexity and power. It is the story of Lidwina of the town of Schiedam in the Netherlands, who died in 1433 at the age of 53.[19]

Several hagiographical accounts of Lidwina exist, incorporating information provided by her confessors; moreover, the town officials of Schiedam, who had her watched for three months, promulgated a testimonial that suggests that Lidwina's miraculous abstinence attracted more public attention than any other aspect of her life. The document solemnly attests to her complete lack of food and sleep and to the sweet odor given off by the bits of skin she supposedly shed.

The accounts of Lidwina's life suggest that there may have been early conflict between mother and daughter. When her terrible illness put a burden on her family's resources and patience, it took a miracle to convince her mother of her sanctity. One of the few incidents that survives from her childhood shows her mother annoyed with her childish dawdling. Lidwina was required to carry food to her brothers at school, and on the way home she slipped into church to say a prayer to the Virgin. The incident shows how girlish piety could provide a respite from household tasks—in this case, as in so many cases, the task of feeding men. We also learn that Lidwina was upset to discover that she was pretty, that she threatened to pray for a deformity when plans were broached for her marriage, and that, after an illness at age fifteen, she grew weak and did not want to get up from her sickbed. The accounts thus suggest that she may have been cultivating illness—perhaps even rejecting food—before the skating accident some weeks later that produced severe internal injuries. In any event, Lidwina never recovered from her fall on the ice. Her hagiographers report that she was paralyzed except for her left hand. She burned with fever and vomited convulsively. Her body putrefied so that great pieces fell off. From mouth, ears, and nose, she poured blood. And she stopped eating.

Lidwina's hagiographers go into considerable detail about her abstinence. At first she supposedly ate a little piece of apple each day, although bread dipped into liquid caused her much pain. Then she reduced her intake to a bit of date and watered wine flavored with spices and sugar; later she survived on watered wine alone—only half a pint a week—and she preferred it when the water came from the river and was contaminated with salt from the tides. When she ceased to take any solid food, she also ceased to sleep. And finally she ceased to swallow anything at all. Although Lidwina's biographers present her abstinence as evidence of saintliness, she was suspected by some during her lifetime of being possessed by a devil instead; she herself appears to have claimed that her fasting was natural. When people accused her of hypocrisy, she replied that it is no sin to eat and therefore no glory to be incapable of eating.[20]

Fasting and illness were thus a single phenomenon to Lidwina. And since she perceived them as redemptive suffering, she urged both on others. We are told that a certain Gerard from Cologne, at her urging, became a hermit and lived in a tree, fed only on manna sent from God. We are also told that Lidwina prayed for her twelve-year-old nephew to be afflicted with an illness so that he would be reminded of God's mercy. Not surprisingly, the illness itself then came from miraculous feeding. The nephew became sick by drinking several drops from a pitcher of unnaturally sweet beer on a table by Lidwina's bedside.

Like the bodies of many other women saints, Lidwina's body was closed to ordinary intake and excreting but produced extraordinary effluvia.[21] The authenticating document from the town officials of Schiedam testifies that her body shed skin, bones, and even portions of intestines, which her parents kept in a vase; and these gave off a sweet odor until Lidwina, worried by the gossip that they excited, insisted that her mother bury them. Moreover, Lidwina's effluvia cured others. A man in England sent for her wash water to cure his ill leg. The sweet smell from her left hand led one of her confessors to confess sins. And Lidwina actually nursed others in an act that she herself explicitly saw as a parallel to the Virgin's nursing of Christ.

One Christmas season, so all her biographers tell us, a certain Catherine, who took care of her, had a vision that Lidwina's breasts would fill with milk, like Mary's, on the night of the Nativity. When she told Lidwina, Lidwina warned her to prepare herself. Then Lidwina saw a vision of Mary surrounded by a host of female virgins; and the breasts of Mary and of all the company filled with milk, which poured out from their open tunics, filling the sky. When Catherine entered Lidwina's room, Lidwina rubbed her own breast and the milk came out, and Catherine drank three times and was satisfied (nor did she want any corporeal food for many days thereafter).[22] One of Lidwina's hagiographers adds that, when the same grace was given to her again, she fed her confessor, but the other accounts say that the confessor was unworthy and did not receive the gift.

Lidwina also fed others by charity and by food multiplication miracles. Although she did not eat herself, she charged the widow Catherine to buy fine fish and make fragrant sauces and give these to the poor. The meat and fish she gave as alms sometimes, by a miracle, went much further than anyone had expected. She gave water and wine and money for beer to an epileptic burning with thirst; she sent a whole pork shoulder to a poor man's family; she regularly sent food to poor or sick children, forcing her servants to spend or use for others money or food she would not herself consume. When she shared the wine in her bedside jug with others it seemed inexhaustible. So pleased was God with her charity that he sent her a vision of a heavenly banquet, and the food she had given away was on the table.

Lidwina clearly felt that her suffering was service—that it was one with Christ's suffering and that it therefore substituted for the suffering of others, both their bodily ills and their time in purgatory. Indeed her body quite literally became Christ's macerated and saving flesh, for, like many other female saints she received stigmata (or so one—but only one—of her hagiographers claims).[23] John Brugman, in the *Vita posterior*, not only underlines the parallel between her wounds and those on a miraculous bleeding host she received; he also states explicitly that, in her stigmata, Christ "transformed his lover into his likeness."[24] Her hagiographers state that the fevers she suffered almost daily from 1421 until her death were suffered in order to release souls in purgatory.[25] And we see this notion of substitution reflected quite clearly in the story of a very evil man, in whose stead Lidwina made confession; she then took upon herself his punishment, to the increment of her own bodily anguish. We see substitution of another kind in the story of Lidwina taking over the toothache of a woman who wailed outside her door.

Thus, in Lidwina's story, fasting, illness, suffering, and feeding fuse together. Lidwina becomes the food she rejects. Her body, closed to ordinary intake and excretion but spilling over in milk and sweet putrefaction, becomes the sustenance and the cure—both earthly and heavenly—of her followers. But holy eating is a theme in her story as well. The eucharist is at the core of Lidwina's devotion. During her pathetic final years, when she had almost ceased to swallow, she received frequent communion (indeed as often as every two days). Her biographers claim that, during this period, only the holy food kept her alive.[26] But much of her life was plagued by conflict with the local clergy over her eucharistic visions and hunger. One incident in particular shows not only the centrality of Christ's body as food in Lidwina's spirituality but also the way in which a woman's craving for the host, although it kept her under the control of the clergy, could seem to that same clergy a threat, both because it criticized their behavior and because, if thwarted, it could bypass their power.[27]

Once an angel came to Lidwina and warned her that the priest would, the next day, bring her an unconsecrated host to test her. When the priest came and pretended to adore the host, Lidwina vomited it out and said that she could easily tell our Lord's body from unconsecrated bread. But the priest swore that the host was consecrated and returned, angry, to the church. Lidwina then languished for a long time, craving communion but unable to receive it. About three and a half months later, Christ appeared to her, first as a baby, then as a bleeding and suffering youth. Angels appeared, bearing the instruments of the passion, and (according to one account) rays from Christ's wounded body pierced Lidwina with stigmata. When she subsequently asked for a sign, a host hovered over Christ's head and a napkin descended onto her bed, containing a miraculous host, which remained and was seen by many people for days after. The priest returned and ordered Lidwina to keep quiet about the miracle but finally agreed, at her insistence, to feed her the miraculous host as communion. Lidwina was convinced that it was truly Christ because she, who was usually stifled by food, ate this bread without pain. The next day the priest preached in church that Lidwina was deluded and that her host was a fraud of the devil. But, he claimed, Christ was present in the bread he offered because it was consecrated with all the majesty of the priesthood. Lidwina protested his interpretation of her host, but she agreed to accept a consecrated wafer from him and to pray for his sins. Subsequently the priest claimed that he had cured Lidwina from possession by the devil, while Lidwina's supporters called her host a miracle. Although Lidwina's hagiographers do not give full details, they claim that the bishop came to investigate the matter, that he blessed the napkin for the service of the altar, and that the priest henceforth gave Lidwina the sacrament without tests or resistance.

As this story worked its way out, its theme was not subversive of clerical authority. The conflict began, after all, because Lidwina wanted a consecrated host, and it resulted in her receiving frequent communion, in humility and piety. According to one of her hagiographers, the moral of the story is that the faithful can always substitute "spiritual communion" (i.e., meditation) if the actual host is not given.[28] But the story had radical implications as well. It suggested that Jesus might come directly to the faithful if priests were negligent or skeptical, that a priest's word might not be authoritative on the difference between demonic possession and sanctity, that visionary women might test priests. Other stories in Lidwina's life had similar implications. She forbade a sinning priest to celebrate mass; she read the heart of another priest and learned of his adultery. Her visions of souls in purgatory especially concerned priests, and she substituted her sufferings for theirs. One Ash Wednesday an angel came to bring ashes for her forehead before the priest arrived. Even if Lidwina did not reject the clergy, she sometimes quietly bypassed or judged them.

Lidwina focused her love of God on the eucharist. In receiving it, in vision and in communion, she became one with the body on the cross. Eating her God, she received his wounds and offered her suffering for the salvation of the world. Denying herself ordinary food, she sent that food to others, and her body gave milk to nurse her friends. Food is the basic theme in Lidwina's story: self as food and God as food. For Lidwina, therefore, eating and not-eating were finally one theme. To fast, that is, to deny oneself earthly food, and yet to eat the broken body of Christ—both acts were to suffer. And to suffer was to save and to be saved.

Lidwina did not write herself, but some pious women did. And many of these women not only lived lives in which miraculous abstinence, charitable feeding of others, wondrous bodily changes, and eucharistic devotion were central; they also elaborated in prose and poetry a spirituality in which hungering, feeding, and eating were central metaphors for suffering, for service, and for encounter with God. For example, the great Italian theorist of purgatory, Catherine of Genoa (d. 1510)—whose extreme abstinence began in response to an unhappy marriage and who eventually persuaded her husband to join her in a life of continence and charitable feeding of the poor and sick—said that the annihilation of ordinary food by a devouring body is the best metaphor for the annihilation of the soul by God in mystical ecstasy.[29] She also wrote that, although no simile can adequately convey the joy in God that is the goal of all souls, nonetheless the image that comes most readily to mind is to describe God as the only bread available in a world of the starving.[30] Another Italian Catherine, Catherine of Siena (d. 1380), in whose saintly reputation fasting, food miracles, eucharistic devotion, and (invisible) stigmata were central,[31] regularly chose to describe Christian duty as "eating at the table of the cross the food of the honor of God and the salvation of souls."[32] To Catherine, "to eat" and "to hunger" have the same fundamental meaning, for one eats but is never full, desires but is never satiated.[33] "Eating" and "hungering" are active, not passive, images. They stress pain more than joy. They mean most basically to suffer and to serve—to suffer because in hunger one joins with Christ's suffering on the cross; to serve because to hunger is to expiate the sins of the world. Catherine wrote:

> And then the soul becomes drunk. And after it . . . has reached the place [of the teaching of the crucified Christ] and drunk to the full, it tastes the food of patience, the odor of virtue, and such a desire to bear the cross that it does not seem that it could ever be satiated. . . . And then the soul becomes like a drunken man; the more he drinks the more he

> wants to drink; the more it bears the cross the more it wants to bear it. And the pains are its refreshment and the tears which it has shed for the memory of the blood are its drink. And the sighs are its food."[34]

And again:

> Dearest mother and sisters in sweet Jesus Christ, I, Catherine, slave of the slaves of Jesus Christ, write to you in his precious blood, with the desire to see you confirmed in true and perfect charity so that you be true nurses of your souls. For we cannot nourish others if first we do not nourish our own souls with true and real virtues. . . . Do as the child does who, wanting to take milk, takes the mother's breast and places it in his mouth and draws to himself the milk by means of the flesh. So . . . we must attach ourselves to the breast of the crucified Christ, in whom we find the mother of charity, and draw from there by means of his flesh (that is, the humanity) the milk that nourishes our soul. . . . For it is Christ's humanity that suffered, not his divinity; and, without suffering, we cannot nourish ourselves with this milk which we draw from charity.[35]

To the stories and writings of Lidwina and the two Catherines—with their insistent and complex food motifs—I could add dozens of others. Among the most obvious examples would be the beguine Mary of Oignies (d. 1213) from the Low Countries, the princess Elisabeth of Hungary (d. 1231), the famous reformer of French and Flemish religious houses Colette of Corbie (d. 1447), and the thirteenth-century poets Hadewijch and Mechtild of Magdeburg. But if we look closely at the lives and writings of those men from the period whose spirituality is in general closest to women's and who were deeply influenced by women—for example, Francis of Assisi in Italy, Henry Suso and Johann Tauler in the Rhineland, Jan van Ruysbroeck of Flanders, or the English hermit Richard Rolle—we find that even to these men food asceticism is not the central ascetic practice. Nor are food metaphors central in their poetry and prose.[36] Food then is much more important to women than to men as a religious symbol. The question is why?

Modern scholars who have noticed the phenomena I have just described have sometimes suggested in an offhand way that miraculous abstinence and eucharistic frenzy are simply "eating disorders."[37] The implication of such remarks is usually that food disorders are characteristic of women rather than men, perhaps for biological reasons, and that these medieval eating disorders are different from nineteenth- and twentieth-century ones only because medieval people "theologized" what we today "medicalize."[38] While I cannot deal here with all the implications of such analysis, I want to point to two problems with it. First, the evidence we have indicates that extended abstinence was almost exclusively a male phenomenon in early Christianity and a female phenomenon in the high Middle Ages.[39] The cause of such a distribution of cases cannot be primarily biological.[40] Second, medieval people did not treat all refusal to eat as a sign of holiness. They sometimes treated it as demonic possession, but they sometimes also treated it as illness.[41] Interestingly enough, some of the holy women whose fasting was taken as miraculous (for example, Colette of Corbie) functioned as healers of ordinary individuals, both male and female, who could not eat.[42] Thus, for most of the Middle Ages, it was only in the case of some unusually devout women that not-eating was both supposedly total and religiously significant. Such behavior must have a cultural explanation.

On one level, the cultural explanation is obvious. Food was important to women religiously because it was important socially. In medieval Europe (as in many countries today) women were associated with food preparation and distribution rather than food consumption. The culture suggested that women cook and serve, men eat. Chronicle accounts of medieval banquets, for example, indicate that the sexes were often segregated and that women were sometimes relegated to watching from the balconies while gorgeous foods were rolled out to please the eyes as well as the palates of men.[43] Indeed men were rather afraid of women's control of food. Canon lawyers suggested, in the codes they drew up, that a major danger posed by women was their manipulation of male virility by charms and potions added to food.[44] Moreover, food was not merely a resource women controlled; it was *the* resource women controlled. Economic resources were controlled by husbands, fathers, uncles, or brothers. In an obvious sense, therefore, fasting and charitable food distribution (and their miraculous counterparts) were natural religious activities for women. In fasting and charity women renounced and distributed the one resource that was theirs. Several scholars have pointed out that late twelfth- and early thirteenth-century women who wished to follow the new ideal of poverty and begging (for example, Clare of Assisi and Mary of Oignies) were simply not permitted either by their families or by religious authorities to do so.[45] They substituted fasting for other ways of stripping the self of support. Indeed a thirteenth-century hagiographer commented explicitly that one holy woman gave up food because she had nothing else to give up.[46] Between the thirteenth and fifteenth centuries, many devout laywomen who resided in the homes of fathers or spouses were able to renounce the world in the midst of abundance because they did not eat or drink the food that was paid for by family wealth. Moreover, women's almsgiving and abstinence appeared culturally acceptable forms of asceticism because what women ordinarily did, as housewives, mothers, or mistresses of great castles, was to prepare and serve food rather than to eat it.

The issue of control is, however, more basic than this analysis suggests. Food-related behavior was central to women socially and religiously not only because food was a resource women controlled but also because, by means of food, women controlled themselves and their world.

First and most obviously, women controlled their bodies by fasting. Although a negative or dualist concept of body does not seem to have been the most fundamental notion of body to either women or men, some sense that body was to be disciplined, defeated, occasionally even destroyed, in order to release or protect spirit is present in women's piety. Some holy women seem to have developed an extravagant fear of any bodily contact.[47] Clare of Montefalco (d. 1308), for example, said she would rather spend days in hell than be touched by a man.[48] Lutgard of Aywières panicked at an abbot's insistence on giving her the kiss of peace, and Jesus had to interpose his hand in a vision so that she was not reached by the abbot's lips. She even asked to have her own gift of healing by touch taken away.[49] Christina of Stommeln (d. 1312), who fell into a latrine while in a trance, was furious at the laybrothers who rescued her because they touched her in order to do so.[50]

Many women were profoundly fearful of the sensations of their bodies, especially hunger and thirst. Mary of Oignies, for example, was so afraid of taking pleasure in food that Christ had to make her unable to taste.[51] From the late twelfth century comes a sad story of a dreadfully sick girl named Alpaïs who sent away the few morsels of pork given her to suck, because she feared that any enjoyment of eating might mushroom madly into gluttony or lust.[52] Women like Ida of Louvain (d. perhaps 1300), Elsbeth Achler of Reute (d. 1420), Catherine of Genoa, or Columba of Rieti (d. 1501), who sometimes snatched up food and ate without knowing what they were doing, focused their hunger on the eucharist partly because it was an acceptable object of craving and partly because it was a self-limiting food.[53] Some of women's asceticism was clearly directed toward destroying bodily needs, before which women felt vulnerable.

Some fasting may have had as a goal other sorts of bodily control. There is some suggestion in the accounts of hagiographers that fasting women were admired for suppressing excretory functions. Several biographers comment with approval that holy women who do not eat cease also to excrete, and several point out, explicitly that the menstruation of saintly women ceases.[54] Medieval theology—profoundly ambivalent about body as physicality—was ambivalent about menstruation also, seeing it both as the polluting "curse of Eve" and as a natural function that, like all natural functions, was redeemed in the humanity of Christ. Theologians even debated whether or not the Virgin Mary menstruated.[55] But natural philosophers and theologians were aware that, in fact, fasting suppresses menstruation. Albert the Great noted that some holy women ceased to menstruate because of their fasts and austerities and commented that their health did not appear to suffer as a consequence.[56]

Moreover, in controlling eating and hunger, medieval women were also explicitly controlling sexuality. Ever since Tertullian and Jerome, male writers had warned religious women that food was dangerous because it excited lust.[57] Although there is reason to suspect that male biographers exaggerated women's sexual temptations, some women themselves connected food abstinence with chastity and greed with sexual desire.[58]

Women's heightened reaction to food, however, controlled far more than their physicality. It also controlled their social environment. As the story of Lidwina of Schiedam makes clear, women often coerced both families and religious authorities through fasting and through feeding. To an aristocratic or rising merchant family of late medieval Europe, the self-starvation of a daughter or spouse could be deeply perplexing and humiliating. It could therefore be an effective means of manipulating, educating, or converting family members. In one of the most charming passages of Margery Kempe's autobiography, for example, Christ and Margery consult together about her asceticism and decide that, although she wishes to practice both food abstention and sexual continence, she should perhaps offer to trade one behavior for the other. Her husband, who had married Margery in an effort to rise socially in the town of Lynn and who was obviously ashamed of her queer penitential clothes and food practices, finally agreed to grant her sexual abstinence in private if she would return to normal cooking and eating in front of the neighbors.[59] Catherine of Siena's sister, Bonaventura, and the Italian saint Rita of Cascia (d. 1456) both reacted to profligate young husbands by wasting away and managed thereby to tame disorderly male behavior. [60] Columba of Rieti and Catherine of Siena expressed what was clearly adolescent conflict with their mothers and riveted family attention on their every move by their refusal to eat. Since fasting so successfully manipulated and embarrassed families, it is not surprising that self-starvation often originated or escalated at puberty, the moment at which families usually began negotiations for husbands for their daughters. Both Catherine and Columba, for example, established themselves as unpromising marital material by their extreme food and sleep deprivation, their frenetic giving away of paternal resources, and their compulsive service of family members in what were not necessarily welcome ways. (Catherine insisted on doing the family laundry in the middle of the night.)[61]

Fasting was not only a useful weapon in the battle of adolescent girls to change their families' plans for them. It also provided for both wives and daughters an excuse for neglecting food preparation and family responsibilities. Dorothy of Montau, for example, made elementary mistakes of cookery (like forgetting to scale the fish before frying them) or forgot entirely to cook and shop while she was in ecstasy. Margaret of Cortona refused to cook for her illegitimate son (about whom she felt agonizing ambivalence) because, she said, it would distract her from prayer.[62]

Moreover, women clearly both influenced and rejected their families' values by food distribution. Ida of Louvain, Catherine of Siena, and Elisabeth of Hungary, each in her own way, expressed distaste for family wealth and coopted the entire household into Christian charity by giving away family resources, sometimes surreptitiously or even at night. Elisabeth, who gave away her husband's property, refused to eat any food except that paid for by her own dowry because the wealth of her husband's family came, she said, from exploiting the poor.[63]

Food-related behavior—charity, fasting, eucharistic devotion, and miracles—manipulated religious authorities as well.[64] Women's eucharistic miracles—especially the ability to identify unconsecrated hosts or unchaste priests—functioned to expose and castigate clerical corruption. The Viennese woman Agnes Blannbekin, knowing that her priest was not chaste, prayed that he be deprived of the host, which then flew away from him and into her own mouth.[65] Margaret of Cortona saw the hands of an unchaste priest turn black when he held the host.[66] Saints' lives and chronicles contain many stories, like that told of Lidwina of Schiedam, of women who vomited out unconsecrated wafers, sometimes to the considerable discomfiture of local authorities.[67]

The intimate and direct relationship that holy women claimed to the eucharist was often a way of bypassing ecclesiastical control. Late medieval confessors and theologians attempted to inculcate awe as well as craving for the eucharist; and women not only received ambiguous advice about frequent communion, they were also sometimes barred from receiving it at exactly the point at which their fasting and hunger reached fever pitch.[68] In such circumstances many women simply received in vision what the celebrant or confessor withheld. Imelda Lambertini, denied communion because she was too young, and Ida of Léau, denied because she was subject to "fits," were given the host by Christ.[69] And some women received, again in visions, either Christ's blood, which they were regularly denied because of their lay status, or the power to consecrate and distribute, which they were denied because of their gender. Angela of Foligno and Mechtild of Hackeborn were each, in a vision, given the chalice to distribute.[70] Catherine of Siena received blood in her mouth when she ate the wafer.[71]

It is thus apparent that women's concentration on food enabled them to control and manipulate both their bodies and their environment. We must not underestimate the effectiveness of such manipulation in a world where it was often extraordinarily difficult for women to avoid marriage or to choose a religious vocation.[72] But such a conclusion concentrates on the function of fasting and feasting, and function is not meaning. Food did not "mean" to medieval women the control it provided. It is time, finally, to consider explicitly what it meant.

As the behavior of Lidwina of Schiedam or the theological insights of Catherine of Siena suggest, fasting, eating, and feeding all meant suffering, and suffering meant redemption. These complex meanings were embedded in and engendered by the theological doctrine of the Incarnation. Late medieval theology, as is well known, located the saving moment of Christian history less in Christ's resurrection than in his crucifixion. Although some ambivalence about physicality, some sharp and agonized dualism, was present, no other period in the history of Christian spirituality has placed so positive a value on Christ's humanity as physicality. Fasting was thus flight not so much *from* as *into* physicality. Communion was consuming—i.e., becoming—a God who saved the world through physical, human agony. Food to medieval women meant flesh and suffering and, through suffering, salvation: salvation of self and salvation of neighbor. Although all thirteenth- and fourteenth-century Christians emphasized Christ as suffering and Christ's suffering body as food, women were especially drawn to such a devotional emphasis. The reason seems to lie in the way in which late medieval culture understood "the female."

Drawing on traditions that went back even before the origins of Christianity, both men and women in the later Middle Ages argued that "woman is to man as matter is to spirit." Thus "woman" or "the feminine" was seen as symbolizing the physical part of human nature, whereas man symbolized the spiritual or rational.[78] Male theologians and biographers of women frequently used this idea to comment on female weakness. They also inverted the image and saw "woman" as not merely below but also above reason. Thus they somewhat sentimentally saw Mary's love for souls and her mercy toward even the wicked as an apotheosis of female unreason and weakness, and they frequently used female images to describe themselves in their dependence on God.[74] Women writers, equally aware of the male/female dichotomy, saw it somewhat differently. They tended to use the notion of "the female" as "flesh" to associate Christ's humanity with "the female" and therefore to suggest that women imitate Christ through physicality.

Women theologians saw "woman" as the symbol of humanity, where humanity was understood as including bodiliness. To the twelfth-century prophet, Elisabeth of Schönau, the humanity of Christ appeared in a vision as a female virgin. To Hildegard of Bingen (d. 1179), "woman" was the symbol of humankind, fallen in Eve, restored in Mary and church. She stated explicitly: "Man signifies the divinity of the Son of Cod and woman his humanity."[75] Moreover, to a number of women writers, Mary was the source and container of Christ's physicality; the flesh Christ put on was in some sense female, because it was his mother's. Indeed whatever physiological theory of reproduction a medieval theologian held, Christ (who had no human father) had to be seen as taking his physicality from his mother. Mechtild of Magdeburg went further and implied that Mary was a kind of preexistent humanity of Christ as the Logos was his preexistent divinity. Marguerite of Oingt, like Hildegard of Bingen, wrote that Mary was the *tunica humanitatis*, the clothing of humanity, that Christ puts on.[76] And to Julian of Norwich, God himself was a mother exactly in that our humanity in its full physicality was not merely loved and saved but even given being by and from him. Julian wrote:

> For in the same time that God joined himself to our body in the maiden's womb, he took our soul, which is sensual, and in taking it, having enclosed us all in himself, he united it to our substance. . . . So our Lady is our mother, in whom we are all enclosed and born of her in Christ, for she who is mother of our saviour is mother of all who are saved in our

saviour; and our saviour is our true mother, in whom we are endlessly born and out of whom we shall never come.[77]

Although male writers were apt to see God's motherhood in his nursing and loving rather than in the fact of creation, they too associated the flesh of Christ with Mary and therefore with woman.[78]

Not only did medieval people associate humanity as body with woman; they also associated woman's body with food. Woman was food because breast milk was the human being's first nourishment—the one food essential for survival. Late medieval culture was extraordinarily concerned with milk as symbol. Writers and artists were fond of the theme, borrowed from antiquity, of lactation offered to a father or other adult male as an act of filial piety. The cult of the Virgin's milk was one of the most extensive cults in late medieval Europe. A favorite motif in art was the lactating Virgin.[79] Even the bodies of evil women were seen as food. Witches were supposed to have queer marks on their bodies (sort of super-numerary breasts) from which they nursed incubi.

Quite naturally, male and female writers used nursing imagery in somewhat different ways. Men were more likely to use images of being nursed, women metaphors of nursing. Thus when male writers spoke of God's motherhood, they focused more narrowly on the soul being nursed at Christ's breast, whereas women were apt to associate mothering with punishing, educating, or giving birth as well.[80] Most visions of drinking from the breast of Mary were received by men.[81] In contrast, women (like Lidwina) often identified with Mary as she nursed Jesus or received visions of taking the Christchild to their own breasts.[82] Both men and women, however, drank from the breast of Christ, in vision and image.[83] Both men and women wove together—from Pauline references to milk and meat and from the rich breast and food images of the Song of Songs—a complex sense of Christ's blood as the nourishment and intoxication of the soul. Both men and women therefore saw the body on the cross, which in dying fed the world, as in some sense female. Again, physiological theory reinforced image. For, to medieval natural philosophers, breast milk was transmuted blood, and a human mother (like the pelican that also symbolized Christ) fed her children from the fluid of life that coursed through her veins.[84]

Since Christ's body itself was a body that nursed the hungry, both men and women naturally assimilated the ordinary female body to it. A number of stories are told of female saints who exuded holy fluid from breasts or fingertips, either during life or after death. These fluids often cured the sick.[85] The union of mouth to mouth, which many women gained with Christ, became also a way of feeding. Lutgard's saliva cured the ill; Lukardis of Oberweimar (d. 1309) blew the eucharist into another nun's mouth; Colette of Corbie regularly cured others with crumbs she chewed.[86] Indeed one suspects that stigmata—so overwhelmingly a female phenomenon—appeared on women's bodies because they (like the marks on the bodies of witches and the wounds in the body of Christ) were not merely wounds but also breasts.

Thus many assumptions in the theology and the culture of late medieval Europe associated woman with flesh and with food. But the same theology also taught that the redemption of all humanity lay in the fact that Christ was flesh and food. A God who fed his children from his own body, a God whose humanity *was* his children's humanity, was a God with whom women found it easy to identify. In mystical ecstasy as in communion, women ate and became a God who was food and flesh. And in eating a God whose flesh was holy food, women both transcended and became more fully the flesh and the food their own bodies were.

Eucharist and mystical union were, for women, both reversals and continuations of all the culture saw them to be.[87] In one sense, the roles of priest and lay recipient reversed normal social roles. The priest became the food preparer, the generator and server of food. The woman recipient ate a holy food she did not exude or prepare. Woman's jubilant, vision-inducing, inebriated eating of God was the opposite of the ordinary female acts of food preparation or of bearing and nursing children. But in another and, I think, deeper sense, the eating was not a reversal at all. Women became, in mystical eating, a fuller version of the food and the flesh they were assumed by their culture to be. In union with Christ, woman became a fully fleshly and feeding self—at one with the generative suffering of God.

Symbol does not determine behavior. Women's imitation of Christ, their assimilation to the suffering and feeding body on the cross, was not uniform. Although most religious women seem to have understood their devotional practice as in some sense serving as well as suffering, they acted in very different ways. Some, like Catherine of Genoa and Elisabeth of Hungary, expressed their piety in feeding and caring for the poor. Some, like Alpaïs, lay rapt in mystical contemplation as their own bodies decayed in disease or in self-induced starvation that was offered for the salvation of others. Many, like Lidwina of Schiedam and Catherine of Siena, did both. Some of these women are, to our modern eyes, pathological and pathetic. Others seem to us, as they did to their contemporaries, magnificent. But they all dealt, in feast and fast, with certain fundamental realities for which all cultures must find symbols—the realities of suffering and the realities of service and generativity.

NOTES

This paper was originally given as the Solomon Katz Lecture in the Humanities at the University of Washington, March 1984. It summarizes several themes that will be elaborated in detail in my forthcoming book, *Holy Feast and Holy Fast: Food Motifs in the Piety of Late Medieval Women*. I am grateful to Rudolph M. Bell, Peter Brown, Joan Jacobs Brumberg, Rachel Jacoff, Richard Kieckhefer, Paul Meyvaert, Guenther Roth, and Judith Van Herik for their suggestions and for sharing with me their unpublished work.

1. Quoted from Suso's letter to Elsbet Stagel in Henry Suso, *Deutsche Schriften im Auftrag der Württembergischen Kommission für Landesgeschichte*, ed. Karl Bihlmeyer (Stuttgart, 1907), 107; trans. (with minor changes) M. Ann Edward in *The Exemplar: Life and Writings of Blessed Henry Suso*, O.P, ed. Nicholas Heller, 2 vols. (Dubuque, Ia., 1962), 1:103.
2. William A. Hammond, *Fasting Girls: Their Physiology and Pathology* (New York, 1879), 6. Quoted in Joan J. Brumberg, "'Fasting Girls': Nineteenth-Century Medicine and the Public Debate over 'Anorexia,'" paper delivered at the Sixth Berkshire Conference on the History of Women, 1–3 June 1984.
3. On the patristic notion that the sin of our first parents was gluttony, see Herbert Musurillo, "The Problem of Ascetical Fasting in the Greek Patristic Writers," *Traditio* 12 (1956): 17, n. 43. For a medieval discussion, see Thomas Aquinas *Summa theologiae* 2–2.148.3. For several examples of very explicit discussion of "eating God" in communion or of "being eaten" by him, see Augustine of Hippo (d. 430). *De civitate Dei*, in *Patrologiae cursus completus: Series latina*, ed. J.-P. Migne [hereafter *PL*], 41, col. 284c; Hilary of Poitiers (d. 367), *Tractatus in CXXV psalmum*, in PL 9, col. 688b-c; idem, *De Trinitate, PL* 10, cols. 246–47; Mechtild of Magdeburg (d. about 1282 or 1297), *Offenbarungen der Schwester Mechtild von Magdeburg oder Das Fliessende Licht der Gottheit*, ed. Gall Morel (Regensburg, 1869; reprint ed., Darmstadt, 1963), 43; Hadewijch (thirteenth century), *Mengeldichten*, ed. J.

Van Mierlo (Antwerp, 1954), poem 16, p. 79; Johann Tauler, sermon 30, *Die Predigten Taulers*, ed. Ferdinand Vetter (Berlin, 1910), 293.

4. See Fritz Curschmann, *Hungersnöte im Mittelalter: Ein Beitrag zur Deutschen Wirtschaftsgeschichte des 8. bis 13. Jahrhunderts* (Leipzig, 1900); Mikhail M. Bakhtin, *Rabelais and His World*, trans. Hélène Iswolsky (Cambridge, Mass., 1968); and Piero Camporesi, *Il pane selvaggio* (Bologna, 1980).
5. For a brief discussion, see P M. J. Clancy, "Fast and Abstinence," *New Catholic Encyclopedia* (New York, 1967), 5:846–50.
6. Lambert, "History of the Counts of Guines," in *Monumenta Germaniae historica: Scriptorum* (hereafter *MGH.SS*), vol. 24 (Hanover, I879), 615.
7. See Peter Brown, "A Response to Robert M. Grant: The Problem of Miraculous Feedings in the Graeco-Roman World," in *The Center for Hermeneutical Studies* 42 (1982): 16–24; Édouard Dumoutet, *Corpus Domini: Aux sources de la piété eucharistique médiévale* (Paris, 1942).
8. Tauler, sermon 31, in *Die Predigten*, 310; trans. E. Colledge and Sister M. Jane, *Spiritual Conferences* (St. Louis, 1961), 258.
9. André Vauchez, *La Sainteté en Occident aux derniers siècles du moyen âge d'après les procès de canonisation et les documents hagiographiques* (Rome, 1981); Donald Weinstein and Rudolph M. Bell, *Saints and Society: The Two Worlds of Western Christendom, 1000–1700* (Chicago, 1982); Richard Kieckhefer, *Unquiet Souls: Fourteenth-Century Saints and Their Religious Milieu* (Chicago, 1984). See also Ernst Benz, *Die Vision: Erfahrungsformen und Bilderwelt* (Stuttgart, 1969), 17–34.
10. For partial (and not always very accurate) lists of miraculous abstainers, see T E. Bridgett, *History of the Holy Eucharist in Great Britain*, 2 vols. (London, 1881), 2:195ff.; Ebenezer Cobham Brewer, *A Dictionary of Miracles: Imitative, Realistic and Dogmatic* (Philadelphia, 1884), 508–10; Peter Browe, *Die Eucharistischen Wunder des Mittelalters* (Breslau, 1938), 49–54; Thomas Pater, *Miraculous Abstinence: A Study of One of the Extraordinary Mystical Phenomena* (Washington, D.C., 1964); and Herbert Thurston, *The Physical Phenomena of Mysticism* (Chicago, 1952), 341–83 and passim. My own list includes ten cases that are merely mentioned in passing in saints' lives or chronicles. In addition, the following women are described in the sources as eating "nothing" or "almost nothing" for various periods and as focusing their sense of hunger on the eucharist: Mary of Oignies, Juliana of Cornillon, Ida of Louvain, Elisabeth of Spalbeek, Margaret of Ypres, Lidwina of Schiedam, Lukardis of Oberweimar, Jane Mary of Maillé, Alpaïs of Cudot, Elisabeth of Hungary, Margaret of Hungary, Dorothy of Montau (or Prussia), Elsbeth Achler of Reute, Colette of Corbie, Catherine of Siena, Columba of Rieti, and Catherine of Genoa. Several others—Angela of Foligno, Margaret of Cortona, Beatrice of Nazareth, Beatrice of Ornacieux, Lutgard of Aywières, and Flora of Beaulieu—experienced at times what the nineteenth century would have called a "hysterical condition" that left them unable to swallow.
11. Possible male exceptions include the visionary monk of Evesham and Aelred of Rievaulx in the twelfth century and Facio of Cremona in the thirteenth; see *The Revelation to the Monk of Evesham Abbey*, trans. Valerian Paget (New York, 1909), 35, 61; *The Life of Aelred of Rievaulx by Walter Daniel*, ed. and trans. F. M. Powicke (New York, 1951), 48ff.; *Chronica pontificum et imperatorum Mantuana* for the year 1256, *MGH.SS* 24: 216. In general those males whose fasting was most extreme—for example, Henry Suso, Peter of Luxembourg (d. 1386), and John the Good of Mantua (d. 1249)—show quite clearly in their *vitae* that, although they starved themselves and wrecked their digestions, they did *not* claim to cease eating entirely nor did they lose their *desire* for food. See Life of Suso in *Deutsche Schriften*, 7–195; John of Mantua, Process of Canonization, in *Acta sanctorum*, ed. the Bollandists [hereafter *AASS*], October: 9 (Paris and Rome, 1868), 816, 840; Life of Peter of Luxemburg, in *AASS*, July: 1 (Venice, 1746), 513, and Process of Canonization, in ibid., 534–39. James Oldo (d. 1404), who began an extreme fast, returned to eating when commanded by his superiors; life of James Oldo, in *AASS*, April: 2 (Paris and Rome, 1865), 603–4.
12. On food multiplication miracles, see Thurston, *Physical Phenomena*, 385–91. According to the tables in Weinstein and Bell, *Saints and Society*, women are 22 percent of the saints important for aiding the poor and 25 percent of those important for curing the sick, although a little less than 18 percent of the total.

13. Elisabeth of Hungary, Rose of Viterbo, Elisabeth of Portugal, Margaret of Fontana, and Flora of Beaulieu; see Jeanne Ancelet-Hustache, *Sainte Elisabeth de Hongrie* (Paris, 1946), 39–42, n. 1; life of Margaret of Fontana, in *AASS*, September: 4 (Antwerp, 1753), 137; and Clovis Brunel, "Vida e Miracles de Sancta Flor," *Analecta Bollandiana* 64 (1946): 8, n. 4. We owe the story, first attached to Elisabeth of Hungary but apocryphal, to an anonymous Tuscan Franciscan of the thirteenth century; see Paul G. Schmidt, "Die zeitgenössische Überlieferung zum Leben und zur Heiligsprechung der heiligen Elisabeth," in *Sankt Elisabeth: Fürstin, Dienerin, Heilige: Aufsätze, Dokumentation, Katalog*, ed. the University of Marburg (Sigmaringen, 1981), 1, 5.
14. Browe, *Die Wunder*, and Antoine Imbert-Gourbeyre, *La Stigmatisation: L'Extase divine et les miracles de Lourdes: Réponse aux libres-penseurs*, 2 vols. (Clermont-Ferrand, 1894), 2:183, 408–9. The work of Imbert-Gourbeyre is on one level inaccurate and credulous, but for my purposes it provides a good index of stories medieval people were willing to circulate, if not of events that in fact happened. See also Caroline W. Bynum, "Women Mystics and Eucharistic Devotion in the Thirteenth Century," *Women's Studies* 11 (1984): 179–214.
15. See Thurston, *Physical Phenomena*, passim; E. Amann, "Stigmatisation," in *Dictionnaire de théologie catholique*, vol. 14, part I (Paris, 1939), col. 2617–19; Imbert-Gourbeyre, "La Stigmatisation"; and Pierre Debongnie, "Essai critique sur l'histoire des stigmatisations au moyen âge," *Études carmélitaines* 21.2 (1936): 22–59. According to the tables in Weinstein and Bell, women provide 27 percent of the wonder-working relics although only 18 percent of the saints. Women also seem to provide the largest number of myroblytes (oil-exuding saints), although more work needs to be done on this topic. Of the most famous medieval myroblytes—Nicolas of Myra, Catherine of Alexandria, and Elisabeth of Hungary—two were women. On myroblytes, see Charles W. Jones, *Saint Nicolas of Myra, Bari and Manhattan: Biography of a Legend* (Chicago, 1978),144–53; J.-K. Huysmans, *Sainte Lydwine de Schiedam* (Paris, 1901), 288–91 (which must, however, be used with caution); and n. 85 below.
16. Frederic C. Tubach, *Index Exemplorum: A Handbook of Medieval Religious Tales* (Helsinki, 1969); see entries for "abstinence," "fasting," "bread," "loaves and fishes," "meat," "host," and "chalice."
17. See, for example, *Tractatus beati Gregorii pape contra religionis simulatores*, ed. Marvin Colker, in *Analecta Dublinensia: Three Medieval Latin Texts in the Library of Trinity College, Dublin* (Cambridge, Mass., 1975), 47, 57; and A. George Rigg, "'Metra de monachis carnalibus': The Three Versions," *Mittellateinisches Jahrbuch* 15 (1980): 134–42, which gives three versions of an antimonastic parody and shows that the adaptation for nuns eliminates most of the discussion of food as temptation.
18. See David Knowles, "The Diet of Black Monks," *Downside Review* 52 [n.s. 33] (1934), 273–90; and Eileen Power, *Medieval English Nunneries, c. 1275 to 1535* (Cambridge, 1922), 161–236.
19. There are four near-contemporary *vitae* of Lidwina, one in Dutch by John Gerlac, two in Latin by John Brugman, and one by Thomas à Kempis. (Both Gerlac and Brugman knew her well: Gerlac was her relative and Brugman her confessor.) See *AASS*, April: 2, 267–360, which gives Brugman's Latin translation of Gerlac's Life with Gerlac's additions indicated in brackets (the *Vita prior*) and Brugman's longer Life (the *Vita posterior*); and Thomas à Kempis, *Opera omnia*, ed. H. Sommalius, vol. 3 (Cologne, 1759), 114–64. See also Huysmans, *Lydwine*.
20. Brugman, *Vita posterior*, 320; see also ibid., 313. It is also important to note that Lidwina at first responded to her terrible illness with anger and despair and had to be convinced that it was a saving imitation of Christ's passion; see *Vita prior*, 280–82, and Thomas à Kempis, Life of Lidwina, in *Opera omnia*, 132–33.
21. For parallel lives from the Low Countries, see Thomas of Cantimpré, Life of Lutgard of Aywières (d. 1246), in *AASS*, June: 4 (Paris and Rome, 1867), 189–210; and G. Hendrix, "Primitive Versions of Thomas of Cantimpré's *Vita Lutgardis,*" *Cîteaux: Commentarii cistercienses* 29 (1978): 153–206; Thomas of Cantimpré, Life of Christina Mirabilis (d. 1224), in *AASS*, July: 5 (Paris, 1868),637–60; Life of Gertrude van Oosten (d. 1358), in *AASS*, January: l (Antwerp, 1643),348–53; and Philip of Clairvaux, Life of Elisabeth of Spalbeek (or Herkenrode) (d. after 1274), *Catalogus codicum hagiographicorum Bibliothecae regiae Bruxellensis*, vol. 1, part 1, ed. the Bollandists, in *Subsidia hagiographica*, vol. 1 (Brussels, 1886), 362–78.

22. *Vita* prior, 283; *Vita posterior*, 344; Thomas à Kempis, Life of Lidwina, 135–36.
23. *Vita posterior*, 331–32, 334–35. See Debongnie, "Stigmatisations," 55–56.
24. *Vita posterior*, 335.
25. See, for example, *Vita prior*, 277, 297. We are told that her relatives and friends benefited especially.
26. See *Vita prior*, 297; see also ibid., 280. And see Thomas à Kempis, Life of Lidwina, 155–56.
27. *Vita prior*, 295–97; *Vita posterior*, 329–35; Thomas à Kempis, Life of Lidwina, 155–56.
28. *Vita posterior*, 330.
29. *Vita mirabile et doctrina santa della beata Caterina da Genova* . . . (Florence, 1568; reprint ed., 1580), 106–7.
30. *Trattato del Purgatorio*, in Umile Bonzi da Genova, *S. Caterina Fieschi Adorno*, 2 vols. (Marietta, 1960), vol. 2, *Edizione critica dei manoscritti cateriniani*, 332–33. Catherine's "works" were compiled after her death, and there is some controversy about their "authorship" but all agree that the treatise on purgatory represents her own teaching.
31. It is quite easy to establish striking parallels between Catherine's behavior and modern descriptions of anorexia/bulimia. For accounts of Catherine's extended inedia, bingeing, and vomiting, see Raymond of Capua, Life of Catherine, in *AASS*, April: 3 (Paris and Rome, 1866), 872, 876–77, 903–7, 960; and the anonymous *I miracoli di Caterina di Jacopo da Siena di Anonimo Fiorentino a cura di Francesco Valli* (Siena, 1936), 5–9, 23–35. On Catherine's eucharistic devotion, see Raymond, Life of Catherine, 904–5, 907, and 909.
32. See, for example, letter 208, *Le lettere de S. Caterina da Siena, ridotte a miglior lezione, e in ordine nuovo disposte con note di Niccolò Tommaseo*, ed. Piero Misciattelli, 6 vols. (Siena, 1913–22), 3: 255–58; letter 11, ibid., 1:44; letter 340, ibid., 5:158–66; and *Catherine of Siena: The Dialogue*, trans. Suzanne Noffke (New York, 1980), 140.
33. See, for example, *Dialogue*, 170; letter 34, *Le lettere*, 1:157; letter 8, ibid., 1:34–38; letter 75, ibid., 2:21–24.
34. Letter 87, ibid., 2:92.
35. Letter 2* (a separately numbered series), in ibid., 6:5–6. Catherine is fond of nursing images to describe God: see letter 86, ibid., 2:81–88; letter 260, ibid., 4:139–40; letter 1*, ibid., 6:1–4; letter 81 (Tommaseo-Misciattelli no. 239) in *Epistolario di Santa Caterina da Siena*, ed. Eugenio Dupré Theseider, vol. 1 (Rome, 1940), 332–33; *Dialogue*, 52, 179–80, 292, and 323–24; and Raymond of Capua, Life of Catherine, 909.
36. On Hadewijch and Mechtild, see n. 3 above. On Mary of Oignies, see Bynum, "Women Mystics and Eucharistic Devotion." For Elisabeth, see Albert Huyskens, *Quellenstudien zur Geschichte der hl. Elisabeth Landgräfin von Thüringen* (Marburg, 1908); and Ancelet-Hustache, *Elisabeth*. For Colette, see the Lives in *AASS*, March: 1 (Antwerp, 1668), 539–619. For an analysis of the extent of food asceticism and food metaphors in the lives and writings of Francis, Suso, Tauler, Ruysbroeck, and Rolle, see my forthcoming book, *Holy Feast and Holy Fast*. Tauler and Rolle (like Catherine of Siena's confessor, Raymond of Capua) were actually apologetic about their inability to fast. And Tauler and Ruysbroeck, despite intense eucharistic piety, use little food language outside a eucharistic context. See also n. 11 above.
37. See, for example, Thurston, *Physical Phenomena*, passim; Benedict J. Groeschel, introduction to *Catherine of Genoa: Purgation and Purgatory* . . . , trans. Serge Hughes (New York, 1979), 11; J. Hubert Lacey, "Anorexia Nervosa and a Bearded Female Saint," *British Medical Journal* 285 (18–25 December 1982): 1816–17. Rudolph M. Bell is at work on a sophisticated study, the thesis of which is that a number of late medieval Italian religious women suffered from anorexia nervosa.
38. There is good evidence for biological factors in women's greater propensity for fasting and "eating disorders." See Harrison G. Pope and James Hudson, *New Hope for Binge Eaters: Advances in the Understanding and Treatment of Bulimia* (New York, 1984), which argues that anorexia and bulimia are types of biologically caused depression, which have a pharmacological cure. See also "Appendix B: Sex Differences in Death, Disease and Diet" in Katharine B. Hoyenga and K. T. Hoyenga, *The Question of Sex Differences: Psychological, Cultural and Biological Issues* (Boston, 1979), 372–90, which demonstrates that, for reasons of differences in metabolism between men and women, women's bodies tolerate fasting better than men's. For a sophisticated discussion of continuities and discontinuities in women's fasting practices, see Joan Jacobs Brumberg, "'Fasting Girls': Reflections on Writ-

ing the History of Anorexia Nervosa," *Proceedings of the Society for Research on Child Development* (forthcoming).

39. See Browe, *Die Wunder*, 49–50; Jules Corblet, *Histoire dogmatique, liturgique et archéologique du sacrement de l'eucharistie*, 2 vols. (Paris, 1885–86), 1:188–91; and n. 10 above.
40. Even for the modern period there is much evidence that confutes a rigidly biochemical explanation of women's inedia. Most researchers agree that incidents of anorexia and bulimia are in fact increasing rapidly, although recent talk of an "epidemic" may be journalistic overreaction. See Hilde Bruch, "Anorexia Nervosa: Therapy and Theory," *American Journal of Psychiatry* 139, no. 12 (December 1982): 1531–38; and A. H. Crisp, et al., "How Common Is Anorexia Nervosa? A Prevalence Study," British *Journal of Psychiatry* 128 (1976): 549–54.
41. Lidwina, Catherine of Siena, and Alpaïs of Cudot (d. 1211) apparently saw their inedia as illness, although all three were accused of demonic possession. See n. 20 above; Catherine of Siena, letter 19 (Tommaseo-Misciattelli no. 92), Epistolario, 1:80–82, where she calls her inability to eat an infermità; Raymond of Capua, Life of Catherine, 906; and Life of Alpaïs, in *AASS*, November: 2.1 (Brussels, 1894), 178, 180, 182–83, and 200.
42. Peter of Vaux, Life of Colette, 576, and account of miracles performed at Ghent after her death, ibid., 594–95. See also the two healing miracles recounted in the ninth-century life of Walburga (d. 779) by Wolfhard of Eichstadt, in *AASS*, February: 3 (Antwerp, 1658), 528 and 540–42. Holy women also sometimes cured people who could not fast; see the Life of Juliana of Cornillon, in *AASS*, April: 1 (Paris and Rome, 1866), 475, and Thomas of Cantimpré, Life of Lutgard, 200–201.
43. Barbara K. Wheaton, *Savouring the Past: The French Kitchen and Table from 1300 to 1789* (Philadelphia, 1983), 1–26.
44. Georges Duby, *The Knight, the Lady and the Priest: The Making of Modern Marriage in Medieval France*, trans. Barbara Bray (New York, 1984), 72, 106.
45. On Mary of Oignies, see Brenda M. Bolton, "*Vitae Matrum*: A Further Aspect of the *Frauenfrage*," in *Medieval Women: Dedicated and Presented to Professor Rosalind M. T. Hill . . .*, ed. D. Baker, Studies in Church History, subsidia 1 (Oxford, 1978), 257–59; on Clare, see Rosalind B. Brooke and Christopher N. L. Brooke, "St. Clare," in ibid., 275–87. And on this point generally see my essay, "Women's Stories, Women's Symbols: A Critique of Victor Turner's Theory of Liminality," in Frank Reynolds and Robert Moore, eds., *Anthropology and the Study of Religion* (Chicago, 1984), 105–25.
46. Life of Christina *Mirabilis*, 654.
47. On this point, see Claude Carozzi, "Douceline et les autres," in *La religion populaire en Languedoc du XIIIe siècle à la moitié du XIVe siècle*, Cahiers de Fanjeaux, no. 11 (Toulouse, 1976), 251–67; and Martinus Cawley, "Lutgard of Aywières: Life and Journal," *Vox Benedictina: A Journal of Translations from Monastic Sources* 1.1 (1984): 20–48.
48. See Vauchez, *La Sainteté*, 406. For Francesca Romana de' Ponziani (d. 1440) and Jutta of Huy (d. 1228), who found the act of sexual intercourse repulsive, see Weinstein and Bell, 39–40, 88–89.
49. Life of Lutgard, 193–95. To interpret these incidents more psychologically, one might say that it is hardly surprising that Lutgard, a victim of attempted rape in adolescence, should feel anesthetized when kissed, over her protests, by a man. Nor is it surprising that the "mouth" and "breast" of Christ should figure so centrally in her visions, providing partial healing for her painful experience of the mouths of men; see Life of Lutgard, 192–94, and Hendrix, "Primitive Versions," 180.
50. Ernest W. McDonnell, *The Beguines and Beghards in Medieval Culture with Special Emphasis on the Belgian Scene* (Rutgers, 1954; reprint ed., New York, 1969), 354–55.
51. James of Vitry, Life of Mary of Oignies, in *AASS*, June: 5 (Paris and Rome, 1867), 551–56.
52. Addendum to the life of Alpaïs, 207–8.
53. See above n. 29; Life of Ida of Louvain, in *AASS*, April: 2, 156–89, esp. 167; Anton Birlinger, ed., "Leben Heiliger Alemannischer Frauen des XIV–XV Jahrhunderts, 1: Dit erst Büchlyn ist von der Seligen Kluseneryn von Rüthy, die genant waz Elizabeth," *Alemannia* 9 (1881): 275–92, esp. 280–83; Life of Columba of Rieti, in *AASS*, May: 5 (Paris and Rome, 1866), 149*–222*, esp. 159*, 162*–164*, 187*, and 200*. This would also appear to be true of Dorothy of Montau (or of Prussia) (d. 1394); see Kieckhefer, *Unquiet Souls*, 22–33.

54. See, for example, Life of Lutgard, 200; Life of Elisabeth of Spalbeek, 378; Peter of Vaux, Life of Colette, 554–55; Life of Columba of Rieti, 188*. This emphasis on the closing of the body is also found in early modern accounts of "fasting girls." Jane Balan (d. 1603) supposedly did not excrete, menstruate, sweat, or produce sputum, tears, or even dandruff; see Hyder E. Rollins, "Notes on Some English Accounts of Miraculous Fasts," *The Journal of American Folk-lore* 34.134 (1921): 363–64.
55. See Charles T. Wood, "The Doctors' Dilemma: Sin, Salvation and the Menstrual Cycle in Medieval Thought," *Speculum* 56 (1981): 710–27.
56. Albert the Great, *De animalibus libri XXVI. nach der Cölner Urschrift*, vol. 1 (Münster, 1916), 682. Hildegard of Bingen, *Hildegardis Causae et Curae*, ed. P. Kaiser (Leipzig, 1903), 102–3, comments that the menstrual flow of virgins is less than that of nonvirgins, but she does not relate this to diet.
57. See, for example, Jerome, letter 22 *ad Eustochium*, PL 22, cols. 404–5, and *contra Jovinianum*, PL 23, cols. 290–312; Fulgentius of Ruspé, letter 3, PL 65, col. 332; John Cassian, *Institutions cénobitiques*, ed. Jean-Claude Guy (Paris, 1965), 206; and Musurillo, "Ascetical Fasting," 13–19. For a medieval preacher who repeats these warnings, see Peter the Chanter, *Verbum abbreviatum*, PL 205, cols. 327–28.
58. See Life of Margaret of Cortona, in *AASS*, February: 3 (Paris, 1865), 313, and Life of Catherine of Sweden, in *AASS*, March: 3 (Paris and Rome, 1865), 504. On the tendency of male writers to depict women as sexual beings, see Weinstein and Bell, 233–36.
59. *The Book of Margery Kempe*, trans. W. Butler-Bowdon (London, 1936), 48–49.
60. Raymond, Life of Catherine, 869; Life of Rita of Cascia, in *AASS*, May: 5 (Paris and Rome, 1866), 226–28. (There is no contemporary life of Rita extant.)
61. Raymond, Life of Catherine, 868–91 passim; Life of Columba, 153*–161*.
62. On Dorothy, see *Vita Lindana*, in *AASS*, October: 13 (Paris, 1883), 505, 515, 523, and 535–43; see also John Marienwerder, *Vita Latina*, ed. Hans Westpfahl, in *Vita Dorotheae Montoviensis Magistri Johannis Marienwerder* (Cologne, 1964), 236–45, and Kieckhefer, *Unquiet Souls*, 22–33. On Margaret, see Father Cuthbert, *A Tuscan Penitent: The Life and Legend of St. Margaret of Cortona* (London, n.d.), 94–95.
63. See above nn. 36, 53, and 61. On Elisabeth, see the depositions of 1235 in Huyskens, *Quellenstudien*, 112–40, and Conrad of Marburg's letter (1233) concerning her life in ibid., 155–60. See also Ancelet-Hustache, *Elisabeth*, 201–6 and 314–18. The importance of Elisabeth's tyrannical confessor, Conrad of Marburg, in inducing her obsession with food is impossible at this distance to determine. It was Conrad who ordered her not to eat food gained by exploitation of the poor; at times, to break her will, he forbade her to indulge in charitable food distribution.
64. On this point, see Bynum, "Women Mystics and Eucharistic Devotion," and idem, *Jesus as Mother: Studies in the Spirituality of the High Middle Ages* (Berkeley, 1982), chap. 5.
65. See Browe, *Die Wunder*, 34.
66. Life of Margaret of Cortona, 341; see also 343, where she recognizes an unconsecrated host.
67. See, for example, Life of Ida of Louvain, 178–79, and account of "Joan the Meatless" in Thomas Netter [Waldensis], *Opus de sacramentis* . . . (Salamanca, 1557), fols. 111v–112r. See also life of Mary of Oignies, 566, where James of Vitry claims that Mary saw angels when virtuous priests celebrated.
68. See Joseph Duhr, "Communion fréquente," *Dictionnaire de spiritualité, ascétique et mystique, doctrine et histoire*, vol. 2 (Paris, 1953), cols. 1234–92, esp. col. 1260.
69. For Imelda, see Browe, *Die Wunder*, 27–28; for Ida of Léau, see Life, in *AASS*, October: 13, 113–14. See also Life of Alice of Schaerbeke (d. 1250), in *AASS*, June: 2 (Paris and Rome, 1867), 473–74; and Life of Juliana of Cornillon, 445–46.
70. Life of Angela of Foligno, in *AASS*, January: 1, 204. For Mechtild's vision, see Bynum, *Jesus as Mother*, 210, n. 129.
71. See Catherine's own account of such a miracle, *Dialogue*, 239.
72. The work of David Herlihy and Diane Owen Hughes, which argues that, in the thirteenth and fourteenth centuries, the age discrepancy between husband and wife increased and the dowry, provided by the girl's family, became increasingly a way of excluding her from other forms of inheritance and from her natal family, suggests some particular reasons for a high level of antagonism between girls and their families in this period. The antagonism would

stem less from the failure of families to find husbands for daughters (as Herlihy suggests) than from the tendency of families to marry girls off early and thereby buy them out of the family when they were little more than children (as Hughes suggests). See David Herlihy, "Alienation in Medieval Culture and Society," reprinted in *The Social History of Italy and Western Europe, 700–1500: Collected Studies* (London, 1978); idem, "The Making of the Medieval Family: Symmetry, Structure and Sentiment," *Journal of Family History* 8.2 (1983): 116–30; and Diane Owen Hughes, "From Brideprice to Dowry in Mediterranean Europe," *Journal of Family History* 3 (1978): 262–96.

73. Vern Bullough, "Medieval Medical and Scientific Views of Women," *Viator* 4 (1973): 487–93; Eleanor McLaughlin, "Equality of Souls, Inequality of Sexes: Women in Medieval Theology," *Religion and Sexism: Images of Woman in the Jewish and Christian Traditions*, ed. R. Ruether (New York, 1974), 213–66; and M.-T d'Alverny, "Comment les théologiens et les philosophes voient la femme?" *Cahiers de civilisation médiévale* 20 (1977): 105–29.
74. See Bynum, *Jesus as Mother*, chap. 4, and idem, "'. . .And Woman His Humanity': Female Imagery in the Religious Writing of the Later Middle Ages," in *New Perspectives on Religion and Gender*, ed. Caroline Bynum, Stevan Harrell, and Paula Richman, to appear.
75. *Die Visionen der hl. Elisabeth und die Schriften der Aebte Ekbert und Emecho von Schönau*, ed. F W. E. Roth (Brunn, 1884), 60; Hildegard, *Liber divinorum operum*, PL 197, col. 885; and idem, *Scivias*, ed. Adelgundis Führkötter and A. Carlevaris, 2 vols. (Turnhout, 1978), 1:225–306, esp. 231 and plate 15.
76. On Mechtild, see *Jesus as Mother*, 229, 233–34, and 244; and on Hildegard, see Barbara Jane Newman, "*O Feminea Forma*: God and Woman in the Works of St. Hildegard (1098–1179)," Ph.D. diss., Yale, 1981, 131–34. And see Marguerite of Oingt, *Speculum*, in *Les Oeuvres de Marguerite d'Oingt*, ed. and trans. A. Duraffour, R. Gardette, and R. Durdilly (Paris, 1965), 98–99.
77. *Julian of Norwich: Showings*, trans. Edmund Colledge and James Walsh (New York, 1978), long text: 292, 294.
78. See *Jesus as Mother*, chap. 4, and Jan van Ruysbroeck, *The Spiritual Espousals*, trans. E. Colledge (New York, n.d.), 43; idem, "Le Miroir du salut éternel," in *Oeuvres de Ruysbroeck l'Admirable*, trans. by the Benedictines of St.-Paul de Wisques, vol. 3 (3rd ed., Brussels, 1921), 82–83; Francis of Assisi, "Salutation of the Blessed Virgin," trans. B. Fahy, in *St. Francis of Assisi: Writings and Early Biographies: English Omnibus of Sources*, ed. Marion Habig (3rd ed., Chicago, 1973), 135–36; and Henry Suso, *Büchlein der Ewigen Weisheit*, in *Deutsche Schriften*, 264.
79. See P. V. Bétérous, "A propos d'une des légendes mariales les plus répandues: Le 'Lait de la Vierge,'" Bulletin *de l'Association Guillaume Budé* 4 (1975): 403–11; Léon Dewez and Albert van Iterson, "La Lactation de saint Bernard: Légende et iconographie," *Cîteaux in de Nederlanden* 7 (1956): 165–89.
80. See Bynum, "'. . . And Woman His Humanity.'"
81. On the lactation of Bernard of Clairvaux, see E. Vacandard, *Vie de saint Bernard, abbé de Clairvaux*, 2 vols. (1895; reprint ed., Paris, 1920), 2:78; and Dewez and van Iterson, "La Lactation." Suso received the same vision; see Life, in *Deutsche Schriften*, 49–50. Alanus de Rupe (or Alan de la Roche, d. 1475), founder of the modern rosary devotion, tells a similar story of himself in his Revelations; see Heribert Holzapfel, *St. Dominikus und der Rosenkranz* (Munich, 1903), 21. See also Alb. Poncelet, "Index miraculorum B. V. Mariae quae saec. VI-XV latine conscripta sunt," *Analecta Bollandiana* 21 (1902): 359, which lists four stories of sick men healed by the Virgin's milk. Vincent of Beauvais, *Speculum historiale* (Venice, 1494), fol. 80r, tells of a sick cleric who nursed from the Virgin. Elizabeth Petroff, *Consolation of the Blessed* (New York, 1979), 74, points out that the Italian women saints she has studied nurse only from Christ, never from Mary, in vision. A nun of Töss, however, supposedly received the "pure, tender breast" of Mary into her mouth to suck because she helped Mary rear the Christchild; Ferdinand Vetter, ed., *Das Leben der Schwestern zu Töss beschrieben von Elsbet Stagel* (Berlin, 1906), 55–56. And Lukardis of Oberweimar nursed from Mary; see her Life in *Analecta Bollandiana* 18 (1899): 318–19.
82. See, for example, Life of Gertrude van Oosten (or of Delft), 350. For Gertrude of Helfta's visions of nursing the Christchild, see *Jesus as Mother*, 208, n. 123. There is one example of a man nursing Christ; see McDonnell, *Beguines*, 328, and Browe, *Die Wunder*, 106.

83. See *Jesus as Mother*, chap. 4, and nn. 35 and 49 above. Clare of Assisi supposedly received a vision in which she nursed from Francis; see "Il Processo di canonizzazione di S. Chiara d'Assisi," ed. Zeffirino Lazzeri, *Archivum Franciscanum Historicum* 13 (1920), 458, 466.
84. See Mary Martin McLaughlin, "Survivors and Surrogates: Children and Parents from the Ninth to the Thirteenth Centuries," *The History of Childhood*, ed. L. DeMause (New York, 1974): 115–18, and *Jesus as Mother*, 131–35.
85. See above nn. 15 and 82. Examples of women who exuded healing oil, in life or after death, include Lutgard of Aywières, Christina *Mirabilis*, Elisabeth of Hungary, Agnes of Montepulciano, and Margaret of Città di Castello (d. 1320). See Life of Lutgard, 193–94; Life of Christina *Mirabilis* 652–54; Huyskens, *Quellenstudien*, 51–52; Raymond of Capua, Life of Agnes, in *AASS*, April: 2, 806; Life of Margaret of Città di Castello, *Analecta Bollandiana* 19 (1900): 27–28, and see also *AASS*, April: 2, 192.
86. Life of Lutgard, 193; Life of Lukardis of Oberweimar, *Analecta Bollandiana* 18 (1899): 337–38; Peter of Vaux, Life of Colette, 563, 576, 585, and 588.
87. For a general discussion of reversed images in women's spirituality, see Bynum, "Women's Stories, Women's Symbols."

13

The Appetite as Voice

JOAN JACOBS BRUMBERG

The symptoms of disease never exist in a cultural vacuum. Even in a strictly biomedical illness, patient responses to physical discomfort and pain are structured in part by who the patient is, the nature of the care giver, and the ideas and values at work in that society. Similarly, in mental illness, basic forms of cognitive and emotional disorientation are expressed in behavioral aberrations that mirror the deep preoccupations of a particular culture. For this reason a history of anorexia nervosa must consider the ways in which different societies create their own symptom repertoires and how the changing cultural context gives meaning to a symptom such as noneating.[1]

In this chapter I suggest a link between the emergence of anorexia nervosa in the nineteenth century and the cultural predispositions of that era. Just as the incidence of anorexia nervosa today is related to powerful contemporary messages about body image and dieting, there is a cultural context—albeit a somewhat different one—that helps to explain the Victorian anorectic. Again, this is not to say that cultural ideas directly cause the disease. At the outset I acknowledged the etiological complexities and the limitations of historical study; anorexia nervosa is a multidetermined disorder that involves individual biological and psychological factors as well as environmental influences. As a historian, I cannot resolve the problem of causation nor can I chart individual psychopathologies. Historical study does, however, illuminate the larger meanings of food and eating in Victorian society and in that process posits a certain set of cultural preoccupations that had particular impact on adolescent women among the bourgeoisie. In effect, by supplying something of the female "food vocabulary" of a distant era, I hope to explain how there could have been anorexia nervosa before there was Twiggy.

The medical literature of the nineteenth century provides few clues to the meaning of anorexic behavior in that period. Physicians reported the characteristic cry of the anorectic as "I will not eat," but they rarely provided the text of the critical subordinate clause, "I will not eat because . . ." The medical literature supplied few accounts by Victorian anorectics, *in their own words,* of why they refused their food and why they deprived their bodies as they did. The Victorian anorectic's understanding of her own behavior remains something of a mystery.

Victorian anorectics did present somatic complaints about gastric discomfort and difficulty in swallowing. Because nineteenth-century doctors emphasized physical diagnosis and somatic treatment, they probably reinforced presentations of this type of distress.[2] Parents also found it easier to accept physical rather than emotional reasons as an explanation of their daughter's emaciation and food-refusing behavior. Yet in cases of anorexia nervosa, biomedical reasons for noneating were quickly undercut, since the diagnosis itself meant that there was no organic disease. Most physicians avoided explanations of etiology and concentrated instead on curing the primary symptoms, noneating and emaciation.

Among the few careful medical reports that include any discussion of motivation is Stephen MacKenzie's 1895 account of his anorexic patient's refusing her food "on account of her mother talking to her about being so fat." A decade earlier Jean-Martin Charcot discovered a rose-colored ribbon wound tightly around the waist of a patient with anorexia nervosa. Questioning revealed the girl's preoccupation with the size of her body: the ribbon was a measure that her waist was not to exceed. Max Wallet's 189S discussion of two cases of anorexia nervosa suggested the same theme and implicated peers. A seventeen-year-old refused her food because of a "fear of being seen as a bit heavy," and a fifteen-year-old stopped eating when she "got the idea that she was too fat after seeing her friends forcing themselves to lose weight."[3]

These behaviors were likely to be dismissed by physicians as the flirtatious "coquetries" and simpleminded "scruples" of female adolescence.[4] In effect, nineteenth-century medicine did not relate anorexia nervosa to the cultural milieu that surrounded the Victorian girl. The ideas of Victorian women and girls about appetite, food, and eating, as well as the cultural categories of fat and thin, were not mentioned as contributing to the disease. Only in the twentieth century has medicine come to understand that society plays a role in shaping the form of psychological disorders and that behavior and physical symptoms are related to cultural systems. Throughout the nineteenth century most doctors gave and accepted formulaic explanations of anorexia nervosa (for example, their patients "craved sympathy" or experienced a "perversion of the will") without providing any substantive discussion of why appetite and food were at issue. These explanations said more about the doctors' general views on adolescence, gender, and hysteria than they did about the specific mentality of patients with anorexia nervosa.

Given the attention paid to anorexia nervosa in late-nineteenth-century Anglo-American medicine, the failure of physicians to document the anorectic's explanations, however mundane or bizarre they might have been, is a provocative omission. This lapse raises a number of questions about the state of doctor-patient relations and the history of diagnostic and therapeutic techniques in the late nineteenth century. What were the dynamics of doctor, patient, and mother within the Victorian examining room? What expectations did the doctor have of his young female patients? A sensitivity to the relationship between culture and symptomatologies prompts additional questions. What motivated young women in the late nineteenth century to persistently refuse food in the face of familial coaxing and professional medical supervision? What role did food and eating play in female identity in the Victorian era?

IN THE EXAMINING ROOM

By the 1870s physical examinations included visual observation and manual manipulation of the body, combined with a few rudimentary tests of body temperature, blood, and urine. Because manual examinations were a progressive innovation done only by better-trained physicians, some patients were probably still unfamiliar and uneasy with the latest information-gathering techniques: listening to the body through a stethoscope, manipulation of the body parts, and tactile probing of the body. In cases involving young women, professionally knowledgeable and socially correct doctors did the examination in the presence of the girl's mother as well as a clinical clerk. The clerk recorded information while the physician listened to, poked, and thumped the patient, who remained partially dressed in her underclothing.[5]

The nineteenth-century physician's new faith in the verifiable external signs and sounds of illness shaped the interaction in the examining room. The doctor was more interested in what the body revealed than in anything the patient had to say about her illness. Educated physicians came to regard the process of history taking as secondary to the process of physical examination. Doctors were assured that patient accounts of illness were more often than not prejudiced, ignorant, and unreliable; personal and family narratives were rarely objective and they almost always revealed the ignorance of lay people about medical phenomena. In this atmosphere of suspicion about all patient accounts, volatile adolescent girls were considered particularly unreliable informants.[6]

As a consequence, the professionally correct doctor turned to the girl's mother, in her authoritative role as parent, for information about the patient's medical history and current symptoms. Social convention supported this strategy; as long as an unmarried girl resided at home, her parents unquestionably had authority over her. Consequently, the doctor, who was in the employ of the parents, dealt with the young woman as a child. In the Victorian examining room, the mother was not only a monitor of the physical examination of her daughter's body but a check on the substance of the conversation between doctor and patient.

In this scenario, which assumes that doctor, mother, and patient all played out their expected social roles, the examining room reproduced the situation in the home. The doctor and the mother were the primary conversants; again, two adults, a male and a female, were focused on the girl's wasting body and her refusal of food. Again, she was told that she ought to eat. Her response, shaped by the nature of the medical investigation and parental expectations, was to say that she could not—eating hurt in some vague, nonspecific way. When the examination showed no organic problem to sustain this interpretation, the doctor made his diagnosis: anorexia nervosa.

It is unlikely that the doctor ever dismissed the mother and tried to see the patient alone in order to search out what was troubling the girl and causing her to refuse food. Propriety worked against such a scenario, as did the conception of the patient as a dependent person and the doctor's lack of interest in girlish narratives. Adolescent patients must have sensed the doctors' disinterest in their point of view. A recovered anorectic told her physician that, during treatment, "I saw that you wished to shut me up."[7]

In an era that valued demure behavior in all women, it is not inconceivable that the anorexic girl honored social conventions by respecting her mother's authority and keeping silent. It is also possible that the partially dressed young woman was so embarrassed by her situation and so intimidated by her doctor that she could not speak. Another explanation, culled directly from the medical record, suggests that the patient responded to questions in a diffident manner. Published case reports repeatedly said that girls with anorexia nervosa were "sullen," "sly," and "peevish," implying that they were as parsimonious with their words as with their food.[8] Refusal to sustain conversation with either one's parents or one's doctor went hand in hand with refusal to eat. The anorexic girl used both her appetite and her body as a substitute for rhetorical behavior.

When the doctor had ascertained that the patient had no physical reason not to eat, his forbearance might ebb. At that point an authoritarian regimen of overfeeding, weighing, and isolation was usually instituted. This regimen became the primary basis of the doctor's relationship with his patient. Conversation, when it occurred, centered on the amount of food taken and weight and strength gained. Both doctor and patient acted as if the girl's illness was strictly physical (rather than emotional), despite the fact that differential diagnosis established exactly the opposite. The physician maintained an exclusive focus on the issue of the girl's body and her need to add flesh. To do otherwise was to pander to the sympathies of a hysterical adolescent.

Anorexic patients did sometimes talk to less authoritative female medical personnel and to their peers. Although the evidence is undeniably scanty, a few examples reveal that the Victorian anorectic shared aspects of her compulsion to starve with individuals she perceived as less threatening and more sympathetic than the doctor. If she did not speak to them directly about why she refused her food, she left telltale pieces of evidence that provided some explanation of her behavior. This evidence was rarely uncovered by the supervising doctor.

For example, ward nurses and nursing nuns were in close and intimate contact with patients—feeding them regularly, washing their bodies, and supervising their waking hours. In the 1890s a French physician ousted a high-strung mother from the home where her anorexic daughter was being treated and sent in a nun, *une religieuse*, to care for the emaciated fifteen-year-old. At first, the "new attitude of her caretakers" terrified the girl and she became even more recalcitrant about eating and said that she wished to die. After three months of an enforced dietary regimen of arrowroot, bread, eggs, and beef tea, the girl left the recuperative home "fat" and capable of a normal, active life. At discharge the nun disclosed that she had in her possession a series of letters written by the former patient, which "constituted a peculiarly interesting witness from the point of view of causation of her malady." The letters, addressed to an older male relative, disclosed that the patient's food refusal was generated by her romantic and "singular passion" for this man who, in the young girl's presence, had explicitly admired another woman who was "extremely lean." In the effort to please him, the girl began to starve herself, walk excessively, and lace herself very tightly. Yet she never once told her doctor about her passion for her relative or her desire to be thin.[9]

The nurse functioned as detective in another case involving a twenty-year-old at St. George's Hospital in London. None of the consulting doctors could find an organic reason why their hysterical patient refused to eat, but they continued to work diligently to ease the stomach pains of which she complained. The nurse on the ward, however, regarded the girl as a malingerer and told the doctors, "On December 6th, whilst the girl was apparently suffering . . . the Queen [Victoria] passed the hospital, on her way to open Blackfriar's Bridge; [the girl] rose in bed to watch her out of the window, having been thought utterly unable to move, owing to pain." On yet another occasion, when friends were admitted for hospital visiting hours, the nurse found the supposedly debilitated girl "sitting up in bed, trying on a new coloured frock."

This same patient, who told the doctors that she could not and would not eat, engaged in surreptitious relationships with other patients in order to get bits of their food, which she would eat on the sly. The nurse at St. George's Hospital found a note from the girl indicating that she did eat secretly:

> My dear Mrs. Evans—I was very sorry you should take the trouble of cutting me such a nice peice [*sic*] of bread and butter yesterday. I would of taken it, but all of them saw you send it, and they would of made enough to have talked about, but I should be very glad if you will cut me a nice peice [*sic*] of crust and put it in a peice [*sic*] of paper and send it or else bring it, so as they do not see it, for they all watch me very much.[10]

The nurse's information on this patient provided no real explanation of why the girl would not eat in the presence of her caretakers, but it did confirm the physician's belief that hysterical adolescents were by definition deceptive. This attitude surely affected the doctors' interactions with anorexic patients. If it was assumed that the patient was by nature duplicitous, then any explanation she gave would be suspect.

Sensing the doctor's loyalties to her parents and his suspicious attitude toward her, the anorectic usually chose not to disclose her private preoccupations to an unsympathetic male authority figure. When she spoke, it was almost always of bodily ills: pain after eating, a sour stomach, difficulty in swallowing, flatulence. Deference, fear, and anger all combined to keep her essentially mute. When her bodily preoccupations were rooted in ideas that the doctor might find childish, inappropriate, or distasteful, her silence became confirmed. Furthermore, there was always the distinct possibility of misunderstanding or embarrassment when girls told personal things to men or boys. The bourgeois world of the nineteenth century was still very much sex segregated.[11] Consequently, enormous emotional risks were involved in baring one's soul to the doctor. Most adult men did not understand the language of girlhood sentiment and knew neither its vocabulary nor its symbols. The silence of the Victorian anorectic was in keeping with her provocative resort to symbolic rather than rhetorical behavior.

THE IRRATIONAL APPETITE

In the late nineteenth century, adolescent girls demonstrated an array of health problems that involved eating and appetite disturbances. These problems lent confusion rather than clarity to the process of making the diagnosis for anorexia nervosa. In effect, there was a wide spectrum of "picky eating" and food refusal, ranging from the normative to the pathological. Anorexia nervosa was the extreme—but it was not altogether alien, given the range of behaviors that doctors saw in adolescent female patients. As one astute twentieth-century physician wrote about the origins of anorexia nervosa in the era of William Gull, "the conditions of life were well staged for such a disturbance."[12]

The health of young women was definitely influenced by a general female fashion for sickness and debility.[13] The sickly wives and daughters of the bourgeoisie provided the medical profession with a ready clientele. In Victorian society unhappy women (and men) had to employ physical complaints in order to be permitted to take on the privileged "sick role." Because the most prevalent diseases in this period were those that involved "wasting," it is no wonder that becoming thin, through noneating, became a focal symptom. Wasting was in style.

Among women, invalidism and scanty eating commonly accompanied each other. The partnership was familiar enough to become the subject of a satirical novel. In *The*

Female Sufferer; or, Chapters from Life's Comedy, Augustus Hoppin satirized the indolent existence of an upper-class invalid who, while ever so ill, managed to run a vigorous social life from her sick chamber. The stylized eating of this "nervous exhaustionist" was central to the author's portrait. Delicate foods such as "tidbits of fruit and jelly," "a snip of a roll," "a wren's leg on toast," were taken only to "appease the cravings of her exhausted nerves"—but not because she was hungry. At times, however, the debilitated patient would become voracious for "dainty" items such as wedding cake, peaches and cream, and freshly cut melon. According to Hoppin, another characteristic type among female sufferers was the woman supposedly perishing of starvation or "pining away." "Well! dying of inanition is doing something, isn't it?" asked one of the admirers who surrounded the sick couch. Another replied, "Inanition being merely action begun, demands too much exertion for [the lady] to finish."[14]

Adolescent girls simply followed and imitated the behavioral styles of adult women. As a consequence, mothers were urged to take action against their daughters' fondness for wasting and debility. In *Eve's Daughters; or, Common Sense for Maid, Wife, and Mother*, Marion Harland told parents:

> Show no charity to the faded frippery of sentiment that prates over romantic sickliness. Inculcate a fine scorn for the desire to exchange her present excellent health for the estate of the pale, drooping, human-flower damsel; the taste that covets the "fascination" of lingering consumption; the "sensation" of early decease induced by the rupture of a blood-vessel over a laced handkerchief held firmly to her lily mouth by agonized parent or distracted lover. All this is bathos and vulgarity . . . Bid her leave such balderdash to the pretender to ladyhood, the low-minded *parvenu*, who, because foibles are more readily imitated than virtues, and tricks than graces, copies the mistakes of her superiors in breeding and sense, and is persuaded that she has learned "how to do it."[15]

Harland, an American, called the "cultivation of fragility" a "national curse."

Of the conditions that affected girls most frequently, dyspepsia and chlorosis both incorporated peculiar eating and both could be confused with anorexia nervosa. Dyspepsia, a form of chronic indigestion with discomfort after eating, was widespread in middle-class adults and in their daughters. Physicians saw the adolescent dyspeptic frequently; advice writers suggested how she should be managed at home; health reformers used her existence to argue for changes in the American diet; and even novelists considered her enough of a fixture on the domestic scene to include her in their portraits of social life.[16] The dyspeptic had no particular organic problem; her stomach was simply so sensitive that it precluded normal eating. Whereas dyspeptic women could be extremely thin, some, according to doctors' reports, were corpulent. Yet dyspepsia sometimes looked much like anorexia nervosa. For example, a physician described his young dyspeptic patients as persons "who enter upon a strict regimen which they follow only too well. By auto-observation and auto-suggestion, by constantly noticing and classifying their foods, and rejecting all kinds that they think they cannot digest, they finally manage to live on an incredibly small amount."[17]

Chlorosis, a form of anemia named for the greenish tinge that allegedly marked the skin of the patient, was the characteristic malady of the Victorian adolescent girl. Although chlorosis was never precisely defined and differentiated, it was unequivocally regarded as a disease of girlhood rather than boyhood. Its symptoms included lack of energy, shortness of breath, dyspepsia, headaches, and capricious or scanty appetite; sometimes the menses stopped. Chlorotic girls tended to lose some weight as a result of poor eating and aversion to specific foods, particularly meat.[18] (Today iron-deficiency anemia corresponds to the older diagnosis of chlorosis.)

Doctors of the Victorian era fostered the notion that all adolescent girls were potentially chlorotic: "Every girl passes as it were through the outer court of chlorosis in her progress from youth to maturity . . . Perhaps, no girl escapes it altogether."[19] In contrast to anorexia nervosa, treatment for this popular disease was relatively easy: large doses of iron salts and a period of rest at home. As a result, parents were not afraid of chlorosis. In fact, it was accepted as a normal part of adolescent development. Many doctors and families were also fond of tonics to stimulate the appetite, restore the blush to the cheek, and cure latent consumption. "Young Girls Fading Away" was the headline of a well-known advertisement for Dr. William's Pink Pills for Pale People, a medicine aimed at the chlorotic market.[20] A vast amount of patent medicine was sold to families that assumed chlorosis in an adolescent whenever her energy, spirits, or appetite waned. In cases that were eventually diagnosed as anorexia nervosa, the patient in the earliest stages may well have been regarded as dyspeptic or chlorotic. Because clinical descriptions of the confirmed dyspeptic, the chlorotic, and the anorectic had many features in common, we must assume that the diagnoses occasionally overlapped.[21]

Taken together, these conditions suggest that young women presented unusual eating and diminished appetite more often than any other group in the population. Apparently, it was relatively normal for a Victorian girl to develop poor appetite and skip her meals, "affect daintiness" and eat only sweets, or express strong food preferences and dislikes.[22] A popular women's magazine told its readership that in adolescence "digestive problems are common, the appetite is fickle, and evidences of poor nourishment abound."[23] Between 1850 and 1900 the most frequent warning issued to parents of girls had to do with forestalling the development of idiosyncrasies, irregularities, or strange whims of appetite because these were precursors of disease as well as signs of questionable moral character.

Ideas about female physiology and sexual development underlay the physician's expectations and his clinical treatment. Doctors believed that women were prone to gastric disorders because of the superior sensitivity of the female digestive system. Using the machine metaphor that was popular in describing bodily functions, they likened a man's stomach to a quartz-crushing machine that required coarse, solid food. By contrast, the mechanisms of a woman's stomach could be ruined if fed the same materials. The female digestive apparatus required foods that were soft, light, and liquid.[24] (Dyspepsia in women could result from the choice of inappropriate foods that required considerable chewing and digestion.)

To the physician's mind, a young woman caught up in the process of sexual maturation was subject to vagaries of appetite and peculiar cravings. "The rapid expansion of the passions and the mind often renders the tastes and appetite capricious," wrote a midcentury physician.[25] Therefore, even normal sexual development had the potential to create a disequilibrium that could lead to irregular eating such as the kind reported in dyspepsia and chlorosis. But physicians reported on eating behavior that was far more bizarre. In fact, the adolescent female with "morbid cravings" was a stock figure in the medical and advice literature of the Victorian period. Stories circulated of "craving damsels" who were "trash-eaters, oatmeal-chewers, pipe-chompers, chalk-lickers, wax-nibblers, coal-scratchers, wall-peelers, and gravel-diggers." The clinical literature also provided a list of "foods" that some adolescent girls allegedly craved: chalk, cinders, magnesia, slate pencils, plaster, charcoal, earth, spiders, and bugs.[26] Modern medicine associates iron-deficiency anemia with eating nonnutritive items, such as pica. For the Victorian physician, nonnutritive eating constituted proof of the fact that the adolescent girl was essentially out of control and that the process of sexual maturation could generate voracious and dangerous appetites.

In this context physicians asserted that even normal adolescent girls had a penchant for highly flavored and stimulating foods. A reputable Baltimore physician, for example, described three girlfriends who constantly carried with them boxes of pepper and salt, taking the condiments as if they were snuff.[27] The story was meant to imply that the girls were slaves of their bodily appetites. Throughout the medical and advice literature an active appetite or an appetite for particular foods was used as a trope for dangerous sexuality. Mary Wood-Allen warned young readers that the girl who masturbated "will manifest an unnatural appetite, sometime desiring mustard, pepper, vinegar and spices, cloves, clay, salt, chalk, charcoal, etc."[28]

Because appetite was regarded as a barometer of sexuality, both mothers and daughters were concerned about its expression and its control. It was incumbent upon the mother to train the appetite of the daughter so that it represented only the highest moral and aesthetic sensibilities. A good mother was expected to manage this situation before it escalated into a medical or social problem. Marion Harland's Mamie, the prototypical adolescent, developed at puberty "morbid cravings of appetite and suffered after eating things that never disagreed with her before." Mamie's mother was cautioned about the possibility that a disturbance of appetite could precipitate an adolescent decline. Mothers were urged to be vigilant: "If Mamie has not a rational appetite, create a digestive conscious [*sic*] that may serve her instead." Mothers were expected to educate, if not tame, their adolescent daughter's propensity for "sweetmeats, bonbons, and summer drinks" as well as for "stimulating foods such as black pepper and vinegar pickle."[29] "Inflammatory foods" such as condiments and acids, thought to be favored by the tumultuous female adolescent, were strictly prohibited by judicious mothers. Adolescent girls were expressly cautioned against coffee, tea, and chocolate; salted meats and spices; warm bread and pastry; confectionery; nuts and raisins; and, of course, alcohol.[30] These sorts of foods stimulated the sensual rather than the moral nature of the girl.

No food (other than alcohol) caused Victorian women and girls greater moral anxiety than meat. The flesh of animals was considered a heat-producing food that stimulated production of blood and fat as well as passion. Doctors and patients shared a common conception of meat as a food that stimulated sexual development and activity. For example, Lucien Warner, a popular medical writer, suggested that meat eating in adolescence could actually accelerate the development of the breasts and other sex characteristics; at the same time, a restriction on the carnivorous aspects of the diet could moderate premature or rampant sexuality as well as overabundant menstrual flow. "If there is any tendency to precocity in menstruation, or if the system is very robust and plethoric, the supply of meat should be quite limited. If, on the other hand, the girl is of sluggish temperament and the menses are tardy in appearance, the supply of meat should be especially generous."[31] Meat eating in excess was linked to adolescent insanity and to nymphomania.[32] A stimulative diet of meat and condiments was recommended only for those girls whose development of the passions seemed, somehow, "deficient."

By all reports adolescent girls ate very little meat, a practice that certainly contributed to chlorosis or iron-deficiency anemia. In fact, many openly disdained meat without being necessarily committed to the ideological principles of the health reformers who espoused vegetarianism.[33] Meat avoidance, therefore, is the most apt term for this pattern of behavior. According to E. Lloyd Jones, adolescent girls "are fond of biscuits, potatoes, etc. while they avoid meat on most occasions, and when they do eat meat, they prefer the burnt outside portion." Another doctor confirmed the same problem in a dialogue between himself and a patient. "Oh, I like pies and preserves but I can't bear meat," the young woman reportedly told the family physician. A "disgust for meat in any form" characterized many of the adolescent female patients of a Pennsylvania practitioner of this period.[34]

When it became necessary to eat meat (say, if prescribed by a doctor), it was an event worthy of note. For many, meat eating was endured for its healing qualities but despised as a moral and aesthetic act. For example, eighteen-year-old Nellie Browne wrote to tell her mother that a delicate classmate [Laura] had, like her own sister Alice, been forced to change her eating habits:

> I am very sorry to hear Alice has been so sick. Tell her she must eat meat if she wishes to get well. Laura eats meat *three* times a day.—She says she cannot go without it.—If Laura *can* eat *meat, I am sure Alice can.* If Laura needs it *three* times a day, Alice needs it *six.* (Italics in original.)[35]

After acknowledging the "common distaste for meat" among his adolescent patients, Clifford Allbutt wrote, "Girls will say the entry of a dish of hot meat into the room makes them feel sick."[36]

The repugnance for fatty animal flesh among Victorian adolescents ultimately had a larger cultural significance. Meat avoidance was tied to cultural notions of sexuality and decorum as well as to medical ideas about the digestive delicacy of the female stomach. Carnality at table was avoided by many who made sexual purity an axiom. Proper women, especially sexually maturing girls, adopted this orientation with the result that meat became taboo. Contemporary descriptions reveal that some young women may well have been phobic about meat eating because of its associations:

> There is the common illustration which every one meets a thousand times in a lifetime, of the girl whose stomach rebels at the very thought of fat meat. The mother tries persuasion and entreaty and threats and penalties. But nothing can overcome the artistic development in the girl's nature which makes her revolt at the bare idea of putting the fat piece of a dead animal between her lips.[37]

In this milieu food was obviously more than a source of nutrition or a means of curbing hunger; it was an integral part of individual identity. For women in particular, how one ate spoke to issues of basic character.

"A WOMAN SHOULD NEVER BE SEEN EATING"

In Victorian society food and femininity were linked in such a way as to promote restrictive eating among privileged adolescent women. Bourgeois society generated anxieties about food and eating—especially among women. Where food was plentiful and domesticity venerated, eating became a highly charged emotional and social undertaking. Displays of appetite were particularly difficult for young women who understood appetite to be both a sign of sexuality and an indication of lack of self-restraint. Eating was important because food was an analogue of the self. Food choice was a form of self-expression, made according to cultural and social ideas as well as physiological requirements. As the anthropologist Claude Lévi-Strauss put it, things must be "not only good to eat, but also good to think."[38]

Female discomfort with food, as well as with the act of eating, was a pervasive subtext of Victorian popular culture.[39] The naturalness of eating was especially problematic among upwardly mobile, middle-class women who were preoccupied with establishing their own good taste.[40] Food and eating presented obvious difficulties because they implied digestion and defecation, as well as sexuality. A doctor explained that one of his anorexic patients "refused to eat for fear that, during her digestion, her face should grow red and appear less pleasant in the eyes of a professor whose lectures she attended after her meals."[41] A woman who ate inevitably had to urinate and move her bowels. Concern about these bodily indelicacies explains why constipation was incorporated into the ideal of Victorian femininity. (It was almost always a symptom in anorexia nervosa.) Some women "boasted that the calls of Nature upon them averaged but one or two demands per week."[42]

Food and eating were connected to other unpleasantries that reflected the self-identity of middle-class women. Many women, for good reason, connected food with work and drudgery. Food preparation was a time-consuming and exhausting job in the middle-class household, where families no longer ate from a common soup pot. Instead, meals were served as individual dishes in a sequence of courses. Women of real means and position were able to remove themselves from food preparation almost entirely by turning over the arduous daily work to cooks, bakers, scullery and serving maids, and butlers. Middle-class women, however, could not achieve the same distance from food.[43]

Advice books admonished women "not to be ashamed of the kitchen," but many still sought to separate themselves from both food and the working-class women they hired to do the preparation and cooking. A few women felt the need to make alienation from food a centerpiece of their identity. A young "lady teacher," for example, "regard[ed] it as unbecoming her position to know anything about dinner before the hour for eating arrived . . . [She was] ashamed of domestic work, and graduate[d] her pupils with a similar sense of false propriety."[44] Similarly, in the 1880s in Rochester, New York, a schoolgirl was chastised by her aunt for describing (with relish) in her diary the foods she had eaten during the preceding two weeks.[45]

Food was to be feared because it was connected to gluttony and to physical ugliness. In advice books such as the 1875 *Health Fragments; or, Steps toward a True Life* women were cautioned to be careful about what and how much they ate. Authors George and Susan Everett enjoined: "Coarse, gross, and gluttonous habits of life degrade the physical appearance. You will rarely be disappointed in supposing that a lucid, self-respectful lady is very careful of the food which forms her body and tints her cheeks." Sarah Josepha

Hale, the influential editor of *Godey's Lady's Book* and an arbiter of American domestic manners, warned women that it was always vulgar to load the plate.[46]

Careful, abstemious eating was presented as insurance against ugliness and loss of love. Girls in particular were told: "Keep a great watch over your appetite. Don't always take the nicest things you see, but be frugal and plain in your tastes."[47] Young women were told directly that "gross eaters" not only developed thick skin but had prominent blemishes and broken blood vessels on the nose. Gluttony also robbed the eyes of their intensity and caused the lips to thicken, crack, and lose their red color. "The glutton's mouth may remind us of cod-fish—never of kisses." A woman with a rosebud mouth was expected to have an "ethereal appetite." A story circulated that Madam von Stein "lost Goethe's love by gross habits of eating sausages and drinking strong coffee, which destroyed her beauty." Women such as von Stein, who indulged in the pleasures of the appetite, were said to develop "a certain unspiritual or superanimal expression" that conveyed their base instincts. "[We] have never met true refinement in the person of a gross eater," wrote the Everetts.[48]

Indulgence in foods that were considered stimulating or inflammatory served not only as an emblem of unchecked sensuality but sometimes as a sign of social aggression. Women who ate meat could be regarded as acting out of place; they were assuming a male prerogative. In *Daniel Deronda* (1876) George Eliot described a group of local gentry, all men, who came together after a hunt to take their meal apart from the women. As they ate, the men took turns telling stories about the "epicurism of the ladies, who had somehow been reported to show a revolting masculine judgement in venison." Female eating was a source of titillation to men precisely because they understood eating to be a trope for sexuality. Furthermore, women who asked baldly for venison were aggressive if not insatiable. What most bothered the local gentry was the women's effrontery to "ask . . . for the fat—a proof of the frightful rate at which corruption might go in women, but for severe social restraint."[49]

Because food and eating carried such complex meanings, manners at the table became an important aspect of a woman's social persona. In their use of certain kinds of conventions, nineteenth-century novelists captured the crucial importance of food and eating in the milieu of middle-class women. Because they understood the middle-class reverence for the family meal, writers such as Jane Austen and Anthony Trollope saw the meal as an arena for potential individual and collective embarrassment. These novelists provided numerous examples of young women whose lives and fortunes hung on the issue of dinner-table decorum. For example, in Austen's *Mansfield Park* (1814) the heroine, Fanny Price, was horrified at the prospect of having her well-bred suitor eat with her family and "see all their deficiencies." Fanny was concerned not only about her family's lower standard of cookery, but about her sister's mortifying tendency to eat "without restraint." In Trollope's *Ralph the Heir* (1871) the family of Mr. Neefit, a tradesman, invited Ralph Newton, a gentleman, to their family table after some degree of preparation and nervousness on the part of the wife and daughter. Newton, who was halfheartedly courting the daughter at the request of her socially ambitious father, ultimately concluded that the young woman was attractive enough but the roughness of her father was unbearable. He found particularly galling the manner in which Mr. Neefit ate his shrimp.[50] Manners at table were often a dead giveaway of one's true social origins. This convention for marking the social distance between the classes was utilized by Mark Twain in the famous scene in *The Prince and the Pauper* (1881) where the prince's impersonator drinks from his fingerbowl.[51]

Women's anxieties about how to eat in genteel fashion were widespread and conveyed by novelists in a number of different ways. In Elizabeth Gaskell's *Cranford* (1853), the middle-class ladies of the town were made uncomfortable by the presentation of foods that were difficult to eat—in this case, peas and oranges. One woman "sighed over her delicate young peas [but] left them on one side of her plate untasted" rather than attempt to stab them or risk dropping them between the two prongs of her fork. She knew that she could not do "the ungenteel thing"—shoveling them with her knife. So, too, oranges presented difficulties for the decorous middle-class women of Cranford:

> When oranges came in, a curious proceeding was gone through. Miss Jenkyns did not like to cut the fruit; for, as she observed, the juice all ran out nobody knew where; sucking (only I think she used some more recondite word) was in fact the only way of enjoying oranges; but then there was the unpleasant association with a ceremony frequently gone through by little babies; and so, after dessert, in orange season, Miss Jenkyns and Miss Matty used to rise up, possess themselves each of an orange in silence, and withdraw to the privacy of their own rooms to indulge in sucking oranges.[52]

In fact, secret eating was not unknown among those who subscribed to the absurd dictum that "a woman should never be seen eating."[53] This statement, attributed by George Eliot to the famed poet Lord Byron, was the ultimate embodiment of Victorian imperatives about food and gender.

Over and over again, in all of the popular literature of the Victorian period, good women distanced themselves from the act of eating with disclaimers that pronounced their disinterest in anything but the aesthetics of food. "It's very little I eat myself," a proper Trollopian hostess explained, "but I do like to see things nice."[54] Apparently, Victorian girls adopted the aesthetic sensibilities of their mothers, displaying extraordinary interest in the appearance and color of their food, in the effect of fine china and linen, and in agreeable surroundings. A 1904 study of the psychology of foods in adolescence reported that boys most valued companionship at table, whereas girls emphasized "ceremony" and "appointments."[55] Attention to the aesthetics of eating seemed to minimize the negative implications of participating in the gustatory and digestive process.

But Victorian women avoided connections to food for a number of other reasons. The woman who put soul over body was the ideal of Victorian femininity. The genteel woman responded not to the lower senses of taste and smell but to the highest senses—sight and hearing—which were used for moral and aesthetic purposes.[56] One of the most convincing demonstrations of a spiritual orientation was a thin body—that is, a physique that symbolized rejection of all carnal appetites. To be hungry, in any sense, was a social faux pas. Denial became a form of moral certitude and refusal of attractive foods a means for advancing in the moral hierarchy.

Appetite, then, was a barometer of a woman's moral state. Control of eating was eminently desirable, if not necessary. Where control was lacking, young women were subject to derision. "The girl who openly enjoys bread-and-butter, milk, beefsteak and potatoes, and thrives thereby, is the object of many a covert sneer, or even overt jest, even in these sensible days and among sensible people."[57] Given the intensity of concern about control of appetite, it is not surprising that some women found strong attraction in cultural figures whose biographies exemplified the triumph of spirit over flesh. Two figures representing the Romantic and medieval traditions became especially relevant to how young women thought about these issues: Lord Byron and Catherine of Siena. Both spoke to the moral desirability of being thin.

Known to the Victorian reading public as the author of the immensely popular epic poems *Childe Harold* and *Don Juan,* Lord Byron (1788–1824) remained an important cultural figure whose life and work stood, even as late as the third quarter of the century, as a symbol of the power of the Romantic movement.[58] Young women who shared the Romantic sensibility found Byron's poetry inspirational. *Childe Harold,* which detailed a youth's struggle for meaning, spoke to the inner reaches of the soul and helped its readers transcend the "tawdry world." For many, such as Trollope's Lizzie Eustace, Byron was "the boy poet who understood it all."[59]

Although Byron's tempestuous love life served to titillate some and revolt others, the poet's struggles with the relation of his body to his mind were of enormous interest to women. Byron starved his body in order to keep his brain clear. He existed on biscuits and soda water for days and took no animal food. According to memoirs written by acquaintances, the poet had a "horror of fat"; to his mind, fat symbolized lethargy, dullness, and stupidity. Byron feared that if he ate normally he would lose his creativity. Only through abstinence could his mind exercise and improve. In short, Byron was a model of exquisite slenderness and his sensibilities about fat were embraced by legions of young women.[60]

Adults, especially physicians, lamented Byron's influence on youthful Victorians. In addition to encouraging melancholia and emotional volatility, Byronism had consequence for the eating habits of girls. In Britain "the dread of being fat weigh[ed] like an incubus" on Romantic youngsters who consumed vinegar "to produce thinness" and swallowed rice "to cause the complexion to become paler."[61] According to American George Beard, "our young ladies live all their growing girlhood in semi-starvation" because of a fear of "incurring the horror of disciples of Lord Byron."[62] Byronic youth, in imitation of their idol, disparaged fat of any kind, a practice which advice writers found detrimental to their good health. "If plump, [the girl] berates herself as a criminal against refinement and aesthetic taste; and prays in good or bad earnest, for a spell of illness to pull her down."[63] Other doctors besides Beard spoke of the popular Romantic association between scanty eating, a slim body, and "delicacy of mind." Beard, however, did not let the blame for modern eating habits rest entirely on the Romantics. He decried the influence of Calvinist doctrine as well. Cultivated people, he said, eat too little because of the old belief that "satiety is a conviction of sin."[64]

Women attuned to the higher senses did find inspiration for their abstemious eating in the austerities of medieval Catholics—particularly Catherine of Siena. Although Protestant Victorian writers presented Catherine's asceticism as a dangerous form of self-mortification, there was also widespread admiration for the spiritual intensity that drove her fasts. Victorian writers used the biography of Saint Catherine to demonstrate how selfhood could be lost to a higher moral or spiritual purpose. This message was considered particularly relevant to girls, in that self-love was supposed to be a distinguishing characteristic of the female adolescent.[65] Saint Catherine's biography was included in inspirational books for girls, and two prominent women of the period demonstrated serious interest in her life. Josephine Butler, an articulate English feminist, published a full-length biography in 1879; and Vida Scudder, a Wellesley College professor of English and a Christian socialist, published her letters in 1895.[66] Because she provided a vivid demonstration of a woman who placed spiritual over bodily concerns, Catherine of Siena was of enormous interest to Anglo-American women.

This lingering ascetic imperative did not go unnoticed by one of the period's most astute observers of religious behavior, William James. In *The Varieties of Religious Experience* the Harvard philosopher and psychologist noted quite correctly that old religious habits of "misery" and "morbidness" had fallen into disrepute. Those who pursued "hard and painful" austerities were regarded, in the modern era, as "abnormal." Yet James noted that young women were the most likely to remain tied to the dying tradition of religious asceticism. Although he understood that ascetic behavior had many sources (what he called "diverse psychological levels"), he did mark girls as the group most likely to embrace "saintliness." "We all have some friend," James wrote, "perhaps more often feminine than masculine, and young than old, whose soul is of this blue-sky tint, whose affinities are rather with flowers and birds and all enchanting innocencies than with dark human passions."[67] Girls seemed to be the most interested in the tenets of what James called saintliness: conquering the ordinary desires of the flesh, establishing purity, and taking pleasure in sacrifice.[68]

Those who were ascetic in girlhood tried to act as well as look like saints. In *The Morgesons* (1862) Veronica, an adolescent invalid and dyspeptic, defined her saintliness through a diet of tea and dry toast. She cropped her hair short in the manner of a penitent; constantly washed her hands with lavender water, as if she were taking a ritual absolution; and on her bedroom wall hung a picture of the martyred Saint Cecilia with white roses in her hair.[69] Although Veronica was a Protestant, she revered Saint Cecilia for her spirituality. Many novelists linked asceticism to physical beauty as well as to spiritual perfection. In short, beautiful women were often "saintlike," a relationship that implied the inverse as well—the "saintlike" were beautiful. Trollope, for example, spoke of a young gentleman who "declared to himself at once that she was the most lovely young woman he had ever seen. She had dark eyes, and perfect eyebrows, and a face which, either for colour or lines of beauty, might have been taken for a model for any female saint or martyr."[70]

By the last decades of the nineteenth century, a thin body symbolized more than just sublimity of mind and purity of soul. Slimness in women was also a sign of social status. This phenomenon, noted by Thorstein Veblen in *The Theory of the Leisure Class,* heralded the demise of the traditional view that girth in a woman signaled prosperity in a man. Rather, the reverse was true: a thin, frail woman was a symbol of status and an object of beauty precisely because she was unfit for productive (or reproductive) work. Body image rather than body function became a paramount concern.[71] According to Veblen, a thin woman signified the idle idyll of the leisured classes.

By the turn of the twentieth century, elite society already preferred its women thin and frail as a symbol of their social distance from the working classes. Consequently, women with social aspirations adopted the rule of slenderness and its related dicta about parsimonious appetite and delicate food. Through restrictive eating and restrictive clothing (that is, the corset), women changed their bodies in the name of gentility.

Women of means were the first to diet to constrain their appetite, and they began to do so before the sexual and fashion revolutions of the 1920s and the 1960s. In the 1890s Veblen noted that privileged women "[took] thought to alter their persons, so as to conform more nearly to the instructed taste of the time."[72] In effect, Veblen documented the existence of a critical gender and class imperative born of social stratification. In bourgeois society it became incumbent upon women to control their appetite in order to encode their body with the correct social messages.[73] Appetite became less of a biological drive and more of a social and emotional instrument.

Historical evidence suggests that many women managed their food and their appetite in response to the notion that sturdiness in women implied low status, a lack of gentility, and even vulgarity. Eating less rather than more became a preferred pattern for those who were status conscious. The pressure to be thin in order to appear genteel came from many quarters, including parents. "The mother, also, would look upon the sturdy frame and ruddy cheeks as tokens of vulgarity."[74] Recall that Eva Williams, admitted to London Hospital in 1895 for treatment of anorexia nervosa, told friends that it was her mother who complained about her rotundity.

A controlled appetite and ill health were twin vehicles to elevated womanhood. Advice to parents about the care of adolescent daughters regularly included the observation that young women ate scantily because they denigrated health and fat for their declassé associations. In 1863 Hester Pendleton, an American writer on the role of heredity in human growth, lamented the fact that the natural development of young women was being affected by these popular ideas. "So perverted are the tastes of some persons," Pendleton wrote, "that delicacy of constitution is considered a badge of aristocracy, and daughters would feel themselves deprecated by too robust health."[75] Health in this case meant a sturdy body, a problem for those who cultivated the fashion of refined femininity. One writer felt compelled to assert: "Bodily health is never pertinently termed 'rude.' It is not coarse to eat heartily, sleep well, and to feel the life throbbing joyously in heart and limb."[76]

Consequently, to have it insinuated or said that a woman was robust constituted an insult. This convention was captured by Anthony Trollope in *Can You Forgive Her?* (1864). After a late-night walk on the grounds of the Palliser estate, the novel's genteel but impoverished young heroine, Alice Vavasour, was criticized by a male guest for her insensitivity to the physical delicacy of her walking companion and host, Lady Glencora Palliser. The youthful and beautiful Lady Glencora caught cold from the midnight romp, but Alice did not. The critical gentleman immediately caught the social implications of the fact that Alice was not unwell from the escapade, and he used her health against her: "Alice knew that she was being accused of being robust . . . but she bore it in silence. Ploughboys and milkmaids are robust, and the accusation was a heavy one."[77] The same associations were relevant thirty years later in the lives of middle-class American girls. Marion Harland observed that the typical young woman "would be disgraced in her own opinion and lost caste with her refined mates were she to eat like a ploughboy."[78]

In the effort to set themselves apart from plowboys and milkmaids—that is, working and rural youth—middle-class daughters chose to pursue a body configuration that was small, slim, and essentially decorative. By eating only tiny amounts of food, young women could disassociate themselves from sexuality and fecundity and they could achieve an unambiguous class identity. The thin body not only implied asexuality and an elevated social address, it was also an expression of intelligence, sensitivity, and morality. Through control of appetite Victorian girls found a way of expressing a complex of emotional, aesthetic, and class sensibilities.

By 1900 the imperative to be thin was pervasive, particularly among affluent female adolescents. Albutt wrote in 1905, "Many young women, as their frames develop, fall into a panic fear of obesity, and not only cut down on their food, but swallow vinegar and other alleged antidotes to fatness."[79] The phenomenon of adolescent food restriction was so widespread that an advice writer told mothers, "It is a circumstance at once fortunate and notable if [your daughter] does not take the notion into her pulpy brain that a healthy appetite for good substantial food is 'not a bit nice,' 'quite too awfully vulgar you know.'"[80]

Because food was a common resource in the middle-class household, it was available for manipulation. Middle-class girls, rather than boys, turned to food as a symbolic language, because the culture made an important connection between food and femininity and because girls' options for self-expression outside the family were limited by parental concern and social convention. In addition, doctors and parents expected adolescent girls to be finicky and restrictive about their food. Young women searching for an idiom in which to say things about themselves focused on food and the body. Some middle-class girls, then as now, became preoccupied with expressing an ideal of female perfection and moral superiority through denial of appetite. The popularity of food restriction or dieting, even among normal girls, suggests that in bourgeois society appetite was (and is) an important voice in the identity of a woman. In this context anorexia nervosa was born.

NOTES

1. The best statement in the extensive literature on the relation of culture to obsessive-compulsive disorder and anorexia nervosa is Albert Rothenberg, "Eating Disorder as a Modern Obsessive-Compulsive Syndrome," *Psychiatry* 49 (February 1986), 45–53. Another historian interested in the changing symptomatology of hysteria is Edward Shorter. See his "The First Great Increase in Anorexia Nervosa," *Journal of Social History* 21 (Fall 1987), 69–96.
2. This point is made by Laurence Kirmayer, "Culture, Affect and Somatization, Part II," *Transcultural Psychiatric Research Review* 21 (1984), 254.
3. London Hospital Physician's Casebooks, MS 107 (1897); quoted in Pierre Janet, *The Major Symptoms of Hysteria* (New York, 1907), p. 234; Max Wallet, "Deux cas d'anorexie hystérique," in *Nouvelle iconographie de la Salpêtrière,* ed. J.-M. Charcot (Paris, 1892), p. 278.
4. See Janet, *Major Symptoms,* p. 234, for these terms. Janet did, however, have a more sophisticated interpretation based on his idea of "body shame."
5. My description of late-nineteenth-century examinations is drawn from clinical case records and from Joel Stanley Reiser, *Medicine and the Reign of Technology* (Cambridge, Mass., 1978), and Charles E. Rosenberg, "The Practice of Medicine in New York a Century Ago," *Bulletin of the History of Medicine* 41 (May–June 1967), 223–253; D. W. Cathell, *The Physician Himself and What He Should Add to His Scientific Acquirements* (Baltimore, 1882).
6. I have tried to incorporate, and move beyond, the perspective on male doctor–female patient relations provided in Barbara Ehrenreich and Deirdre English, *Witches, Midwives and Nurses* (New York, n.d.) and *Complaints and Disorders: The Sexual Politics of Sickness* (Old Westbury, N.Y., 1973).

   At the outset I asked, Can the absence of a patient voice be attributed solely to misogyny and authoritarianism on the part of doctors? I think not. Without discounting the relevance of both these well-known attributes of nineteenth-century medical men, I posit that the silence stemmed from a complex of medical and social factors that shaped interactions between doctors and patients in such a way that doctors really had no coherent, firsthand information to give. In a system of medicine that emphasized physical diagnosis, somatic complaints were primary. Traditional ideas about hysteria in women and girls prevailed, providing the rationale for medical and moral therapy.

7. J.-M. Charcot, *Clinical Lectures on Diseases of the Nervous System* (London, 1889), p. 214.
8. See for example John Ogle, "A Case of Hysteria: 'Temper Disease,'" *British Medical Journal* (July 16, 1870), 59; William Gull, "Anorexia Nervosa (Apepsia Hysterica, Anorexia Hysterica)," *Transactions of the Clinical Society of London* 7 (1874), 22–28; Thomas Stretch Dowse, "Anorexia Nervosa," *Medical Press and Circular* 32 (August 3, 1881), 95–97, and ibid. (August 17, 1881), 147–148; W. J. Collins, "Anorexia Nervosa," *Lancet* (January 27, 1894), 203.
9. Charles Féré, *The Pathology of Emotions* (London, 1899), pp. 79–80. The home was designated a Maison de Santé.
10. Ogle, "A Case of Hysteria," pp. 57–58.
11. The strongest statement on the separation of male and female spheres is Carroll Smith-Rosenberg, "The Female World of Love and Ritual: Relations between Women in Nineteenth-Century America," *Signs 1* (1975), 1–29. In addition to this generative article see Nancy Cott, *The Bonds of Womanhood* (New Haven, 1977); Ann Douglas, *The Feminization of American Culture* (New York, 1977); Mary Ryan, *Cradle of the Middle Class* (New York, 1982). Ellen Rothman's *Hands and Hearts: A History of Courtship in America* (New York, 1984) provides some rethinking of the Smith-Rosenberg thesis.
12. John Ryle to Parkes Weber, January 27, 1939, PP/FDW F. Parkes Weber Papers, Wellcome Institute, London.
13. Ann Douglas Wood, "The Fashionable Diseases: Women's Complaints and Their Treatment in Nineteenth Century America," *Journal of Interdisciplinary History* 4 (1973), 25–52; John S. Haller, Jr., and Robin Haller, *The Physician and Sexuality in Victorian America* (Urbana, Ill., 1974). See also Judith Walzer Leavitt, ed., *Women and Health in America: Historical Readings* (Madison, Wis., 1984).
14. Augustus Hoppin, *A Fashionable Sufferer; or, Chapters from Life's Comedy* (Boston, 1883), pp. 16–17, 35, 55, 135.
15. Marion Harland, *Eve's Daughters; or, Common Sense for Maid, Wife, and Mother*, with an introduction by Sheila M. Rothman (Farmingdale, N.Y., 1885; reprint ed., 1978), pp. 135, 153.
16. See for example Elizabeth Stoddard, *The Morgesons* (New York, 1862). Stoddard's novel depended in large part on the contrast between two adolescent sisters in a wealthy New England family. Thirteen-year-old Veronica was in a premature adolescent decline and remained at home, isolated from society; sixteen-year-old Cassandra was vigorous and "about in the world." Cassandra served as narrator of the story and presented many intimate details of the behavior of her ailing and dyspeptic sister.

    Stoddard never told the reader what was exactly wrong with the invalid Veronica, but she appeared to combine both dyspepsia and anorexia. Cassandra explained: "Delicacy of constitution the doctor called the disorder. She had no strength, no appetite, and looked more elfish than ever. She would not stay in bed, and could not sit up, so father had a chair made for her, in which she could recline comfortably." Much of the trouble with Veronica revolved around her lack of appetite, complicated by a simultaneous interest in food preparation and in the consumption habits of others. From her chair she directed the family maid to cook elaborate dishes that she would not eat.

    For long periods of time Veronica Morgeson ate only a single kind of food. "As we began our meal," Cassandra recounted, "Veronica came in from the kitchen with a plate of toasted crackers. She set the plate down, and gravely shook hands with me, saying she had concluded to live entirely on toast, but supposed I would eat all sorts of food as usual." Although Veronica ate virtually nothing at the family table, she asked many questions about food and cast aspersions on her sister's normal appetite. Whenever Cassandra said she was hungry, Veronica's eyes "sparkled with disdain." Above and beyond her physical problems with digestion, Veronica Morgeson took some strange emotional delight in her denial of hunger. Her reclusive existence, her listless appetite, and her sensitive stomach all implied that her body was subordinated to other, higher, more spiritual concerns.
17. Bernard Hollander, *Nervous Disorders of Women* (London, 1916), p. 77.
18. For the general history of chlorosis see Frank Panettiere, "What Ever Happened to Chlorosis?" *Alaska Medicine* 15 (May 1973), 68–70; Eugene Stransky, "On the History of Chlorosis,"

Episteme 8: 1 (1974), 26–46; and Ronald E. McFarland, "The Rhetoric of Medicine: Lord Herbert's and Thomas Carew's Poems of Green Sickness," *Journal of the History of Medicine* 30 (July 1975), 250–258. For more perceptive in-depth studies of Great Britain see Karl Figlio, "Chlorosis and Chronic Disease in Nineteenth Century Britain: The Social Constitution of Somatic Illness in a Capitalist Society," *Social History* 3 (1978), 167–197; and lrvine Loudon, "Chlorosis, Anemia, and Anorexia Nervosa," *British Medical Journal* 281 (December 20–27, 1980), 1669–75; and idem, "The Diseases Called Chlorosis," *Psychological Medicine* 14 (1984), 27–36. For the United States see R. P. Hudson, "The Biography of Disease: Lessons Learned from Chlorosis," *Bulletin of the History of Medicine* 51 (1977), 448–463; S. R. Huang, "Chlorosis and the Iron Controversy: An Aspect of Nineteenth-Century Medicine," Ph.D. diss., Harvard University, 1978; Joan Jacobs Brumberg, "Chlorotic Girls, 1870–1920: An Historical Perspective on Female Adolescence," *Child Development* 53 (December 1982), pp. 1468–77; A. C. Siddall, "Chlorosis—Etiology Reconsidered," *Bulletin of the History of Medicine 56* (1982), 254–260.

19. J. H. Montgomery, *Clinical Observations on Cases of Simple Anemia or Chlorosis Occurring in Young Women in the Decade Following Puberty* (Erie, Pa., 1919), n.p.
20. *Elmira Daily Gazette and Free Press,* March 26, 1898. On women and the patent medicine business see Sarah Stage, *Female Complaints: Lydia Pinkham and the Business of Women's Medicine* (New York, 1979).
21. Loudon, "Chlorosis," posits that after 1850 chlorosis incorporated at least three different kinds of disorders common in young women, each involving loss of weight, anemias, and amenorrhea. He argues that chlorosis was a functional disorder closely related to anorexia nervosa: they were two "closely related conditions, each a manifestation of the same type of psychological reaction to the turbulence of puberty and adolescence" (p. 1675). Loudon is correct that the two diseases had much in common, and his observation probably means that there was more anorexia nervosa than heretofore thought. Anorexia nervosa was usually a class-specific diagnosis, however. Given the family dynamic that was part of the disorder, working-class girls were unlikely to develop anorexia nervosa. Moreover, if a similar pattern of noneating developed in a working-class girl, family reverberations were different and medicine was more likely to call the disorder chlorosis or depression.
22. G. Stanley Hall, *Adolescence,* vol. 1 (New York, 1904), pp. 252–253.
23. Lillie A. Williams, "The Distressing Malady of Being Seventeen Years Old," *Ladies' Home Journal* (May 1909), p. 10.
24. James Henry Bennet, *Nutrition in Health and Disease* (London, 1877), pp. 59–60, 170. Among others who noted that children and women had more easily disturbed digestive systems are Elizabeth Blackwell, *The Laws of Life* (New York, 1852), and Anna Brackett, *The Education of American Girls* (New York, 1874).
25. Edward Smith, *Practical Dietary for Families, Schools, and the Labouring Classes* (London, 1864), p. 141.
26. These descriptions are from Harland, *Eve's Daughters,* pp. 111–113; see also Charles E. Simon, "A Study of Thirty-One Cases of Chlorosis," *American Journal of Medical Sciences* 113 (April 1897), 399–423; Lucien Warner, *A Popular Treatise on the Functions and Diseases of Women* (New York, 1875), p. 70; E. H. Ruddock, *The Lady's Manual of Homeopathic Treatment* (New York, 1869), p. 32.
27. Simon, "Thirty-One Cases," pp. 413–414.
28. Mary Wood-Allen, *What a Young Girl Ought to Know* (Philadelphia, 1905), p. 89.
29. Harland, *Eve's Daughters,* pp. 111–115. The notion of an adolescent decline had some foundation. Among adolescents tuberculosis was a particularly serious threat with a high mortality rate. Most people agreed that once a child passed beyond infancy and early childhood, adolescence stood as the next critical juncture in the life course. Adolescence required caution, whether one was male or female, but physical decline seemed to occur more often among girls. For a summary of a number of different statements about the vulnerability of the female adolescent see Nellie Comins Whitaker, "The Health of American Girls," *Popular Science Monthly* 71 (September 1907), 240.
30. Harvey W. Wiley, *Not By Bread Alone* (New York, 1915), pp. 245, 248–250, 256; Brackett, *Education of American Girls,* pp. 25–26.
31. Warner, *A Popular Treatise,* p. 54.
32. On meat eating and sexual excess see Vern Bullough and Martha Voight, "Women, Men-

struation and Nineteenth Century Medicine,' *Bulletin of the History of Medicine* 47 (1973), 66–82. According to many physicians, flesh eating contributed to a "neurotic temperament." T. S. Clouston, superintendent of the Edinburgh asylum, espoused a widely held view: "I have found . . . a large proportion of the adolescent insane h[ave] been flesh-eaters, consuming and having a craving for much animal food"; see his "Puberty and Adolescence Medico-Psychologically Considered," *Edinburgh Medical Journal* 26 (July 1880), 17. This article was reprinted in the *American Journal of Insanity* in April 1881.

33. See Albert J. Bellows, *The Philosophy of Eating* (New York, 1869), for a typical statement about meat by a health reformer. And see Stephen Nissenbaum, *Sex, Diet and Debility in Jacksonian America: Sylvester Graham and Health Reform* (Westport, Conn., 1980) on nineteenth-century vegetarianism.
34. E. L. Jones, *Chlorosis: The Special Anemia of Young Women* (London, 1897), p. 39; Charles Meigs, *Females and Their Diseases* (Philadelphia, 1848), p. 361; Montgomery, *Clinical Observations.* Susan Williams, *Savory Suppers and Fashionable Feasts: Dining in Victorian America* (New York, 1985), notes that rare or "underdone meat" was "out of fashion" and particularly "disgusting" to women and children (p. 239).
35. Nellie Browne to her mother [April 1859?], Sarah Ellen Browne papers, Schlesinger Library. Jane Hunter called this letter to my attention.
36. J. Clifford Allbutt, *A System of Medicine,* vol. 5 (New York, 1905), p. 517.
37. "The Antagonism between Sentiment and Physiology in Diet," *Current Literature* 42 (February 1907), 222. In reporting on anorexia nervosa, Pierre Janet described the phenomenon of *la crainte d'engraisser*—literally, the fear of taking on grease.
38. Quoted in Claude Fischler, "Food Preferences, Nutritional Wisdom, and Sociocultural Evolution," in *Food, Nutrition and Evolution,* ed. Dwain Watcher and Norman Kretchmer (New York, 1981), p. 58.
39. I borrow the term "subtext" from literary criticism, particularly from the semioticians. Food in nineteenth-century fiction and culture serves as a set of signs and symbols with communicative power. See Roland Barthes, "From Work to Text," in *Textual Strategies: Perspectives in Post-Structuralist Criticism,* ed. J. Harari (Ithaca, N.Y., 1979), pp. 73–81.
40. Williams, *Savory Suppers,* chap. 1, describes a pattern of middle-class concern over eating and correct eating behaviors. See also Jocelyne Kolb, "Wine, Women, and Song: Sensory Referents in the Works of Heinrich Heine," Ph.D. diss., Yale University, 1979, p. 8.
41. Janet, *Major Symptoms,* p. 234. Kolb, "Wine, Women, and Song," writes, "In the early nineteenth century, when Heine was writing, the mere mention of food in a lyrical passage was generally shocking enough to achieve ironic distance" (p. 71). The Romantic conception of food as an emblem of both the positive and negative sides of sensuality continued into the late nineteenth century.
42. Harland, *Eve's Daughters*, p. 81.
43. Although middle-class women were frequently assisted by a household servant, resulting in some reduction of time spent in the kitchen, they spent more and more energy planning meals, purchasing food, and determining ways to make eating an aesthetic experience. See Ruth Schwartz Cowan, *More Work for Mother: The Ironies of Household Technology from the Open Hearth to the Microwave* (New York, 1983).
44. George Everett and Susan Everett, *Health Fragments; or, Steps Toward a True Life* (New York, 1875), p. 35.
45. Fannie Munn Field diary, December 17, 1886, Box 9:18, Munn-Pixley Papers, Department of Rare Books and Special Collections, Rush Rhees Library, University of Rochester, Rochester, N.Y. Susan Williams kindly brought this example to my attention.
46. Everett and Everett, *Health Fragments*, p. 25; Sarah Josepha Hale, *Receipts for the Million* (Philadelphia, 1857), p. 509.
47. [Society for Promoting Christian Knowledge], *Talks to Girls by One of Themselves* (London, 1894), p. 104.
48. Everett and Everett, *Health Fragments,* pp. 26, 29; Leslie A. Marchand, *Byron: A Portrait* (New York, 1970), p. 386.
49. George Eliot, *Daniel Deronda* (New York, n.d.), p. 104.
50. Jane Austen, *Mansfield Park* (New York, 1963), pp. 311–312; Anthony Trollope, *Ralph the Heir* (London, 1871), p. 195.
51. Mark Twain, *The Prince and the Pauper* (New York, 1881).

52. Elizabeth Gaskell, *Cranford* (New York, 1906), pp. 41, 53. Oranges were problematic for American eaters also; see Williams, *Savory Suppers,* pp. 108–109.
53. Eliot, *Daniel Deronda,* p. 104. The actual text of Byron's statement is, "A woman should never be seen eating or drinking, unless it be lobster salad and champagne, the only truly feminine and becoming viands" (Marchand, *Byron,* p. 133). In William Makepeace Thackeray's *History of Pendennis* (London, 1848–50) the character of Blanche is delineated by her peculiar appetite and secret eating: "When nobody was near, our little sylphide, who scarcely ate at dinner more than six grains of rice . . . was most active with her knife and fork, and consumed a very substantial portion of mutton cutlets: in which piece of hypocrisy it is believed she resembled other young ladies of fashion." (Quoted in Ann Alexandra Carter, "Food, Feasting, and Fasting in the Nineteenth Century British Novel," Ph.D. diss., University of Wisconsin, 1978, p. 3.)
54. Anthony Trollops, *Can You Forgive Her?* (New York, 1983), p. 70.
55. Sanford Bell, "An Introductory Study of the Psychology of Foods," *Pedagogical Seminary* 9 (1904), 88–89. In *Adolescence,* vol. 2 (New York, 1907), pp. 14–15, G. Stanley Hall noted that appetite varied a great deal in adolescence in response to "psychic motives." Hall did not differentiate between male and female adolescents, although Bell certainly did.
56. Kolb, "Wine, Women, and Song," suggests that this idea of higher and lower senses was inherited from the eighteenth century—specifically from Chevalier de Jancourt, who wrote on the subject in the famed *Encyclopédie.*
57. Harland, *Eve's Daughters*, p. 153.
58. On Byron's influence among the Victorians see Donald David Stone, *The Romantic Impulse in Victorian Fiction* (Cambridge, Mass., 1980).
59. Lizzie Eustace in Anthony Trollope's *Eustace Diamonds,* quoted in Stone, *Romantic Impulse*, p. 51.
60. On Byron's life and his struggles with food and eating see Edward John Trelawny, *Records of Shelly, Byron and the Author*, ed. David Wright (London, 1973), pp. 11, 35–36, 86, 97–98, 245. Mary Jacobus pointed out to me that Trelawny, notoriously "pro-Shelley," is a less than totally disinterested source of information on Byron's dieting. It is significant that Shelley too was a picky eater and a vegetarian, contributing further to the romance of undereating.
61. J. Milner Fothergill, *The Maintenance of Health* (London, 1874), pp. 80–81. This report of girls swallowing vinegar is not anomalous. See Brumberg, "Chlorotic Girls."
62. George Beard, *Eating and Drinking* (New York, 1871), p. 104.
63. Harland, *Eve's Daughters*, p. 124.
64. Beard, *Eating and Drinking*, p. v. Beard connected the American propensity for scanty eating to health reformers of an evangelical bent. He spoke of the "vast army of Jeremiahs who have gone up and down the land, predicting that gluttony will be our ruin" (p. iv). The links between parsimonious eating and religiosity are also suggested in Anthony Trollope, *Rachel Ray* (London, 1880). The plot of this novel is interesting because of the contrast between Mrs. Ray, a loving mother who likes tea and buttered toast, and her austere daughter, Mrs. Prime, an asexual, pious, churchgoing widow who likes "her tea to be stringy and bitter" and "her bread stale" (p. 5). The older women who influence Rachel Ray are defined by their appetite and eating behavior.
65. See for example Clouston, "Puberty and Adolescence," p. 14.
66. H. Davenport Adams, *Childlife and Girlhood of Remarkable Women* (New York, 1895), chap. 5; Josephine Butler, *Catherine of Siena* (London, 1878); Vida Scudder, *Letters of St. Catherine of Siena* (New York, 1905).
67. William James, *The Varieties of Religious Experience,* with an introduction by Reinhold Niebuhr (New York, 1961), pp. 79, 221, 238–240. James observed that as a consequence of secularization painful austerities or asceticism were not considered abnormal. "A strange moral transformation has within the past century swept over our Western world. We no longer think we are called on to face physical pain with equanimity. It is not expected of a man that he should either endure it or inflict much of it, and to listen to the recitals of cases of it makes our flesh creep morally as well as physically . . . The result of this historical alteration is that even in the Mother Church herself, where ascetic discipline has such a fixed traditional prestige as a factor of merit, it has largely come into desuetude, if not discredit. A believer who flagellates or 'macerates' himself today arouses more wonder and fear than

emulation" (p. 238). On asceticism see also Emile Durkheim, *The Elementary Forms of Religious Life*, trans. Robert Nisbet (London, 1976), pp. 299–321.

68. In the twentieth century many ideological followers of G. S. Hall spoke of the idealism of adolescents, that is, their search for moral purity. Asceticism in adolescence is rarely discussed, however. An interesting article that makes this connection and also distinguishes between adaptive and pathological asceticism is S. Louis Mogul, "Asceticism in Adolescence and Anorexia Nervosa," *Psychoanalytic Study of the Child* 35 (1980), 155–175. Mogul is reliant on Anna Freud, *The Ego and the Mechanisms of Defense* (London, 1937).
69. See note 16 above and Stoddard, *The Morgesons*, pp. 30, 57, 61, 140.
70. Trollope, *Ralph the Heir*, p. 29.
71. Thorstein Veblen, *The Theory of the Leisure Class* (New York, 1967).
72. Ibid., pp. 145–149.
73. My argument about the ways in which cultural and class concerns are encoded in the body follows from Michel Foucault, *History of Sexuality*, vol. 1 (New York, 1980), and *Madness and Civilization* (New York, 1965). The discipline of the body and its relation to social theory is explored by Brian Turner, "The Discourse of Diet," *Theory, Culture and Society* 1: 1 (1982), 23–32.
74. Hester Pendleton, *Husband and Wife; or, The Science of Human Development through Inherited Characteristics* (New York, 1863), p. 66.
75. Ibid., pp. 65–66.
76. Harland, *Eve's Daughters,* p. 134.
77. Trollope, *Can You Forgive Her?* p. 297.
78. Harland, *Eve's Daughters,* p. 111.
79. Allbutt, *A System of Medicine,* vol. 3, p. 485.
80. Harland, *Eve's Daughters,* p. 111.

14

Conflict and Deference

MARJORIE DEVAULT

Studies of contemporary couples suggest that equal sharing of family work is quite rare (e.g. Hood 1983; Hertz 1986; Hochschild 1989). In most families, women continue to put family before their paid jobs, and take primary responsibility for housework and child care. Indeed, some studies suggest that husbands require more work than they contribute in families: Heidi Hartmann (1981a) shows that working women with children and husbands spend more time on housework than single mothers, whether their husbands "help" with the work or not. Some analysts suggest that women's responsibility for housework persists because men's careers are typically more lucrative than women's, but Berk's study of the determinants of couples' family work patterns (1985) shows that these decisions do not depend on economics alone. In two-paycheck families, women continue to do more household work than men, even when such a pattern is not economically rational. Berk's explanation is that the "production of gender"—of a sense that husband and wife are acting as "adequate" man and woman—takes precedence over the most economically efficient production of household "commodities." Her conclusion, arrived at through statistical analysis of a large sample of households, is consistent with clues from the speech of informants in this study, who connected cooking with "wife" and asserted the importance of "a meal made by (a) mother."

This chapter explores the effect of these gendered expectations on the expression of conflict and on relations of service and deference. In order to take account of the diversity of families in this period of change, we must consider several different patterns, ranging from families attempting to share household work to those where relations of male dominance and female submission are enacted and enforced physically, through violence and abuse. I will suggest that expectations of men's entitlement to service from women are powerful in most families, that these expectations often thwart attempts to construct truly equitable relationships and sometimes lead to violence. I do not mean to argue that husbands are all tyrannical or that marriages aiming at egalitarianism do not represent significant change. Rather, my aim is to identify the varied but powerful effects of taken-for-granted beliefs and expectations about gender, and to begin to confront these expectations as barriers to change.

PROBLEMS OF SHARING

Even when husbands and wives are committed to the idea of sharing household responsibilities, the character of family work contributes to an asymmetry in effort and attention to household needs. Perhaps because the household routine is such a coherent whole, it often seems easiest for one person to take responsibility for its organization, even when others share the actual work. This person—traditionally, the housewife—is the one who keeps an entire plan in mind. In the households I studied, three men shared cooking with their wives, and all three reported that for the most part they take directions from their wives. Ed, a psychologist who has begun to do more and more cooking for his family, still commented:

> It's not my domain. I have the minute to minute decisions, because I'm the one who's here, but she's really the one who decides things. I just carry out the decisions. It feels more like my domain now, because I've been doing so much more than I used to, but she's still in charge. She's the organized one.

He perceives her as "organized"; in fact, it is her activity that produces this perception, since even with his increased participation in carrying out tasks, she is the one who does the organizing. Similarly, Robin, whose husband Rick does virtually all the cooking, reported that she was "the supervisor":

> He handles the greater portion of, you know, taking care of stuff around the house. I tell him. I say, "OK, we need this, this, and this done," and that's what he does. Like I say, "You better go to the store, we need some milk," or "We need the laundry done," and he'll do it. He's the employee, and I'm the supervisor.

Rick agreed. When I asked if "keeping track of things" was a large part of his work, he responded:

> You're talking to the wrong person. Robin tells me when I've got to remember stuff. She's—if she could be a computer, she'd have the greatest memory bank in the world. I'm scatterbrained. And I don't remember a lot of things, I'll let things run down and just let things go completely out . . . She'll keep a list in her head for a week.

Though he is quite comfortable and skilled at doing the discrete tasks of cooking, he seems not to have learned the practices associated with responsibility for feeding; she, rather than he, does the work of keeping things in mind.

The fact that standards and plans for housework are typically unarticulated also makes it difficult for husbands and wives to share work in the family. The houseworker defines the work as part of an overall design for the household routine, but this design is only partially conscious. Phyllis, a white single mother, complained, "Once you've got the whole system in your head, it's very hard to translate that into collective work, I think." She has tried to share the work with her daughters, but finds it difficult:

> We once made schedules. There was probably something in the paper about that. And I tried to make—you know, we would all take a chore, and write it down and do it. And you know something? When you work all day, and come home, it's almost easier to do it than to have to supervise other people doing something they don't want to do.

Like a manager, Phyllis is responsible for planning and overseeing a range of activities involving others. At first glance, the problem—to get the children to do some of the work—seems similar to the problem of supervision in the workplace. However, a wife or mother lacks the formal authority of a manager. Further, since family work is mostly invisible, household members learn not to expect to share it, and the woman in charge of a household may not even be able to specify fully the tasks to be divided. The invisibility of monitoring, for example, can make it difficult to share the work of provisioning, because one never knows whether another is thinking about what is needed. A woman whose husband shares some housework commented:

> I'll be surprised. Like just the other day, we had just a little bit of milk left. And I know he saw that, because he handled the milk carton. But he went to the store and he didn't buy milk.

Tasks such as planning and managing the sociability of family meals are also invisible, and since maintaining their invisibility is part of doing the work well, people are often unable, or reluctant, to talk explicitly about them.

Serious attempts to share housework require a great deal of communication among household members, because the overall design for the work must be developed, and then continually revised, collectively. Sometimes these attempts at equality provoke conflict over the definition of the work. The problems involved were evident in the comments of a black professional couple, Ed and Gloria, who have been moving slowly toward a more equal division of household work during the course of their marriage. They have had to redefine some activities. Gloria enjoys gardening and yardwork, for example, and used to consider them part of her housework. Her husband Ed is not so interested, and argues that these activities are "hobbies"; he explained, "I say, 'Look, don't be dumping on me because I'm not doing that, because it's not my hobby.'" When they told me about their routine, both of their accounts emphasized that their activities and decisions depended on a variety of factors, and had to be made practically from moment to moment—both kept repeating, "It depends." They must talk about these things, or find some other way of making decisions together. For example, he reported:

> She does more cooking, but every now and then she'll say that she wants me to do the cooking. And if I'm really uptight, I'll say, "Look, I'll go out and buy it [laughing]. You're working, we can spend the money." Otherwise I'll go ahead and do it.

She described unspoken decisions about cleaning up after dinner:

> That depends on how tired I am or how tired I feel he is. If I've had a hard week or a hard day I just leave, and I don't care, I just walk out. And either Ed does it or it's there in the morning. And then it depends on how much time I have. If I have to go straight out, then

> I don't know what happens. But if I have some time then I might clean up some dishes. And then if I'm feeling up, or if I think he's down, then I'll clean up after supper.

Making these decisions requires sensitivity, and when there are problems, explicit talk:

> We just feel each other out. We know each other so well that we can read each other. I know when he's uptight and he knows when I'm uptight. We don't really talk about it. The only time we talk is when we're not reading each other very well, and then one person starts to feel dumped on. Then we talk, we say "What's going on?" and we try to work it out.

Such comments illustrate the extent to which housework is typically hidden from view. When one person takes responsibility for the work, others rarely think about it. Even the one who does it—because so much of her thought about it is never shared—may not be fully aware of all that is involved. Her work can come to seem like a natural expression of caring.

When couples begin to share the work of care, its "workful" unnaturalness—the effort behind care—is necessarily exposed. The underlying principles of housework must be made visible. The work must be seen as separable from the one who does it, instead of in the traditional way as an expression of love and personality. Some couples do attempt to discuss the effort required to produce the kind of family life they desire. But as the next section will show, many people accept with little thought pervasive cultural expectations that connect relations of service with the definitions of "husband" and "wife," and "mother" and "father."

HUSBANDS AND WIVES

When I talked with women about their household routines, many of them spoke of their partners' preferences as especially compelling considerations. In discussions about planning meals, several women mentioned their husbands' wage work as activity that conferred a right to "good food": "He works hard," or "He is in a very demanding work situation." (And their more detailed comments indicate that they use the phrase "good food" here to mean "food that he likes.") These comments are consistent with the findings from studies of power within marriage, which suggest that paid employment brings power and influence within the family (Blood and Wolfe 1960). Family members easily recognize the importance of paid work, and Charles and Kerr (1988) found that many women rationalized their sole responsibility for cooking in terms of their husbands' wage work. However, beliefs about work and domestic service do not operate in the same way for men and women. While men who work are said to deserve service, women who work for pay are (at most) excused from the responsibility of providing that service; they are rarely thought entitled to service themselves (Murcott 1983). Studies of the impact of women's employment on family patterns suggest that men's work at home increases only slightly when their wives take jobs. Employed wives typically manage household tasks by redefining the work and doing less than they would if they were home. I will argue here that women's service for husbands is based on more than the importance of men's paid work. Women's comments about feeding reveal powerful, mostly unspoken beliefs about relations of dominance and subordination between men and women, and especially between husbands and wives. They show that women learn to think of service as a proper form of relation to men, and learn a discipline that defines "appropriate" service for men.

Rhian Ellis (1983) suggests that many incidents of domestic violence are triggered by men's complaints about the preparation and service of meals. She notes, for example, the expectation in some working-class subcultures that a wife will serve her husband a hot meal immediately when he returns from work, and she cites examples from several studies of battered women who report being beaten when they violated this expectation. Typical accounts portray husbands who return home hours late, sometimes in the middle of the night, and still expect to be served well and promptly. In other studies, researchers report that conflict can arise from men's complaints about the amount or quality of food they are served. Some of the researchers who report such incidents remark on the fact that violence can be triggered by such "trivial" concerns, but Ellis suggests that the activities of cooking and serving food in particular ways are in fact quite significant because they signal a wife's acceptance of a subservient domestic role and deference to her husband's wishes. In these situations, men insist on enforcing exaggerated relations of dominance and subordination within the family. We will see below that some of the patterns taken for granted in families without explicit violence are based on similar assumptions about women's deference to men's needs: assumptions that women should work to provide service and that men are entitled to receive it. Though battering represents an extreme version of inequity between husbands and wives, it highlights the significance of the observations that will follow, and suggests a vicious circle: the idea that some version of womanly deference is "normal" may contribute to an ideology of male entitlement that supports violence against wives and mothers.

The households I studied were not, to my knowledge, ones in which violence was prevalent. Only a few of the people I interviewed spoke of any discord with their spouses. But their reports of daily routines suggested that implicit marital "bargains" were often based on taken-for-granted notions of men's entitlement to good food and domestic service. It was clear, for example, that in most households, wives are very sensitive to husbands' evaluations of their cooking. Teresa, a young Chicana woman, described the pressure of learning to cook when she was first married and her husband and brother-in-law, who ate with them regularly, were both "judging" her:

> It was really a lot of strain, to make two men happy, who were judging you. You know, "You don't make it half as good as my mother did." So that kind of pushed me a little to learn fast.

After seven years of marriage, her husband still has a great deal of influence over her routine. He buys most of the meat for their household, since they both agree that he knows more about butchering and can purchase better meat than she does. He has high standards and communicates them clearly. When he goes shopping with her, Teresa reported, "I go through the canned food aisle, and he'll say, 'Why are you taking so many cans?'" Teresa does not think of her husband as particularly demanding. She not only accepts her husband's preferences, but also thinks of her occasional failures to satisfy them as "cheating":

> I do cheat a little and I—like beans take about an hour to make—so if I forget that I've run out of beans and I don't make any, I'll just open a can of beans and just warm them up.

But when he tastes them—I don't know, there must be something about the taste—he'll say, "These are canned beans, right?" No matter how hard I try to hide the can.

Teresa told me proudly that he likes the way she cooks and that he has gained so much weight that people tease him about what a good cook she mast be. But she is still anxious about judgments of her cooking:

Maybe that's why I still don't like cooking, because I know that every time I serve something on the table I'm going to be judged and criticized, and you know, "This tastes awful."
[MD: Do you really get that?]
No, no. But I'm afraid of it, right.

She describes a more relaxed kind of meal on the days when her husband works a second job and she eats casually with the children:

Now Saturday, my husband works on Saturdays, and that's the day that my kids are home also. So Saturday would be hot dog day . . . To them it's a big treat to have hot dogs. And to me it's another treat, because I don't have to go to the whole trouble of doing or preparing a whole dinner . . . [She explains that her husband works until 10 or 11 at night.] So that means the whole day, we don't have a father to yell over us. That's what Felix says, he says, "Oh, we don't have a dad to yell over us." So we'll have some kind of Campbell's soup, some kind of vegetable or chicken or something, out of a can.

Some writers have suggested that men are especially authoritarian in Hispanic families, and the pattern that Teresa describes—with "a dad to yell over" the family—is consistent with such an idea. However, I found instances of a similar attitude among women in all of the class and ethnic groups included in this study. While none of these women—including Teresa—described their own attitudes in terms of service or deference to husbands (in fact, many took some trouble to explain to me that they were not mere servants in their own households), they spoke of accepting husbands' demands in a matter-of-fact tone that illustrates the force of male preference. For example, one affluent white woman explained:

My husband doesn't like a prepared salad dressing. So I make my own. And he now is on a kick of having me make orange juice, rather than buying the frozen concentrate. So I'm going to have to go out and get an orange juice squeezer.

Her tone suggests that she takes for granted that his preferences should determine her work: that he is "on a kick" means that she will buy new equipment and adopt a new method for preparing juice. The same idea appears in one of Donna's comments, as she talked about the foods that she and her husband like:

I used to have pork chops three or four times a week. And then he just said, "I don't want pork chops and that's it." And I haven't bought them since.

She tells a small story here that conveys the drama involved in such mundane matters.

By casting her report in this way, she indicates how easily (and thus, we assume, how legitimately in her mind) he puts forward his claims: "he just said . . . And I haven't bought them since." Susan, remembering times when she had to scrimp and save, explains how things are different now: "If the man wants a steak, he gets it. Period. No questions asked." Again, the sense is of men's entitlement to make claims and have them met within the family.

A number of women referred to husbands as the sources for elaborated standards. For example, a recently divorced, white professional woman commented on how the change had affected her cooking; before the divorce, meals for her husband and children had been rather elaborate:

> The expectations of supper, you know, the big meal of the day. I knew that I had to have meat and potatoes, and you know, the usual fanfare. But you know, that really required me to be home, by five o'clock in the afternoon, to get all of this ready by six. But I knew that was expected of me.

After the divorce, food was "not a priority," and her routines became "very simple." She is concerned about nutrition, and carefully monitors her children's eating, but she no longer prepares elaborate meals:

> After my husband left, things got very simple with food. I found, you know, I didn't have to be in the kitchen, and shopping all of the time. I had always flirted with the idea. But you know, being married, you're just a slave to the kitchen. And once I got out of that, I just had more choices. I mean I had more flexibility in what I could do and couldn't do.

She does not explain exactly why her choices expanded; we do not know precisely how her husband influenced her decisions before the divorce. But she describes a shift from "the usual fanfare" to a "very simple" routine, and her language is clearly that of constraint and permission: her husband's presence—whether explicitly or through her own expectation—dictated what she "could do and couldn't do."

The fact that many women seem unaware of this tone of deference toward husbands, or at least are unable to articulate its basis, does not detract from its force. Indeed, for those who value some form of egalitarianism in their marriages, the requirement to serve a husband—which might be resisted if it were more explicit—is not necessarily diminished by its invisibility. One white woman, whose husband is an executive, provides an example. When she is home alone during the day, she eats very casually, and she talked about how difficult it is, when her husband occasionally works at home, to prepare a "really decent lunch." When I asked why he could not eat the same lunch that she does every day, she was unable to explain except on the basis of an inarticulate "feeling." She attempts to explain her concern as an issue of nutrition, but her talk about nonfattening food is peppered with references to meals that are "really decent" in some other way:

> It would be hard, if I had to feed him every day, to think up really decent lunches. I eat yogurt, and peanut butter and jelly, all kinds of fattening things that I shouldn't, and every day I say I shouldn't be doing this. But it's hard to think of a well-rounded, man-sized meal.

I asked what she might prepare for him, and she gave an example:

> All right. Yesterday, we thought our girls were coming out the night before and I had bought some artichokes for them, so I cooked them anyway. So I scooped out the center and made a tuna salad and put it in the center, on lettuce, tomatoes around. And then, I had made zucchini bread . . . So that was our lunch. If I had to do it every day I would find it difficult.
>
> [MD: When you make a distinction between the kind of lunch you would have and what you'd fix for him, what's involved in that? Is it because of what he likes, or what?]
>
> I just feel he should have a really decent meal. He would not like—well, I do terrible things and I know it's fattening. Like I'll sit down with yogurt and drop granola into it and it's great. Well, I can't give him that for lunch.
>
> [MD: Why not?]
>
> He doesn't, he wouldn't like it, wouldn't appreciate it. Or peanut butter and jelly, for instance, it's not enough of a lunch to give him.

This woman's discussion is quite confusing: when she is alone, she does "terrible things" that her husband "wouldn't appreciate." Yet she went on to report that in fact, he likes peanut butter and jelly and would probably enjoy a peanut butter sandwich. Still, it is "not enough of a lunch to give him." In fact, her explanations—the references to nutrition concerns and what he "appreciates" or not—obscure what seems the real point, that "not enough of a lunch," means simply that she has not done enough to prepare it. Her example shows how much trouble she takes to prepare a "decent" meal and serve it attractively. A "man-sized meal" may not be so much larger in quantity, but should be a meal she has worked on for him.

What we see in all of these comments are specific versions of a socially produced sense of appropriate gender relations, a sense that certain activities are associated with the very fundamental cultural categories "man"/"husband"/"father" and "woman"/"wife"/"mother." (Haavind [1984] discusses similar interpretive frames for marital exchanges among Norwegian couples.) These associations are learned early and enforced through everyday observation of prevailing patterns of gender relations; they are rarely justified or even articulated explicitly, but explicit statement is hardly necessary. For most people, these understandings have become part of a morally charged sense of how things should be, so that even those who strive for some version of equity are prey to their pervasive effects.

MEN WHO COOK

Perhaps the one who cooks tries, "naturally," to please the one he or she cooks for, regardless of gender. That is, in homes where men did as much cooking as women, perhaps the inequalities of service and deference found in typical family settings would be less pronounced. The possibility is one that is difficult to assess, precisely because gender is so strongly associated with activity. In the sample of households I studied, only three men cooked more than occasionally. Of these three, only one (Rick, a white transportation worker) has taken on almost all of the cooking, as women have typically done in traditional families; one (Ed, a black psychologist) prepares the dinners most evenings, but usually by finishing the preparation of foods his wife organizes during the previous weekend; and the third (a white professional worker) reported cooking about twice each week. Since even Rick reported that his wife is the one who organizes their routine, reminding him what needs to be done each day, none of these men has taken on the kind of sole responsibility for family work that has been the most common pattern for women, and none seems to do even half of the coordinative work of planning. In short, they cook, but they are only beginning to share the work of "feeding the family" in the broader sense I have been developing. These few men's reports cannot provide any definitive answers to questions about gender and household activity, but they are worth examining alongside their wives' accounts if only as suggestive pointers toward understanding how men's and women's understandings of family work might differ. Their comments indicate that they have begun to learn and practice household skills such as preparing specific foods, juggling schedules in order to bring the family together for a meal, and improvising with the materials at hand. However, their reports also suggest that they do not feel the force of the morally charged ideal of deferential service that appears in so many women's reports.

When women talked about what they cook, they frequently referred to husbands' and children's preferences as the fixed points around which they designed the meals. Such comments are mostly absent from the reports of these men, except for occasional references to particular foods that young children will not accept. Rick, for example, likes cooking partly because he can be inventive; his explanation emphasizes his own creativity more than the tastes of those he serves:

> I don't use any recipes. It's just by what I want to put in the thing. And if I like it . . . It's just whatever I have up here, what can I think of now?

He reported that his wife and child like his cooking, but only late in the interview, in the context of a longer exchange. When I asked him what kinds of cooking he did best, his answer again suggested freedom from external standards even as he mentioned his attention to his wife's tastes; he explained that he has not mastered baking, and then continued:

> Other than the meats and stuff, it's just my imagination, whatever I want to do with it. And whatever Robin, you know, I might ask her, like the taste of this, do you like it? And if she doesn't like it I won't put it in.

Picking up on this last comment, I attempted to probe for some elaboration of his feelings about the evaluations of others:

> [MD: So what about Robin and Kate? Do they think you're a pretty good cook?]
> Yeah. I surprise them. I surprise myself sometimes [laughing].
> [MD: Yeah. They don't have too many complaints, then?]
> No. If so, I don't know . . . [laughing and shaking his head as if the thought had never occurred to him].

Rick takes on more responsibility than the other men I talked with, and he does so even in the face of some pointed teasing from some of his working-class friends who disapprove. But even though he understands the urgent demands of children who need to be fed, he seems able to set limits for his efforts more comfortably than most of the women I talked with, and has more success saying no to the tasks he simply does not enjoy. Both he and Robin reported that she cooks when they entertain large groups; he explained that she cooks "if it's gonna be any effort out on cooking—you know, let's go, go for it—instead of just fixing one or two things like I do." And in contrast to the many women who spoke of their own "laziness" and "bad habits," his matter-of-fact reporting of his faults, while perhaps more reasonable, is striking in its lack of shame:

> I don't remember a lot of things, I'll let things run down and just let things go completely out. I do that quite frequently. Keeping things in your head—well, I know what I really need. But it's going out and getting it—and having the money. Or just going out and getting it, that you know, if I'm extra tired and I don't feel like going out, I'm just not going to go out. The heck with that. I'll think of something else, or do something else with it, or not use it, whatever.

Like the women I have discussed, Rick takes advantage of the flexibility built into housework to avoid work he dislikes; unlike many women, he seems to do so guiltlessly.

A somewhat different situation provides another version of this attitude. Ed took up cooking not because he enjoys it but because he had to, when his wife's new job and long commute meant that she returned home too late to prepare their meals. As he explained the routine, she plans meals for each day, but he decides when he begins the cooking whether or not to follow her plan, and she accepts whatever he decides. I saw the system in operation when I observed a dinner at their house. Just as he started to serve, Ed stopped and said, "Oh, I forgot—Gloria would have liked us to have a salad." But then, shrugging, "Well, I didn't get to it." The ability simply to dismiss the work that cannot be completed, without the anxieties that plague so many women, springs at least in part from differential cultural expectations: the notion that caring work is optional or exceptional for men while it is obligatory for women.

These comments can only be taken as suggestive, since so few men were interviewed, and since even these represent only a few of the variety of household arrangements that are possible. Still, the language of these few men points toward some rather different understandings of what in the work is "burdensome" or "convenient." Ed, for example, sees as a "burden" an aspect of planning that the women I talked with took so much for granted that few even mentioned it. He reported:

> We occasionally get some fresh vegetables, but we usually have frozen vegetables. Because the fresh stuff, somebody has to be there to consume it, you can't delay. Then it becomes my burden, you know, I have to be thinking, and orchestrating, how to use such a variety of food.

And another man explained to me that he and his wife never use "convenience foods," but then defined as a "convenience" the extra work that she puts in on the weekends:

> One convenience that we will use sometimes is this. You know when you come home from work and you have to cook you don't really have much time. You don't have time to sim-

> mer sauces or anything like that. So sometimes on the weekend Katherine will make up something, or if she has time maybe two or three things, something that can be heated up later in the week and will actually taste better then.

Here, we see how thoroughly women's household work is obscured from view and thus framed as not requiring discussion or negotiation. Even as they began to share this work, all three of these men continued to attribute imbalances in the division of labor to their wives' "personalities," defining much of the extra work that wives did in terms of fortuitous propensities for organization or planning.

Men who care for children alone probably take the work of feeding more seriously, simply because they are forced to take sole responsibility for this work that most men are unaccustomed to doing. In addition to the men who served as informants in this study, I talked briefly with two single fathers who do all of the work to care for their families. Both emphasized the terrific worries produced by the necessity of feeding their children. One talked of an "overwhelming anxiety" that began each day as he finished work and realized that he had to get a meal on the table for his three demanding children; the other wrote me that I should emphasize "the STRUGGLE involved in this feeding work," and the fact that "behind closed doors, dinner is often a nightmare." These are the comments of men who cannot rely on partners even to help out or fill in (much less to plan menus and prepare meals during the weekend), who are forced to learn, as most women do, that "if I don't do it, no one else will." Such comments may also reflect the lack of guidance—and resulting panic—for men who must unlearn cultural expectation, disentangling work and gender, separating "care" from "woman"; they must learn to provide service instead of receiving it.

STRUGGLE AND SILENCE

Overt conflict about who will do housework is surprisingly rare.[1] Informants in only two of the thirty households I studied discussed any sustained, explicit conflict about who would do the work or how it should be done, and results from a large survey are similar: over half of the wives responding reported no difference of opinion over who should do what, and only seven percent reported "a lot" of difference of opinion (Berk 1985:188; see also Haavind 1984). Fairly quickly, most houseworkers develop adjustments that are satisfactory enough to mute potential complaints from household members. The boundaries that produce complaints become the givens of the work, and those who do it find ways to manage around these givens. The perception that routine is chosen provides an interpretive frame for redefining the adjustments that are made.

Susan, who is quite happy with her household routine, told a story from the early days of her marriage, of her first definitions of her work, an episode of resentment, and its resolution:

> When we first got married, I played "Suzy Homemaker." I was young and stupid—what did I know? We lived in the suburbs and I worked in the city, and I had to get up at five every morning to get to work. And then on my days off, I'd get up to fix him breakfast, and you know, put on makeup, all that kind of thing. After a while my sister-in-law kind of pulled me aside, and told me I'd better cool it, or he'd get used to that kind of thing.
>
> I still remember, once I came home after a grueling day, and there was my old man, sit-

> ting in front of the TV with his potato chips. I said, "God, in my next life I hope I come back as a 26-inch Zenith, I'd get more attention!" That was probably our first fight. But it had been brewing for about four months. You know, we were just getting used to our differences.
>
> I like the way I do things, I'm used to it. I just get it all done on Monday and then I don't have to worry about it. If I don't do it, I'm a wreck by Wednesday. It's not that I like this kind of work, but you have to do it.

She thinks of this as something other than conflict over work: they were "getting used to [their] differences." She knows that she "has to do" housework, so she has found a way of doing it that she "likes," or at least, that she is "used to." Now, her husband goes to work before she is up; he has coffee and a doughnut, or buys something to eat on the job, and she sleeps until her daughter wakes up. She says, "If I'm in a pinch, my husband's not beyond doing the laundry, or washing a floor." But it is clear that she accepts responsibility for the housework: it is her domain, and though she would not say that she likes the work, she has accepted what seems to her a satisfactory compromise.

In many households, like this one, such compromises are negotiated with little overt struggle. Some men accept more responsibility, or are less demanding than others; some women are satisfied to take on the family work with little help. But accommodations cannot always be found. When conflict about housework does arise, it can be quite painful, at least partly because it carries so much emotional significance. Women who resist doing all of the work, or resist doing it as their husbands prefer, risk the charge—not only from others, but in their own minds as well—that they do not care about the family. When I talked with Jean, for example, she was engaged in an ongoing struggle to get her husband to share the housework. She spoke of her continuing frustration in two interviews, a year apart, and I felt her ambivalence when she told me:

> In spite of all this, I love him. [Laughing, but then serious] No, I do love him, and I'm willing to make some sacrifices, but there are times when I really just go off half-crazy. Because the pressure is just too much sometimes. I just feel it's not fair. It's not a judicious way to live, a fair and equal way to live.

She seems afraid that any complaint will be heard as her lack of feeling for her husband. She insists, indeed, that she is willing to make "sacrifices," as a loving wife should. His resistance to helping with family work is apparently not subject to such an interpretation.

Since feeding work is associated so strongly with women's love and caring for their families, it is quite difficult for women to resist doing all of the work. Bertie, for example, had been married for over twenty years when I talked with her, and had experienced a long period of difficult change. She told the following story:

> There was a time when I was organized, did things on time, on a schedule. I cooked because I felt a responsibility to cook. I felt guilty if I didn't give my husband a certain kind of meal every day . . . When I made the transition it was hard. For me and him. And he's still going through some problems with objecting to it. But I felt that I had put undue pressure on myself, by trying to do what people used to do, you know . . . when the husband could pay the bills, and the wife took care of the house . . .

> I told my family that there were certain things that I needed, which went neglected for many years. And when I recognized my own needs, there was a problem . . . I had given so much of my life to my husband and children, that he thought that I was wrong, not to give them that much time anymore. But I needed to go back to school, needed to improve myself, I needed time to myself . . . They've come to accept it now. Five, six years ago it was really rough. But now they accept, that you know, I'm a person. I am to be considered a person. I have rights, you know, to myself. It was a rough ride for a while. And I suppose it could have gone in another direction. But it didn't.

Things had changed by the time I talked with Bertie: she was pursuing a degree and spent less time cooking elaborate meals. Her husband and children have not taken on much of the work burden, and typically do without meals when she is away from home in the evening—sometimes the girls will prepare sandwiches—so Bertie continues to be responsible for the bulk of the work, simply doing less than in the past. Yet when she talks about their struggles, she still worries about being "selfish":

> I do take that time now for myself. But I count my study time, and my class time, as my own, you know. You know, so that I don't—I try not to be selfish.

When I asked what "being selfish" meant, she replied:

> I make sure that I have time with them. If my stuff gets to be too much, whatever is necessary, whatever, whatever is important, I try to do. Because we still have the children to raise. So there must be some sharing. They're still there, I can't treat them as though they're not there. Even though they're pretty independent, on their own, they still require a lot of attention. So I have to be careful not to give too much to myself. Because you can fall into that. You know, studying too much. It's hard to describe.

It is, indeed, hard to describe. She claims that she has "rights," like any other person. Yet she will do "whatever is necessary" for her family, and must be "careful not to give too much" to herself. Her talk reveals quite different standards for evaluating her own needs and those of others. Raising such issues within the family requires this intense scrutiny of a woman's own desires.

Bertie's system for accounting her time was not unique. Jean, the other woman who was struggling with these issues, reported considering her situation in similar terms. She identified two blocks of time as "hers," but both were hers in a rather ambiguous sense:

> I feel the only real time I have to myself is usually my lunch hour [at her job]. I consider that my time for myself.

She has to be firm to maintain even this break officially sanctioned by her employer. Her husband, who works at night and wakes up around noon, would like her to come home so that they could spend time together. But, she reported:

> I do kind of resist that, on any kind of a regular basis. Although, in order to keep a marriage going I should maybe not do that. It's hard for me. Because I know there's a need

> there, I feel that too. But this gets into all kinds of other issues, about him not helping. I feel that if he would help out more around the house, I'd be more willing to come home and spend more time with him. But you know, I feel like this is my own time for me.

On Sundays they go to church, partly for the children, and partly because she and her husband enjoy the "social aspect," and she counts this activity as her own time as well:

> So Sunday morning is never a time when I can do chores or anything. And in a way, I mean, I count that as time for me, in a different sort of way.

These women have asserted to their families that they are people too, with rights (and it is surely striking that they feel a need to put forward such a claim). However, they still must calculate which time to claim as "theirs," and the logic of caring for others labels any too-active exercise of their rights as "selfish."

These women's stories help to show why there were only two of them in the group I studied, why conflict about housework is infrequent. They illustrate how, in the family context, a mother's claims for time to pursue her own projects can so easily be framed as a lack of care, and a mother's claim even to be "a person" may be taken as "selfish." If the act of pressing a claim for time off or help from others is so fraught with interpersonal danger, it is perhaps not surprising that so many women choose to accommodate to inequitable arrangements instead of resisting them.

THE LANGUAGE OF CHOICE

Many women spoke of doing work they did not enjoy in order to please their husbands. However, very few of them expressed explicit discomfort about these efforts, and only those described above reported any sustained, overt conflict with their husbands. I was puzzled and a bit dismayed by their complacency about what I saw as inequity. But as I analyzed their reports, I began to see how the organization of family work contributes to their responsiveness to husbands' demands. As I showed in the last chapter, individuals have considerable flexibility in designing household routines, and they choose routines shaped to the idiosyncrasies of those in the family. They find ways to adjust to special demands, and then take their adjustments for granted, often describing them as no particular trouble. These choices, and the sense of autonomy that comes with making them, combine to hide the fact that they are so often choices made in order to please others.

Both deference and a sense that her deferential behavior has been freely chosen can be seen in Donna's comments, for example. Her husband, a mail carrier, is moody and unpredictable, difficult to please and quite openly critical when he does not like the meal she prepares. Although she told me several times that he is "not fussy," she also reported checking with him about every evening's menu before she begins to cook. I remarked that his preferences seemed quite important, and she responded:

> Yeah. I like to satisfy him, you know, because a lot of times I'll hear, 'Oh, you don't cook good,' or something like that.

The possibility of such criticism becomes part of the context within which she plans her work. She thinks ahead about what to prepare, but final decisions depend on his responses. When I asked what she would prepare the night of our interview, she could not answer:

> I haven't talked to him today, so I really don't know . . .
> [MD: Does he always know what he wants?]
> Well—I give him choices. Or he'll say "I don't care." So then it's up to me and I just take out something. Hopefully tonight—I would like to have the pot roast—so maybe he'll say yes. Because he actually bought it the other day, so he might want it.

Donna finds ways to build a routine that provides some shape for her work and still allows accommodation to his day-to-day tastes. She explained, for instance, how she plans her shopping:

> Like I'll ask him, what do you want me to pick up? you know, what kind of meat do you want me to pick up? And he'll go through the paper, and he'll tell me, do this, get this. But as far as really making it out [a menu], I just don't. Because sometimes he might not be in the mood for it, he might not want it, or something like that. So I just leave it up in the freezer.

Her scheme sometimes involves an extra trip to the store:

> Then if he wants something, then I'll just go to the store and get what he wants. It's really kind of day by day. I find it easier that way. I couldn't sit there and write what I'm having for dinner every day. I just can't do that.

When I asked why not:

> I don't know, I figure maybe it's just me. I just can't sit there and write, well, we're going to have this and this and this. And then that day you might not have a taste for it. And then you'll want something else. That's the way I look at it.

She "finds it easier" to plan meals day by day, and she presents this as just her way, a personal inclination. It seemed clear to me that her strategy was shaped by her husband's demands, in response to his moodiness and in order to avoid his sharp words of criticism. Within the constraints of their relationship, she does make choices in order to avoid trouble. Interpreting her accommodations as choices freely made, she translates his peculiarities into a general observation: "you might not have a taste for it. And then you'll want something else." And thus she presents the result of her strategizing as her own belief: "the way I look at it."

Even when family members are not so demanding, the pattern of choosing to adjust to others is common. Another woman explained how she has chosen a breakfast routine that lets her sleep a bit later instead of eating with the family:

> During the week I usually get their food to the table and then I make a lunch [for her husband]. It's more pragmatic. I could get up earlier and do that, but I choose to stay in bed and avoid sitting at the table.

Such comments stress autonomy and choice; however, it is clear that these women's decisions are not so freely made as they suggest. When husbands decide to press their claims, these become the fixed points around which adjustments and "choices" are made. One white woman, married to a journalist, reported a more conflictual negotiation over the breakfast routine, and explained how she has adjusted her morning schedule to accommodate her husband's ideas about breakfast:

> Breakfast has turned into more of a social occasion than I perhaps would care for. For my husband it's a real social affair, and we got into huge fights years ago. He always from the day we got married expected me to get up and fix his breakfast, no matter what time he was going anywhere. Then we lived overseas and we had two maids, and I couldn't see any point in getting up just to sit with him—I didn't even have to *make* the breakfast. Well, that was a "dreadful, dreadful thing." Finally he got over that, and I don't mind getting up, you see, all right, that's a personal thing . . . So I usually fix breakfast for the two of us. Which is nice—but I would like to be able to read the newspaper, myself.

On this issue, her husband is adamant—it is a "dreadful, dreadful thing" not to have breakfast together—and she has adjusted to this "personal" preference. But she also describes her adjustment in somewhat contradictory terms: she "doesn't mind getting up," it's "all right," even "nice," but still, she would rather read the newspaper.

The choices that women talk about are not entirely illusory: in many ways, houseworkers can choose to do the work as they like. They adopt different general strategies: some maintain that they "couldn't live with" a regular routine, while others describe themselves as "disciplined" and "big on rules." To some extent, people even choose not to do the kinds of work they dislike. One woman, who would like to "just forget about" cooking, has simplified her food routine so that her work is quite automatic: she prepares meals that are "very easy to cook, and very quick also." And Sandra, who enjoys cooking and prepares elaborate meals for her husband (in addition to simpler, early meals for the children), thinks of her efforts as "compensation" for the cleaning that she does not enjoy, and often does not do. Still, these real choices—some of which certainly do ease the burdens of housework—seem also to provide a rationale for deference: women emphasize their freedoms and minimize their adjustments to others.

As women make choices about housework, their decisions include calculations about when to press their own claims and when to defer to others. The choice to do something in the way one prefers oneself is made to fit among the more compelling demands of others, especially husbands. The houseworker comes to understand her work in terms of a compromise that seems fair: since she is free to choose in some ways, it is only fair to defer in others. Most women seem only partly conscious of this logic. They, and others, notice the choices but not the deference. In their talk about decision making they tend to conflate benefits for the household group all told with their own more specific interests and preferences.

Such calculation can be seen in this affluent white woman's comment about going out to dinner. She explained that she does not really enjoy cooking and "always looks forward to" going out. But she often thinks that her husband is not so interested, and "being sensitive" to him can interfere with her enjoyment:

> If it's just family, and we wait a real long time, I keep thinking of all the things I could be doing instead of waiting, and then I wonder if it's really worth going out for dinner. And I think part of that is being sensitive to my husband's thinking. Because he has to eat out every noon anyway. And sometimes a couple of times a week he has to take people out for dinner. So it's not a great pleasure for him, to go out again just to get me out of the kitchen. So if it takes a real long time, I just feel, why did I do this, you know?

She does not feel this way—and can relax and enjoy herself—when she eats out with her son; knowing her husband's feeling changes her own, to the point that she wonders, "Why did I do this, you know?" Donna, managing a household with little extra money, worries about the expense of eating out, but talks about the decision to go out in much the same way:

> To me, I can go to the store and pay for two days', meals, or go to a restaurant and pay for one meal. And a lot of times the kids don't finish their meal . . . I was going to suggest for Easter, going out to eat. But maybe I'm better off just getting something and having it here.

Her own preferences disappear into those of the household group: though she might enjoy a holiday, the children might not eat and money spent would be wasted. In the end she concludes, "I'm better off" doing the work of cooking at home.

These are issues that are difficult for many women to discuss; their talk is often hesitant, sometimes contradictory. But my point is not that they are confused or disingenuous about their positions relative to others. Rather, I mean to show that they are in a situation not easily described in terms of either autonomy or control. These women do make choices about their work, though many of those choices are made within a structure of constraints produced by others. The work itself is defined in terms of service to others, and husbands' demands are given special force through cultural assumptions about appropriate relations between husbands and wives. What a husband insists on typically becomes a requirement of the work, and a woman who arranges the routine to satisfy her own preferences as well as his may simply be making her work more difficult. The fact that so many women frame these accommodations as "choices" means that they are less likely to make choices more obviously in their own behalf when the interests of family members conflict. In such situations, women seem to assume that they have made enough choices, and often come to define deference as equity.[2]

DOMINANCE AND SUBORDINATION IN EVERYDAY LIFE

I have argued throughout this book that the work of "feeding a family" is skilled practice, a craft in which many women feel pride and find much satisfaction. Such a view suggests that wives care for family members not because they are coerced or compelled by despotic husbands, but because they believe the work of care is valuable and important. The discussion in this chapter, however, suggests as well that the work of caring—however valuable or valued by those who do it—is implicated in subtle but pervasive ways in relations of inequality between men and women. Some husbands insist quite explicitly that their wives display subordination by providing domestic service. For most men, however, such coercion is as unnecessary as it would be unpleasant. In most households, wives display deference to husbands simply because catering to a man is built into a cultural definition of "woman" that includes caring activity and the work of feeding. For many women and men, patterns of "womanly" service for men simply "feel right." In some cases, the recognition of a husband's claim to service is quite direct ("Like I said, whatever he wants, I'll make it the way he likes, and everything he likes with it."); in other reports, references to "really decent," "man-sized" meals point toward a more diffuse sense that a husband, because he is a man, deserves special service. Both kinds of statements show how the everyday activities of cooking, serving, and eating become rituals of dominance and deference, communicating relations of power through non-verbal behavior (Henley 1977).

Many women take pleasure in preparing food that pleases, in serving family members, in rewarding a husband for his work at a difficult job. Many think of the craft of attentive service as work they choose. But few women are themselves the recipients of a similarly attentive service in return. We might assume that men who cook like to please the ones they cook for. But they do not talk about preparing a meal that is "enough for a woman." Indeed, they cannot talk (or think) this way: the idea of a "woman-sized meal" is so dissonant with prevailing cultural meanings that it sounds quite wrong. Here, we see how categories of expression interact with people's everyday family activity. The gender inequalities inherent in language and in a multitude of nonverbal behaviors are woven into the fabric of social relations produced as people go about the mundane affairs of everyday life. Even when fathers cook, their activity—however similar to that of mothers who cook—is framed differently. There are no terms within which men think of cooking as service for a woman, no script suggesting that husbands should care for wives through domestic work. Some women are beginning to insist on more equal relations, and some husbands are beginning to struggle at taking equal responsibility for family work. But these attempts are made without a cultural imagery to support them, and in opposition to established understandings about appropriate activities for men and women.

Because of the expectation that women will be responsible for caring work, their own independent activities are likely to conflict with requirements for family service. When wives and mothers assert their rights to pursue individual projects, they often discover the limits of choice and the force of cultural expectation. When women resist—by demanding help with housework or a respite from serving others—they challenge a powerful consensual understanding of womanly character by suggesting that women's care for others is effort rather than love. Many have trouble speaking plainly about the limits of caring work, and many find that in the long run it is easier to do all the work required than to press claims for an equitable division of labor.

The invisibility of the work that produces "family," the flexibility underlying perceptions of "choice" about the work, and the association between caring work and the supposedly "natural" emotions of a loving wife and mother all tend to suppress conflict over housework. Since the work itself is largely unrecognized, and often misidentified as merely "love" rather than also effort, redefinition is required before questions of dividing the work can be discussed. Those who have benefited from the work often have trouble recognizing it as such, and indeed, have little incentive to do so. Further, many women find that they can make enough choices and adjustments in some areas that accommodation in others seems preferable to sustained conflict. Those who insist on negotiating new household patterns must confront their own and others' sense that they do so out of "selfishness" or insufficient concern for others. Even as they struggle for more equitable arrangements, these women carefully ration ("count") the time and attention they give to their own needs, while attempting to provide "whatever" their families require. Their demands for themselves are painfully visible within the family, while their accommodations to others remain largely unacknowledged.

The claim that caring is work, or that this work should be shared by all those who are able to do it, must be made against powerful beliefs about the naturalness and importance of family life, and about men's and women's dispositions and roles. For a woman to provoke and sustain conflict in this area is to risk the charge that she is unnatural or unloving. The costs of conflict are high. Conversely, when a husband complains, or even hints at complaint, his claims carry with them the weight of generations of traditional practice and a body of expert advice about housekeeping and family life based on the assumption that women will serve others. As women adjust and accommodate, choosing deference to others and fitting their own projects into frames established by others, their actions contribute to traditional assumptions about woman's "nature." Thus, we see that in addition to its constructive, affiliative aspect, the work of care—as presently organized—has a darker aspect, which traps many husbands and wives in relations of dominance and subordination rather than mutual service and assistance.

NOTES

A portion of this chapter first appeared in "Conflict over Housework: A Problem That (Still) Has No Name," in *Research in Social Movements, Conflicts and Change: A Research Annual,* edited by Louis Kriesberg (Greenwich, CT: JAI Press, 1990), pp. 189–202. Reprinted by permission.

1. This is not to say that other kinds of family conflicts are not sometimes expressed through the provision of food or reactions to it. I was less interested in these emotional dynamics associated with food than with the practices of organizing, preparing, and serving it, and I did not ask interviewees to speculate about the significance of food in conflicts other than those focused on family work issues. Charles and Kerr (1988) provide somewhat more information about food practices as expressions of interpersonal struggles, especially between parents and young children.
2. Haavind, in an analysis of "Love and Power in Marriage," based on studies of Norwegian couples, makes a similar observation about understandings of "choice" that accompany marriage based on romantic love: "If you have the right to marry whom you please, the responsibility for how it works out is also yours. Therefore, it is difficult to share our disappointment with anyone outside our own marriage" (1984:161).

REFERENCES

Berk, Sarah Fenstermaker. 1985. *The Gender Factory: The Apportionment of Work in American Households.* New York: Plenum Press.

Blood, Robert O. and Donald M. Wolfe. 1960. *Husbands and Wives.* New York: Free Press.

Charles, Nickie and Marion Kerr. 1988. *Women, Food and Families.* Manchester: Manchester University Press.

Ellis, Rhian. 1983. The Way to a Man's Heart: Food in the Violent Home. In Anne Murcott, ed., *The Sociology of Food and Eating,* 164–171. Aldershot: Gower Publishing.

Haavind, Hanne. 1984. Love and Power in Marriage. In Harriet Hotter, ed. *Patriarchy in a Welfare Society,* 136–167. Oslo: Universitetsforlaget.

Hartmann, Heidi I. 1981. The Family as the Locus of Gender, Class, and Political Struggle: The Example of Housework. *Signs* 6:366–94.

Hertz, Rosanna. 1986. *More Equal than Others: Women and Men in Dual Career Marriages.* Berkeley: University of California Press.

Hochschild, Arlie, with Anne Machung. 1989. *The Second Shift: Working Parents and the Revolution at Home.* New York: Viking.

Hood, Jane C. 1983. *Becoming a Two-Job Family.* New York: Praeger.

Henley, Nancy M. 1977. *Body Politics: Power, Sex and Nonverbal Communication.* New York: Simon and Schuster.

Murcott, Anne. 1983. "It's a Pleasure to Cook for Him": Food, Mealtimes and Gender in some South Wales Households. In Eva Garmarnikow et al. eds. *The Public and the Private*, 78–90. London: Heinemann.

Food, Body, and Culture

If culture inscribes bodies, it is food that leaves the clearest mark, and that mark is most often read on women's bodies. Of central concern in Europeanized Western cultures is fat. Several articles here demonstrate how people in various cultures define and feel about fat. Hortense Powdermaker reviews the anthropological literature to show that obesity is a Western problem and that in many non-Western cultures, fat symbolizes beauty and fertility. No volume on foodways would be complete without a piece by Hilde Bruch, a brilliant and humane pioneer in the psychological understanding of eating problems. Here Bruch discusses psychosocial dimensions of body image and its relationship to self-awareness. Susan Bordo looks at the meaning of food and body to women today, showing how women's subordination is continually recreated through their relationship to the body. Massara, Sobo, and Hughes examine three cultures—Puerto Ricans in the United States, rural Jamaicans, and African-Americans—that celebrate the large female body and associate it with power, well-being, fertility, and a good family life.

15

An Anthropological Approach to the Problem of Obesity

HORTENSE POWDERMAKER

I am the only person participating in the Symposium who has never done any research on the problem of obesity, and who is decidedly not an expert on the subject. My role here as an anthropologist is to attempt to set the problem in the context of the culture. In this day of specialists we cultural anthropologists are the specialists in a holistic approach. In our studies of primitive or preliterate tribal societies, we have asked questions concerning relationships between different elements of culture. How are the functioning of the family, the economic and class organization, the political system, the religious and magical beliefs, the values that men live by related to each other and integrated in that abstraction we call culture? In this paper I shall give a cultural approach to the problem of obesity, raise questions, and, quite tentatively, offer some hypotheses. These will provide some understanding of the complexities of the problem and a basis for future research.

In setting the problem of obesity in the frame of the culture of contemporary society, my focus will be on the roles of food and of physical activity in our value systems. Incidentally, I wonder why so much of the education designed to reduce the incidence of obesity is centered on food rather than on activity. Is it assumed that food habits may be modified more easily than those of physical activity?

My basic questions are concerned with the symbolism of fatness and thinness in our society and the relationship of each to other symbols and to our values. I would be interested in differences in the symbols and in the relative strength of the same symbols in class, ethnic, religious, sex, and age groups, and among individuals. I would assume that there might be conflicting values concerning fatness and thinness, about eating and physical activity as there are in many other areas of our life, and that some of this conflict might stem from the fact that we live in a rapidly changing society, where traditional values linger beside new ones. I would also be interested in the cultural study of people who are not obese as well as those who are, i.e., some kind of control group in which variables are limited. As an anthropologist, I am naturally interested in a comparative approach, i.e., the symbolism of obesity and thinness in other cultures and the many-sided role of food and eating in them, assuming that this comparative knowledge would illuminate the problem in our society.

Beginning with the last point, let me summarize briefly some relevant facts from preliterate, tribal societies. In a large number of these societies the economy was a subsistence one, whether characterized by food gathering, hunting, fishing, agriculture, raising cattle, or some combination of these activities. A major part of all activity was concerned with the production of food. Tools were crude—a wooden hoe and a stone axe. The only means of transportation was by foot or canoe. Food-growing plots were often several miles from the village; the clearing of the dense bush in tropical and semi-tropical parts of the world by the men was a strenuous job, as was also the planting and weeding by the women. Strenuous physical activity was the norm for men and for women, whatever the type of economy. But although everyone worked hard and long in the production of food, hunger was a common experience. Famines and periods of scarcity were not unusual. Seasonal changes, plagues, pests, and many other natural causes tended to produce alternate periods of shortage and relative plenty. It is, therefore, not difficult to understand that gluttony, one of the original sins in our society, was an accepted and valued practice for these tribal peoples whenever it was possible. In anticipating a feast a Trobriand Islander in the Southwest Pacific says, "We shall be glad, we shall eat until we vomit."[1] A South African tribal expression is, "We shall eat until our bellies swell out and we can no longer stand."[2]

The function of food and eating was, and still is, not restricted to the biological aspects. Food is the center of a complex value system and an elaborate ideology centers about it. Religious beliefs, rituals, prestige systems, etiquette, social organization, and group unity are related to food. Throughout the Pacific, in Africa, and in most other parts of the tribal world, kinship groups work together in the production of food. Distribution of food is part of traditional obligations between people related biologically and through marriage ties, between clans, and between chiefs and their subjects. The accumulation of food, particularly for ritual occasions, is a major way of obtaining prestige. At all significant events in the individual's life history—birth, puberty, marriage, death—there must be a feast, and the amount of food reflects the prestige of those giving it. Less formal but of equal significance is the relationship of food-giving to hospitality, valued even more among tribal peoples than among ourselves.

The importance of food is not limited to relations among the living. It plays a significant role in relationships with dead ancestors and gods. Offerings of food are made to them, so that they will grant the requests of the living and protect them from sickness and other misfortunes. The spirits of the dead and the ancestral gods presumably have to eat and, among some tribes, observe the same eating etiquette as do the living. In Haiti the gods are very demanding, and providing their food becomes a means of controlling and manipulating them, for the gods depend on men for their strength. In the same country death is symbolized in many instances as being "eaten" by evil gods and, in a modern context of a railroad accident, the locomotive is said to be a machine that eats people. This oral aggression of evil gods (and, presumably, the locomotive, too) is regarded as being motivated by the desire to acquire strength through being fed.[3] The function of food in magical and religious practices throughout the world is well known, and food taboos are part of many religious rituals in both tribal and modern societies. We could go on almost indefinitely describing the social role of food.[4, 5]

But we turn now to the more personal role of food for the individual. The infant's first relationship with his mother is a nutritive one. In primitive societies it is fairly common for a child to be nursed at his mother's breast for several years. For the infant in all societies, suckling and eating appear to be among the earliest sensory experiences and pleasures. The psychoanalysts call it the oral stage. We tend to agree with them that early infantile experiences have lasting effects. In some tribal societies such as the one I studied in the Southwest Pacific, the stomach is the seat of the emotions. "Bel belong me hot" is the pidgin English way of expressing deep feeling, whether occasioned by anger, sexual desire, or eating well.[6] The same concept appears in Africa and other parts of the world.

Given the scarcity of food and the ever-present fear of famine in many tribal societies, the significant social role of food, and the lasting impact of the infant's first sensory satisfactions, it is not surprising to find that stoutness or some degree of obesity is often regarded with favor. This is particularly true for the concept of female attractiveness. Among the Banyankole, a pastoral people in East Africa, when a girl began to prepare for marriage at the age of eight, she was not permitted to play and run about, but kept in the house and made to drink large quantities of milk daily so that she would grow fat. By the end of a year she could only waddle. "The fatter she grew the more beautiful she was considered and her condition was a marked contrast to that of the men, who were athletic and well-developed." The royal women, the king's mother and his wives, vied with each other as to who should be the stoutest. They took no exercise, but were carried in litters when going from place to place.[7]

Among the Bushmen of South Africa, the new moon is spoken of as a man because of its slenderness, and the full moon is a woman because of its roundness. Masculine and feminine endings are given to the same roots to denote sex: male endings for strong, tall, slender things and female for weak, small, round ones.[8] Today in a mining community on the Copperbelt of Northern Rhodesia where I have done fieldwork, in one popular song a young man sings,

> "Hullo, Mama,* the beautiful one, let us go to town;
> You will be very fat, you girl, if you stay with me."
> (* "Mama" is a term of address for a woman.)

The standard of beauty for a woman here was not the fatness which we mentioned earlier, but rather a moderate plumpness.

Summarizing briefly for tribal pre-literate societies, we note that hunger was common and that a high proportion of men's and women's energy was spent in producing enough food to stay alive; that food was not only a biological necessity, but that its social and psychological functions were also very significant. The giving of food was a prominent part of all relationships: between kindred, between clans, with dead ancestors, and with gods. Food played a role in ritual, magic and witchcraft, and in hospitality. The accumulation of food was a mark of great prestige. Fatness was a mark of beauty and desirability in women.

We turn now to our contemporary society. It is characterized by an economy of plenty as compared to the economy of scarcity in tribal societies. We eat too much. We have too much of many things. According to the population experts, there are too many people in the world, due to the decline in mortality rates. A key theme in this age of plenty—people, food, things—is consumption. We are urged to buy more and more things and new things:

food, cars, refrigerators, television sets, clothes, etcetera. We are constantly advised that prosperity can be maintained only by ever-increasing consumption. This is in sharp contrast to our own not too distant past, when saving and thrift were among the prized virtues and emphasis was on production rather than consumption.

Another important change in our modern industrial society is that physical activity is almost non-existent in most occupations, particularly those in the middle and upper classes. We think of the ever-increasing white-collar jobs, the managerial and professional groups, and even the unskilled and skilled laborers in machine and factory production. For some people there are active games in leisure time, probably more for males than females. But, in general, leisure time activities tend to become increasingly passive. We travel in automobiles, we sit in movies, we stay at home and watch television. Most people live too far away to walk to their place of work. Walking for pleasure is very rare. Former President Truman's daily walk is regarded as one of his peculiarities. The trend for those who are advised to take exercise and who also have the necessary wealth is a passive form-massage, the electric table which vibrates the body, and other electrical devices.

But while people may exercise less and live in an economy of plenty, they are becoming increasingly aware of the problem of obesity. There is a continuing enlargement of our knowledge of nutrition, of the relationship between obesity and certain diseases, and to health and longevity in general, and a wide popularization of this knowledge. This past month we had a "Nutrition Week," and every day our mass media—newspapers, radio and television—carry information about food and its relationship to health, disease and physical attractiveness.

Our standards of beauty, particularly for the female, have undergone a great change from tribal societies and from our own past. The slender, youthful-looking figure is now desired by women of all ages. The term "matronly," with its connotation of plumpness, is decidedly not flattering. Although the female body is predisposed to proportionately more fat and the male to more muscle,[9] the plump or stout woman's body is considered neither beautiful nor sexually attractive. Our guess is that a hundred years ago the term "matronly" was not unflattering. The role of a wife today as an active sex mate, as compared to her role in our more Puritanical past with its emphasis on motherhood rather than on the pleasures of sexual experiences, may be significant in this context. For this and for other reasons the contemporary cult of youthfulness appears to be stronger among women than among men. At almost any middle and upper class gathering of middle-aged men and women, a large proportion of the latter will have dyed their hair, while most of the men will have the symbolic grey hair of aging. It is generally assumed that physical attractiveness is more important for the female than for the male in their respective search for a mate. Success, wealth, and vigor are significant eligibility criteria for potential husbands and fathers. Of course, sex appeal is important for men, too, but it seems not to be so much associated with seeming youthfulness as it is for women.

However, the cult of youthfulness is not confined to women. As science enables us all to live longer and longer, men and women want to remain young longer and longer. This is not a new desire. The quest for the fountain of youth is one of the well-known themes in mythology. The desire to remain healthy and "fit" as long as possible seems quite normal to us. Yet our excessive need to *look* young may also be related to other trends in our culture. Middle-aged people often find it difficult to get jobs, and they are faced with enforced, and sometimes unwanted, retirement at a fixed age. The cult of youthfulness may also have some connection with our apparent concern about sexual potency and sexual pleasure. Many books and articles discuss these as a difficult problem, and their large sale presumably indicates considerable anxiety about sexuality in our culture. Do people with this kind of anxiety have more, or less, difficulty in dieting and keeping their bodies young-looking?

We have indicated a number of strong trends in our culture which run counter to obesity. The desire for health, for longevity, for youthfulness, for sexual attractiveness is indeed a powerful motivation. Yet obesity is a problem. Otherwise we would not be participating in this Symposium. We ask, then, what cultural and psychological factors might be counteracting the effective work of nutritionists, physicians, beauty specialists, and advertisements in the mass media? We have a number of hypotheses. We think there may be considerable ambivalence for many people in regard to being fat or thin, to over-eating or to dieting. This ambivalence could, in turn, come from conflicting patterns in our culture.

I have a hypothesis that, consciously or unconsciously, our symbolism for a maternal woman is on the plump or obese side. There is the figure of a pregnant woman and, as already indicated, the infantile satisfactions gained from food given by a mother or mother-surrogate. The image for mother and for mate may be in conflict.

Then, too, food is a very significant symbol in our prestige system. The kind of food, the quantity, and the manner in which it is served are among the important criteria of social class. In most tribal societies, even those with a highly stratified social system, everyone—royalty and commoners—ate the same kind of food, and if there was famine everyone was hungry. In our society there are sharp distinctions. Although there are probably relatively few people today who know sustained hunger because of poverty, poor people eat differently from rich people. Fattening, starchy foods are common among the former, and in certain ethnic groups, particularly those from southern Europe, women tend to be fat. Obesity for women is therefore somewhat symbolic for lower class. In our socially mobile society this is a powerful deterrent. The symbolism of obesity in men has been different. The image of a successful middle-aged man in the middle and upper classes has been with a pouch," or "bay-window," as it was called a generation ago. We are all familiar with pictures of this type, resplendent with gold watch chain across the large stomach. Today this particular male class-symbolism is changing, probably because of the increased knowledge of the relationship of obesity to heart-malfunctioning and to other diseases.

Although slenderness becomes increasingly a symbol of social status, the food of the wealthy is still rich and plentiful, and their dinner parties are often, quite literally, a sign of conspicuous consumption. With the ever-increasing diversity of foods, food has become not only a matter of social status, but also a mark of one's personality and taste. More and more people are becoming gourmets, and with the declining number of servants, the hostess—and often the host, too—display their individual style and taste in cooking.[10] We become more personally interested in food as we become more aware of the problems connected with overeating.

The giving of food to people who are in trouble is a still widely prevailing folk custom and is reflected in our radio "soap operas." When someone is having marital or financial problems or when there is illness in the family, a good neighbor brings in food and says, "You must eat to keep up your strength." The same correlation of eating with strength runs through many food advertisements, particularly those designed to reach young, growing children in the television audience. It would be interesting to do an analysis of the mass media advertisements of food which are directed toward children. It would be equally desirable to analyze the advertisements concerning reducing foods, pills, and other products directed toward adults.

Our symbols for fatness or thinness are not clear-cut, as old and new patterns mingle. We have the beliefs that fat people are good-natured, contented, likable, funny, and also that they are foolish, "greasy," and greedy. There is the well-known image from Shakespeare's *Julius Caesar*, in which Caesar prefers his followers to be fat, and fears those who are lean and hungry.

> Let me have men about me that are fat.
> Sleek-headed men, and such as sleep o'nights:
> Yond Cassius has a lean and hungry took;
> He thinks too much; such men are dangerous.
> *Julius Caesar*, Act 1, Scene 2.

A study of heroes and heroines and villains in our mass media, in terms of their fatness and thinness, might be revealing. I cannot offhand remember any fat movie villains, male or female. But this would be interesting to check.

While the family in our society is no longer an economic unit for the production of food, as it was in primitive society, the family meal remains one of the few times when the family is united and drawn together. Parents still are the givers of food, and most of us are aware of the intense interest with which young siblings watch mother cut a pie and their anxiety over whether the slices are even. This is true in homes where food is plentiful, and obviously food is a symbol for the mother's favoring or not favoring one child more than another.

Eating well, a full stomach, is still one of our main ways of achieving a state of euphoria. A really good dinner sets all of us up. This is probably connected with the fact that one of the earliest forms of security and of sensory pleasure is connected with the intake of food and that about it are centered the first human relations. The eating of food and the giving of it thus remains a symbol of love, affection, and friendliness, as well as a source of pleasure in itself.

It is often stated and rather commonly believed that indulgence in overeating is a conscious or unconscious compensation for frustration or neurotic problems. We ask a further question: why do some people seek this form of compensation rather than another form? Is there, for instance, one type of person who tends to be alcoholic and another to overeat? A number of studies have indicated a comparatively low rate of alcoholism among Jews.[11, 12] They show that sobriety is a strong moral virtue among orthodox and pious Jews, and that drunkenness is associated with the outgroup, the Gentiles. The Jewish norms of moderate drinking and sobriety are bound up with the ceremonial and ritual observances, with their religious beliefs, and with the value of remaining separate from Christians. It is assumed that Jews have the same proportion of neurotic and other problems that could lead to alcoholism as do Christians. The norms favored by any group for meeting problems are part of its culture and are internalized in childhood. It would be interesting to find out whether overeating and obesity are more common among orthodox Jews than among reformed Jews and Christians of the same class. We think, too, that there could be regional as well as religious differences in attitudes toward obesity. One suspects that there would be considerable difference between the South and New England.

We have a number of other questions concerning possible correlations of cultural and psychological factors with obesity. Is the ability to diet, and to diet consistently, related to belief in a measure of control over one's fate? Is it related to the strength of the belief in science? Is obesity correlated with orientations toward asceticism versus sensory pleasures? Has there been any study of obesity among monks and nuns? Do people who value sensory pleasures in general, such as those derived from perfumes, from physical contacts, from sexual experience, demonstrate an ability to diet more, or less, successfully than others? The degree of emphasis on sensory pleasure may be culturally determined, may vary from one historical period to another in the same culture, and from one class and ethnic group. And within each group there can be variations due to genetic idiosyncratic factors in the life history of individuals.

There are time limits to the number of questions we can raise. We have tried to indicate some of the cultural factors underlying the problem of obesity. Our society, with its economy of plenty and lack of physical activity, as compared to the economy of scarcity and the hard physical work in tribal societies, provides increasing opportunities for people to eat more food and to become obese. At the same time, other cultural factors, such as the knowledge of nutrition and of the relationship of obesity to disease and longevity and the popularization of the knowledge, our cult of youthfulness and the emphasis on the beauty of the slender body, particularly for the female, our class stereotypes, all tend to keep people from taking advantage of the opportunities to gorge on food. Yet there are many who overeat. We have hypotheses that this may be related to our deeply imbedded desire for the euphoria which comes from a full stomach, with other sensory indulgences or a lack of them, with conflicting imagery about a motherly woman versus a sex mate, with the use of food as a status symbol and as an expression of personality tastes, and with cultural norms about food and standards of beauty in different religious, class, ethnic, and regional groups. We have asked a number of questions relating to possible cultural correlations, for which there is no data. Mainly we have tried to show some of the intricate and complex ramifications of eating and of obesity in the tribal societies of the past characterized by too little food, and in our contemporary culture characterized by too much food.

NOTES

Presented as part of a Symposium on *Prevention of Obesity,* sponsored by the American Heart Association and held at The New York Academy of Medicine, May 26, 1959.

I am much indebted to my friend and colleague, Dr. George Grosser, for his critical reading of the manuscript of this paper and for his helpful suggestions.

1. Malinowski, B. *Argonauts of the Western Pacific.* New York, E. P. Dutton & Co., 1922, p. 171.
2. Knopf, A. *Das Volk der Xosa-Kaffern.* Berlin, 1889, p. 88.
3. Bourguignon, E. Persistence of folk belief: some notes on cannibalism and zombis in Haiti, *J. Amer. Folklore* 72: 42, 1959.
4. Richards, A. I. *Hunger and Work in a Savage Tribe.* Glencoe, Ill., The Free Press, 1948.
5. Radcliffe-Brown, A. B. *The Andaman Islanders.* Cambridge Univ. Pr., 1922.

6. Powdermaker, H. *Life in Lesu: The Study of a Melanesian Society in New Ireland.* New York, W. W. Norton & Co., 1933, pp. 232–34.
7. Roscoe, J. *The Northern Bantu.* Cambridge Univ. Pr., 1915, p. 38. Ibid., *The Banyankole,* 1923, pp. 116–17, 120.
8. Schapera, I. *The Khoisan Peoples of South Africa: Bushmen and Hottentots.* London, Routledge and Keegan Paul, 1930, p. 427.
9. Scheinfeld, A. *Women and Men.* New York, Harcourt, Brace & Co., 1943, p. 147.
10. Riesman, D., Glazer, N. and Denney, R. *The Lonely Crowd.* New York, Doubleday & Co., 1953, pp. 168–69.
11. Bales, R. F. The "Fixation Factor" in Alcohol Addiction: An Hypothesis derived from a Comparative Study of Irish and Jewish Social Norms. Doctoral dissertation. *Arch. Widener Libr.,* Harvard Univ., 1944.
12. Snyder, C. R. Culture and Jewish Sobriety: The Ingroup-Outgroup Factor. In *The Jews, Social Patterns of an American Group.* M. Sklare, edit. Glencoe, Ill., The Free Press, 1958.

16

Body Image and Self-Awareness

HILDE BRUCH

The expression "body image" is so widely used in the psychiatric evaluation of patients that it is surprising to note how vague the concept is. The need for the formulation of a "body schemata" was felt by neurologists who observed that patients with various lesions in the central nervous system experienced their body as markedly changed and distorted. Much experimental work with changing theoretical formulations has been carried out by neurologists and psychologists. This has been ably summarized by Fisher and Cleveland (p. 6) and more recently by Shontz (18). Yet with all these efforts the concept of the body image has not yet been sufficiently developed, either theoretically or empirically, to enable one to measure it or its components with any degree of assurance.

In spite of these limitations it is a concept of definite clinical usefulness for summarizing the multitude of attitudes patients may experience and express about their body. It was Schilder who first extended the concept beyond what had been observed in neurological patients by including clinical psychiatric considerations (16). In *The Image and Appearance of the Human Body,* published in English in 1935 (17), he explored body-image phenomena in relation to normal and abnormal behavior. He spoke of body image as "the picture of our own body which we form in our mind, that is to say the way in which the body appears to ourselves." This is a plastic concept which is built from all sensory and psychic experiences and it is constantly integrated in the central nervous system. Underlying it is the concept of Gestalt, which sees life and personality as a whole.

Schilder conceived of the body image as something more than a postural model arising from the changing body movement and positions, but as representing an integrated pattern of all the immediate organic and psychic experiences. However, he emphasized the importance of motility, i.e., that we would not know much about our bodies unless we moved them. As a child moves around in his environment, sensations stemming from multiple perceptual and muscular feedback are integrated into a dynamically developing body image; thus motility plays an essential role, not only in defining the boundaries of the self, but in differentiating one's self from the total perceptual environment. The inactivity so characteristic of obese people thus appears to be related to their often disturbed body concept (12). Schilder emphasized also the continuous interaction of various factors, and felt that body image in some way preceded and determined the body structure. He felt it was an unsolved problem how the reflection in the psychic sphere was related to predisposition and constitutional elements and to life experiences.

A child's body perception is modified and gradually extended in the course of his growth so as to conform to the current body structure. At the same time his capacities for conceptualization and for experiencing and interpreting reality undergo great changes, as has been studied by Piaget (14). The child also absorbs the attitudes of others toward his body and its parts. He may develop a body concept that is pleasing and satisfying, or he may come to view his body and its parts as unpleasant, dirty, shameful, or disgusting (10). In earlier studies on body interest and hypochondriasis in children, Levy (11) had recognized that the attitude of parents was integrated into a child's body concept and that derogatory attitudes had a very strong effect. He gave the example of a fat girl whose mother always pointed out her obesity with disgust. This mother was quite hostile to the child and made that particular body configuration a mark of her disapproval.

Ideally there should be no discrepancy between the body structure, body image, and social acceptance. Obese people live under the pressure of derogatory social environments. Such a continuous insult to a person's physical personality may result in a cleavage between body structure and the desired and socially acceptable image. According to Schilder, psychosis may arise from such discrepancies.

Own Observations

My own observations were made over a period of 35 years on obese and anorexic patients, on severely disturbed as well as normally functioning individuals. I was greatly influenced by Schilder's work but gradually developed some independent formulations. In particular I came to the conclusion that correct or incorrect interpretation of enteroceptive stimuli and the sense of control over and ownership of the body needed to be included in the concept of body awareness or body identity. The study of obese and anorexic patients brought strikingly into the open the extent to which social attitudes toward the body, the concept of beauty in our society, and our preoccupation with appearance enter into the picture. The obsession of the Western world with slimness, the condemnation of any degree of overweight as undesirable and ugly, may well be considered a distortion of the social body concept, but it dominates present day living.

A young person whose constitutionally or experientially determined body structure does not conform to the socially acceptable image finds himself under enormous pressure and constant criticism. For an understanding of the disturbances of the body image of patients with a deviant body size, biological, psychic, and social forces must be conceived of as in constant interaction. The deviant body size itself is related to or even the result of disturbances in hunger awareness, or of other bodily sensations. Thus evaluation of enteroceptive awareness needs to be included in a study of the body image. Various terms have been used to indicate such inclusions: body concept, body identity, body percept, etc. This new concept includes the correctness or error in cognitive awareness of the bodily self, the accuracy in recognizing stimuli coming from without or within, the sense of control over one's own bodily functions, the affective reaction to the reality of the body configuration, and one's rating of the desirability of one's body by others. The range of attitudes expressed by patients talking about their concept and awareness of their body goes far beyond the earlier definition of body image. In evaluation of all of these experiences it is not always possible to make sharp distinctions.

Therapeutically it is important to help an individual come to a more realistic awareness of his functioning self, bodily or otherwise, and to integrate various experiences into a functioning whole with changing impulses, whereby it is important that he comes to experience them as being under his own control. The therapeutic encounter offers the chance of witnessing how a patient develops, how he gradually becomes freer in discussing his manifest concepts and feelings about his body, and how they are related to other aspects of his self-awareness. Gradually he comes to understand how the distortion of his structural as well as his functional body concept is closely interwoven with his experiences and interactions with significant people throughout his life.

DISTURBANCES IN SIZE AWARENESS

Best known among the disturbances in size awareness is the anorexic's denial, his vigorous defense of his emaciated body as not too thin but as just right, as normal. As a matter of fact, this complete denial of his starved appearance is pathognomic for true anorexia nervosa. Patients with severe weight loss for other reasons will not only readily admit that they are too thin, but often actively complain about the weight loss.

During the initial phase, when they reject anything coming from the outside, in particular all efforts to help them, this denial of thinness appears to be part of the general negativistic picture. However, even after a good therapeutic relationship has developed, when they appear to be actively interested in understanding the background of their condition, they will complain, with a certain bewilderment, that they cannot "see" how thin they are. A woman of 20, who seemed to be making good progress, admitted "I really cannot see how thin I am. I look into the mirror and still cannot see it; I know I am thin because when I feel myself I notice that there is nothing but bones."

Another girl of 19, also doing well in therapy, showed her physician two photographs taken on the beach, one when she was 15 years old and of normal weight, and the other when 17, with her weight down to 70 lbs. She asked him whether he could see a difference and admitted that she had trouble seeing a difference, though she knew there was one, and that she had been trying to correct this. When she looks at herself in a mirror then she sometimes can see that she is too skinny, "But I can't hold onto it." She may remember it for an hour but then begins to feel again that she is much larger. "I feel inwardly that I am larger than that—no matter what I tell myself. Even last summer I felt large, that was when I was at my lowest (67 lbs), but I felt I was very large." When first admitted to the hospital she felt that all the people around her were "too large," and even that the building was too large. She would look at herself in the mirror several times a day to help her maintain a realistic image, but each time it got larger. During a discussion her therapist used a comparison with a pumping mechanism, an analogy she eagerly took up. Regardless of what she saw in the mirror, there was this inner mechanism that kept on inflating her self-image. Only through repeated reality testing (looking into the mirror) could she "let the air go out again." A physician father described his daughter's attitude: "She appears to become profoundly depressed after a heavy meal and then feels as though she weighs 400 lbs. Conversely, when she is not eating and her weight is falling she appears in good spirits."

A realistic body-image concept is a precondition for recovery in anorexia nervosa. Many patients will gain weight for a variety of reasons but no real or lasting cure is achieved without correction of the body-image misperception. The resolution of the denial was studied through self-image confrontation in one case by Gottheil (8). After initial denial this patient began to see how thin she was, more strikingly on video tape than by looking into the mirror. After repeated self-confrontation it became more difficult for her to maintain the denial, and a change in her body image occurred so that thinness became ugly rather than comforting to her. The same observation is made with patients in long-term treatment without such direct confrontation with the video tape. Recovered patients invariably will express shocked amazement when they look at photographs from the time of their illness. When a new anorexic is admitted to the hospital they often will tell the personnel that it is useless to confront the new patient with her scrawny picture, or make her look at her ugly mirror image. Remembering how they felt, they will say, "It's no use—she cannot 'see' how thin she is."

Fat people, in contrast, vary widely in the way they perceive themselves. It seems that people who become obese as adults rather than in early childhood have a more realistic perception than those who suffer from lifelong obesity. Misperception in childhood obesity extends even to the mother. More than one mother whose fat child is referred for treatment from the school will say with indignation, "They say he is too fat. He doesn't look that way to me." Here, too, little therapeutic progress can be made unless the abnormal appearance is perceived and acknowledged. I recall a 42-year-old man, highly successful in his career, who had been overweight since adolescence, who in spite of circulatory difficulties did not follow a diet, denying that there was anything wrong with his size. One day he reported that he had stood in front of the mirror, taking in all details of how fat he was, and that he finally acknowledged "All of this is me." Until then he had done what many fat people do, simply avoided looking into a mirror or being photographed. Characteristic is the often repeated cartoon of the fat lady looking into the mirror and seeing in it a slender, glamorous, beauty queen.

Misconceptions about size appear also as the opposite, an exaggerated interpretation of any curve and excess weight as grotesquely fat. This is the background of the patients who eventually become anorexic, but also of the much larger group whom I have called "thin fat people." One later anorexic girl described this process. She had experienced any bodily change during puberty with intense discomfort. She began to deny that she had breasts or rounded buttocks, and maintained these denials over the years, long before her anorexic syndrome started. In a way she had developed a negative phantom, not seeing and accepting her budding body.

Not infrequently patients who have been severely obese for a long time will not "see" themselves as thinner even after weight reduction of considerable proportion; they carry the image of their former size like a phantom with them. This phenomenon was studied at the Rockefeller Institute in a group of superobese patients following weight reduction. These patients expressed concern over the alteration of their body size and experienced something like increased permeability of ego boundaries (7). Their figure drawings following weight loss showed larger waist diameters, with the belt extending over the body line and body area. In spite of the weight loss they reported persistent feelings of largeness. Characteristic is the acute sense of amazement of formerly fat people when someone sits down next to them on a bus, or the surprise of formerly fat girls when they are looked at or whistled at with admiration, instead of being stared at as monstrosities.

Misjudgments may occur also in concepts of the body structure. A bright 14-year-old boy, extremely obese (225 lbs), was phobic and would not go out on the street alone. He spoke with much embarrassment about the reasons for his fears, particularly his fear of losing weight.

> I really am afraid of any injury. I thought my body was like a thin layer of skin full of jelly. If you got hurt, the jelly would come out. I thought if you got hurt, *all* of it would come tumbling out—that there would be just a puddle of jelly or pea soup and a balloon of skin left around my bones. I did not really believe in what we learned in biology about bones and muscle. I thought it was only jelly hardened. I thought everything would just flow out—that I would become empty. I was so afraid of this emptiness; that was what led to the real stuffing.

He also had memories of having seen pictures of victims of concentration camps, which had filled him with real horror—that's what he would look like if he did not eat or if he got injured and the jelly ran out.

A student nurse who was hospitalized for an acute schizophrenic episode, and who had grown quite fat, was observed to eat ravenously whenever she had had an argument or felt threatened. Eventually she gave an explanation. She was afraid that the hostility of others and their angry words would rattle around inside her and keep on wounding her. By stuffing herself with food she would cover her sore inside, like with a poultice, and she would not feel the hurt so much. Similar distortions were reported by Petro about an 18-year-old anorexic girl (13). As she developed tender feelings toward the analyst, a desire came over her to sit snugly in his lap, to feel love and protective safety. She experienced it as a gradual shrinking of her body, turning into a baby, and being cuddled and held with infinite care. The patient's minute body softened up and finally merged with the analyst's body. On other occasions she would experience the analyst's body penetrating or engulfing or becoming one with her's in an indescribable way. This way her voidness was filled up, her separate existence ceased, and she was not able to feel who was in whom—and actually this distinction became meaningless.

I should like to add an artist's description of the fat person's misperception of herself. It is taken from *An American Home* by Helen Eustis (5).

> There was Mrs. Harrington, a small boned woman who wore tiny high heeled shoes with platform soles to increase her height, and there was her maid, Pearl, a young woman with dark brown skin. She had a small pretty face, but she was obese to the point of deformity, and her features were nearly lost in the largeness of her cheeks and chins, just as her personality was nearly lost in the neuroses of her size and race. In a dim, mixed-up way, Pearl felt that inside her, underneath her layers of flesh and her dark skin, there lived a Mrs. Harrington, too, loving luxury, waiting to be born and enjoy the world, beautiful, slim, desirable and even white.

MISPERCEPTION OF BODILY FUNCTIONS

The fat person's plaintive statement, "Everything I eat turns to fat," or its exaggeration, "I just have to look at chocolate cake and I gain 10 pounds," probably contains a kernel of truth; namely, that there are some metabolic disturbances facilitating fat deposition, but mainly it reflects his inability to judge correctly how much he eats. However much, there is always room for more. Others completely disassociate their size and what they eat. They act and talk as if they and their body were separate entities, and as if food they sneak (what they eat when nobody else watches) will not make them fat.

Conversely, the anorexic will complain of feeling "full" after a few bites of food, or even a few drops of fluid. I have the impression that this sense of fullness really is a phantom phenomenon, a projection of memories of formerly experienced sensations. An 18-year-old girl, intelligent and articulate, but obsessed with her size and with food, felt so little differentiated from others that she would assume the identity of whomever she was around and by watching others eat, in a way, "have people eat for her," and feel "full" after that, without having eaten at all. After having starved for some time, she reported, "I keep my mind eternally preoccupied with what size I am, always hoping it will become smaller. If I must eat—that takes too much mental energy to decide what, how much, and why must I. Every day I wake up in a prison, actually enjoying the confinement."

Another young anorexic patient "imitated eating—bite by bite—even the choice of food." She observed slim women or tall boys, not "voluptuous females" and would imitate what they ate. She is quite definite that she will *not eat* by self-choice.

A 29-year-old anorexic who had been sick for over 11 years and had several times been in critical condition, claimed that she had existed on fluids only for over six months, because food would make her "weak." She described the "weird ideas" she had in relation to food, "that it is the idea that solid food remains in the state I see it in after I eat it, and that it becomes part of me and thus has power over me." Even in talking about this she feels a sudden blankness, like from an electric shock, which she explains as "I suddenly saw an idea—my eating people in the same way as food."

Another frightening thought, "a depressed body thought," was:

> How will I look if I eat? I always feel full right then—it is as if I have really eaten, not just thought about it. Then I see myself—all of me—or sometimes just the hands or feet, just the breast or just the stomach—grossly distorted into a balloon like protruding hideous things. It's not that I fear food as much as I fear the irrational feeling that somehow the food almost has the power over me that a person would—it is almost as if it (the food) could make me eat it.

Other functions, too, are misinterpreted by fat and thin people. I shall give a few examples of the distortions in the assessment of their strength. It had been recognized during the study of obese children that the accumulation of fat, the increased body size, was not just the passive sequel of a positive energy balance but that it expressed a child's inner picture of what he wanted to be. His great weight gave him a certain feeling of safety and strength. The heavy layers of fat act like a wall behind which the child finds protection against a threatening outside world. Overt aggressive behavior is rare in obese children. Usually the eating itself seems to satisfy their aggressive impulses, and their size is a way of self-assertion. Many fat children have a dread of being small and skinny; although they are unhappy about their large size, they find security in it. This was graphically expressed by a girl of 6 who had grown obese when she was 3 years old, following a summer during which her high-strung and nervous mother had cared for her in place of her placid nurse (3).

This shy and awkward child revealed her problems and conflicts in a series of drawings. She was jealous of her younger sister, who was pretty and well-liked and enjoyed playing with other children. She sensed her mother's annoyance and disgust with her and was aware of her own hostile impulses toward her mother. She expressed this in her drawing of her mother when she said, "I make a ruffle around her dress, so she trips—and shall I make a cage around her?"

The "cage" had been previously used to express her hostility to a certain doctor whom she hated because he had "pricked" her. The drawings for this doctor and her mother were very similar in the symbolic use of form and color. She had finished the doctor's picture by cross-hatching the body, remarking, "I make a cage out of his dress and he will die."

She was very reluctant to draw a picture of herself. She would draw pictures of a "fairy," emptier and cruder than her other drawings. When asked to tell something about the fairy, she went into a singsong: "That's a girl—any girl—5500 years old. She does not like herself. I don't like myself. She is a fairy and she does not like to be a fairy. She wants to be a real girl."

The theme of the "girl who does not like herself" was accompanied by resigned remarks that there was nothing one could do to change her. She tried repeatedly to draw a picture of a girl who could like herself. She ended up by completely covering a picture that had been quite pleasing in the beginning, with scribblings of different colored crayons, finishing with a vigorous black scribble across the face. This was done with much display of emotion; "It's myself. I put every color in the box on it. I was not good in school last year. I put black on her face, and now you can't see her any more. *I hide her. She does not want to get hurt. We put her behind a wall.* A black face, nobody likes her."

In discussing this picture she repeated: "I don't like myself," and explained: "because I am fat. I eat too much, my mother says." This was followed by an enthusiastic description of all the things she likes to eat: "I like spaghetti, best. And do you know what else? I want them all in one meal: spaghetti, hominy, rice, noodles, and grits."

Fear of losing weight, of becoming thin, is common in fat youngsters. This fear is often shared by the mothers who in sly ways will refeed their children after they have reduced. Sophisticated obese adults will rarely express this explicitly, but do so rather indirectly by developing symptoms of weakness and unbearable tension, or by getting depressed while on a reducing regimen.

Anorexia nervosa patients experience something like the exact opposite. A 17-year-old girl, who went on a diet and lost weight when she learned that her boyfriend at college was dating other girls, felt for the first time "that she was getting results." Gradually she developed the feeling that her body had magical qualities. "My body could do anything—it could walk forever and ever and not get tired. I separated my mind and my body. My mind was tricky but my body was honest. It knew exactly what to do and I knew exactly what I could do. I felt very powerful on account of my body—my only weakness was my mind." At that time she started *compulsive walking rituals,* would walk for many miles even though she was increasingly cachectic. She would walk whether it was hot or raining, or a thunderstorm threatened. She felt "I have will power to walk as far as I want anytime—no matter what the weather is." She felt the same about her weight, "This was something I could control. I still don't know what I look like or what size I am, but I know my body can take anything." She was condescending toward her physicians who were increasingly alarmed about her condition, for she knew that she was immortal.

A 16-year-old anorexic girl, when at her lowest weight, was afraid of being "strong." Her ideal was to be weak, ethereal, and thin so that she could accept everybody's help without feeling guilty. Her deepest desire was to be blind so that she could show how noble she was in the face of suffering, and then she would be respected by everyone for this nobility. There was no realistic awareness of what it would be like to be blind, of not being able to see. In spite of this desire she was extremely active and perfectionistic, and would not permit herself to go to sleep until she had done calisthenics to the point of her muscles hurting.

Another anorexic girl, from an upper-class background, felt being fat indicated that one was exploitative. "If you are thin they don't think that you are rich and that your life is too easy. Being fat is like the kings in the Middle Ages; they are just rich and powerful and do nothing and everybody works for them. Looking and being exhausted shows that one does a lot, and without that I feel so undeserving."

Changes in this functional awareness are necessary milestones on the road to recovery. To quote from a patient who was doing well, "I took a walk—not to wear myself out or to prove 'I could make it' but just to enjoy the bright blue sky and the pretty yellow flowers. I seemed to do it without this 'double track' thinking." Another girl reported with real joy, "Last night I ate a chocolate bar and I enjoyed it. I told the nurse what kind I wanted and when she brought it, that's what I ate. There was none of the old fear 'I shouldn't do it' or that I wouldn't be able to stop once I had eaten one."

A 19-year-old girl who had become severely anorexic when away at college, with her weight dropping from 110 lbs in September to 70 lbs in January, made a serious suicidal attempt in February when her weight had risen to 75 lbs because she was afraid of getting "too fat" again. When there was further increase in weight she gave up all control as "useless," and in September her weight had risen to 140 lbs. Asked about this, she said, "It looks the same to me—I'm just fat." The only way she could imagine feeling proud again would be by being very slim, and she considered this a hopeless goal that she couldn't attain again.

MISPERCEPTION OF SEXUAL ROLE

It is generally assumed that obese people have difficulties with sex and that they are confused about their sexual role. Anorexia nervosa is conceived of by some as expressing a rejection of sex or of pregnancy. It is true that many obese people, particularly those who become psychiatric patients, express uneasiness about their sexual identity, but it is misleading to generalize from their problems to all obese people.

Among the many obese patients I have observed there was no instance of transsexualism, with the desire of changing his or her sex, nor delusional misconception of being a member of the other sex. There were a few instances of homosexuality or lesbianism, and one of transvestism. The incidence is not higher, maybe even lower, than among my nonobese patients. The great majority did not misperceive their anatomical structures, though anorexic girls tended to exclude awareness of curves and breasts from their body images. There was, however, marked confusion about the gender role. Rorschach records of obese children revealed, besides a preoccupation with the problem of "size," definite evidence of gender confusion, expressed by their tendency of interpreting symmetrical figures as "a man and a woman," or "a boy and a girl," a finding not observed in other patients (2). This type of interpretation was given more often by obese boys than girls.

Men who had been obese since childhood, or who had been obese children but had outgrown it, and who functioned adequately often expressed doubts about their masculinity; they were convinced of not being right, of being too soft, not masculine enough. This was more marked when there had been breast development during adolescence, even though it had disappeared. Such men are exceedingly careful about maintaining their weight at a normal level, though one aspect of their feeling "not right" is the ease with which they tend to gain weight when they relent their vigilance.

From direct contact with mothers of fat children it was learned that many had openly expressed a wish for a girl child, and some had raised their sons as if they were daughters. Some mothers were able to follow advice to discontinue this and to permit their sons to be boys. The most seriously disturbed boy in this group had a mother who did not change in her attitude. He became openly psychotic in adolescence, pulled out his body hair in pubescence "in order not to be like an animal," began to bleach his hair, used cosmetics, and said in an artificially high-pitched voice that he wanted to be good and would not have any dealings with the dirty habits of other youngsters.

Many other obese adolescents will speak about not wanting to be like the rough boys. One spoke openly of his fear of growing up, of his fear of manhood, and he related this to his hatred for his father. "It is not that I want to be a woman—I don't want to be a man, the way my father is." He felt his father wanted him to be brave and to accomplish things, but since he felt his father wanted this for *his own* gratification, he would not do any of the things expected of a young man.

Men who become obese later in life seem to feel differently. Often they consider their girth a sign of their imposing power and virility. This was described as a cultural trait for German men, with upper-class men tending to be heavy as a sign of their status and power (15).

On the whole, obese girls and women are much more open and articulate in discussing their dissatisfaction with their sex or their concepts of what they would want to be. In psychoanalytic literature there are repeated references to the large body being equated with the phallus. Even during intensive analytic work, not one of my patients made such a statement spontaneously, though they would pick it up if it was suggested to them. Many obese girls and women openly expressed how, throughout their lives, they had felt they had been destined to be "more than a woman." Bigness means to be like father.

> Consciously I fight being fat, but unconsciously I accept it. It has to do with the fact that father was fat. He eats all the things he is not supposed to eat, and he is big all over. Mother will put him on a diet which he follows for a few days and then he gives it up. As long as I remember he was big all over. He was quite slim when he was younger but when he became an important man he grew fat and has maintained his weight ever since. He has all the attributes of being busy and important—he could not be thin at the same time. I always felt just like him and I accepted it that it is the thing to be when one is so important. Mother was always so careful about what she ate. When she gained one or two pounds, she would go on a diet. That proved to me "people with nothing else to do, they can keep on diet-

ing." *I* had more important things to do, I am like my father. I always thought of mother as useless and unimportant.

This idea can be found in the thinking of many fat women, that only unimportant people can waste time and energy on dieting. "The littleness of it, so picayune. You have to be so careful—that is not living," or "I hate the small, chattering magpie kind of woman, so small—such small souls. When I am with skinny girls I feel like a mountain with lots of gnats around. They make sweetness and littleness important." Or, "Being a woman means doing things as a minor study. Being big is my real aspiration. That means not to be a woman but to be somewhat more than a man—a hero."

The openly expressed desire of such a woman was not to be man, but, what the Greeks called the "third sex," to be both a woman *and* a man. Several had elaborate fantasies about their lives in both roles. One condition for this was *being big*, in spite of all the criticism and social ostracism their fatness implied. This is even more openly expressed by anorexic girls, whether they had been previously overweight or not. One girl with active fantasies about sex would stand in front of the mirror and she saw in the reflection not the scrawny creature she was at that time but the most attractive young woman, and in her mind's eye she experienced her viewing self, the body outside the mirror, as a young man who was going to seduce this girl. In a way she had established the absolute of independence, of needing nothing from the outside, of carrying out a love affair with herself.

Several of the anorexic girls had had active fantasies as children about being a boy—a young prince or pageboy. One carried on a continuous fantasy, silently entertaining herself at the dinner table where her parents and older sister were engaged in conversation, feeling that what she had to say was not important or would not be listened to. Puberty development put a shocking end to this self-image of being the long-legged prince who walked along with a striding gait. Even after she had become anorexic she could recapture this sense of being the prince by going on walks where she would take long striding steps.

Of particular interest are observations of patients who alternate between being fat and thin. A young divorced woman entered the psychiatric service after 3 months of compulsive eating during which she had gained 50 lbs, and had become completely incapacitated. Compulsive eating had been a problem for her for the previous eight years, but never to the same degree. She had been married at age 19, after a year and a half of college, and was divorced a few months later. She then went to live abroad where she stayed for some time. After her return she married again and this, too, ended in divorce. She began to eat almost without stopping, hoarded and hid food in her apartment, became untidy and seclusive, and appeared to be unable to care for herself. She was quite articulate, sometimes overly dramatic, in describing her experiences, and some of the alternatives she described were, I am sure, artificially constructed. She considered her fat self "unreal" and felt as soon as she had lost weight that she was cured, that she was "real" again. As this *real, thin self* she feels transparent, admired, acclaimed, active, capable of giving, well-groomed, interested in men, having many heterosexual experiences. But *when fat* she feels *unreal*, dirty, slovenly, a nothing, passive, always on the receiving end, gorging herself like in a stupor, only interested in women, even becoming involved in homosexual affairs. In working with her it became apparent that her inner perception was as confused during the thin, starving phase as during the slovenly, fat period. What was revealed was dissatisfaction with her sex, her size, her capacities, and her ability to relate to others.

EMOTIONAL, AFFECTIVE, AND SOCIAL ASPECTS

Body image disturbances are experienced also in the affective sphere; this is probably the aspect of which one is most aware, whether one is satisfied with or unhappy about one's body and appearance. I have tried, in the preceding discussion, to separate cognitive and perceptive awareness of one's body from the affective attitude, yet comments on the feeling tone invariably crept in. Similarly, it will not be possible to speak of one's emotional attitude toward his body without attention to the social value system. Our social climate praises slenderness to such an extent that it is astounding that not all fat people suffer from disgust and self-hatred for being fat. Many, in particular adolescents and those who have been fat since childhood, speak with real anguish about their terrible fate and their hatred for their ugly and loathsome bodies. There is a tendency to consider the social stigma, the feeling of being a member of a despised minority, as the main cause for the psychological problems and low self-esteem of the obese.

The experience of self-hatred and contempt for being fat is frequent enough, yet it appears to be not only determined by social attitudes; it is also closely interwoven with psychological and interpersonal experiences early in life. It is not surprising that a young man whose mother had said proudly "the family thinks the bigger he is the better he is" developed a basically self-accepting attitude, even though severely overweight since childhood. Entirely different is the self-image of a young girl who, under constant criticism, had become so discouraged that she had withdrawn from all activities and was so apathetic that she gave the impression of being stupid, if not feeble-minded. Speaking of her childhood she would say "as far back as I can remember I was too fat and my mother always harped on it. Everything was done to make me thinner; that was the only thing that mattered. That's why I hate myself and my body. Fat people just disgust me."

I have seen occasionally a fat adolescent in psychiatric consultation who could say with reassurance, "Yes, I know it is not fashionable but that's the way I'm built, and my mother can say what she wants. I have to go through life my own way." The great majority of obese youngsters who come to psychiatric attention express their dislike and disgust in many dramatic ways. If one would rely on first information only, one might be inclined to say that childhood and adolescent obesity invariably leads to hateful, negative appraisal of and disgust for one's own body, but with which one is deeply identified. In play sessions with children Homburger (Erikson) (9) observed that obese children constructed buildings which resembled the outline of their bodies, a finding not observed in any other group. Mixed with this expressed disgust is guilt for having only oneself to blame, and a sense of shame for being weak-willed. Some openly equate the guilt about overeating with guilt about masturbation. Being fat is a public display of their transgression, a demonstration of self-indulgence and lack of control and will power. Yet with extended contact (during therapy), the information often changes. Gradually it becomes apparent that dislike of fatness is only the surface response; it is a concrete symbol of everything one dislikes in oneself and of what one considers contemptible and bad. Many will state that their fatness is only what shows on the outside; their deeper shame is their conviction of being awkward and ugly on the inside.

Quite often an obese youngster will "advertise" his low self-esteem, his contempt for his ugly body, by the way he dresses and presents himself. Most of my observations were made before "hippie" dress had become the fashion. A boy of 14, who made it a point to look disheveled and neglected, explained: "You can tell me from my cover." Occasionally he would appear well-dressed and well-groomed, looking entirely different, not even as grossly obese as usual. This was on days when he planned to visit his wealthy grandfather. He feared that the doorman would not admit him in his disheveled garb. The 16-year-old daughter of a beautiful woman, repeatedly on the "best dressed" list, usually appeared in gray or brown sack-like dresses, bought at the cheapest stores. Even after she had lost weight, the color and fit of her dresses would announce her mood.

Anorexics, too, often express their body awareness through their clothing. Whether it is fashionable or not, they will hold their dresses in with tightly-drawn wide belts. The belt seems to serve as a control; as long as it can be closed at the same hole they feel reassured of not having gained an ounce. With all the denial of the scrawniness of their figures, many seem to dislike the skeleton-like look of their arms; summer or winter, they will wear long-sleeved dresses, covering their ugly-looking arms. As one expressed it, though she did not feel "too thin," she felt that her arms made her look like "a praying mantis."

In working with such people a much more complex attitude of their self-concept gradually evolves. Many speak quite openly of feeling like a dual person, a fat one and a thin one. "I seem to be two people; one wants to be thin, but the other *wants* to be fat." When asked about the personality of the fat and thin person, it turns out that many fat people conceive of the thin person as one who is doing everything that is expected of a young girl, who is conforming, studying, socially active, well-groomed and well-dressed, but who deep down is a selfish and undesirable person who lacks inner qualities and values and who relies on looks, is ruthless, and babbles socially, "having a line." In contrast, the fat girl, the one who procrastinates with her studies, who is unhappy about moping around at home, who is the constant butt of her parents' and friends' belittling remarks, is deep down a worthwhile person, kind and considerate, honest in dealing with people and truly valuing her friends. The deepest pain is felt because the family and society at large praise and admire the thin one, regardless of her selfish and ruthless traits, and devalue and despise the fat one with the much more valuable human qualities.

A brilliant young college girl described her whole childhood as a time of continuous humiliation. "I always felt I was dull and clumsy—even as a child I was considered 'too big.'" She had become fat only after puberty but the conviction of not being desirable, of not having "sex appeal," had preceded the weight gain. "It was not the heaviness—there was something about me, the bigness, the clumsiness." But with all her complaining, there was also reluctantly admitted pride in being big. When she enrolled in college, she was rather disappointed that nobody paid special attention to her for being so big and heavy. Somehow the feeling of "specialness" had been a comfort in this continuous shame and embarrassment about her bodily existence.

One might say that anorexia nervosa by its very existence proves that the hateful self-contempt is not really related to the excess weight, but to some deep inner dissatisfaction. Not one of the anorexic patients whom I have come to know over the years had set out to reach this state of pitiful emaciation. All they had wanted to achieve was to feel better about themselves. Since they had felt that "being too fat" was the cause of their despair, they were determined to correct it. Whatever weight they reached in this struggle for self-respect and respect from others, it was "not right" for giving them inner reassurance, and so the downhill course continued. One girl had suffered all her life from the feeling of being "ugly," of not being a "beauty." She was overweight and started a reducing program, at first rather sensibly, reducing from 140 to 110 lbs between ages 13 and 16. But then, when things were still "not right," she went on a starvation regimen, to a low of 67 lbs. She explained it as "about being fat I could do something, but not about being ugly." She knew that her skeleton-like appearance was hideous, if not from her own inspection then from the reaction of others. Her argument was "What if my weight went back to normal and I were still ugly, what would I do then?" At the same time she knew that what was covered under the term "ugly" applied as much to psychological attributes and to disappointment in her own achievements and behavior as to the physical evidence.

Another anorexic girl said of herself, "I was trying to be what other people thought I was. I am an unknown quantity—it all began with mother's obsession with social success."

A systematic study of obese people's emotional attitudes toward their body was conducted by Stunkard and Mendelson (19) and Stunkard and Burt (20). They found that almost half of the obese individuals who had become obese before or at the time of puberty, but who were post-adolescent at the time of the study, had severely disturbed body-image concepts, along with greatly impaired interpersonal and heterosexual adjustment. In striking contrast not one of the patients who had become obese as adults had the same degree of disturbance in these areas. These findings are in agreement with my own observations that it is the distorting and undermining interpersonal experiences early in life that result in this self-concept of being monstrous and grotesque.

OWNERSHIP OF THE BODY AND ITS CONTROL

It was only through long-range therapeutic work that the meaning of the continuous complaint of "not being right" was gradually recognized through the discovery that these patients experienced their bodies as not being truly their own, as being under the influence of others. They felt they had no control over their bodies and its functions. It had always been known that fat people suffered from "lack of will power," and this term has often been used in a reproachful way. In the course of my studies it became clear that this lack of will power really represents one of the basic issues and is related to their inability to perceive their bodily needs. Fat people tend to talk about their bodies as external to themselves. They do not feel identified with this bothersome and ugly thing they are condemned to carry through life, and in which they feel confined or imprisoned. An English writer, Cyril Connolly, himself a fat man, expressed this graphically: "Imprisoned in every fat man a thin one is wildly signaling to be let out" (4). Another English writer, Kingsley Amis, paraphrased this: "Outside every fat man there is an even fatter man trying to close in" (1).

Anorexia nervosa patients will express this not being identified with their bodies even more definitely. I quote from the history of a 20-year-old woman who had become anorexic at age 15, reducing from 90 to 55 lbs. She made what looked like a spontaneous recovery but became depressed when she went to college at age 18. She also began to suffer from bouts of compulsive eating which terrified her because she was afraid of getting fat. Her weight fluctuated between 103 and 110 lbs, but from her continuous preoccupation with her weight one might have concluded that she was monstrously fat.

In the course of treatment she gradually recognized that her whole life had been an effort to fit into the idealized image her parents had planned for her, particularly her mother, whose own childhood had been deprived and who had wanted her daughter to have all the things she had missed. The dominant feeling at college, which led to the depressive withdrawal, was the question, "Do *I* have to go to college just because mother never went?" Toward the end of treatment she became concerned with the question of what had led to this particular illness. She had never thought of it in an active way, namely that she herself had stopped eating. She had always thought of it as an outside event, as "*It* happened to me." Once she had recognized the importance of this, she raised the question "Do you think I was hurting my parents by not eating? Father is so overconcerned about anything that has to do with physical health. I realize now I was hurting them by not eating. The more they worried about it the more I was hurting them." It had never occurred to her that it was she and her body that were undergoing this ordeal of starvation.

A young fat girl from a similarly overprivileged background had one favorite daydream, namely that she would reduce to such an extent that she would look pitiful and her father would be worried and would plead with her to eat. There was a feeling of triumph that it would serve him right, "that will truly hurt him when I look miserable and scrawny and he cannot make me eat." Again there was no awareness that she and her body would be involved and suffer.

This was even more clearly expressed by a male anorexic who had been sick since age 12 and had successfully resisted all treatment efforts, weighing less than 50 lbs at age 18. Throughout this time all his struggles and fights had been directed against efforts to *make* him eat. He protested, "It's my life. If I don't want to eat why should my mother make a fuss over it." He had gradually developed a real fear of the scale. "I feel I get evaluated by it and then I am panicky. If I gain, *they* are so proud; and if I lose, my mother blows her head off. It is always somebody else's business."

In connection with a visit to his parents he used the expression "after all, I am their property." When this was singled out for discussion he at first wanted to dismiss it, and said that he had used the wrong expression. But then he took it up in detail, saying "Everything I have and own comes from them."

It was only after considerable therapeutic progress that he began to recognize his body and its functions as his own, and could let go of his long-standing symptoms. When he was transferred to an open ward he expressed the feeling: "I am free—I *own my body*—I am not supervised any more by nurses or by mother." His attitude toward his weight and what he ate underwent a complete change. "I have controlled myself so long, now I want to enjoy food." Or at another time, "Now if I lose weight it makes me feel sick that I am losing something that is *mine*." This young man, whose treatment will be discussed in Chapter 17, made a complete recovery.

Probably the most dramatic misperception of her body, and dissatisfaction and despair about it, was expressed by a young woman whose history was presented in Chapter 4. I shall repeat here a few sentences relevant for the development of her body image. As a child she had been under relentless pressure to fulfill her mother's image of what she should be. Unfortunately this image stood in contrast to the child's constitutional make-up. The mother had been in her late thirties when she became pregnant, at a time when she no longer expected a child. She had accepted that she would become a mother in the hope of having a child like her sister's plump and cuddly baby. Her newborn baby looked long and stringy and the pediatrician jokingly remarked, "Put some fat on those bones." The mother took this recommendation literally and a situation developed in which she shoveled food into the infant, who would reject it; she would shovel it back, forcing the child to swallow some more. Initially it looked as if the mother was winning this contest; the little girl grew fat but then was blamed for being clumsy and ugly.

This girl was unable to exercise any control over her eating, she was either monstrously fat or maintained her weight at an artificially low level. This she could accomplish only by forcing her parents never to have extra food in the home except what she needed to maintain this low weight. She reproached her parents for having created her as a monster; she now was waiting for them to change so that they could recreate her in a better image. She expressed in exaggerated degree the feeling of having no sense of ownership of her body or any effective self-awareness. She suffered from enormous self-hatred for being fat, though she could not "see" herself either as fat or as thin, and was completely unable to control her food intake or any other aspect of her life. She experienced herself as controlled from the outside, lacking initiative and autonomy, without a personality, the sad product of her parents' mistakes, "a nothing," and the despised body was, she felt, not really hers.

REFERENCES

Amis, K., *One Fat Englishman,* Harcourt, Brace & World, Inc., New York, 1963.

Bruch, H., Obesity in childhood and personality development, *Amer. J. Ortho-psychiat.*, 11: 467–474, 1941.

Bruch, H., Food and emotional security, *The Nervous Child*, 3: 165–173, 1944.

Connolly, C., *The Unquiet Grave,* Harper & Bros., New York, 1945.

Eustis, H., An American home, pp. 47–64, in *The Captains and the Kings Depart,* Harper & Bros., New York, 1949.

Fisher, S., and Cleveland, S. E., *Body Image and Personality,* Dover Publications, Inc., New York, 1968.

Glucksman, M. L., Hirsch, J., McCully, R. S., Barron, B. A., and Knittle, J. L., The response of obese patients to weight reduction, II. A quantitative evaluation of behavior, *Psychosom. Med.*, 30: 359–375, 1968.

Gottheil, E., Backup, C. E., and Cornelison, F. S., Denial and self-image confrontation in a case of anorexia nervosa, *J. Nerv. Ment. Dis.*, 148: 238–250, 1969.

Homburger (Erikson), E., Traumatische Konfigurationen im Spiel, *Imago*, 23: 447–462, 1937.

Kolb, L. C., Disturbances of the body image, pp. 749–769, in *American Handbook of Psychiatry, Vol. 1*, S. Arieti, ed., Basic Books, Inc., New York, 1959.

Levy, D. M., Body interest in children and hypochondriasis, *Amer. J. Psychiat.*, 12: 295–311, 1932.

Nathan, S., and Pisula, D., Psychological observations of obese adolescents during starvation treatment, *J. Amer. Acad. Child Psychiat.*, 9: 722–740, 1970.

Petro, A., Body image and archaic thinking, *Int. J. Psychoanal.*, 40: 1–9, 1959.

Piaget, J., *The Construction of Reality in the Child,* Basic Books, Inc., New York, 1954.

Pflanz, M., Mediziniseh-soziologische Aspekte der Fettsucht, *Psyche*, 16: 579–591, 1963.

Schilder, P., *Das Koerperschema: Ein Beitrag zur Lehre vom Bewusstsein des eigenen Koerpers,* Julius Springer, Berlin, 1923.

Schilder, P., *The Image and Appearance of the Human Body,* Psyche Monographs, 4, London, 1935.

Shontz, F. C., *Preceptual and Cognitive Aspects of Body Experience,* Academic Press, New York, 1969.

Stunkard, A., and Mendelson, M., Obesity and the body image: I. Characteristics of disturbances in the body image of some obese persons, *Amer. J. Psychiat.*, 123: 1296–1300, 1967.

Stunkard, A., and Burt, V., Obesity and the body image: II. Age at onset of disturbances in the body image, *Amer. J. Psychiat.*, 123: 1443–1447, 1967.

17

Anorexia Nervosa

Psychopathology as the Crystallization of Culture

SUSAN BORDO

1996 PREFATORY NOTE

In 1983, preparing to teach an interdisciplinary course called "Gender, Culture, and Experience," I felt the need for a topic that would enable me to bring feminist theory alive for a generation of students that seemed increasingly suspicious of feminism. My sister, Binnie Klein, who is a therapist, suggested that I have my class read Kim Chernin's The Obsession: Reflections on the Tyranny of Slenderness. *I did, and I found my Reagan-era students suddenly sounding like the women in the consciousness-raising sessions that had first made me aware of the fact that my problems as a woman were not mine alone. While delighted to have happened on a topic that was so intensely meaningful to them, I was also disturbed by what I was reading in their journals and hearing in the privacy of my office. I had identified deeply with the general themes of Chernin's book. But my own disordered relations with food had never reached the point of anorexia or bulimia, and I was not prepared for the discovery that large numbers of my students were starving, binging, purging, and filled with self-hatred and desperation. I began to read everything I could find on eating disorders. I found that while the words and diaries of patients were enormously illuminating, most of the clinical theory was not very helpful. The absence of cultural perspective—particularly relating to the situation of women—was striking.*

As a philosopher, I was also intrigued by the classically dualistic language my students often used to describe their feelings, and I decided to incorporate a section on contemporary attitudes toward the body in my metaphysics course. There, I discovered that although it was predominantly my female students who experienced their lives as a perpetual battle with their bodies, quite a few of my male students expressed similar ideas when writing about running. I found myself fascinated by what seemed to me to be the cultural emergence of a set of attitudes about the body which, while not new as ideas, *were finding a special kind of embodiment in contemporary culture, and I began to see all sorts of evidence for this cultural hypothesis. "Anorexia Nervosa: Psychopathology as the Crystallization of Culture," first published in 1985, was the result of my initial exploration of the various cultural axes to which my students' experiences guided me in my "Gender, Culture, and Experience" and metaphysics courses. Other essays followed, and ultimately a book,* Unbearable Weight: Feminism, Western Culture, and the

Body, *further exploring eating disorders through other cultural interconnections and intersections: the historically female disorders, changes in historical attitudes toward what constitutes "fat" and "thin," the structural tensions of consumer society, the post-modern fascination with re-making the self.*

Since I began this work in 1983, my then-tentative intuitions have progressively been validated, as I have watched body practices and attitudes that were a mere ripple on the cultural scene assume a central place in the construction of contemporary subjectivity. In 1995, old clinical generalizations positing a distinctive class, race, family and "personality" profile of the woman most likely to develop an eating disorder no longer hold, as images of the slender, tight body become ever-more widely deployed, asserting their homogenizing power over other cultural ideals of beauty, other cultural attitudes toward female appetite and desire. The more generalized obsession with control of the body which I first began to notice in the early eighties now supports burgeoning industries in exercise equipment, diet products and programs, and cosmetic surgery—practices which are engaged in by greater numbers and more diverse groups of people all the time. On television, "infomercials" hawking stomach-flatteners, miracle diet plans, and wrinkle-dissolving cosmetics have become as commonplace as aspirin ads. As the appearance of our bodies has become more and more important to personal and professional success, the incidence of eating disorders has risen, too, among men. All of this has led to an explosion of written material, media attention and clinical study, much of it strongly bearing out my observations and interpretations. I have not, however, incorporated any new studies or statistics into the piece reprinted here. With the exception of a few endnotes, it appears substantially as it did in its original version.

> Historians long ago began to write the history of the body. They have studied the body in the field of historical demography or pathology; they have considered it as the seat of needs and appetites, as the locus of physiological processes and metabolisms, as a target for the attacks of germs or viruses; they have shown to what extent historical processes were involved in what might seem to be the purely biological "events" such as the circulation of bacilli, or the extension of the lifespan. But the body is also directly involved in a political field; power relations have an immediate hold upon it; they invest it, mark it, train it, torture it, force it to carry out tasks, to perform ceremonies, to emit signs.
>
> *Michel Foucault*, Discipline and Punish

> I believe in being the best I can be,
> I believe in watching every calorie . . .
>
> *Crystal Light television commercial*

EATING DISORDERS, CULTURE, AND THE BODY

Psychopathology, as Jules Henry has said, "is the final outcome of all that is wrong with a culture."[1] In no case is this more strikingly true than in that of anorexia nervosa and bulimia, barely known a century ago, yet reaching epidemic proportions today. Far from being the result of a superficial fashion phenomenon, these disorders, I will argue, reflect and call our attention to some of the central ills of our culture—from our historical heritage of disdain for the body, to our modern fear of loss of control over our future, to the disquieting meaning of contemporary beauty ideals in an era of greater female presence and power than ever before.

Changes in the incidence of anorexia[2] have been dramatic.[3] In 1945, when Ludwig Binswanger chronicled the now famous case of Ellen West, he was able to say that "from a psychiatric point of view we are dealing here with something new, with a new symptom."[4] In 1973, Hilde Bruch, one of the pioneers in understanding and treating eating disorders, could still say that anorexia was "rare indeed."[5] Today, in 1984, it is estimated that as many as one in every 200–250 women between the ages of thirteen and twenty-two suffer from anorexia, and that anywhere from 12 to 33 percent of college women control their weight through vomiting, diuretics, and laxatives.[6] The New York Center for the Study of Anorexia and Bulimia reports that in the first five months of 1984 it received 252 requests for treatment, as compared to the 30 requests received in all of 1980.[7] Even correcting for increased social awareness of eating disorders and a greater willingness of sufferers to report their illnesses, these statistics are startling and provocative. So, too, is the fact that 90 percent of all anorectics are women, and that of the 5,000 people each year who have part of their intestines removed as an aid in losing weight 80 percent are women.[8]

Anorexia nervosa is clearly, as Paul Garfinkel and David Garner have called it, a "multidimensional disorder," with familial, perceptual, cognitive, and, possibly, biological factors interacting in varying combinations in different individuals to produce a "final common pathway."[9] In the early 1980s, with growing evidence, not only of an overall increase in frequency of the disease, but of its higher incidence in certain populations, attention has begun to turn, too, to cultural factors as significant in the pathogenesis of eating disorders.[10] Until very recently, however, the most that could be expected in the way of cultural or social analysis, with very few exceptions, was the (unavoidable) recognition that anorexia is related to the increasing emphasis that fashion has placed on slenderness over the past fifteen years.[11] This, unfortunately, is only to replace one mystery with another, more profound than the first.

What we need to ask is *why* our culture is so obsessed with keeping our bodies slim, tight, and young that when 500 people were asked what they feared most in the world, 190 replied, "Getting fat."[12] In an age when our children regularly have nightmares of nuclear holocaust, that as adults we should give *this* answer—that we most fear "getting fat"—is far more bizarre than the anorectic's misperceptions of her body image, or the bulimic's compulsive vomiting. The nightmares of nuclear holocaust and our desperate fixation of our bodies as arenas of control—perhaps one of the few available arenas of control we have left in the twentieth century—are not unconnected, of course. The connection, if explored, could be significant, demystifying, instructive.

So, too, we need to explore the fact that it is women who are most oppressed by what Kim Chernin calls "the tyranny of slenderness," and that this particular oppression is a post-1960s, post-feminist phenomenon. In the fifties, by contrast, with middle-class women once again out of the factories and safely immured in the home, the dominant ideal of female beauty was exemplified by Marilyn Monroe—hardly your androgynous, athletic, adolescent body type. At the peak of her popularity, Monroe was often described as "femininity incarnate," "femaleness embodied"; last term, a student of mine described her as "a cow." Is this merely a change in what size hips, breasts, and waist are considered attractive, or has the very idea of incarnate femaleness come to have a different meaning, different associations, the capacity to stir up different fantasies and images, for the culture of the eighties? These are the sorts of questions that need to be addressed if we are to achieve a deep understanding of the current epidemic of eating disorders.

The central point of intellectual orientation for this essay is expressed in its subtitle. I take the psychopathologies that develop within a culture, far from being anomalies or aberrations, to be characteristic expressions of that culture; to be, indeed, the crystallization of much that is wrong with it. For that reason they are important to examine, as keys to cultural self-diagnosis and self-scrutiny. "Every age," says Christopher Lasch, "develops its own peculiar forms of pathology, which express in exaggerated form its underlying character structure."[13] The only aspect of this formulation with which I would disagree, with respect to anorexia, is the idea of the expression of an underlying, unitary cultural character structure. Anorexia appears less as the extreme expression of a character structure than as a remarkably overdetermined *symptom* of some of the multifaceted and heterogeneous distresses of our age. Just as anorexia functions in a variety of ways in the psychic economy of the anorexic individual, so a variety of cultural currents or streams converge in anorexia, find their perfect, precise expression in it.

I will call those streams or currents "axes of continuity": *axes* because they meet or converge in the anorexic syndrome; *continuity* because when we locate anorexia on these axes, its family resemblances and connections with other phenomena emerge. Some of these axes represent anorexia's *synchronicity* with other contemporary cultural practices and forms—bodybuilding and jogging, for example. Other axes bring to light *historical* connections: for instance, between anorexia and earlier examples of extreme manipulation of the female body, such as tight corseting, or between anorexia and long-standing tradition and ideologies in Western culture, such as our Greco-Christian traditions of dualism. The three axes that I will discuss in this essay (although they by no means exhaust the possibilities for cultural understanding of anorexia) are the *dualist axis,* the *control axis,* and the *gender/power axis.*[14]

Throughout my discussion, it will be assumed that the body, far from being some fundamentally stable, acultural constant to which we must *contrast* all culturally relative and institutional forms, is constantly "in the grip," as Foucault puts it, of cultural practices. Not that this is a matter of cultural *repression* of the instinctual or natural body. Rather, there is no "natural" body. Cultural practices, far from exerting their power *against* spontaneous needs, "basic" pleasures or instincts, or "fundamental" structures of body experience, are already and always inscribed, as Foucault has emphasized, "on our bodies and their materiality, their forces, energies, sensations, and pleasures."[15] Our bodies, no less than anything else that is human, are constituted by culture.

Often, but not always, cultural practices have their effect on the body as experienced (the "lived body," as the phenomenologists put it) rather than the physical body. For example, Foucault points to the medicalization of sexuality in the nineteenth century, which recast sex from being a family matter into a private, dark, bodily secret that was appropriately investigated by such specialists as doctors, psychiatrists, and school educators. The constant probing and interrogation, Foucault argues, ferreted out, eroticized and solidified all sorts of sexual types and perversions, which people then experienced (although they had not done so originally) as defining their bodily possibilities and pleasures. The practice of the medical confessional, in other words, in its constant foraging for sexual secrets and hidden stories, actually *created* new sexual secrets—and eroticized the acts of interrogation and confession, too.[16] Here, social practice changed people's *experience* of their bodies and their possibilities. Similarly, as we shall see, the practice of dieting—of saying no to hunger—contributes to the anorectic's increasing sense of hunger as a dangerous eruption from some alien part of the self, and to a growing intoxication with controlling that eruption.

The *physical* body can, however, also be an instrument and medium of power. Foucault's classic example in *Discipline and Punish* is public torture during the Ancien Régime, through which, as Dreyfus and Rabinow put it, "the sovereign's power was literally and publicly inscribed on the criminal's body in a manner as controlled, scenic and well-attended as possible."[17] Similarly, the nineteenth-century corset caused its wearer actual physical incapacitation, but it also served as an emblem of the power of culture to impose its design on the female body.

Indeed, female bodies have historically been significantly more vulnerable than male bodies to extremes in both forms of cultural manipulation of the body. Perhaps this has something to do with the fact that women, besides *having* bodies, are also *associated* with the body, which has always been considered woman's "sphere" in family life, in mythology, in scientific, philosophical, and religious ideology. When we later consider some aspects of the history of medicine and fashion, we will see that the social manipulation of the female body emerged as an absolutely central strategy in the maintenance of power relations between the sexes over the past hundred years. This historical understanding must deeply affect our understanding of anorexia and of our contemporary preoccupation with slenderness.

This is *not* to say that I take what I am doing here to be the unearthing of a long-standing male conspiracy against women or the fixing of blame on any particular participants in the play of social forces. In this I once again follow Foucault, who reminds us that although a perfectly clear logic, with perfectly decipherable aims and objectives, may characterize historical power relations, it is nonetheless "often the case that no one was there to have invented" these aims and strategies, either through choice of individuals or through the rational game plan of some presiding "headquarters."[18] We are not talking, then, of plots, designs, or overarching strategies. This does not mean that individuals do not *consciously* pursue goals that in fact advance their own position. But it does deny that in doing so they are consciously directing the overall movement of power relations or engineering their shape. They may not even know what that shape is. Nor does the fact that power relations involve domination by particular groups—say, of prisoners by guards, females by males, amateurs by experts—entail that the dominators are in anything like full control of the situation or that the dominated do not sometimes advance and extend the situation themselves.[19] Nowhere, as we shall see, is this collaboration in oppression more clear than in the case of anorexia.

THE DUALIST AXIS

I will begin with the most general and attenuated axis of continuity, the one that begins with Plato, winds its way to its most lurid expression in Augustine, and finally becomes metaphysically solidified and scientized by Descartes. I am referring, of course, to our dualistic heritage: the view that human existence is bifurcated into two realms or substances: the bodily or material, on the one hand; the mental or spiritual, on the other.

Despite some fascinating historical variations which I will not go into here, the basic imagery of dualism has remained fairly constant. Let me briefly describe its central features; they will turn out, as we will see, to comprise the basic body imagery of the anorectic.

First, the body is experienced as *alien*, as the not-self, the not-me. It is "fastened and glued" to me, "nailed" and "riveted" to me, as Plato describes it in the *Phaedo*.[20] For Descartes, the body is the brute material envelope for the inner and essential self, the thinking thing; it is ontologically distinct from that inner self, is as mechanical in its operations as a machine, is, indeed, comparable to animal existence.

Second, the body is experienced as *confinement and limitation:* a "prison," a "swamp," a "cage," a "fog"—all images that occur in Plato, Descartes, and Augustine—from which the soul, will, or mind struggles to escape. "The enemy ["the madness of lust"] held my will in his power and from it he made a chain and shackled me," says Augustine.[21] In the work of all three philosophers, images of the soul being "dragged" by the body are prominent. The body is "heavy, ponderous," as Plato describes it; it exerts a downward pull.[22]

Third, the body is *the enemy,* as Augustine explicitly describes it time and again, and as Plato and Descartes strongly suggest in their diatribes against the body as the source of obscurity and confusion in our thinking. "A source of countless distractions by reason of the mere requirement of food," says Plato; "liable also to diseases which overtake and impede us in the pursuit of truth; it fills us full of loves, and lusts, and fears, and fancies of all kinds, and endless foolery, and in very truth, as men say, takes away from us the power of thinking at all. Whence come wars, and fightings, and factions? Whence but from the body and the lusts of the body."[23]

And, finally, whether as an impediment to reason or as the home of the "slimy desires of the flesh" (as Augustine calls them), the body is the locus of *all that threatens our attempts at control.* It overtakes, it overwhelms, it erupts and disrupts. This situation, for the dualist, becomes an incitement to battle the unruly forces of the body, to show it who is boss. For, as Plato says, "Nature orders the soul to rule and govern and the body to obey and serve."[24]

All three—Plato, Augustine, and, most explicitly, Descartes—provide instructions, rules, or models of how to gain control over the body, with the ultimate aim—for this is what their regimen finally boils down to—of learning to live without it.[25] By that is meant: to achieve intellectual independence from the lure of the body's illusions, to become impervious to its distractions, and, most important, to kill off its desires and hungers. Once control has become the central issue for the soul, these are the only possible terms of victory, as Alan Watts makes clear:

> Willed control brings about a sense of duality in the organism, of consciousness in conflict with appetite . . . But this mode of control is a peculiar example of the proverb that nothing fails like success. For the more consciousness is individualized by the success of the will, the more everything outside the individual seems to be a threat—including . . . the uncontrolled spontaneity of one's own body. . . . Every success in control therefore demands a further success, so that the process cannot stop short of omnipotence.[26]

Dualism here appears as the offspring, the by-product, of the identification of the self with control, an identification that Watts sees as lying at the center of Christianity's ethic of anti-sexuality. The attempt to subdue the spontaneities of the body in the interests of control only succeeds in constituting them as more alien and more powerful, and thus more needful of control. The only way to win this no-win game is to go beyond control, to kill off the body's spontaneities entirely—that is, to cease to *experience* our hungers and desires.

This is what many anorectics describe as their ultimate goal. "[I want] to reach the point," as one puts it, "when I don't need to eat at all."[27] Kim Chernin recalls her surprise when, after fasting, her hunger returned: "I realized [then] that my secret goal in dieting must have been the intention to kill off my appetite completely."[28]

It is not usually noted, in the popular literature on the subject, that anorexic women are as obsessed with *hunger* as they are with being slim. Far from losing her appetite, the typical anorectic is haunted by it—in much the same way that Augustine describes being haunted by sexual desire—and is in constant dread of being overwhelmed by it. Many describe the dread of hunger, "of not having control, of giving in to biological urge," to "the craving, never satisfied thing,"[29] as the "original fear" (as one puts it),[30] or, as Ellen West describes it, "the real obsession." "I don't think the dread of becoming fat is the real . . . neurosis," she writes, "but the constant desire for food . . . [H]unger, or the dread of hunger, pursues me all morning . . . Even when I am full, I am afraid of the coming hour in which hunger will start again." Dread of becoming fat, she interprets, rather than being originary, served as a "brake" to her horror of her own unregulatable, runaway desire for food.[31] Bruch reports that her patients are often terrified at the prospect of taking just one bite of food, lest they never be able to stop.[32] (Bulimic anorectics, who binge on enormous quantities of food—sometimes consuming up to 15,000 calories a day[33]—indeed *cannot* stop.)

These women experience hunger as an alien invader, marching to the tune of its own seemingly arbitrary whims, disconnected from any normal self-regulating mechanisms. Indeed, it would not possibly be so connected, for it is experienced as coming from an area *outside* the self. One patient of Bruch's says she ate breakfast because "my stomach wanted it," expressing here the same sense of alienation from her hunger (and her physical self) as Augustine's when he speaks of his "captor," "the law of sin that was in my member."[34] Bruch notes that this "basic delusion," as she calls it, "of not owning the body and its sensations" is a typical symptom of all eating disorders. "These patients act," she says, "as if for them the regulation of food intake was outside [the self]."[35] This experience of bodily sensations as foreign is, strikingly, not limited to the experience of hunger. Patients with eating disorders have similar problems in identifying cold, heat, emotions, and anxiety as originating in the self.[36]

While the body is experienced as alien and outside, the soul or will is described as being trapped or confined in this alien "jail," as one woman describes it.[37] "I feel caught in my body," "I'm a prisoner in my body":[38] the theme is repeated again and again. A typical fantasy, evocative of Plato, imagines total liberation from the bodily prison: "I wish I could get out of my body entirely and fly!"[39] "Please dear God, help me . . . I want to get out of my body, I want to get out!"[40] Ellen West, astute as always, sees a central meaning of her self-starvation in this "ideal of being too thin, of being *without a body*."[41]

Anorexia is not a philosophical attitude; it is a debilitating affliction. Yet, quite often a highly conscious and articulate scheme of images and associations—virtually a metaphysics—is presented by these women. The scheme is strikingly Augustinian, with evocations of Plato. This does not indicate, of course, that anorectics are followers of Plato or Augustine, but that the anorectic's metaphysics makes explicit various elements, historically grounded in Plato and Augustine, that run deep in our culture.[42] As Augustine often speaks of the "two wills" within him, "one the servant of the flesh, the other of the spirit," who "between them tore my soul apart," so the anorectic describes a "spiritual struggle," a "contest between good and evil," often conceived explicitly as a battle between mind or will and appetite or body.[43] "I feel myself, quite passively," says West, "the stage on which two hostile forces are mangling each other."[44] Sometimes there is a more aggressive alliance with mind against body: "When I fail to exercise as often as I prefer, I become guilty that I have let my body 'win' another day from my mind. I can't wait 'til this semester is over . . . My body is going to pay the price for the lack of work it is currently getting. I can't wait!"[45]

In this battle, thinness represents a triumph of the will over the body, and the thin body (that is to say, the nonbody) is associated with "absolute purity, hyperintellectuality and transcendence of the flesh. My soul seemed to grow as my body waned; I felt like one of those early Christian saints who starved themselves in the desert sun. I felt invulnerable, clean and hard as the bones etched into my silhouette."[46] Fat (that is to say, becoming *all* body) is associated with the taint of matter and flesh, "wantonness,"[47] mental stupor and mental decay.[48] One woman describes how after eating sugar she felt "polluted, disgusting, sticky through the arms, as if something bad had gotten inside."[49] Very often, sexuality is brought into this scheme of associations, and hunger and sexuality are psychically connected. Cherry Boone O'Neill describes a late-night binge, eating scraps of leftovers from the dog's dish:

> I started slowly, relishing the flavor and texture of each marvelous bite. Soon I was ripping the meager remains from the bones, stuffing the meat into my mouth as fast as I could detach it.
>
> [Her boyfriend surprises her, with a look of "total disgust" on his face.]
>
> I had been caught red-handed . . . in an animalistic orgy on the floor, in the dark, alone. Here was the horrid truth for Dan to see. I felt so evil, tainted, pagan . . . In Dan's mind that day, I had been whoring after food.[50]

A hundred pages earlier, she had described her first romantic involvement in much the same terms: "I felt secretive, deceptive, and . . . tainted by the ongoing relationship" (which never went beyond kisses).[51] Sexuality, similarly, is "an abominable business" to Aimee Liu; for her, staying reed-thin is seen as a way of avoiding sexuality, by becoming "androgynous," as she puts it.[52] In the same way, Sarah, a patient of Levenkron's, connects her dread of gaining weight with "not wanting to be a 'temptation' to men."[53] In Liu's case, and in Sarah's, the desire to appear unattractive to men is connected to anxiety and guilt over earlier sexual abuse. Whether or not such episodes are common to many cases of anorexia,[54] "the avoidance of any sexual encounter, a shrinking from all bodily contact," is, according to Bruch, characteristic of anorectics.[55]

THE CONTROL AXIS

Having examined the axis of continuity from Plato to anorexia, we should feel cautioned against the impulse to regard anorexia as expressing entirely modern attitudes and fears. Disdain for the body, the conception of it as an alien force and impediment to the soul, is very old in our Greco-Christian traditions (although it has usually been expressed most forcefully by male philosophers and theologians rather than adolescent women!).

But although dualism is as old as Plato, in many ways contemporary culture appears *more* obsessed than previous eras with the control of the unruly body. Looking now at contemporary American life, a second axis of continuity emerges on which to locate anorexia. I call it the *control axis.*

The young anorectic, typically, experiences her life as well as her hungers as being out of control. She is a perfectionist and can never carry out the tasks she sets herself in a way that meets her own rigorous standards. She is torn by conflicting and contradictory expectations and demands, wanting to shine in all areas of student life, confused about where to place most of her energies, what to focus on, as she develops into an adult. Characteristically, her parents expect a great deal of her in the way of individual achievement (as well as physical appearance), yet have made most of the important decisions for her.[56] Usually, the anorexic syndrome emerges, not as a conscious decision to get as thin as possible, but as the result of her having begun a diet fairly casually, often at the suggestion of a parent, having succeeded splendidly in taking off five or ten pounds, and then having gotten hooked on the intoxicating feeling of accomplishment and control.

Recalling her anorexic days, Aimee Liu recreates her feelings:

> The sense of accomplishment exhilarates me, spurs me to continue on and on. It provides a sense of purpose and shapes my life with distractions from insecurity. . . . I shall become an expert [at losing weight]. . . . The constant downward trend [of the scale] somehow comforts me, gives me visible proof that I can exert control.[57]

The diet, she realizes, "is the one sector of my life over which I and I alone wield total control."[58]

The frustrations of starvation, the rigors of the constant physical activity in which anorectics engage, the pain of the numerous physical complications of anorexia: these do not trouble the anorectic. Indeed, her ability to ignore them is further proof to her of her mastery of her body. "This was something I could control," says one of Bruch's patients. "I still don't know what I look like or what size I am, but I know my body can take anything."[59] "Energy, discipline, my own power will keep me going," says Liu. "Psychic fuel, I need nothing and no one else, and I will prove it. . . . Dropping to the floor, I roll. My tailbone crunches on the hard floor. . . . I feel no pain. I will be master of my own body, if nothing else, I vow."[60] And, finally, from one of Bruch's patients: "*You make of your own body your very own kingdom where you are the tyrant, the absolute dictator.*"[61]

Surely we must recognize in this last honest and explicit statement a central modus operandi for the control of contemporary bourgeois anxiety. Consider compulsive jogging and marathon-running, often despite shin splints and other painful injuries, with intense agitation over missing a day or not meeting a goal for a particular run. Consider the increasing popularity of triathlon events such as the Iron Man, whose central purpose appears to be to allow people to find out how far they can push their bodies—through long-distance swimming, cycling, and running—before they collapse. Consider lawyer Mike Frankfurt, who runs ten miles every morning: "*To run with pain is the essence of life.*"[62] Or consider the following excerpts from student journals:

> The best times I like to run are under the most unbearable conditions. I love to run in the hottest, most humid and steepest terrain I can find. . . . For me running and the pain associated with it aren't enough to make me stop. I am always trying to overcome it and the biggest failure I can make is to stop running because of pain. Once I ran five of a ten-mile run with a severe leg cramp but wouldn't stop—it would have meant failure.[63]

> When I run I am free. . . . The pleasure is closing off my body—as if the incessant pounding of my legs is so total that the pain ceases to exist. There is no grace, no beauty in the running—there is the jarring reality of sneaker and pavement. Bright pain that shivers and splinters sending its white hot arrows into my stomach, my lung, but it cannot pierce my mind. I am on automatic pilot—there is no remembrance of pain, there is freedom—I am losing myself, peeling out of this heavy flesh. . . . Power surges through me.[64]

None of this is to dispute that the contemporary concern with fitness has nonpathological, nondualist dimensions as well. Particularly for women, who have historically suffered from the ubiquity of rape and abuse, from the culturally instilled conviction of our own helplessness, and from lack of access to facilities and programs for rigorous physical training, the cultivation of strength, agility, and confidence clearly has a positive dimension. Nor are the objective benefits of daily exercise and concern for nutrition in question here. My focus, rather, is on a subjective stance, become increasingly prominent, which, although preoccupied with the body and deriving narcissistic enjoyment from its appearance, takes little pleasure in the *experience* of embodiment. Rather, the fundamental identification is with mind (or will), ideals of spiritual perfection, fantasies of absolute control.

Not everyone, of course, for whom physical training is a part of daily routine exhibits such a stance. Here, an examination of the language of female body-builders is illustrative. Body-building is particularly interesting because on the surface it appears to have the opposite structure to anorexia: the body-builder is, after all, building the body *up*, not whittling it down. Body-building develops strength. We imagine the body-builder as someone who is proud, confident, and perhaps most of all, conscious of and accepting of her physicality. This is, indeed, how some female body-builders experience themselves:

> I feel . . . tranquil and stronger [says Lydia Cheng]. Working out creates a high everywhere in my body. I feel the heat. I feel the muscles rise, I see them blow out, flushed with lots of blood. . . . My whole body is sweating and there's few things I love more than working up a good sweat. That's when I really feel like a woman.[65]

Yet a sense of joy in the body as active and alive is *not* the most prominent theme among the women interviewed by Trix Rosen. Many of them, rather, talk about their bodies in ways that resonate disquietingly with typical anorexic themes.

There is the same emphasis on will, purity, and perfection: "I've learned to be a stronger person with a more powerful will . . . pure concentration, energy and spirit." "I want to be as physically perfect as possible." "Body-building suits the perfectionist in me."

"My goal is to have muscular perfection."[66] Compulsive exercisers—whom Dinitia Smith, in an article for *New York* magazine, calls "The New Puritans"—speak in similar terms: Kathy Krauch, a New York art director who bikes twelve miles a day and swims two and a half, says she is engaged in "a quest for perfection." Mike Frankfurt, in describing his motivation for marathon running, speaks of "the purity about it." These people, Smith emphasizes, care little about their health: "They pursue self-denial as an end in itself, out of an almost mystical belief in the purity it confers."[67]

Many body-builders, like many anorectics, unnervingly conceptualize the body as alien, not-self:

> I'm constantly amazed by my muscles. The first thing I do when I wake up in the morning is look down at my "abs" and flex my legs to see if the "cuts" are there. . . . My legs have always been my most stubborn part, and I want them to develop so badly. Every day I can see things happening to them. . . . I don't flaunt my muscles as much as I thought I would. I feel differently about them; they are my product and I protect them by wearing sweaters to keep them warm.[68]

Most strikingly, body-builders put the same emphasis on *control*: on feeling their life to be fundamentally out of control, and on the feeling of accomplishment derived from total mastery of the body. That sense of mastery, like the anorectic's, appears to derive from two sources. First, there is the reassurance that one can overcome all physical obstacles, push oneself to any extremes in pursuit of one's goals (which, as we have seen, is a characteristic motivation of compulsive runners, as well). Second, and most dramatic (it is spoken of time and again by female body-builders), is the thrill of being in total charge of the shape of one's body. "Create a masterpiece," says *Fit* magazine. "Sculpt your body contours into a work of art." As for the anorectic—who literally cannot *see* her body as other than her inner reality dictates and who is relentlessly driven by an ideal image of ascetic slenderness—so for the body-builder a purely mental conception comes to have dominance over her life: "You visualize what you want to look like . . . and then create the form." "The challenge presents itself: to rearrange things." "It's up to you to do the chiseling; you become the master sculptress." "What a fantasy, for your body to be changing! . . . I keep a picture in my mind as I work out of what I want to look like and what's happened to me already."[69] Dictation to nature of one's own chosen design for the body is the central goal for the body-builder, as it is for the anorectic.

The sense of security derived from the attainment of this goal appears, first of all, as the pleasure of control and independence. "Nowadays," says Michael Sacks, associate professor of psychiatry at Cornell Medical College, "people no longer feel they can control events outside themselves—how well they do in their jobs or in their personal relationships, for example—but they can control the food they eat and how far they can run. Abstinence, tests of endurance, are ways of proving their self-sufficiency."[70] In a culture, moreover, in which our continued survival is often at the mercy of "specialists," machines, and sophisticated technology, the body acquires a special sort of vulnerability and dependency. We may live longer, but the circumstances surrounding illness and death may often be perceived as more alien, inscrutable, and arbitrary than ever before.

Our contemporary body-fetishism expresses more than a fantasy of self-mastery in an increasingly unmanageable culture, however. It also reflects our alliance *with* culture against all reminders of the inevitable decay and death of the body. "Everybody wants to live forever" is the refrain from the theme song of *Pumping Iron*. The most youth-worshipping of popular television shows, *Fame,* opens with a song that begins, "I want to live forever." And it is striking that although the anorectic may come very close to death (and 15 percent do indeed die), the dominant experience throughout the illness is of *invulnerability.*

The dream of immortality is, of course, nothing new. But what is unique to modernity is that the defeat of death has become a scientific fantasy rather than a philosophical or religious mythology. We no longer dream of eternal union with the gods; instead, we build devices that can keep us alive indefinitely, and we work on keeping our bodies as smooth and muscular and elastic at forty as they were at eighteen. We even entertain dreams of halting the aging process completely. "Old age," according to Durk Pearson and Sandy Shaw, authors of the popular *Life Extension,* "is an unpleasant and unattractive affliction."[71] The mega-vitamin regime they prescribe is able, they claim, to prevent and even to reverse the mechanisms of aging.

Finally, it may be that in cultures characterized by gross excesses in consumption, the "will to conquer and subdue the body" (as Chernin calls it) expresses an aesthetic or moral rebellion.[72] Anorectics initially came from affluent families, and the current craze for long-distance running and fasting is largely a phenomenon of young, upwardly mobile professionals (Dinitia Smith calls it "Deprivation Chic").[73] To those who are starving *against* their wills, of course, starvation cannot function as an expression of the power of the will. At the same time, we should caution against viewing anorexia as a trendy illness of the elite and privileged. Rather, its most outstanding feature is powerlessness.

THE GENDER/POWER AXIS

Ninety percent of anorectics are women. We do not, of course, need to know that particular statistic to realize that the contemporary "tyranny of slenderness" is far from gender-neutral. Women are more obsessed with their bodies than men, less satisfied with them,[74] and permitted less latitude with them by themselves, by men, and by the culture. In a 1984 *Glamour* magazine poll of 33,000 women, 75 percent said they thought they were "too fat." Yet by Metropolitan Life Insurance Tables, themselves notoriously affected by cultural standards, only 25 percent of these women were heavier than their optimal weight, and a full 30 percent were *below* that weight.[75] The anorectic's distorted image of her body—her inability to see it as anything but too fat—although more extreme, is not radically discontinuous, then, from fairly common female misperceptions.

Consider, too, actors like Nick Nolte and William Hurt, who are permitted a certain amount of softening, of thickening about the waist, while still retaining romantic-lead status. Individual style, wit, the projection of intelligence, experience, and effectiveness still go a long way for men, even in our fitness-obsessed culture. But no female can achieve the status of romantic or sexual ideal without the appropriate *body*. That body, if we use television commercials as a gauge, has gotten steadily leaner since the mid-1970s.[76] What used to be acknowledged as an extreme required only of high fashion models is now the dominant image that beckons to high school and college women. Over and over, extremely slender women students complain of hating their thighs or their stomachs (the anorectic's most dreaded danger spot); often, they express concern and anger over frequent teasing by their boyfriends. Janey, a former student, is 5' 10" and weighs

132 pounds. Yet her boyfriend Bill, also a student of mine, calls her "Fatso" and "Big Butt" and insists she should be 110 pounds because (as he explains in his journal for my class) "that's what Brooke Shields weighs." He calls this "constructive criticism" and seems to experience extreme anxiety over the possibility of her gaining any weight: "I can tell it bothers her yet I still continue to badger her about it. I guess that I think that if I continue to remind her things will change faster."[77] This sort of relationship, in which the woman's weight has become a focal issue, is not at all atypical, as I have discovered from student journals and papers.

Hilda Bruch reports that many anorectics talk of having a "ghost" inside them or surrounding them, "a dictator who dominates me," as one woman describes it; "a little man who objects when I eat" is the description given by another.[78] The little ghost, the dictator, the "other self" (as he often described) is always male, reports Bruch. The anorectic's *other* self—the self of the uncontrollable appetites, the impurities and taints, the flabby will and tendency to mental torpor—is the body, as we have seen. But it is also (and here the anorectic's associations are surely in the mainstream of Western culture) the *female* self. These two selves are perceived as at constant war. But it is clear that it is the male side—with its associated values of greater spirituality, higher intellectuality, strength of will—that is being expressed and developed in the anorexic syndrome.[79]

What is the meaning of these gender associations in the anorectic? I propose that there are two levels of meaning. One has to do with fear and disdain for traditional female roles and social limitations. The other has to do, more profoundly, with a deep fear of "the Female," with all its more nightmarish archetypal associations of voracious hungers and sexual insatiability.

Adolescent anorectics express a characteristic fear of growing up to be mature, sexually developed, and potentially reproductive women. "I have a deep fear," says one, "of having a womanly body, round and fully developed. I want to be tight and muscular and thin."[80] Cherry Boone O'Neill speaks explicitly of her fear of womanhood.[81] If only she could stay thin, says yet another, "I would never have to deal with having a woman's body; like Peter Pan I could stay a child forever."[82] The choice of Peter Pan is telling here—what she means is, stay a *boy* forever. And indeed, as Bruch reports, many anorectics, when children, dreamt and fantasized about growing up to be boys.[83] Some are quite conscious of playing out this fantasy through their anorexia; Adrienne, one of Levenkron's patients, was extremely proud of the growth of facial and body hair that often accompanies anorexia, and especially proud of her "skinny, hairy arms."[84] Many patients report, too, that their father had wanted a boy, were disappointed to get "less than" that, or had emotionally rebuffed their daughter when she began to develop sexually.[85]

In a characteristic scenario, anorexia develops just at the outset of puberty. Normal body changes are experienced by the anorectic, not surprisingly, as the takeover of the body by disgusting, womanish fat. "I grab my breasts," says Aimee Liu, "pinching them until they hurt. If only I could eliminate them, cut them off if need be, to become as flat-chested as a child again."[86] The anorectic is exultant when her periods stop (as they do in *all* cases of anorexia[87] and as they do in many female runners as well). Disgust with menstruation is typical: "I saw a picture at a feminist art gallery," says another woman. "There was a woman with long red yarn coming out of her, like she was menstruating. . . . I got that *feeling*—in that part of my body that I have trouble with . . . my stomach, my thighs, my pelvis. That revolting feeling."[88]

Some authors interpret these symptoms as a species of unconscious feminist protest, involving anger at the limitations of the traditional female role, rejection of values associated with it, and fierce rebellion against allowing their futures to develop in the same direction as their mothers' lives.[89] In her portrait of the typical anorexic family configuration, Bruch describes nearly all of the mothers as submissive to their husbands but very controlling of their children.[90] Practically all had had promising careers which they had given up to care for their husbands and families full-time, a task they take very seriously, although often expressing frustration and dissatisfaction.

Certainly, many anorectics appear to experience anxiety about falling into the lifestyle they associate with their mothers. It is a prominent theme in Aimee Liu's *Solitaire*. Another woman describes her feeling that "[I am] full of my mother ... she is in me even if she isn't there" in nearly the same breath as she complains of her continuous fear of being "not human ... of ceasing to exist."[91] And Ellen West, nearly a century earlier, had quite explicitly equated becoming fat with the inevitable (for an elite woman of her time) confinements of domestic life and the domestic stupor she associates with it:

> Dread is driving me mad ... the consciousness that ultimately I will lose everything; all courage, all rebelliousness, all drive for doing; that it—my little world—will make me flabby, flabby and fainthearted and beggarly.[92]

Several of my students with eating disorders reported that their anorexia had developed after their families had dissuaded them from choosing or forbidden them to embark on a traditionally male career.

Here anorexia finds a true sister-phenomenon in the epidemic of female invalidism and "hysteria" that swept through the middle and upper-middle classes in the second half of the nineteenth century.[93] It was a time that, in many ways, was very like our own, especially in the conflicting demands women were confronting: the opening up of new possibilities versus the continuing grip of the old expectations. On the one hand, the old preindustrial order, with the father at the head of a self-contained family production unit, had given way to the dictatorship of the market, opening up new, nondomestic opportunities for working women. On the other hand, it turned many of the most valued "female" skills—textile and garment manufacture, food processing—out of the home and over to the factory system.[94] In the new machine economy, the lives of middle-class women were far emptier than they had been before.

It was an era, too, that had been witnessing the first major feminist wave. In 1840, the World Anti-Slavery Conference had been held, at which the first feminists spoke loudly and long on the connections between the abolition of slavery and women's rights. The year 1848 saw the Seneca Falls Convention. In 1869, John Stuart Mill published his landmark work "On the Subjection of Women." And in 1889 the Pankhursts formed the Women's Franchise League. But it was an era, too (and not unrelatedly, as I shall argue later), when the prevailing ideal of femininity was the delicate, affluent lady, unequipped for anything, but the most sheltered domestic life, totally dependent on her prosperous husband, providing a peaceful and comfortable haven for him each day after his return from his labors in the public sphere.[95] In a now famous letter, Freud, criticizing John Stuart Mill, writes:

> It really is a still-born thought to send women into the struggle for existence exactly as men. If, for instance, I imagine my gentle sweet girl as a competitor it would only end in my telling her, as I did seventeen months ago, that I am fond of her and that I implore her to withdraw from the strife into the calm uncompetitive activity of my home.[96]

This is exactly what male doctors *did* do when women began falling ill, complaining of acute depression, severe headaches, weakness, nervousness, and self-doubt.[97] Among these women were such noted feminists and social activists as Charlotte Perkins Gilman, Jane Addams, Elizabeth Cady Stanton, Margaret Sanger, British activist Josephine Butler, and German suffragist Hedwig Dohm. "I was weary myself and sick of asking what I am and what I ought to be," recalls Gilman,[98] who later went on to write a fictional account of her mental breakdown in the chilling novella *The Yellow Wallpaper.* Her doctor, the famous specialist S. Weir Mitchell, instructed her, as Gilman recalls, to "live as domestic a life as possible. Have your child with you all the time.... Lie down an hour every day after each meal. Have but two hours intellectual life a day. And never touch pen, brush or pencil as long as you live."[99]

Freud, who favorably reviewed Mitchell's 1887 book and who advised that psychotherapy for hysterical patients be combined with Mitchell's rest cure ("to avoid new psychical impressions"),[100] was as blind as Mitchell to the contribution that isolation, boredom, and intellectual frustration made to the etiology of hysteria. Nearly all of the subjects in *Studies in Hysteria* (as well as the later *Dora*) are acknowledged by Freud to be unusually intelligent, creative, energetic, independent, and, often, highly educated. (Berthe Pappenheim—"Anna O."—as we know, went on after recovery to become an active feminist and social reformer.) Freud even comments, criticizing Janet's notion that hysterics were "psychically insufficient," on the characteristic coexistence of hysteria with "gifts of the richest and most original kind."[101] Yet Freud never makes the connection (which Breuer had begun to develop)[102] between the monotonous domestic lives these women were expected to lead after they completed their schooling, and the emergence of compulsive daydreaming, hallucinations, dissociations, and hysterical conversions.

Charlotte Perkins Gilman does make that connection. In *The Yellow Wallpaper* she describes how a prescribed regime of isolation and enforced domesticity eventuates, in her fictional heroine, in the development of a full-blown hysterical symptom, madness, and collapse. The symptom, the hallucination that there is a woman trapped in the wallpaper of her bedroom, struggling to get out, is at once a perfectly articulated expression of protest and a completely debilitating idée fixe that allows the woman character no distance on her situation, no freedom of thought, no chance of making any progress in leading the kind of active, creative life her body and soul crave.

So too for the anorectic. It is indeed essential to recognize in this illness the dimension of protest against the limitations of the ideal of female domesticity (the "feminine mystique," as Betty Friedan called it) that reigned in America throughout the 1950s and early 1960s—the era when most of their mothers were starting homes and families. This was, we should recall, the era following World War II, an era during which women were fired en masse from the jobs they had held during the war and shamelessly propagandized back into the full-time job of wife and mother. It was an era, too, when the "fuller figure," as Jane Russell now calls it, came into fashion once more, a period of "mammary madness" (or "resurgent Victorianism," as Lois Banner calls it), which glamorized the voluptuous, large-breasted woman.[103] This remained the prevailing fashion tyranny until the late 1960s and early 1970s.

But we must recognize that the anorectic's protest, like that of the classical hysterical symptom, is written on the bodies of anorexic women, not embraced as a conscious politics—nor, indeed, does it reflect any social or political understanding at all. Moreover, the symptoms themselves function to preclude the emergence of such an understanding. The idée fixe—staying thin—becomes at its farthest extreme so powerful as to render any other ideas or life-projects meaningless. Liu describes it as "all encompassing."[104] West writes: "I felt all inner development was ceasing, that all becoming and growing were being choked, because a single idea was filling my entire soul."[105]

Paradoxically—and often tragically—these pathologies of female protest (and we must include agoraphobia here, as well as hysteria and anorexia) actually function as if in collusion with the cultural conditions that produced them.[106] The same is true for more moderate expressions of the contemporary female obsession with slenderness. Women may feel themselves deeply attracted by the aura of freedom and independence suggested by the boyish body ideal of today. Yet, each hour, each minute spent in anxious pursuit of that ideal (for it does not come naturally to most mature women) is in fact time and energy taken from inner development and social achievement. As a feminist protest, the obsession with slenderness is hopelessly counterproductive.

It is important to recognize, too, that the anorectic is terrified and repelled, not only by the traditional female domestic role—which she associates with mental lassitude and weakness—but by a certain archetypal image of the female: as hungering, voracious, all-needing, and all-wanting. It is this image that shapes and permeates her experience of her own hunger for food as insatiable and out of control, that makes her feel that if she takes just one bite, she will not be able to stop.

Let us explore this image. Let us break the tie with food and look at the metaphor: hungering. . . voracious . . . extravagantly and excessively needful . . . without restraint . . . always wanting . . . always wanting too much affection, reassurance, emotional and sexual contact, and attention. This is how many women frequently experience themselves, and, indeed, how many men experience women. "Please, God, keep me from telephoning him," prays the heroine in Dorothy Parker's classic "A Telephone Call,"[107] experiencing her need for reassurance and contact as being as out of control and degrading as the anorectic does her desire for food. The male counterpart to this is found in Paul Morel in Lawrence's *Sons and Lovers* : "Can you never like things without clutching them as if you wanted to pull the heart out of them?" he accuses Miriam as she fondles a flower. "Why don't you have a bit more restraint, or reserve, or something. . . . You're always begging things to love you, as if you were a beggar for love. Even the flowers, you have to fawn on them."[108] How much psychic authenticity do these images carry in 1980s America? One woman in my class provided a stunning insight into the connection between her perception of herself and the anxiety of the compulsive dieter. "You know," she said, "the anorectic is always convinced she is taking up too much space, eating too much, wanting food too much. I've never felt that way, but I've often felt that I was *too much*—too much emotion, too much need, too loud and demanding, too much *there*, if you know what I mean."[109]

The most extreme cultural expressions of the fear of woman as "too much"—which almost always revolve around her sexuality—are strikingly full of eating and hungering metaphors. "Of woman's unnatural, *insatiable* lust, what country, what village doth not complain?" queries Burton in *The Anatomy of Melancholy.* [110] "You are the true hiennas," says Walter Charleton, "that allure us with the fairness of your skins, and when folly hath brought us within your reach, you leap upon us and *devour* us."[111]

The mythology/ideology of the devouring, insatiable female (which, as we have seen, is the image of her female self the anorectic has internalized) tends historically to wax and wane. But not without rhyme or reason. In periods of gross environmental and social crisis, such as characterized the period of the witch-hunts in the fifteenth and sixteenth centuries, it appears to flourish.[112] "All witchcraft comes from carnal lust, which is in women *insatiable,* say Kramer and Sprenger, authors of the official witch-hunters handbook, *Malleus Malificarum.* For the sake of fulfilling the "*mouth* of the womb . . . [women] consort even with the devil."[113]

Anxiety over women's uncontrollable hungers appears to peak, as well, during periods when women are becoming independent and are asserting themselves politically and socially. The second half of the nineteenth century, concurrent with the first feminist wave discussed earlier, saw a virtual flood of artistic and literary images of the dark, dangerous, and evil female: "sharp-teethed, devouring" Sphinxes, Salomes, and Delilahs, "biting, tearing, murderous women." "No century," claims Peter Gay, "depicted woman as vampire, as castrator, as killer, so consistently, so programmatically, and so nakedly as the nineteenth."[114] No century, either, was so obsessed with sexuality—particularly female sexuality—and its medical control. Treatment for excessive "sexual excitement" and masturbation in women included placing leeches on the womb,[115] clitoridectomy, and removal of the ovaries (also recommended for "troublesomeness, eating like a ploughman, erotic tendencies, persecution mania, and simple 'cussedness'").[116] The importance of female masturbation in the etiology of the "actual neurosis" was a topic in which the young Freud and his friend and colleague Wilhelm Fliess were especially interested. Fliess believed that the secret to controlling such "sexual abuse" lay in the treatment of nasal "genital spots"; in an operation that was sanctioned by Freud, he attempted to "correct" the "bad sexual habits" of Freud's patient Emma Eckstein by removal of the turbinate bone of her nose.[117]

It was in the second half of the nineteenth century, too, despite a flurry of efforts by feminists and health reformers,[118] that the stylized "S-curve," which required a tighter corset than ever before, came into fashion.[119] "While the suffragettes were forcefully propelling all women toward legal and political emancipation," says Amaury deRiencourt, "fashion and custom imprisoned her physically as she had never been before."[120] Described by Thorstein Veblen as a "mutilation, undergone for the purpose of lowering the subject's vitality and rendering her permanently and obviously unfit for work," the corset indeed did just that.[121] In it a woman could barely sit or stoop, was unable to move her feet more than six inches at a time, and had difficulty in keeping herself from regular fainting fits. (In 1904, a researcher reported that "monkeys laced up in these corsets moped, became excessively irritable and within weeks sickened and died"!)[122] The connection was often drawn in popular magazines between enduring the tight corset and the exercise of self-restraint and control. The corset is "an ever present monitor," says one 1878 advertisement, "of a well-disciplined mind and well-regulated feelings."[123] Today, of course, we diet to achieve such control.

It is important to emphasize that, despite the practice of bizarre and grotesque methods of gross physical manipulation and external control (clitoridectomy, Chinese foot-binding, the removal of bones of the rib cage in order to fit into the tight corsets), such control plays a relatively minor role in the maintenance of gender/power relations. For every historical image of the dangerous, aggressive woman there is a corresponding fantasy—an ideal femininity, from which all threatening elements have been purged—that women have mutilated themselves *internally* to attain. In the Victorian era, at the same time that operations were being performed to control female sexuality, William Acton, Richard von Krafft-Ebing, and others were proclaiming the official scientific doctrine that women are naturally passive and "not very much troubled with sexual feelings of any kind."[124] Corresponding to this male medical fantasy was the popular artistic and moral theme of woman as ministering angel; sweet, gentle, domestic, without intensity or personal ambition of any sort.[125] Peter Gay suggests, correctly, that these ideals must be understood as a reaction-formation to the era's "pervasive sense of manhood in danger," and he argues that few women actually fit the "insipid goody" (as Kate Millett calls it) image.[126] What Gay forgets, however, is that most women *tried* to fit—working classes as well as middle were affected by the "tenacious and all-pervasive" ideal of the perfect lady.[127]

On the gender/power axis the female body appears, then, as the unknowing medium of the historical ebbs and flows of the fear of woman as "too much." That, as we have seen, is how the anorectic experiences her female, bodily self: as voracious, wanton, needful of forceful control by her male will. Living in the tide of cultural backlash against the second major feminist wave, she is not alone in constructing these images. Christopher Lasch, in *The Culture of Narcissism,* speaks of what he describes as "the apparently aggressive overtures of sexually liberated women" which convey to many males the same message—that women are *voracious, insatiable,*" and call up "early fantasies of a possessive, suffocating, *devouring* and castrating mother."[128]

Our contemporary beauty ideals, by contrast, seemed purged, as Kim Chernin puts it, "of the power to conjure up memories of the past, of all that could remind us of a woman's mysterious power."[129] The ideal, rather, is an "image of a woman in which she is not yet a woman": Darryl Hannah as the lanky, newborn mermaid in *Splash;* Lori Singer (appearing virtually anorexic) as the reckless, hyperkinetic heroine of *Footloose;* The Charley Girl; "Cheryl Tiegs in shorts, Margaux Hemingway with her hair wet; Brooke Shields naked on an island";[130] the dozens of teenage women who appear in Coke commercials, in jeans commercials, in chewing gum commercials.

The images suggest amused detachment, casual playfulness, flirtatiousness without demand, and lightness of touch. A refusal to take sex, death, or politics too deadly seriously. A delightfully unconscious relationship to her body. The twentieth century has seen this sort of feminine ideal before, of course. When, in the 1920s, young women began to flatten their breasts, suck in their stomachs, bob their hair, and show off long, colt-like legs, they believed they were pursuing a new freedom and daring that demanded a carefree, boyish style. If the traditional female hourglass suggested anything, it was confinement and immobility. Yet the flapper's freedom, as Mary McCarthy's and Dorothy Parker's short stories brilliantly reveal, was largely an illusion—as any obsessively cultivated sexual style must inevitably be. Although today's images may suggest androgynous independence, we need only consider who is on the receiving end of the imagery in order to confront the pitiful paradox involved.

Watching the commercials are thousands of anxiety-ridden women and adolescents (some of whom may well be the very ones appearing in the commercials) with anything *but* an unconscious relation to their bodies. They are involved in an absolutely contradictory state of affairs, a totally no-win game: caring desperately, passionately, obsessively about attaining an ideal of coolness, effortless confidence, and casual freedom. Watching the commercials is a little girl, perhaps ten years old, whom I saw in Central Park, gazing raptly at her father, bursting with pride: "Daddy, guess what? I lost two pounds!" And watching the commercials is the anorectic, who associates her relentless pursuit of thinness with power and control, but who in fact destroys her health and imprisons her imagination. She is surely the most startling and stark illustration of how cavalier power relations are with respect to the motivations and goals of individuals, yet how deeply they are etched on our bodies, and how well our bodies serve them.

NOTES

This essay was presented as a public lecture at Le Moyne College, was subsequently presented at D'Youville College and Bennington College, and was originally published in the *Philosophical Forum* 17, no. 2 (Winter 1985). I wish to thank all those in the audiences at Le Moyne, D'Youville, and Bennington who commented on my presentations, and Lynne Arnault, Nancy Fraser, and Mario Moussa for their systematic and penetrating criticisms and suggestions for the *Forum* version. In addition, I owe a large initial debt to my students, particularly Christy Ferguson, Vivian Conger, and Nancy Monaghan, for their observations and insights.

1. Jules Henry, *Culture Against Man* (New York: Alfred A. Knopf, 1963).
2. When I wrote this piece in 1983, the term *anorexia* was commonly used by clinicians to designate a general class of eating disorders within which intake-restricting (or abstinent) anorexia and bulimia-anorexia (characterized by alternating bouts of gorging and starving and/or gorging and vomiting) are distinct subtypes (see Hilde Bruch, *The Golden Cage: The Enigma of Anorexia Nervosa* [New York: Vintage, 1979], p. 10; Steven Levenkron, *Treating and Overcoming Anorexia Nervosa* [New York: Warner Books, 1982], p. 6; R. L. Palmer, *Anorexia Nervosa* [Middlesex: Penguin, 1980], pp. 14, 23–24; Paul Garfinkel and David Garner, *Anorexia Nervosa: A Multidimensional Perspective* [New York: Brunner/Mazel, 1982], p. 4). Since then, as the clinical tendency has been increasingly to emphasize the differences rather than the commonalities between the eating disorders, bulimia has come to occupy its own separate classificatory niche. In the present piece I concentrate largely on those images, concerns, and attitudes shared by anorexia and bulimia. Where a difference seems significant for the themes of this essay, I will indicate the relevant difference in a footnote rather than overcomplicate the main argument of the text. This procedure is not to be taken as belittling the importance of such differences, some of which I discuss in "Reading the Slender Body."
3. Although throughout history scattered references can be found to patients who sound as though they may have been suffering from self-starvation, the first medical description of anorexia as a discrete syndrome was made by W. W. Gull in an 1868 address at Oxford (at the time he called the syndrome, in keeping with the medical taxonomy of the time, *hysteric apepsia*). Six years later, Gull began to use the term *anorexia nervosa*; at the same time, E. D. Lesegue independently described the disorder (Garfinkel and Garner, *Anorexia Nervosa,* pp. 58–59). Evidence points to a minor "outbreak" of anorexia nervosa around this time (see Jacobs Brumberg, *Fasting Girls* (Cambridge: Harvard University Press, 1988), a historical occurrence that went unnoticed by twentieth-century clinicians until renewed interest in the disorder was prompted by its reemergence and striking increase over the past twenty years (see note 11 of "Whose Body Is This?" for sources that document this increase). At the time I wrote the present piece, I was not aware of the extent of anorexia nervosa in the second half of the nineteenth century.
4. Ludwig Binswanger, "The Case of Ellen West," in Rollo May, ed., *Existence* (New York: Simon and Schuster, 1958), p. 288. He was wrong, of course. The symptom was not new, and we now know that Ellen West was not the only young woman of her era to suffer from anorexia. But the fact that Binswanger was unaware of other cases is certainly suggestive of its infrequency, especially relative to our own time.

5. Hilde Bruch, *Eating Disorders* (New York: Basic Books, 1973), p. 4.
6. Levenkron, *Treating and Overcoming Anorexia Nervosa,* p. 1; Susan Squire, "Is the Binge-Purge Cycle Catching?" *Ms.* (Oct. 1983).
7. Dinitia Smith, "The New Puritans," *New York Magazine* (June 11, 1984): 28.
8. Kim Chernin, *The Obsession: Reflections on the Tyranny of Slenderness* (New York: Harper and Row, 1981), pp. 63, 62.
9. Garfinkel and Garner, *Anorexia Nervosa,* p. xi. Anorectics characteristically suffer from a number of physiological disturbances, including amenorrhea (cessation of menstruation) and abnormal hypothalamic function (see Garfinkel and Garner, *Anorexia Nervosa,* pp. 58–59, for an extensive discussion of these and other physiological disorders associated with anorexia; also Eugene Garfield, "Anorexia Nervosa: The Enigma of Self-Starvation," *Current Contents* [Aug. 6, 1984]: 8–9). Researchers are divided, with arguments on both sides, as to whether hypothalamic dysfunction may be a primary cause of the disease or whether these characteristic neuroendocrine disorders are the result of weight loss, caloric deprivation, and emotional stress. The same debate rages over abnormal vasopressin levels discovered in anorectics. Touted in tabloids all over the United States as the "explanation" for anorexia and key to its cure. Apart from such debates over a biochemical predisposition to anorexia, research continues to explore the possible role of biochemistry in the self-perpetuating nature of the disease, and the relation of the physiological effects of starvation to particular experiential symptoms such as the anorectic's preoccupation with food (see Bruch, *The Golden Cage,* pp. 7–12; Garfinkel and Garner, *Anorexia Nervosa,* pp. 10–14).
10. Initially, anorexia was found to predominate among upper-class white families. There is, however, widespread evidence that this is now rapidly changing (as we might expect; no one in America is immune to the power of popular imagery). The disorder, it has been found, is becoming more equally distributed, touching populations (e.g., blacks and East Indians) previously unaffected, and all socioeconomic levels (Garfinkel and Garner, *Anorexia Nervosa,* pp. 102–3). There remains, however, an overwhelming disproportion of women to men (Garfinkel and Garner, *Anorexia Nervosa*, pp. 112–13).
11. Chernin's *The Obsession,* whose remarkable insights inspired my interest in anorexia, remains *the* outstanding exception to the lack of cultural understanding of eating disorders.
12. Chernin, *The Obsession*, pp. 36–37. My use of the expression "our culture" may seem overly homogenizing here, disrespectful of differences among ethnic groups, socioeconomic groups, subcultures within American society, and so forth. It must be stressed here that I am discussing ideology and images whose power is *precisely* the power to homogenize culture. Even in pre-mass-media cultures we see this phenomenon: the nineteenth-century ideal of the "perfect lady" tyrannized even those classes who could not afford to realize it. With television, of course, a massive deployment of images becomes possible, and there is no escape from the mass shaping of our fantasy lives. Although they may start among the wealthy elite ("A woman can never be too rich or too thin"), media-promoted ideas of femininity and masculinity quickly and perniciously spread their influence over everyone who owns a TV or can afford a junk magazine or is aware of billboards. Changes in the incidence of anorexia among lower-income groups (see note 10, above) bear out this point.
13. Christopher Lasch, *The Culture of Narcissism* (New York: Warner Books, 1979), p. 88.
14. I choose these three primarily because they are where my exploration of the imagery, language, and metaphor produced by anorexic women led me. Delivering earlier versions of this essay at colleges and conferences, I discovered that one of the commonest responses of members of the audience was the proffering of further axes; the paper presented itself less as a statement about the ultimate meaning or causes of a phenomenon than as an invitation to continue my "unpacking" or anorexia as a crystallizing formation. Yet the particular axes chosen have more than a purely autobiographical rationale. The dualist axes serve to identify and articulate the basic body imagery of anorexia. The control axis is an exploration of the question "Why now?" The gender/power axis continues this exploration but focuses on the question "Why women?" The sequence of axes takes us from the most general, most historically diffuse structure of continuity—the dualist experience of self—to ever narrower, more specified arenas of comparison and connection. At first the connections are made without regard to historical context, drawing on diverse historical sources to exploit their familiar coherence in an effort to sculpt the shape of the anorexic experience. In this section, too, I want to suggest that the Greco-Christian tradition provides a particularly fertile soil for the development of anorexia. Then I turn to the much more specific context of

American fads and fantasies in the 1980s, considering the contemporary scene largely in terms of popular culture (and therefore through the "fiction" of homogeneity), without regard for gender difference. In this section the connections drawn point to a historical experience of self common to both men and women. Finally, my focus shifts to consider, not what connects anorexia to other general cultural phenomena, but what presents itself as a rupture from them, and what forces us to confront how ultimately opaque the current epidemic of eating disorders remains unless it is linked to the particular situation of women.

The reader will notice that the axes are linked thematically as well as through their convergence in anorexia: the obsession with control is linked with dualism, and the gender/power dynamics discussed implicitly deal with the issue of control (of the feminine) as well.

15. Michel Foucault, *The History of Sexuality.* Vol. 1: *An Introduction* (New York: Vintage, 1980), p. 155.
16. Foucault, *History of Sexuality*, pp. 47–48.
17. Hubert L. Dreyfus and Paul Rabinow, *Michel Foucault: Beyond Structuralism and Hermeneutics* (Chicago: University of Chicago Press, 1983), p. 112.
18. Foucault, *History of Sexuality,* p. 95.
19. Michel Foucault, *Discipline and Punish* (New York: Vintage, 1979), p. 26.
20. Plato, *Phaedo,* in *The Dialogues of Plato,* ed. and trans. Benjamin Jowett, 4th ed., rev. (Oxford: Clarendon Press, 1953), 83d.
21. St. Augustine, *The Confessions,* trans. R. S. Pine-Coffin (Middlesex: Penguin, 1961), p. 164.
22. *Phaedo* 81d.
23. *Phaedo* 66c. For Descartes on the body as a hindrance to knowledge, see *Conversations with Burman* (Oxford: Clarendon Press, 1976), p. 8, and *Passions of the Soul* in *Philosophical Works of Descartes,* 2 vols., trans. Elizabeth S. Haldane and G. R. T. Ross (Cambridge: Cambridge University Press, 1969), vol. 1, p. 353.
24. *Phaedo* 80a.
25. Indeed, the Cartesian "Rules for the Direction of the Mind," as carried out in the *Meditations* especially, are actually rules for the transcendence of the body—its passions, its senses, the residue of "infantile prejudices" of judgment lingering from that earlier time when we were "immersed" in body and bodily sensations.
26. Alan Watts, *Nature, Man, and Woman* (New York: Vintage, 1970), p. 145.
27. Bruch, *Eating Disorders,* p. 84.
28. Chernin, *The Obsession,* p. 8.
29. Entry in student journal, 1984.
30. Bruch, *The Golden Cage,* p. 4.
31. Binswanger, "The Case of Ellen West," p. 253.
32. Bruch, *Eating Disorders,* p. 253.
33. Levenkron, *Treating and Overcoming Anorexia Nervosa,* p. 6.
34. Bruch, *Eating Disorders,* p. 270; Augustine, *Confessions,* p. 164.
35. Bruch, *Eating Disorders,* p. 50.
36. Bruch, *Eating Disorders,* p. 254.
37. Entry in student journal, 1984.
38. Bruch, *Eating Disorders,* p. 279.
39. Aimee Liu, *Solitaire* (New York: Harper and Row, 1979), p. 141.
40. Jennifer Woods, "I Was Starving Myself to Death," *Mademoiselle* (May 1981): 200.
41. Binswanger, "The Case of Ellen West," p. 251 (emphasis added).
42. Why they should emerge with such clarity in the twentieth century and through the voice of the anorectic is a question answered, in part, by the following two axes.
43. Augustine, *Confessions,* p. 165; Liu, *Solitaire,* p. 109.
44. Binswanger, "The Case of Ellen West," p. 343.
45. Entry in student journal, 1983.
46. Woods, "I Was Starving Myself to Death," p. 242.
47. Liu, *Solitaire,* p. 109.
48. "I equated gaining weight with happiness, contentment, then slothfulness, then atrophy, then death." (From case notes of Binnie Klein, M.S.W., to whom I am grateful for having provided parts of a transcript of her work with an anorexic patient.) See also Binswanger, "The Case of Ellen West," p. 343.
49. Klein, case notes.
50. Cherry Boone O'Neill, *Starving for Attention* (New York: Dell, 1982), p. 131.

51. O'Neill, *Starving for Attention,* p. 49.
52. Liu, *Solitaire,* p. 101.
53. Levenkron, *Treating and Overcoming Anorexia Nervosa,* p. 122.
54. Since the writing of this piece, evidence has accrued suggesting that sexual abuse may be an element in the histories of many eating-disordered women (see note 2 in "Whose Body Is This?").
55. Bruch, *The Golden Cage,* p. 73. The same is not true of bulimic anorectics, who tend to be sexually active (Garfinkel and Garner, *Anorexia Nervosa,* p. 41). Bulimic anorectics, as seems symbolized by the binge-purge cycle itself, stand in a somewhat more ambivalent relationship to their hungers than do abstinent anorectics. See "Reading the Slender Body," in this volume, for a discussion of the cultural dynamics of the binge-purge cycle.
56. Bruch, *The Golden Cage,* p. 33.
57. Liu, *Solitaire,* p. 36.
58. Liu, *Solitaire,* p. 46. In one study of female anorectics, 88 percent of the subjects questioned reported that they lost weight because they "liked the feeling of will power and self-control" (G. R. Leon, "Anorexia Nervosa: The Question of Treatment Emphasis," in M. Rosenbaum, C. M. Franks, and Y. Jaffe, eds., *Perspectives on Behavior Therapy in the Eighties* [New York: Springer, 1983], pp. 363–77).
59. Bruch, *Eating Disorders,* p. 95.
60. Liu, *Solitaire,* p. 123.
61. Bruch, *The Golden Cage,* p. 65 (emphasis added).
62. Smith, "The New Puritans," p. 24 (emphasis added).
63. Entry in student journal, 1984.
64. Entry in student journal, 1984.
65. Trix Rosen, *Strong and Sexy* (New York: Putnam, 1983), p. 108.
66. Rosen, *Strong and Sexy,* pp. 62, 14, 47, 48.
67. Smith, "The New Puritans," pp. 27, 26.
68. Rosen, *Strong and Sexy,* pp. 61–62.
69. Rosen, *Strong and Sexy*, pp. 72, 61. This fantasy is not limited to female body-builders. John Travolta describes his experience training for *Staying Alive* : "[It] taught me incredible things about the body . . . how it can be reshaped so you can make yourself over entirely, creating an entirely new you. I now look at bodies almost like pieces of clay that can be molded." ("Travolta: 'You Really Can Make Yourself Over,'" *Syracuse Herald-American,* Jan. 13, 1985.)
70. Smith, "The New Puritans," p. 29.
71. Durk Pearson and Sandy Shaw, *Life Extension* (New York: Warner, 1982), p. 15.
72. Chernin, *The Obsession,* p. 47.
73. Smith, "The New Puritans," p. 24.
74. Sidney Journard and Paul Secord, "Body Cathexis and the Ideal Female Figure," *Journal of Abnormal and Social Psychology* 50: 243–46; Orland Wooley, Susan Wooley, and Sue Dyrenforth, "Obesity and Women—A Neglected Feminist Topic," *Women's Studies Institute Quarterly* 2 (1979): 81–92. Student journals and informal conversations with women students have certainly borne this out.
75. "Feeling Fat in a Thin Society," *Glamour* (Feb. 1984): 198.
76. The same trend is obvious when the measurements of Miss America winners are compared over the past fifty years (see Garfinkel and Garner, *Anorexia Nervosa,* p. 107). Some evidence has indicated that this tide is turning and that a more solid, muscular, athletic style is emerging as the latest fashion tyranny.
77. Entry in student journal, 1984.
78. Bruch, *The Golden Cage,* p. 58.
79. This is one striking difference between the abstinent anorectic and the bulimic anorectic: in the binge-and-vomit cycle, the hungering female self refuses to be annihilated, is in constant protest. And, in general, the rejection of femininity discussed here is *not* typical of bulimics, who tend to strive for a more "female"-looking body as well.
80. Entry in student journal, 1983.
81. O'Neill, *Starving for Attention,* p. 53.
82. Entry in student journal, 1983.
83. Bruch, *The Golden Cage,* p. 72; Bruch, *Eating Disorders,* p. 277. Others have fantasies of androgyny: "I want to go to a party and for everyone to look at me and for no one to know

whether I was the most beautiful slender woman or handsome young man" (as reported by therapist April Benson, panel discussion, "New Perspectives on Female Development," third annual conference of the Center for the Study of Anorexia and Bulimia, New York, 1984).

84. Levenkron, *Treating and Overcoming Anorexia Nervosa,* p. 28.
85. See, for example, Levenkron's case studies in *Treating and Overcoming Anorexia Nervosa,* esp. pp. 45, 103; O'Neill, *Starving for Attention,* p. 107; Susie Orbach, *Fat Is a Feminist Issue* (New York: Berkley, 1978), pp. 174–75.
86. Liu, *Solitaire,* p. 79.
87. Bruch, *The Golden Cage,* p. 65.
88. Klein, case study.
89. Chernin, *The Obsession,* pp. 102–3; Robert Seidenberg and Karen DeCrow, *Women Who Marry Houses: Panic and Protest in Agoraphobia* (New York: McGraw-Hill, 1983), pp. 88–97; Bruch, *The Golden Cage,* p. 58, Orbach, *Fat Is a Feminist Issue,* pp. 169–70. See also my discussions of the protest thesis in "Whose Body Is This?" and "The Body and the Reproduction of Femininity" in this volume.
90. Bruch, *The Golden Cage,* pp. 27–28.
91. Bruch, *The Golden Cage,* p. 12.
92. Binswanger, "The Case of Ellen West," p. 243.
93. At the time I wrote this essay, I was unaware of the fact that eating disorders were frequently an element of the symptomatology of nineteenth-century "hysteria"—a fact that strongly supports my interpretation here.
94. See, among many other works on this subject, Barbara Ehrenreich and Deirdre English, *For Her Own Good* (Garden City: Doubleday, 1979), pp. 1–29.
95. See Martha Vicinus, "Introduction: The Perfect Victorian Woman," in Martha Vicinus, ed., *Suffer and Be Still: Women in the Victorian Age* (Bloomington: Indiana University Press, 1972), pp. x–xi.
96. Ernest Jones, *Sigmund Freud: Life and Work* (London: Hogarth Press, 1956), vol. 1, p. 193.
97. On the nineteenth-century epidemic of female invalidism and hysteria, see Ehrenreich and English, *For Her Own Good*; Carroll Smith-Rosenberg, "The Hysterical Woman: Sex Roles and Conflict in Nineteenth-Century America," *Social Research* 39, no. 4 (Winter 1972): 652–78; Ann Douglas Wood, "The 'Fashionable Diseases': Women's Complaints and Their Treatment in Nineteenth Century America," *Journal of Interdisciplinary History* 4 (Summer 1973).
98. Ehrenreich and English, *For Her Own Good,* p. 2.
99. Ehrenreich and English, *For Her Own Good,* p. 102.
100. Sigmund Freud and Josef Breuer, *Studies on Hysteria* (New York: Avon, 1966), p. 311.
101. Freud and Breuer, *Studies on Hysteria,* p. 141; see also p. 202.
102. See especially pp. 76 ("Anna O."), 277, 284.
103. Marjorie Rosen, *Popcorn Venus* (New York: Avon, 1973); Lois Banner, *American Beauty* (Chicago: University of Chicago Press, 1983), pp. 283–85. Christian Dior's enormously popular full skirts and cinch-waists, as Banner points out, are strikingly reminiscent of Victorian modes of dress.
104. Liu, *Solitaire,* p. 141.
105. Binswanger, "The Case of Ellen West," p. 257.
106. This is one of the central themes I develop in "The Body and the Reproduction of Femininity," the next essay in this volume.
107. Dorothy Parker, *Here Lies: The Collected Stories of Dorothy Parker* (New York: Literary Guild of America, 1939), p. 48.
108. D. H. Lawrence, *Sons and Lovers* (New York: Viking, 1958), p. 257.
109. This experience of oneself as "too much" may be more or less emphatic; depending on such variables as race, religion, socioeconomic class, and sexual orientation. Luise Eichenbaum and Susie Orbach (*Understanding Women: A Feminist Psychoanalytic Approach* [New York: Basic Books, 1983]) emphasize, however, how frequently their clinic patients, nonanorexic as well as anorexic, "talk about their needs with contempt, humiliation, and shame. They feel exposed and childish, greedy and insatiable" (p. 49). Eichenbaum and Orbach trace such feelings, moreover, to infantile experiences that are characteristic of all female development, given a division of labor within which women are the emotional nurtures and physical caretakers of family life. Briefly (and this sketch cannot begin to do justice to their rich and complex analysis): mothers unwittingly communicate to their daughters that feminine needs

are excessive and bad and that they must be contained. The mother does this out of a sense that her daughter will have to learn the lesson in order to become properly socialized into the traditional female role of caring for others—of feeding others, rather than feeding the self—and also because of an unconscious identification with her daughter, who reminds the mother of the "hungry, needy little girl" in herself, denied and repressed through the mother's *own* "education" in being female: "Mother comes to be frightened by her daughter's free expression of her needs, and unconsciously acts toward her infant daughter in the same way she acts internally toward the little-girl part of herself. In some ways the little daughter becomes an external representation of that part of herself which she has come to dislike and deny. The complex emotions that result from her own deprivation through childhood and adult life are both directed inward in the struggle to negate the little-girl part of herself and projected outward onto her daughter" (p. 44). Despite a real desire to be totally responsive to her daughter's emotional needs, the mother's own anxiety limits her capacity to respond. The contradictory messages she sends out convey to the little girl "the idea that to get love and approval she must show a particular side of herself. She must hide her emotional cravings, her disappointments and her angers, her fighting spirit. . . . She comes to feel that there must be something wrong with who she really is, which in turn must mean that there is something wrong with what she needs and what she wants. . . . This soon translates into feeling unworthy and hesitant about pursuing her impulses" (pp. 48–49). Once she has grown up, of course, these feelings are reinforced by cultural ideology, further social training in femininity, and the likelihood that the men in her life will regard her as "too much" as well, having been schooled by their own training in masculine detachment and autonomy.

(With boys, who do not stir up such intense identification in the mother and who, moreover, she knows will grow up into a world that will meet their emotional needs [that is, the son will eventually grow up to be looked after by his future wife, who will be well trained in the feminine arts of care], mothers feel much less ambivalent about the satisfaction of needs and behave much more consistently in their nurturing. Boys therefore grow up, according to Eichenbaum and Orbach, with an experience of their needs as legitimate, appropriate, worthy of fulfillment.)

The male experience of the woman as "too much" has been developmentally explored, as well, in Dorothy Dinnerstein's ground-breaking *The Mermaid and the Minotaur: Sexual Arrangements and Human Malaise* (New York: Harper and Row, 1976). Dinnerstein argues that it is the woman's capacity to call up memories of helpless infancy, primitive wishes of "unqualified access" to the mother's body, and "the terrifying erotic independence of every baby's mother" (p. 62) that is responsible for the male fear of what he experiences as "the uncontrollable erotic rhythms" of the woman. Female impulses, a reminder of the autonomy of the mother, always appear on some level as a threatening limitation to his own. This gives rise to a "deep fantasy resentment" of female impulsivity (p. 59) and, on the cultural level, "archetypal nightmare visions of the insatiable female" (p. 62).

110. Quoted in Brian Easlea, *Witch-Hunting, Magic, and the New Philosophy* (Atlantic Highlands, N. J.: Humanities Press, 1980), p. 242 (emphasis added).
111. Quoted in Easlea, *Witch-Hunting,* p. 242 (emphasis added).
112. See Peggy Reeve Sanday, *Female Power and Male Dominance* (Cambridge: Cambridge University Press, 1981), pp. 172–84.
113. Quoted in Easlea, *Witch-Hunting,* p. 8.
114. Peter Gay, *The Bourgeois Experience: Victoria to Freud.* Vol. 1: *Education of the Senses* (New York: Oxford University Press, 1984), pp. 197–201, 207.
115. Chernin, *The Obsession,* p. 38.
116. Ehrenreich and English, *For Her Own Good,* p. 124.
117. See Jeffrey Masson's controversial *The Assault on Truth: Freud's Suppression of the Seduction Theory* (Toronto: Farrar Strauss Giroux, 1984) for a fascinating discussion of how this operation (which, because Fliess failed to remove half a meter of gauze from the patient's nasal cavity, nearly killed her) may have figured in the development of Freud's ideas on hysteria. Whether or not one agrees fully with Masson's interpretation of the events, his account casts light on important dimensions of the nineteenth-century treatment of female disorders and raises questions about the origins and fundamental assumptions of psychoanalytic theory that go beyond any debate about Freud's motivations. The quotations cited in this essay can be found on p. 76; Masson discusses the Eckstein case on pp. 55–106.
118. Banner, *American Beauty,* pp. 86–105. It is significant that these efforts failed in large part

because of their association with the women's rights movement. Trousers like those proposed by Amelia Bloomer were considered a particular badge of depravity and aggressiveness, the *New York Herald* predicting that women who wore bloomers would end up in "lunatic asylums or perchance in the state prison" (p. 96).

119. Banner, *American Beauty,* pp. 149–50.
120. Amaury deRiencourt, *Sex and Power in History* (New York: David McKay, 1974), p. 319. The metaphorical dimension here is as striking as the functional, and it is a characteristic feature of female fashion: the dominant styles always decree, to one degree or another, that women *should not take up too much space,* that the territory we occupy should be limited. This is as true of cinch-belts as it is of foot-binding.
121. Quoted in deRiencourt, *Sex and Power in History,* p. 319.
122. Kathryn Weibel, *Mirror, Mirror: Images of Women Reflected in Popular Culture* (New York: Anchor, 1977), p. 194.
123. Christy Ferguson, "Images of the Body: Victorian England," philosophy research project, Le Moyne College, 1983.
124. Quoted in E. M. Sigsworth and T. J. Wyke, "A Study of Victorian Prostitution and Venereal Disease," in Vicinus, ed., *Suffer and Be Still,* p. 82.
125. See Kate Millett, "The Debate over Women: Ruskin vs. Mill," and Helene E. Roberts, "Marriage, Redundancy, or Sin: The Painter's View of Women in the First Twenty-Five Years of Victoria's Reign," both in Vicinus, ed., *Suffer and Be Still.*
126. Gay, *The Bourgeois Experience,* p. 197; Millett, "Debate over Women," in Vicinus, ed., *Suffer and Be Still,* p. 123.
127. Vicinus, "Introduction," p. x.
128. Lasch, *The Culture of Narcissism,* p. 343 (emphasis added).
129. Chernin, *The Obsession,* p. 148.
130. Charles Gaines and George Butler, "Iron Sisters," *Psychology Today* (Nov. 1983): 67.

18

Que Gordita

EMILY MASSARA

CULTURE AND THE ETIOLOGY OF OBESITY

Based on the findings of the present study, the distribution of medically defined obesity in the Philadelphia Puerto Rican community shows women with an average overall incidence of obesity which is 19.4 percent higher than for men (see Table 10). Moreover, the results indicate that older women, in the forty to eighty age bracket, have a 66.7 percent greater prevalence of medical obesity than younger women under twenty-five. As is elucidated in the focused studies of "heavy" Puerto Rican women, a constellation of cultural factors, described in a more holistic framework in the foregoing chapters, encourage and/or sanction weight gain in women as they advance through the life cycle. Indeed, the data indicate a dramatic 47.3 percent increase in the incidence of female obesity from 13.3 percent among women under twenty-five to 60.6 percent among women in the twenty-six to thirty-nine age group, and is associated with a period in the life cycle of women when weight gain has many positive connotations.

Anticipated by the wife and encouraged by the husband, a woman's weight increase upon marriage serves as a visible sign, particularly to her family, that he is adequately providing for her. As such, her weight tangibly confirms the husband's newly-gained status of manhood, achieved as breadwinner for his family. For a woman whose weight falls within the demonstrated wide normative weight range, an increase of fifteen to twenty pounds during the early years of marriage may be viewed as enhancing her shapeliness, vitality and health. From the time of marriage, several cultural concepts and patterns surrounding food and the role of the "good wife" increase a woman's exposure to food, thereby enhancing its significance in her social and emotional life. Desiring to show her new husband that she is a "good cook," she prepares generous amounts of traditional Puerto Rican foods, high in carbohydrate and fat content, which afford the feeling of "fullness," a satiety that compliments her culinary skills. Reinforcing her focus on food, a husband may exert his right to be a "finicky" eater by requesting special dishes, or by requiring her to cook for him again after they have eaten elsewhere because he does not feel "satisfied" unless she cooks for him. One of the ways in which a husband may actively contribute to weight gain is by insisting that his wife accompany him in an "extra meal" should he return home late in the evening for dinner. As her family expands, food assumes increasing importance as a means of demonstrating the nurtrance and caring so central to the definition of the "good wife" and mother. Moreover, with the growth of a family, more extensive involvement with food results from the characteristic social organization of the dinner meal whereby a woman becomes involved with its preparation and serving two or three times. And, finally, increased food intake may result from snacking. Not infrequently, increase in food consumption follows a pattern of skipping or having a minimal breakfast consisting merely of coffee. As a result, the largest bulk of additional calories is consumed in a relatively short period of time.

Weight gain associated with pregnancy in the early years of marriage is facilitated by the traditional belief that eating restraint or an unsatisfied desire for food will jeopardize the well-being of the unborn child.

The anthropometric findings clearly indicate that a woman's weight increase may exceed initial expectations as she proceeds through the life cycle. As the ethnographic materials show, however, the resulting "heaviness" is of little negative social consequence, and may indeed be favored as an appropriate weight for an older woman. The demonstrated wide normative weight range, whereby, according to one cultural standard, "excessive" weight is defined at +85.5 percent beyond medical "ideals," both reflects and sanctions the lack of negative valuation on "heaviness" in women. Thus, for instance, 35.9 percent of the sample classified a female body type of +45 percent above medical norms as "plump," while 15.3 percent considered this body size to be "regular," only 10.5 percent considered it to be "too heavy." Indeed, as the qualitative materials show, the "heavy" woman may tend to her daily round with complacent unselfconsciousness about her body size. Moreover, while no longer conforming to cultural "ideals," whose weights converge with medical "ideals," concern about weight for its effect on her appearance is incompatible with the important self-concept of the "good wife" and mother whose primary concern is the nurturance and care of her family.

As a result of these findings, a significant conclusion of the present study is that the "milder" and even "moderate" forms of medical obesity, in contrast to the more "extreme" forms present in the community, need not be the result of a pattern of eating which is stress-induced. Indeed, weight, which by cultural standards, might be considered to fall within the lower range of medical obesity, may be aspired to for its positive connotations of "tranquility," health and a lack of problems in life.

At the same time, as is poignantly illustrated in the study of the sociocultural matrix surrounding the weight increase of Ramona Santiago, the constellation of cultural factors facilitating weight gain, combined with a culturally patterned constellation of factors inducing stress, can lead to a more "severe" form of obesity resulting from eating in response to stress. Indeed, an equally significant conclusion of the present study is that, in some instances, unrestrained eating serves as a subterfuge for the expression of negative feelings in the family context. As such, this form of eating behavior is related to a wider culturally patterned complex, often described as "hiding" true feelings "inside" or in the "heart," whereby women channel socially unacceptable feelings inward. The resulting inconsistencies between perceived and expressed feelings often exact psychic costs manifested in "suffering," depression, the need for "extra sleep," "bad nerves" or weight gain. Nevertheless, by refraining from direct expression of proscribed emotions, women are able to function in a manner befitting the "good wife" and mother.

The tendency to associate weight gain with positive social situations, as well as the general bias in favor of weight gain and "heaviness," influences individual perceptions about cause of weight gain. As was clearly evidenced in the case studies, although women might at times describe their eating behavior as being "out of control," more often they would attribute weight gain to factors beyond their control, and, on balance, maintain that they did not eat "too much." Thus, it would be difficult to document the extent of a wider pattern of eating in response to "nerves," due to the ambiguous perceptions about the role played by "nerves" in weight gain. For example, influenced by the predominant association of weight gain with "tranquility," many informants who cited "nerves" as a cause of weight gain specified that it was the tranquilizers taken for the "nerves" which were, in fact, responsible.

Thus, regardless of whether or not stress is a factor involved in the etiology of obesity, perhaps the most valuable finding of the present study is the complex and powerful effect of cultural factors on the etiology of weight gain among women in the Puerto Rican community. In particular, cultural values and patterns surrounding the domains of family, sex role organization, food and health shape perceptions about the cause, significance and effects of weight gain, which, in turn, contribute to the etiology of obesity. And, in those instances where stress contributes to the development of obesity, the materials from the studies of the sociocultural matrix related to weight gain highlight the important role played by culture in creating and channelling stress. Indeed, it is the culturally sanctioned, indirect patterns of dealing with socially unacceptable feelings which, exacerbating perceptions of helplessness, may provoke eating patterns leading to additional weight gain.

Moreover, while the social process of migration is clearly implicated in the etiology of obesity of many women, of perhaps more interest, especially for the purposes of the present study, are the findings concerning the manner in which cultural values influence weight gain associated with the migration process. For instance, based on the assumption that heavier is healthier, an increased appetite and weight gain may be a conscious goal, particularly as an antidote to illness associated with the colder climate in the migrant environment (see "Ana Reyes"). In addition, the uncharacteristic abandon in eating behavior exhibited in the life histories of many of the women who migrated before they were married provides additional evidence that food constitutes a domain characterized by a relaxation of the typical controls expected of women.

Perhaps it is no coincidence that all of the "heavy" women in the present study have developed diseases, such as hypertension and diabetes, which are scientifically correlated with medically defined obesity. Nevertheless, the relationship of these illnesses to "heaviness" tends not to be socially recognized, or, if acknowledged, to be ignored. Indeed, with a diabetic or hypertensive condition under control through medication, it is difficult, particularly for younger women, to be concerned about the alleged secondary effects of these serious medical diseases or the eventual development of cardiovascular disease. For judging health by traditional concepts, an extremely "heavy" woman manifesting a "good appearance" characterized by energy, vitality and a "zest" for life exudes the impression of good health. Moreover, in accordance with cultural definitions of health, weight gain and a good appetite may be viewed as a sign of good health; whereas weight loss tends to be associated with malnutrition and poor health.

And, finally, weights deemed to be "heavy" by cultural standards may have significant social effects. As is indicated by some of the case materials, culturally-defined obesity may be perceived by some women as being advantageous in countering the physical and psychological imbalance in the structure of male/female relationships based on the assumption of male superiority and strength. For example, in the view of some of the "heavy" women, "heaviness" is valued for its role in assuring an "asexual" self-preservation, which thereby affords a wider social life outside of the home without jeopardizing her status as a "good wife," and thus, her husband's "dignity" as a "man."

With respect to medical treatment, the present study clarifies the incongruous assumptions often held by physician and patient about the definition, cause and nature of obesity and thus perceptions about the need for weight reduction and the most efficacious means for doing so. Clearly, the study suggests the need for the physician to maintain communication and cooperation of a woman's husband and/or other close family members. Indeed, throughout the period of supervised weight reduction, it would seem that the obese female patient would benefit from a relationship characterized by personal encouragement and close supervision. Techniques of behavioral modification might be of value in long-term weight reduction by decreasing food intake without attempting to change basic diet. In so doing, a patient would be able to function in roles in which involvement in food is essential for positive self-concepts and constitutes an integral part of her social life. Many women, particularly those whose weight is the result of eating in response to unresolved stress, would benefit from exercise programs. Not only would exercise facilitate muscle tone during the process of weight reduction, but it would provide a constructive outlet for pent up feelings which might otherwise be expressed in uncontrolled eating behavior.

PROBLEMS AND PROSPECTS FOR FURTHER STUDY

With regard to methodology, the "emic"/"etic" approach adopted in the present study, whereby western measurements and standards were used as a basis for measuring and quantifying "emic" classifications and standards of obesity, provided valuable insights. For instance, by this method, in combination with results from more qualitative study concerning cultural valuations on "thinness" and weight loss, "heaviness" and weight gain, I was not only able to document the wide normative weight range, but to explain its full significance in contributing to obesity in the life cycle of women. Clearly, this style of research would have valuable application in other ethnic communities, particularly those characterized by a high incidence of medical obesity.

The findings of the rate of medical obesity among men in the Puerto Rican community suggest that a study focusing on weight gain among men would also be worthy of research. In addition, research on cultural factors contributing to obesity in children, and the effect of acculturation on cultural weight valuations, warrant study.

APPENDIX I. QUESTIONNAIRE

I. Basic Census Information

M___ F ___ 1. Sex

___ 2. Age

___ 3. Marital Status:

___ a. Married

___ b. Widowed

___ c. Separated

___ d. Divorced

___ e. Single

___ 4. Number of Children

___ 5. Highest grade of education completed

___ 6. Occupation

___ a. Manual work

___ b. Office work

___ c. Self-employed

___ d. Unemployed

___ e. Housewife

P.R. ___ U.S. ___ 7. Where were you born?

___ 8. (If born in Puerto Rico) How old were you whey you came to the U.S.?

19

The Sweetness of Fat

Health, Procreation, and Sociability in Rural Jamaica

ELISA J. SOBO

In the United States there is a well-known saying that you can't be too rich or too thin, but in rural Jamaica, amassing wealth and keeping slim have antisocial connotations. Ideally, relatives provide for each other, sharing money and food. Because kin share wealth, no one gets rich; because kin feed each other, no one becomes thin. Cultural logic has it that people firmly tied into a network of kin are always plump and never wealthy.

Especially when not well liked, thin individuals who are neither sick nor poor are seen by their fellow villagers as antisocial and *mean* or *stingy.*[1] These individuals do not create and maintain relationships through gift-giving and exchange. They hoard rather than share their resources. Their slender bodies bespeak their socially subversive natures: thinness indicates a lack of nurturant characteristics and of moist, procreative vitality—things on which a community's reproduction depends.

Rural Jamaicans' negative ideas about thinness are linked with their ideas about health. As Sheets-Johnstone points out, "The concept of the body in any culture and at any time is shaped by medical beliefs and practices" (1992: 133). Notions concerning health can profoundly influence the interactive and symbolic communications made through our bodies. These notions greatly influence the ideal standards set for bodies and affect the ways we experience, care for, and shape (or try to shape) our bodies and those of others (Browner 1985; Ehrenreich and English 1979; Nichter and Nichter 1987; Payer 1988).

Importantly, notions about health are—in a very tangible way—notions about body ideals, and they have social meaning. Health traditions do not exist in isolation from other realms of culture, such as gender relations and economy (Farmer 1988; Jordanova 1980; Martin 1987), nor are they isolated from extracultural influences, such as ecology and global political conditions (Farmer 1992; Vaughan 1991). Often, ideas about the body and its health are put forward as rationalizations or ideological supports for conditions, such as class and gender inequalities or personal maladjustments (e.g., Kleinman

1980; Laws et al. 1985; Lock 1989; Scheper-Hughes 1992). In this chapter, I describe the traditional health beliefs that inform understandings of body shape in rural Jamaica, and I trace the connections between these ideas and Jamaican understandings about sociability (see also Sobo 1993b).

For rural Jamaicans, the ideal body is plump with vital fluids, and maintaining the flow of substances through the body is essential for good health. Taylor (1992) argues that an emphasis on maintaining a continuous, unimpeded flow through the body is common among those who value reciprocity and emphasize the obligation kin have to share with each other, which Jamaicans do. Sickness occurs when the flow is blocked or otherwise "anomic" (Taylor's term, 1988); individual pathologies are homologous with social pathologies, caused by disturbances in the flow of mutual support and aid.

Taylor shows that health-related symbolism "establishes implicit connections between the bodily microcosm and the social macrocosm" (1988: 1343). "Liquids are especially privileged vehicles of this symbolism," he says, "because they possess the capacity to flow, and thus to mediate between distinct realms of being . . . attenuating the opposition between self and other" (1988: 1344). In rural Jamaica, people are physically linked by bodily liquids—fluids like semen and the blood that flows from mother to fetus during gestation. They also are linked through food that is shared. Both vital bodily fluids and foods fatten the body, making plumpness an index of the quality and extent of one's social relations as well as an index of good physical health (see Cassidy 1991).

The concept of the body-in-relation may seem foreign to U.S. or Western European readers who tend to view the body like they view the self—as autonomous, individual, and independent. Their bodies serve primarily as vehicles for the expression of the individual self, and so of self-directed denial, control, and mastery (Becker 1990: 1–10). Jamaicans, however, recognize the body's shape as an index of aspects of the social network in which a person is (or is not) enmeshed and of those individual traits that affect that person's social connectedness, such as the ability and willingness to give (see Cassidy 1991).

Influenced by British interests, much of the anthropological literature on Jamaica deals with kinship and social structure (e.g., Blake 1961; Clarke 1957; Douglass 1992; Smith 1988). Some studies examine the cultural construction of kinship, but none examine the ethnophysiology of blood ties and most overlook the body as such, despite its necessary role in procreation. Some works concerned with Jamaican family planning include descriptions of the reproductive body (e.g., Brody 1981; MacCormack 1985), but the health-related significance of blood and the physical intricacies of consanguineal and other consubstantial kin ties (and of their behavioral ramifications) are left unexplored. Pan-Caribbean ethnomedical notions about blood are discussed by Laguerre (1987), but the social and cultural meanings of body morphology and of bodily components (and the sharing thereof) have received little attention.[2]

METHODS AND SETTING

Research for this chapter was carried out in a coastal village of about eight hundred people in the parish of Portland, where I lived for a year in 1988 and 1989 (see Sobo 1993b for a full account of the research). Data were collected through participant-observation and interviews that took place in community settings and in private yards. I also solicited drawings of the body's inner workings from participants.

Like most Jamaicans, the majority of the villagers were impoverished descendants of enslaved West Africans.[3] Many engaged in small-scale gardening, yet few could manage on this alone. To supplement their meager incomes, people also took in wash, hired themselves out for odd jobs, engaged in part-time petty trade like selling oranges, and relied on relatives for help.

Jamaican villages typically consist of people brought together by ancestry, or by proximity to a shop or postal agency. In some cases, they are organized around an estate where village members sell their labor. Households are often matrifocal (see Sargent and Harris 1992: 523; Smith 1988: 7–8), and nonlegal conjugal unions and visiting relationships (in which partners reside separately) are common. Houses are generally made of wood planks and zinc sheeting; often they lack plumbing and electricity. People build their houses as far apart as possible, but they are usually still within yelling distance of a neighbor.

BODY BASICS

Jamaicans value large size, and they *build* the body by eating. Different foods turn into different bodily components as needed, either for growth or to replenish substances lost through work and other activities. Comestibles that do not so much build the body but serve to make people feel full are called *food*. In common Jamaican usage, *food* means only tubers—belly-filling starches not seen as otherwise nutritious.

Blood is the most vital and the most meaning-invested bodily component. It comes in several types. When unqualified by adjective or context, the word *blood* means the red kind, built from thick, dark liquid items such as soup, stout, and porridge and from reddish edibles such as tomatoes. Red wine, also referred to as tonic wine, can be used to build blood, and blood is sometimes called wine. Some think that the blood of *meat-kind*, such as pork or beef, is directly incorporated into human blood; others say that meat's juices build blood. Wild hog meat, redder than regular pork, is supernutritious and vitality boosting because wild hogs feed mainly on red-colored roots, said to be beneficial blood-builders. People point out that meat-kind left sitting out or from which all vital fluid has drained (as when cooked for a long time in soup) loses its nutritive value and serves only as *food* to *fill belly*.

Sinews, another type of blood, comes from okra, fish eyes, and other pale slimy foods, such as egg white or the gelatinous portions of boiled cow skin or hoof. *Sinews* refers to, among other substances, the joint lubricant that biomedical specialists call synovial fluid, which resembles egg white. Sinews is essential for smooth joint movements and steady nerves. The functioning of the eyes depends on sinews too: the eyes are filled with it and glide left and right and open and shut with its aid. Sinews, also associated with procreation, is found in sexual effluvia and breast milk. Many call sinews *white blood*, as opposed to red.

People have less elaborate ideas about what edibles other bodily components are made of. Vitamins, contained in the strengthening tablets and tonics that are popular and easily available, build and fatten. Some Jamaicans argue that meat-kind builds muscles. Most agree that corn meal builds flesh. A few suggest that milk builds bones, at least in children but not necessarily in adults whose bones have already developed.

The most important part of the inner body is the *belly*, where blood is made. This big cavity or bag extends from just below the breast to the pelvis. The belly is full of bags and tubes, such as the *baby bag* and the *urine tube*. A main conduit leads from the top of the body through the belly to the bottom, with tributary bags and tubes along its length. Sometimes, tube and bag connections are not tightly coupled. A substance improperly propelled can meander off course, slide into an unsuitable tube or bag, lodge, and cause problems.

FOOD SHARING AND SOCIAL RELATIONS

In reviewing the social significance and health benefits of big size cross-culturally, Cassidy (1991) found that socially dominant individuals who are enmeshed in sound relationships are usually large. Bigness tends to ensure reproductive success and survival in times of scarcity, and plumpness is generally considered attractive. According to Brink (1989), such is the case in many of the West African societies from which people were taken to Jamaica as slaves. In these societies, those who can afford to do so seclude their adolescent girls in special "fattening rooms" and, after a period of ritual education and heavy eating, the girls emerge fat, attractive, and nubile.

In Jamaica, where a respected adult is called a *big man* or a *big woman,* good relations involve food sharing, and people on good terms with others are large. Weight loss signals social neglect. A Jamaican seeing someone grow thin wonders about the sorts of life stresses that have caused the weight loss (rather than offering congratulations for it and attributing it to a "good" diet, as many middle- and upper-class people in the United States do).

In the ideal Jamaican world, mothers feed their children, kin feed kin, and lovers feed each other. Men involved with women put on pounds from the meals their women serve them. Likewise, women display the status of their relations with their measurements; the breadth of the *backside* is particularly symbolic. Villagers noticed when a woman named Meg began to *mauger down* (get thin, grow meager) and lose her once-broad bottom; they knew—and they broadcast—that her affair with a rich old man had ended as she apparently no longer received food or resources from him.

Food sharing is a part of good social relations, and it, as well as other kinds of sharing, ends when people fall out. People with *something between them* (i.e., strife) both cease to give gifts and refuse to receive them. For instance, they refuse food from each other (often because they fear being poisoned; Sobo 1992). A disruption in the flow of goods and services signals the disintegration of a relationship.[4] Sister Penny knew that her relationship with Mister Edward was in trouble on the day he refused and sent back the dinner that she regularly prepared and had her daughter carry down the road to him at his mother's house, where he lived.

Good relationships and good eating go hand in hand, but plumpness depends on more than mere food—it depends on pleasant household conditions. Living in a household where *the conditions* (that is, the group dynamics) are harmonious and agreeable ensures both physical and mental vitality. No matter what they eat, unhappy people who live where *the conditions* are unpleasant lack energy, and they *draw down* (get thin) as fat *melts off*.

When a young woman named Any lost weight and grew lackadaisical, villagers knew that she and her live-in boyfriend were having problems. Indeed, Amy's young man had taken up with his sister's boyfriend's sister. Amy's declining physical state and lethargy indicated this change in the conditions. Even with plenty to eat, a person in her position would lack energy and pull down mauger because, as one woman commenting on the situation explained, "people with worries can't fat."

SWEETNESS, RIPENESS, AND DECAY

Fatness at its best is associated with moistness, fertility, and *kindness* (a sociable and giving nature) as well as with happiness, vitality, and bodily health in general. People know that drinks and warm, moist, cooked food can fatten them, while cold rice, *overnight food* (leftovers), and *dryers* such as store-bought crackers usually cannot. Fatness connotes fullness and juicy ripeness, like that of ripe fruit *well sweet* and soon to burst. Young boys fill out when they approach adulthood; young girls plump up in late adolescence as a prelude to childbearing. Men often call pubescent girls *soon ripe,* and they allude to sex with *ripe* girls through talk of harvesting.

Jamaicans call pleasing things *sweet.* When someone unexpectedly laughs or smiles, they are commonly asked, "Is what sweet you so?" People associate sweet goodness with fatness too. Men often describe plump women whom they find attractive as sweet. Good food also is sweet. Something sweet is ready to eat or ripe for enjoyment. As it approaches maturity, fruit swells and sweetens. A dream of fruit at its fullest, sweetest stage of development means that the time is ripe for whatever project the dreamer had in mind. Ripeness connotes urgent readiness (as for sexual relations). It can also mean ill-mannered precociousness, just as *green* (unripe, unprocessed) can describe naiveté. Unruly, disrespectful children are put down: "You too ripe!"

Overripe fruit rots and its sweetness sours. After it swells and ripens it declines, coming to resemble feces—soft, dark, fetid, and sometimes maggot infested. Overripe fruit is never eaten. By picking fruit just as it *turns* (from green), Jamaicans avoid the possibility of contamination with rot.

Ideas about decay give expletives power. The curse *rhatid* expresses, as a homonym, the connection between rotted matter and problems worthy of wrath, pronounced "rhat" or "rhot" (see Cassidy 1982: 175). The negative connotations of rot and decay make *bumbo clot, rhas clot,* and *blood clot* among the most insulting epithets available, for these phrases describe the cloth *(clot)* diapers used to sop up dangerous waste that seeps out from the bowels of the body. *Bumbo clot* loosely refers to the diapers once used to catch the fecal and other matter that oozes from corpses when they are moved (morticians generally take care of this now), while *blood clots* serve as menstrual rags. *Rhas clots* do either.

Ideas about decay also fuel subversive banter. While playing bingo in the back room of a shop on the main road, which only more rebellious characters do as it brings disapproval, one rowdy woman named Pet denounced another boisterous player, Glory, for not having bathed. Glory, who had bathed, retorted: "You stink like ripe banana" (in other words, "you stink like feces"; picture overripe banana flesh). Pet playfully drew power from this complaint, warning Glory that most of the bellyful of ripe fruit that she had lately eaten was ready for gaseous rectal expulsion.

All that gets taken into the body, whether to build or fill belly, must get used or expelled because unincorporated excess begins to swell and decay. This knowledge leads people to associate superfluous or unutilized food, fat, health, and such with filth and the inevitable process of decomposition that accompanies death. Some Jamaicans speak of "good" fat and "bad" fat, the good being firm like a fit mango and the bad being spongy, soft, hanging slack, and denoting declining fitness as if a person was an overripe fruit, beginning to break down or rot.

WASTE AND WASHOUT

Not all that gets ingested is transformed into specific components like sinews or blood, and some things are not utilized in the body's structure at all. Extra liquids become urine, and solid food turns to *didi* or feces, which move out from the belly cavity through the *tripe* (digestive tubing) and are expelled. People who do not use the toilet often enough literally fill up with waste. Trisha once asked her five-year-old daughter why the white missionaries always had such soft and overfat bellies. The little girl explained that "the tripe them fulla didi." Delighted, Trisha reiterated, "they fulla shit!"

A body that does not efficiently rid itself of excess and rotting waste turns septic inside because too much decomposition then takes place internally (most of waste's decomposition should take place outside of the body). A number of things can cause digestive inefficiency. Too much of a rich, strength-giving food such as cheese or cream can *clide* or clog the inner works, causing sluggish digestion and a backup of food in the belly. So can gluttony, which also reduces the amount of food available to others. Things like coffee grinds or undigested *hard food*—tough edibles such as bones or coco (a very hard tuber)—can also cause problems by settling in the *belly-bottom* and blocking the exit tubes. Held too long, food rots in place, festers, and sickens. The belly might even burst from buildup.

The most popular cure-all, the *washout* or laxative purge, eliminates blockages and harmful waste from the system. A washout once a month—a schedule modeled gynocentrically on the menstrual cycle (Sobo 1993a)—is advised. The importance of keeping clean inside explains one woman's choice of survival essentials (made as she fled her house during the 1988 hurricane), which included only "the ingredients for a washout" plus a blanket and some biscuits. It makes clear why every household medicinal supply includes, if nothing more, a purgative such as Epsom salts, cathartic herbs, or castor oil. And it explains why so many people understand the life-support devices seen in hospitals, such as the nasogastric tube, to be mechanically effecting washout cures.

The importance of keeping clean inside so that proper, balanced flow can be maintained parallels the importance of keeping goods and services flowing through networks of kin and corresponds in a number of ways to the idea that hoarding is bad. An overabundance of perishable resources not passed on will rot. Even money uncirculated is associated with decay, as the traditional association between feces and money (Chevannes 1990) reveals. Hoarding means neglecting one's social network, possibly allowing others to experience avoidable hardship, which can lessen the network's cohesiveness.

THINNESS

Like cleanliness and balance, plumpness is important for good health. Few rural Jamaicans want to reduce. Diet foods and beverages are only seen in bigger towns. People generally assume that they are meant for diabetics, because no one should wish to be thin. Thinness is associated with ideas antithetical to those that "good" fat connotes. Thinness and fatness are to each other as the lean, dry, white meat of a chicken is to its fatty, moist, dark parts—the parts that most eaters prefer. Ideas about infertility and unkindness are linked with the notion of thinness. People taunt others by saying they will dry up and grow thin from antisocial *meanness.* Their observations of the elite and those in power who are light-skinned and whom they see as thin reinforce this belief.

Thin people are understood to lack the vitality associated with moist and juicy "good" fat. Like an erect penis or breasts plumped with milk, like a fat juicy mango, the body seems more vital when full of fluid and large in size. While too much blood or food overburdens the body and can rot and cause sickness, as noted above, *dry* bodies have no vital nature at all (low levels of bodily fluids and fat can lead women to have trouble conceiving). A slim person, especially a slim woman, is called *mauger*—meager and powerless—as if not alive at all and, like a mummy or an empty husk, far beyond that powerfully dangerous state of decay. A thin, dry body reveals a person's non-nurturant nature and his or her lack of social commitment.

BEING SKINNY, BEING MEAN

Kindness involves altruistic, kinlike sharing. Kind people give what is asked for and also offer things. They treat others as if family. A mean person, like a stranger (not kin), never shares and always refuses requests. Everyone hates a person who is *near* or *exact,* such as someone who never cooks extra dinner—someone stingy with food and so with their sociability. People concerned with their reputations are free with what money they have, buying drinks and putting on a show of kindness so that others cannot call them *mean.*

Mean people use very little salt in their cooking. Salt costs money, and it is associated with (among other things) imported foods, healing, good spiritual forces, and heaviness. It affects food's flavor, and most Jamaicans declare that they simply will not eat *fresh* (unsalted) food because it tastes bad. Like their cooking styles, their bodies give mean people away. Those who are *near* have a *cubbitch hole* or dent of covetousness at the *neck-back;* in other words, they are thin. Jamaicans say that mean people's bodies "dry down," "dry out," and "come skin and bone."

Vy and her brother, both in their early twenties, laughed about the mean old woman with whom their mother sent them to live fifteen years ago when she had no money and could not *keep* (support) or care for them herself. The old woman's thin body and flat cooking betrayed her nearness. Some stingy people draw down mauger or slim because, on top of not feeding others, they starve themselves, Vy explained.

When the woman did share food, it was *fresh* (unsalted) and otherwise ill prepared. She doled out small portions not big enough to *fill belly* but only, Vy said, to "nasty up me teeth" (to dirty the mouth without satisfying hunger). The woman would boil soup from the same piece of dried fish every day for a week, removing the piece each evening to use the following day. The soup carried nothing of substance and lost its salted taste by day three. It served as a sign of this woman's lack of a will to nurture, as did her thin, husklike form. A body with more vital juices inside would have housed a more social and giving person.

Store-bought snack food is also associated with thinness. The procurement and ingestion of packaged ready-to-eat foods (e.g., cookies, cheese puffs, sweet buns) do not involve the regular division of labor and tradition of sharing. Money is instrumentally exchanged for packaged food so that no reciprocity expectations ensue. Not having been prepared by a loving and morally obligated other and especially when ingested alone and away from home, store-bought foods—not coincidentally called *dryers*—represent the antisocial. Accordingly, those who eat dryers in lieu of home-cooked meals will be thin.

KINSHIP AND SOCIABILITY

Jamaican kinship ideally involves a sense of interdependence and obligation, which ensues from shared bodily substance. Thin people have little bodily substance to spare. But sharing one's substance and having children generally are required to attain full adulthood because they affirm one's link to a social network.[5]

Through procreation, a person puts the parental gift of lifeblood from which his or her body was built back into social circulation, confirming and tightening his or her bonds to kin and community. Those who do not reproduce their social networks through procreative relationships physically subvert the social and moral order, as do those who act selfishly, hoarding resources or keeping to themselves so that others cannot ask for favors and so entangle them socially. The bodies of those elite individuals who are thinner than the poor reveal (to the poor) how they came to enjoy their *uplift:* through stingy meanness and by divorcing themselves from a social network of kin.

Jamaicans ascertain kinship through blood ties, in which both kinds of blood figure. Male and female sexual fluids are parallel types of white blood, and mingling white blood is key to conception. Pregnancy occurs when male semen meets a *ripe* female egg (sometimes equitably referred to as a sperm) or the female equivalent of semen—female white blood—in a woman's reproductive tract. Eggs mature and are discharged either monthly or upon sexual excitation. They burst when overripe and no longer fertilizable (as if decaying fruit), and they are excreted through the *tube* (the vagina).

During gestation, parents feed and so build and *grow* the fetus with their bodily substances. At first the fetus develops through simple accretion. After a few months, the *babymother's* (red) blood and food that she has ingested get eaten by the fetus and are then transformed into its bodily components. The fetus also eats semen ejaculated into its mother during its gestation. If a woman has many lovers, whosoever has invested the most sperm should receive credit for paternity.

The white blood that lovers share in nonreproductive sex only mixes temporarily. But in reproductive sex, blood mingles permanently in an offspring, creating an indirect link between parents. Siblings are related by virtue of having incorporated blood from the same source (or sources, when both parents are shared); other kin links, such as those between cousins, are similarly traced to a consanguineal origin.

The creation of kinship is not always seen as fixed or finished at birth. Older, more traditional Jamaicans say that babies continue the consubstantiation process that creates kin even after birth by drinking their mothers' milk, which is a kind of white blood. Accordingly, kinship ties can be created between a baby and an otherwise unrelated wet nurse. While biological motherhood is central in establishing maternity rights, traditional Jamaicans subscribe to an attenuated form of what Watson (cited in Meigs 1987: 120) refers to as "nurture kinship"—kinship that can be altered after birth by what Meigs calls "postnatal acts."

A child not related by virtue of received blood can become as if related to someone by virtue of that person's caretaking efforts. Feeding children is important outside of the womb as in it; food sharing can be a source of, as well as an index of, relatedness. A woman who feeds a child after it is born can claim motherhood: she *grows* it, as the genetic mother did when she fed it in her womb. Likewise, a man who spends his money on food for a child can claim fatherhood. Following the ethnophysiological model of parenthood, kinship can be established and claimed by a caretaker who puts great amounts of effort into raising, growing, feeding, and so building a child. Food from a caretaker that is taken into and made part of a child's body works like incorporated blood to create and maintain kin ties.

A big family provides, ideally at least, a big network of unselfconscious, altruistic support. People are physically driven to lend this support simply because they originate from or share *one blood* and are thus of *one accord,* sharing *one heart* and *one love.* Kin terms express and add an enduring, unconditional quality to relationships, and so cloak instrumental dealings that would otherwise seem self-centered, competitive, antisocial, and thus distasteful, given the ideology of sociable reciprocity.

Fecundity represents reciprocation—blood for blood. This reciprocity is practical, tangible, and homologous with the idealized reciprocal exchange of resources between kin. A daughter often sends one of her own children to live with and help her mother, giving a life (in the form of her young child) in return for having been given her own life. My neighbor Estelle could hardly wait for her grandson to *come up* in age so that she could send for him to come live with her and help with those tasks she could not manage.

Procreation confirms male and female virility and fecundity, demonstrating that one has life inside to give and that one's vital essence or life blood will persist. Thin people have little nurturant, willingly sociable capacity, while plump ones—including those with pregnant bellies—are full with it. Those who do not reproduce only serve to work. Their blood disappears from the social circles that individuals, joining together in the culturally recommended fashion, create and re-create.

THE INFERTILE MULE

Mules do not reproduce. Mules just work. Like mules, women are known for their carrying capabilities; women carry wood, water, and so on, and mules serve as pack animals. Donkeys and old women are traditionally associated (Chevannes 1990). One saying about women holds that God made two kinds of donkeys, but one kind cannot talk (Hurston 1990: 76). Nonfecund women are called mules: unlike fertile women and donkeys (the more common beast of burden), mules do not breed.

Mules (and so nonfecund women) are traditionally associated with prostitutes (Chevannes 1990). In addition to defying female role expectations and posing an antisocial challenge (whether intended or not), nonmonogamous and nonfecund women fulfill the culturally informed sexual but not the procreative needs of men (nor do they fulfill their own culturally constructed childbearing needs). Although as sexual and subsistence workers nonfecund women, like prostitutes and mules, are useful, those who do not provide society with children are said to "have no use" in the end. This phrase implies that the ultimate purpose of the individuals—and of sex, which ideally ends in *discharge* (the release of sperm and eggs)—is to re-create society.

Mules do not reproduce and so cannot establish any enduring form of community. They do not share their blood with other mules—they do not establish relationships confirmed by offspring. They are stubborn and uncooperative. Mules represent particularly antisocial and selfish beings uninterested in growing up to establish new bonds and unable to return the gift of life by fulfilling the social obligation of creation.

The dependence of those with no outside ties or offspring remains like children's dependence on parents. Nonfecund women can never really achieve social adulthood because they, like mules, do not overtly signal childhood's end by having babies and establishing bonds—blood bonds—with the larger society outside their immediate family circles (infertile men can, of course, claim to have fathered children not really their own).

Those who do not have children are generally not well articulated into a social network. They have no *blood ties* to any *babyfathers'* households (so no paternal grandmothers, aunts, or uncles are bound to help them with the children). Furthermore, without children to *response* for, their claims on the resources of their own natal networks are weak. (Such a situation does not always entail suffering. Those with small networks and no offspring are minimally obligated to others.)

Having children expresses a readiness to fulfill obligations to society and one's relatives. It demonstrates a willingness to share resources and also one's body substance in order to maintain and regenerate the social environment into which one was born. People who do not create and maintain network ties tacitly release themselves from their social group's hold.

PLUMPNESS AND SEXUAL FLUIDS

The thin body lacks vital fluids, which can cause embarrassment and shame because this means a person can be cast as infertile and antisocial. That is why it *cut* Dara when a girl who hated her asserted that Dara had to wear two *suits* or layers of clothing to look presentable—to pad her mauger frame. Dara was slim, but Jamaicans denigrated for thinness were never extraordinarily slender in my culturally conditioned opinion. Even the women pictured in a British hardware company's Jamaican give-away calendar would be judged fat by mainstream U.S. standards. But two teenaged male participants in this research laughed at their picture, calling them *mauger dogs.*

People are sensitive to such insults. Some prepubescent girls buy special so-called *anorexa* pills to help them put on weight. Pharmacists sold the Anorexal brand until it was withdrawn from the market; now most carry a brand of anorexa pill called Peritol. The pill, an antihistamine (cyproheptadine) that enhances the appetite, can be bought with no questions asked.

As a child matures and begins to develop a healthy libido or *nature,* the body swells with sexual fluids. A good shape is a matter of firmness and proportion. Men should be muscular, never flabby. Women's bottoms should be broad (my fictive brother boasted that Jamaican women have the broadest bottoms in the world) and breasts should *stand up* high. A sagging or diminutive bosom indicates a lack of plumping juices. Thin, tomboyish girls with short hair and girls whose hair stays short because it is dry and brittle and breaks off are insultingly called *dry-head pickney,* which alludes to their immaturity (*pickney* means child or children) and, moreover, to their lack of vital moistness and so to their potential roles as infertile subversives.

Sexual fluid, like fatness itself, is good, but here again too much can be harmful and balance must be maintained. Orgasm releases body-swelling sexual fluids. Celibate women or those with poor lovers may become sick through the retention of too much *sweet water.* (Wet dreams help keep males well.) A young woman, Jean, once pointed to her friend Titia's crotch, calling "punani fat, eee?" Jean's joke about the size of Titia's *punani* or pubic area implied its disuse. If Titia did not have sex soon, Jean suggested, a buildup of overripe sexual juices would cause bodily harm.

A buildup of unreleased semen or sweet water can cause *teenager bumps* or pimples as excess sexual fluid tries to work its way out through the pores. The association between pimples and unused sexual discharge is a logical extension of the notion that a good thing like money or vital fat can go bad if not properly dispersed, and of the idea that overfull, overripe fruit bursts and sours. Some say teenager bumps are more common among youths whose caretakers can afford to provide them with expensive rich foods but who also insist that they remain celibate, in keeping with the behavioral expectations for those of high station (regarding the maintenance of respect, see Wilson 1973).

DISCHARGE AND DECAY

Bodily equilibrium is essential, so sexual fluids must be discharged now and again. Women receive men's discharge during intercourse. Like their own sexual fluids, sperm taken in fattens women, making them sexually appealing and attractive. A teenaged girl's increasing plumpness is as much a perceived result of her becoming sexually active as it is a positive result of her own growing moistness and fertility. Some people, to support their claim for the health-enhancing value of sperm, say that prostitutes and other women who perform oral sex get fat.

But prostitution and oral sex are not condoned (Sobo 1993a). The reasons for prostitution's poor standing have to do with its instrumental nature and the popular belief that prostitutes seek to avoid pregnancy. Conception does not have to happen during sex, but the possibility of conceiving should exist; the interpersonal flow of sexual fluids, like resources, should not be blocked. With oral sex, conception cannot occur; furthermore, the mouth is meant for food, not sexual fluids. So in situations where the morality of oral sex and prostitution are called into question, people will say that the toxicity of discharge improperly or excessively taken in makes women who have oral sex or who prostitute themselves *pull down* or lose weight.

Semen can cause problems even for women who participate only in socially sanctioned sex acts. When conception does not happen, discharge can lodge in inaccessible spaces. It rots quickly, growing toxic and polluting. Menstruation serves as a purifying *washout* for the female reproductive system (Sobo 1992).[6]

DRY SPINE

Men also need to be careful about semen, albeit for different reasons. Men emphasize the importance of intercourse and ejaculation for *clearing the line*. If this is not done, excess amounts of undischarged semen and sinews harden up in the spine, causing back pain and sexual problems. Drinking beer helps promote the flow of fluids through the penis. It is assumed that men will find partners and so need never masturbate to clear the line. Women who stay celibate too long may get *nerves* problems, as the *nerves* system contains sinews. Discharging cleanses the body, promoting as well as signaling good health. People—mostly men—use this knowledge to justify frequent sex and to coerce others into partnership. Asymmetric gender power relations find support in traditional health beliefs.

Jamaicans believe that men, as opposed to women, always want sex and cannot easily control their urges. Male sexual behavior is justified as an inherent part of their biology. However, sex taxes the male constitution. Men lose much more sinews when discharging than do women. Like any resource, semen must be shared. But people with too many demands on their energy or resources inevitably end up feeling *sucked out*, *eaten out*, or even *dried out*. A man who runs out of sinews comes down with *dry spine*, a condition involving ejaculatory dysfunction.

Dry spine can occur if a man involves himself with an abnormally libidinous or *white liver* woman—a woman who *works* her men far too hard, that is, bone-dry as well as to the bone. The livers of sexually hungry women may be described as white because they are full of sinews, because of the symbolic link between egoism and lightness, or because such women conform to the stereotyped view of sexually ravenous white female tourists. White liver women are feared by men insecure about their power to satisfy women sexually. Such men experience performance anxiety, and impotent men or those who simply *can't do the work* of sex are vindictively called soft, which harms their reputations.

Men with dry spine will *draw down* thin as their stores of vital juice are dried out and depleted. They grow infertile (like mules) and are sexually *useless*. Some have trouble urinating: not a drop of sinews remains to help ease urine's passage through the genital tube. Squatting like women helps. Ma Bovie, telling me about a man so damaged by an overenthusiastic partner that he could hardly walk, offered a story about a cock with light, dry, stingy meat. Rooster meat should be dark, moist, fatty, and rich. Hindsight told her that this cock must have suffered a terrible case of dry spine: she remembered that his hen wives demanded "plenty" sex.

Dry spine can reverse itself over time, but recovery demands total sexual abstinence so that the body can renew its store of sinews. Sufferers can take lime juice because, as it draws out heat, it dissipates the libido or *nature*. One woman surmised that the pope must drink lots of lime juice, for having nature is only natural and health requires expurgating it rather than simply repressing and not spending it. But anyone who has gone too far in expressing his nature and has dried his sinews or juice stores must add plenty of slimy okra, gelatinous cow foot, and other sinews-building foods to his diet. Drinks made from condensed milk and soursop (*Annona muricate*) juice or from the seaweed Irish moss (*Gracilaria spp.*), which resembles semen, and coconut *jelly* (the white, gel-like meat of young coconuts) are especially helpful to *put it back*.

A healthy libido or nature, which demonstrates one's physical and social vitality, depends on healthy blood. *Roots* tonic, made mainly from plant roots gathered during the full moon (when they are supposedly plumpest and most powerful), energizes the body by building, cleansing, and mobilizing the blood. A man with dry spine takes root tonic once every morning and night to replenish and enrich his blood stores. After a year of this treatment, his spine and his sex life should be fine. His body will have grown moist and full with vital fluid. Many men whose spines are fine take the roots regularly anyhow, swearing by its effects on potency. Women take it too, for general health enhancement.

BAD SHAPE

Fluid loss in women is most noticeable in the bosom, which, after a woman has had children or after a miscarriage or abortion, sags flat and low. A sagging bosom indicates a declining physical condition—a *used up*, postreproductive body. Too soft, with no babies in her belly and no life-giving milk to make firm her breasts, a woman with a "bad" shape is, like an overripe fruit, on her way to decay.

Cow's milk and other foods that build sinews increase a woman's store of breast milk and so plump the bosom. Sperm also increases the amount of white blood women carry. A nursing woman should only have intercourse with lovers whose sperm helped grow the nursing child. Otherwise, the child will get sick from drinking breast milk that contains foreign sperm and from contact with or the ingesting of any foreign sweat that remains on the mother's breast after sex with the nonfather. The baby will fail to thrive.

A woman with a small, slow child risks being accused of having an *outside man*. People may surmise that economic instability forced the woman into the liaison. This will shame her, and it will shame her *babyfather* by indicating that he failed to provide for his child.

Through early weaning, women with new lovers can avoid harming their babies, their reputations, or their conjugal relationships. Early weaning also helps women maintain figures that are attractive to men, as prolonged nursing *tires* their bodies.

Bosoms sag with use and age, but they also fall after abortions or miscarriages, or when a body is overworked and so depleted of its stores of blood. Women not blessed with firm breasts risk being accused of taking purgatives to *wash away baby* (abortion is not condoned: Sobo 1995). They risk being labeled prostitutes, who overwork their bodies with sexual labor exchanged directly for instrumental gain and which is therefore antisocial labor. The body shape associated with a separation between sex and reproduction, and with aging and death, is thin and flaccid, not firmly plump like the sociable one.

SOCIABLE PLUMPNESS

Meanings attributed to personal appearances are context-specific, and circumstances such as personal vendettas affect which meanings get linked with whose bodies. The good as well as the bad can be highlighted. For example, a thin individual can escape ridicule if his or her svelte shape is caused by working long and hard for the benefit of others. As a general rule, however, and especially when a slender person's behavior gives others cause to disparage him or her, the thin person is cast in a bad light as lacking the willingness and the capability for giving life. The individual is branded selfish and mean, and people point to his or her body's dry, husklike nature as a confirmation of these antisocial, nonprocreative leanings. Like the person too rich, the person too thin—whether through circumstances or choice—is seen to shirk his or her social duties to share with and nurture kin. The thin individual is seen to contribute little to society, and the shape of his or her body is used to bear witness to this.

The condition of a person's relationships is inferred by others when they observe and comment on the state of his or her body. In turn, people try to mold the shapes of their bodies in order to affect the inferences other people make. Bordo (1990) explores mainstream U.S. dieting and body sculpting through exercise with this notion (and typical U.S. ideas about self-control) in mind. While mainstream Americans prefer regimes that lead to thinness, Jamaicans attempt to fatten their bodies (and those of others whom they *care* or *response* for). Both types of manipulations are efforts to construct and promote oneself as a sociable, desirable individual within a given cultural context.

A fuller understanding of the interactive dimension of body sculpting and the reading of bodily shape opens one avenue to the study of the cultural aspects of ideas about nutrition and the standards for physical beauty and health. The study of traditional health beliefs and social and moral ideals exposes much of the logic behind body shapes and the regimes people attempt to adhere to in order to affect them.

The life-affirming, prosocial associations of the plump body in Jamaica are expressed in the traditional saying that "What don't fat, kill; what don't kill, fat." Foods or events either fatten or bring death. People try to stay fat because the plump body is healthily dilated with vital lifeblood and because it suggests to others that one is kind, sociable, and happy to fulfill obligations to kin and community. Thinness is ultimately linked with death, but the fat person's body is richly fertile, and the fat person is judged a nurturant and constructive member of a thriving network of interdependent kin.

NOTES

This chapter is part of a larger study of Jamaican health traditions and their uses; ideas and information presented in this chapter are discussed more fully in my book, *One Blood* (Sobo 1993b) and in various articles (1993a, 1992, 1995). The research and much of the writing were carried out with the guidance of F. G. Bailey. Tom Csordas, Mark Nichter, Nicole Sault, William Wedenoja, and Drexel Woodson provided thoughtful comments and suggestions.

1. Many of the institutional and structural barriers to class mobility remain invisible to most of the people that they hinder, and blame for impoverishment is frequently placed on fellow villagers (Austin 1984).
2. Works dealing with "racial" body characteristics are exceptions (e.g., Henriques 1958; Hoetink 1985; Phillips 1973). Features such as hair texture and skin color are used to determine one's ancestral heritage. Features considered more "Black" or more African are denigrated. Some people bleach their skin and straighten their hair to affect a "whiter" appearance. Color consciousness is greatest among the middle class because of their realistic concern over social mobility. The ways people manipulate the "racial" features of their bodies may be shifting as pride in the African portion of Caribbean heritage increases.
3. For this reason, and because many urban dwellers were born rurally (see Brody 1981: 101) or have rural mind-sets (Brody 1981: 69), and also because many elite, modernized Jamaicans retain traditional beliefs, I often refer to the participants simply as Jamaicans.
4. Drexel Woodson notes that people who have had a falling out—people with *something between them* (strife)—do, in fact, continue to give and receive *something*—ire, enmity, or spleen. The flow of enmity between individuals marks a shift in the relationship's character (D. Woodson, personal communication).
5. Just bearing children does not confer social adulthood; as I argue elsewhere (Sobo 1995, 1993b), responsible parenting and accepting responsibility for the well-being of others are essential for gaining and maintaining adult status. Nonfecund women who establish a name for themselves as excellent foster mothers or those who demonstrate a high level of social commitment through good works or through a service-oriented career, such as that of a doctor, healer, or pastor, can sometimes gain adult standing in the community (but they are not guaranteed it).
6. The menstrual period is often called an *unclean* time, and menstruating women are often denigrated as *unclean*, yet menstruation purifies the body. The association with uncleanness has to do with the *dirt* (unused, rotting semen) that the menstrual flow washes out. Moreover, it has to do with the power women have to carry out *unclean business* (magical manipulations) with their menstrual blood. Women can secretly add the blood of menstruation to others' food. Once ingested and incorporated into the body, menstrual blood obligates the victims just as *blood ties* obligate kin. (For a detailed analysis see Sobo 1992.)

REFERENCES

Austin, D. 1984. *Urban Life in Kingston, Jamaica: The Culture and Class Ideology of Two Neighborhoods*. New York: Gordon and Breach Science Publishers.

Bailey, F. G. 1971. *Gifts and Poison*. New York: Schocken Books.

Becker, A. 1990. "Body Image in Fiji: The Self in the Body and in the Community." Ph.D. diss., Harvard University.

Blake, J. 1961. *Family Structure in Jamaica*. New York: Free Press.

Bordo, S. 1990. "Reading the Slender Body." In *Body Politics*, ed. M. Jacobus, E. F. Keller, and S. Shuttleworth, pp. 83–112. New York: Routledge.

Brink, P. J. 1989. "The Fattening Room among the Annang of Nigeria." *Medical Anthropology* 12:131–143.

Brody, E. 1981. *Sex, Contraception, and Motherhood in Jamaica*. Cambridge: Harvard University Press.

Browner, C. H. 1985. "Traditional Techniques for Diagnosis, Treatment, and Control of Pregnancy in Cali, Colombia." In *Women's Medicine: A Cross-Cultural Study of Indigenous Fertility Regulations*, ed. L. F. Newman, pp. 99–123. New Brunswick, N. J.: Rutgers University Press.

Cassidy, C. M. 1991. "The Good Body: When Bigger Is Better." *Medical Anthropology* 13:181–213.

Cassidy, F. G. 1982. *Jamaica Talk: Three Hundred Years of the English Language in Jamaica*. London: Macmillan Education. First published in 1961.

Chevannes, B. 1990. "Drop-pan and Folk Consciousness." *Jamaica Journal* 22(2):45–50.

Clarke, E. 1957. *My Mother Who Fathered Me: A Study of the Family in Three Selected Communities in Jamaica*. Boston: George Allen and Unwin.

Douglass, L. 1992. *The Power of Sentiment: Love, Hierarchy, and the Jamaican Family Elite*. Boulder, Colo.: Westview Press.

Ehrenreich, B., and D. English. 1979. *For Her Own Good: 150 Years of the Experts' Advice to Women*. London: Pluto Press.

Farmer, P. 1988. "Bad Blood, Spoiled Milk: Bodily Fluids as Moral Barometers in Rural Haiti." *American Ethnologist* 15(1):62–83.

———. 1992. *AIDS and Accusation: Haiti and the Geography of Blame*. Berkeley: University of California Press.

Henriques, F. 1958. *Family and Colour in Jamaica*. London: Macgibbon and Kee.

Hoetink, H. 1985. "'Race' and Color in the Caribbean." In *Caribbean Contours*, ed. S. Mintz and S. Price, pp. 55–84. Baltimore: Johns Hopkins University Press.

Hurston, Z. N. 1990. *Tell My Horse: Voodoo and Life in Haiti and Jamaica*. With a new forward. San Francisco: Harper and Row. First published in 1938.

Jordanova, L. J. 1980. "Natural Facts: A Historical Perspective on Science and Sexuality." In *Nature, Culture, and Gender*, ed. C. P. MacCormack and M. Strathern, pp. 42–69. New York: Cambridge University Press.

Kitzinger, S. 1982. "The Social Context of Birth: Some Comparisons between Children in Jamaica and Britain." In *Ethnography of Fertility and Birth*, ed. C. P. MacCormack, pp. 181–204. San Diego: Academic Press.

Kleinman, A. 1980. *Patients and Healers in the Context of Culture: An Exploration of the Borderland between Anthropology, Medicine, and Psychiatry*. Berkeley: University of California Press.

Laguerre, M. 1987. *Afro-Caribbean Folk Medicine*. South Hadley, Mass.: Bergin and Garvey Publishers.

Laws, S., V. Hay, and A. Eagan. 1985. *Seeing Red: The Politics of Premenstrual Tension*. London: Hutchinson.

Lock, M. 1989. "Words of Fear, Words of Power: Nerves and the Awakening of Political Consciousness." *Medical Anthropology* 11:79–90.

MacCormack, C. P. 1985. "Lay Concepts Affecting Utilization of Family Planning Services in Jamaica." *Journal of Tropical Medicine and Hygiene* 88:281–285.

MacCormack, C. P., and A. Draper. 1987. "Social and Cognitive Aspects of Female Sexuality in Jamaica." In *The Cultural Construction of Sexuality*, ed. P. Caplan, pp. 143–161. New York: Tavistock Publications.

Martin, E. 1987. *The Woman in the Body: A Cultural Analysis of Reproduction*. Boston: Beacon Press.

Mauss, M. 1967. *The Gift*. New York: W. W. Norton.

Meigs, A. S. 1987. "Blood Kin and Food Kin." In *Conformity and Conflict: Readings in Cultural Anthropology*, ed. J. P. Spradley and D. W. McCurdy, pp. 117–124. Boston: Little, Brown.

Nichter, M., and M. Nichter. 1987. "Cultural Notions of Fertility in South Asia and Their Impact on Sri Lankan Family Planning Practices." *Human Organization* 46(1):18–27.

Payer, L. 1988. *Medicine and Culture: Varieties of Treatment in the United States, England, West Germany, and France*. New York: Henry Holt.

Phillips, A. S. 1973. *Adolescence in Jamaica*. Kingston: Jamaica Publishing House.

Sargent, C., and M. Harris. 1992. "Gender Ideology, Child Rearing, and Child Health in Jamaica." *American Ethnologist* 19:523–537.

Scheper-Hughes, N. 1992. *Death Without Weeping: The Violence of Everyday Life in Brazil*. Los Angeles: University of California Press.
Sheets-Johnstone, M. 1992. "The Materialization of the Body: A History of Western Medicine, A History in Process." In *Giving the Body Its Due*, ed. M. Sheets-Johnstone, pp. 132–158. Albany: State University of New York Press.
Smith, R. T. 1988. *Kinship and Class in the West Indies: A Genealogical Study of Jamaica and Guyana*. New York: Cambridge University Press.
Sobo, E. J. 1992. "'Unclean Deeds': Menstrual Taboos and Binding 'Ties' in Rural Jamaica." In *Anthropological Approaches to the Study of Ethnomedicine*, ed. M. Nichter, pp. 101–126. New York: Gordon and Breach.
————. 1993a. "Bodies, Kin, and Flow: Family Planning in Rural Jamaica." *Medical Anthropology Quarterly* 7(1):50–73.
————. 1993b. *One Blood: The Jamaican Body*. Albany: State University of New York Press.
————. 1995. "Abortion Traditions in Rural Jamaica." *Social Science and Medicine*.
STATIN (Statistical Institute of Jamaica). 1982. Census. Unpublished Portland Information, in author's files.
Taylor, C. C. 1988. "The Concept of Flow in Rwandan Popular Medicine." *Social Science and Medicine* 27(12):1343–1348.
————. 1992. "The Harp That Plays by Itself." In *Anthropological Approaches to the Study of Ethnomedicine*, ed. M. Nichter, pp. 127–147. New York: Gordon and Breach.
Vaughn, M. 1991. *Curing Their Ills: Colonial Power and African Illness*. Stanford: Stanford University Press.
Wilson, P. J. 1973. *Crab Antics*. New Haven: University Press.

20

Soul, Black Women, and Food

MARVALENE H. HUGHES

Women's search for personal identities has probably never been pursued so actively in the United States as during the escalation of the Women's Liberation movement. At the same time the sex revolution is in progress, the collective consciousness of personkind seems to be gravitating back into the history of time in search of roots. Alex Haley's *Roots* and the televised saga of *Roots* highlighted in me, and I believe many people, the quest for ancestral cultures and genetic origins. This is indeed a century when history must record the presence of a collective consciousness that presents a mental mirror reflecting back into the ages of time. In these reflections each of us hopes to find missing pieces of our life's puzzles. As a Black woman, some pieces of my puzzle will relate to my African ancestry; some pieces will relate to the Black struggle in U.S. history; and other pieces will relate to my female identity.

One of the most symbolic tools that may be used by Black Americans in our search for roots in food—soul food. The essence of Black culture has been handed down through oral history, generation after generation in the African tradition, through the selection and preparation of soul food. The dominant figure in the cultural translation through food is the Black woman. Her expressions of love, nurturance, creativity, sharing, patience, economic frustration, survival, and the very core of her African heritage are embodied in her meal preparation. In addition, the effects of Euro-American traditions on her African culture may also be transmitted through her cooking traditions.

The roots of plants are basic to the meaning of soul food. Roots represent the state of being grounded and the state of stabilization. The choice of roots as a basic Black food can be traced to the African tradition. "Yams and sweet potatoes have provided an unbroken link in the Black man's diet from sixteenth-century Africa to twentieth-century America."[1] Accessibility to the roots of plants is preserved by possession of seeds. Forced to leave their native land, their home, family, and African tribes, many slaves brought seeds with them. The watermelon seed, for example, now a symbol of the American South, was introduced to this country by enslaved Africans.[2] Similarly, slaves brought okra, which later became a key ingredient for the preparation of gumbo, a New Orleans, French-related dish.[3] Southern Blacks use okra as a special seasoning supplement to "greens," a term used to describe leaf vegetables like collards and turnips.

The determination to hold on to native foods by bringing seeds into this country may be symbolic of the ever-present determination to preserve the African culture through food. Black-eyed peas were introduced to this country in 1674[4] and other seeds, such as the sesame seed, followed. In clinging to the seed, Blacks proclaimed internally (to themselves) and externally (to the world) that they would maintain an African ethnic identity through food. Blacks have sought many avenues to the maintenance of Black identity in U.S. cultures; soul food became one preserver of Black culture.

Food remains one of the Black woman's self-concept expressions. Through her mysterious, spiritual self-confidence and through her arrogance in food preparation, the Black woman gains a sense of pride as she watches her extended family—her man, her children, and maybe her grandparents, sisters, nieces and friends—enjoy the soulful tastes and textures prepared by her skillful hands. Even when she prepares meals as a way of making a living, she takes pride in watching her consumers literally gorge themselves until the fatty tissue forms and finds a permanent resting and growing place. Plumpness is s symbol of the wonderful job which she is performing. Even when her job occupation is food preparation for White America, her success symbol is plumpness. A big body to the Black woman represents health and prosperity. The interrelatedness of the concepts "big" and "beautiful" is African. Bigness represents health and prosperity, but in America thinness is beautiful. Having learned these American values, could it be that the Black woman (and Black man) enjoy making White America fat and "ugly" by its standards?

My mom receives a lot of praise from her nine children and her grandchildren when she is cooking. Like many Black woman of her generation, giving birth to nine children was a planned, pleasant accomplishment. In giving birth to many children, the Black woman sends out messages to the world—"the one thing over which I have control is my body. . . . this is my only opportunity to experience the feeling of plenty. . . . this is my contribution to the future Black world."

When I have the opportunity to visit my mom, she seldom spends time having long heart-to-heart talks unless I sit in the kitchen and join her in her cooking. However, when there are big family gatherings, I learn a lot about my roots as I listen to accounts presented orally by all of the family members. Seldom does my mom talk about the meaning of getting fat and enjoying food, yet almost all of her day is devoted to the kitchen. She is happiest when she satisfies our oral cravings. It is obvious as she mixes together a "handful" of, a "pinch" of, a "dash" of, that her culinary creativity is much like other Black women who defy the rules of measurement and scientific cooking precision. Many Black people of her generation never used recipes or measuring instruments. In fact, Hess and Hess[5] note that the deliberate illiteracy that slavery perpetuated made it necessary for many Black cooks to have recipes read to them. Slaves were accustomed to their African cultural cooking, and they have to learn European techniques in order to become "good" kitchen maids as prescribed by Western tradition.[6] The European cooking techniques underwent many adaptations as the Black woman added her African cultural style. Her recipe variations were passed from generation to generation through oral history in the African tradition, while "proper English housewives . . . kept manuscript books (in the European tradition) of recipes, which were handed down from mother to daughter."[7]

Using her "common senses" of taste, smell, sight, and touch and her soulful intuition, the Black woman tends to reject scientific progress in food in favor of her basic cultural knowledge and intuition. Sometimes this serves to her advantage, especially in a progressive technological era when the effects of food consumption and food preservation practices on health are often determined after the fact. I did not concern myself with preservatives and additives in frozen foods when I was with my mom, for example, because we always had fresh vegetables and fruits. Even now, among the Blacks, I observe many of these fresh food practices.

I believe as we Blacks increase our working hours, the soul-food ritual of home cooking is still observed. I do not fully accept that eating out is less frequent among Blacks because it represents sheer economic deficiencies. This concept of economic deprivation denies the cultural preference for a soulful home-cooked meal. The vegetable trays in our refrigerators are stocked with the closest thing to mom's garden we can find at the grocery store. Local grocers tend to stock our greens and fresh vegetables when we Blacks are ghetto-ized. When our middle-class socioeconomic status takes us to the suburb, we often drive into the "ghetto" in search of soul food—another indication of Black folks' commitment to preserve "soul" in food preparation.

When I recently embarked on my gardening fling, I noticed that my garden was different from my neighbors' garden. I had corn, black-eyed peas, squash, watermelon, string beans ("snap beans" as we learned back home), radishes, beets, green peas ("English beans") and tomatoes (just like mom's). My neighbors shared with me their curiosity about my garden. First, they could not identify by name some of my vegetables. Second, they had no comprehension of the personal fulfillment that I received in producing and cultivating the garden. My garden gave me a sense of pride and took me back to my family roots. When I could dig into the earth, observe the growing results of my cultivation of the soil, and become an active participant in my food production, I became reconnected with my African culture in a private, intimate, historical sense. My garden represented a regeneration of the earth, a spiritual connectedness that allowed me to pay homage to the earth; it provided me the channel to relate my African respect for land and living things to my African spirituality.

Presently, I live in Del Mar, California, where the local grocers have a scanty, occasional supply of collards, turnips, and black-eyed peas. I have never seen pigs feet, pigs tails, chitterlings, and hog jowl in stock in the local stores. Unfortunately, my Black friends who choose to stay in the San Diego Black "ghetto" have the illusion that middle-class Blacks try to deny their cultural food heritage when they move to the suburbs. I firmly believe that whether we live in the suburbs or the ghettos, there is a bond among Black folks—Black food culture is one of the connecting links which transcends geography. At a typical Black potluck dinner, whether in the suburbs or the ghettos, there are ham hocks, pigs feet, neck bones, pigs tails, potato salad, sweet-potato pie, and the "escargot of soul food," chitterlings.

Soul food is an expression of the central core of Black culture. Black recipes, like the Black culture, are handed down from generation to generation by oral African history. Black culture itself has survived through oral history. The word "soul" is a part of Black history that represents a cultural compactness. "Is she soulful?" "Can he dance soulfully?" "Does she make soulful music?" "Can she make good soul food? "Would you believe that he knows how to make soul food?" I have heard all those questions repeatedly, and I have never felt that they were clichés. Having soul is knowing that when the human layers are peeled away there is a hidden, impenetrable gem, the sapphire.

The "human layers" in the Westernized American culture are thick. They embrace a set of values that are filled with contradictions and confusions for Blacks; they signify the polarities of a Western society with an abundance of labels: good versus bad, power versus powerlessness, beauty versus ugliness, Black versus White, smart versus dumb, advantaged versus disadvantaged, educated versus undereducated, privileged versus underprivileged, thinness versus fatness, soul food versus European foods, poverty versus wealth. The word "versus" in the American culture portrays an extremely low tolerance level for differences and pluralisms. It is, therefore, difficult to perceive the true meaning of "melting pot" since the American melting pot is not fluid. Instead of blending cultures, it superimposes the Euro-white-American cultures over other non-Euro-white cultures. In so doing, it tends to place positive, ruling powers on the Euro-American groups and negative, nonruling powers on peoples of color and Judeo identities. While Jewish people have gained power by mastering skills of economic rulership, peoples of color—Blacks and Browns—have not gained equal economic access.

When Blacks accept the labels of Western polarities, I am concerned. I am concerned about how Black people adopt many of the same Western cultural concepts that are used to perpetuate hierarchical social-class structures, and thus, Black oppression. As Black people, we recognize the institutionalization of labels, and we commonly accept that power is rationalized and, indeed, perpetuated through labeling. (By power I mean legislative power, money, educational access, institutional access, and "success" access.) Yet, some Blacks use those same power strategies on each other. A beautiful word, like "soul," for example, is a victim of this culture's labeling. "Is she soulful?" presupposes that there are deselection criteria, or that on a scale of one to ten, you can measure her soulfulness. Black sisters and brothers are *all* soulful; that, for me, is the real essence of the word "soul."

Black food is all soulful, too. All black-eyed peas are soulful whether prepared in a Black person's kitchen, a white kitchen, or a Jewish kitchen. Just as Blacks have adopted Western behaviors, many aspects of the Black culture have been adopted by non-Blacks. One of my white friends prepares black-eyed peas and ham hocks for the New Year's Day party. I once asked him where he learned how to do that. "I am from the deep South," he replied. The Southern Black Belt, where slavery in this country was concentrated, provided a vast cooking (and therefore, cultural) repertoire through the Black woman's domination of the white kitchen. There was and is today an informal Black kitchen network that subtly penetrates white America.

Reflecting the mobility of contemporary society, Blacks today are scattered throughout most of mainland U.S.A. Despite affirmative action legislation, the professional upward mobility of Blacks in the United States has not changed significantly. In fact, the gap between Black and white professional advancement has widened. There remains a heavy domestic professional concentration of Black women (and Black men) either in family kitchens or in commercial food preparation. We cannot dismiss the fact that this domestic cooking role of Black women is a historically acquired role definition assigned her by slave masters. Cooking, professionally, is a position of servitude both economically and politically. Politically, the message the Black cook receives is that it is her duty to nurture white Americans by cooking their meals, taking care of their children, cleaning their homes, and doing the laundry. The kitchen-bound/domestic-bound Black woman is still in slavery. Her economic plight is still destined for poverty.

Economics dictate the kind of food a family can purchase. The Black woman has demonstrated that economics cannot control her soulful creativity in the preparation of foods. Consider some of the classic foods that are labeled soul food: pigs feet, ham hocks, chitterlings, pigs ears, hog jowl, tripe, cracklings, chicken backs, giblets, chicken wings, and oxtails. All of those were initially leftovers or "throw-aways" given to Blacks who wanted to believe they were in a land of plenty. For Blacks in this country, life has been a perpetual process of leftovers: leftover food, leftover jobs, leftover houses, leftover textbooks, leftover schools, leftover big cars, leftover communities, leftover everything. Denied educational and economic opportunities, the Black woman (and the Black man) stood in line for leftovers.

The survival-oriented Black woman trusts her creative skills to "make something out of nothing." She acquired the unique survival ability to cook (and therefore use) all parts of everything. She cooked the portions of a hog that the master (and white America) discarded; she cooked the roots and leaves of plants; she cooked leftovers. Blacks learn to enjoy the leftovers and turn them into special treats. When the Black woman demonstrated how good the leftovers could be, "leftover foods" became a household word in white America.

It is now common practice for food manufacturers to package many of these leftovers—ham hocks, oxtails, and chitterlings—and deliver them to some of the most economically elite communities in America. Eldridge Cleaver refers to this trend as white America's "going slumming" attitude.[8] He cautions us to remain alert to the economic and political realities that not withstanding the fact that the white culture eats chitterlings, the ghetto person wants steak. Eating soul food for white America represents a food cultural plunge, and it epitomizes the acculturation of white America through Black food practices.

Soul food represents a Black spiritual ritual deeply embedded in the culture of Afro-Americans. Black people discovered centuries ago that a spiritual core is vital to survival, and that the development of Black ego and pride-building can take place most forcefully in the Black religious institution. Used to maintaining a personal balance in times of struggles, the Negro spiritual portrays the message throughout history that life will become brighter, happier—someday.

> "Swing low, sweet Chariot,
> Cumin' for to carry me home."

The relationship of Black religion to food practices is not parallel to other religion's practices with food. Catholics, Jews, Moslems, Hindus, Seventh-Day Adventists, and other religious groups practice food doctrines which set apart certain foods. Generally, "unacceptable" religious foods may also dictate special days when foods may or may not be served, or when food preparation is prohibited altogether. Although some Black people in the United States are now affiliated with religious groups that adhere to food selection and food-scheduling principles, in the mainstream of Black religious cultures, this food norm is not practiced.

Eating during a spiritual, religious ritual connotes a special celebration for Blacks. When a pastor is ordained, when there are baptisms, when there are "footwashings" at the Primitive Baptist Church, when special holidays approach, Black church yards are replete with spreads of soul food. Soul food at Black church meetings convert the religious, spiritual festivity into a communal Black gathering. The Black preacher, the symbol of oral traditionalism, has the privilege of receiving the choice foods, such as chicken thighs and legs. For Black people the dark pieces of the chicken are the "choice" parts. Whether on the church ground or in the Black home, the Black preacher is the first to choose his food, and other Blacks yield to his sacred position of spiritual religious wisdom and cultural power.

Perhaps more than any other personality, the Black preacher transmits the oral African history. He has learned the art of translating history orally, and the ring of his voice provides appropriate intonations to connect with the spiritual souls of sisters and brothers in the congregation. In addition, the content of his presentation will have significance.

The Black cultural heritage represents a deep inner spirit to share. Blacks have a unique cultural heritage of communal festivities that abound in food sharings at churches, parties, and funerals. Until the mid-60s, Black people had to pack food whenever they expected not to be at home for meals, as Blacks did not have hotel and restaurant privileges in many geographical areas in America. They became accustomed to sharing with other Blacks, even if they were strangers.

At the central core of Black food celebrations is the *intent of sharing*. When there is a hog killing (usually during fall or winter), the neighbors and friends are included. Women play special assisting roles at hog killings; they usually clean the meat, do the trimmings (shave off the fat), cook some samples, give portions to neighbors, and prepare the meat for storage. Men assume the "masculine" chores like slaughtering the beast, removing the hairy covering, and cutting it into major parts, such as hams and shoulders. The hog-killing ritual is one of the few clearly delineated sex-role activities for Black men and Black women.

The role definition around hog killings relates to the African hunting tradition. African men were hunters and providers; women were nurturing souls for the children, men, and elders. With "mankind," this phenomenon has been basically universal. However, the institution of "mankind" tends to break down in oppressed ethic cultures where preoccupations are with basic survival. Preoccupied with her individual survival and the survival of her children, the Black woman lost her "female" role identity in this culture. By default, her need to attain economic independence endowed her with the role-freeing experience of androgyny. Basic survival (self-preservation) transcends powers of loyalty attributed to sex gender. Self-preservation does not transcend the power of motherhood, however. Motherhood is a spiritual linking bond that connects the Black woman with her child in much the same manner that she is connected with earth. Given the choice of her own survival and the survival of her child, the child often becomes the surviving recipient. Cooking as a function of sex role is less sex stereotyping (more androgynous) among Blacks because of the Black person's frequent occupational involvement with cooking outside the home.

Economic factors and oppressive conditioning have contributed to the androgynous state of Black women. Ladner refers to the absence of protectiveness received by Black women.[9] Instead, there was an implicit message for her to gain independence and autonomy. Using her masculine powers to aspire for personal autonomy, the Black woman always knew that she would not marry an ambitious, business executive or a wealthy heir who would "rescue" her. Whether a domestic worker or teacher, she always thought of herself as a career woman who would contribute to the economic support of the family. Despite her major economic role in the Black family, she would maintain dominance over the Black kitchen and allow the Black man to assume the dominant role in the household. This is in contradiction to the sociological myth of Black family matriarchy. Dominance over her kitchen was her opportunity to reign over the oral, nurturing territory of the household; it was, and still is, her chance to demonstrate to her man and her family that she is endowed with a special gift of culinary creativity.

When the Black man asks, "What's for dinner, Momma?" he is paying her the highest possible compliment. Black men frequently address their partners as "momma," and the psychological connotation is void of a negative psychoanalytic philosophy. The term "momma" means "I really dig you." It is fondly used when the Black woman's food has "stuck to the ribs" or passed the soul food test. Black men use "momma" to express to their women that there is a connecting soulful linkage. Translated, it means, "Your African soul meets mine on a deeper-than-surface human level—You are treating me good—You are nurturing me."

In a typical Black kitchen three solid meals are prepared daily: breakfast, dinner, and supper. Breakfast most often consists of grits, homemade biscuits, ham or bacon, molasses or canned preserves, fresh milk, and fresh eggs. Although bacon and ham are traditionally used for breakfast, it is not unusual to have country-smothered steak, fried chicken, or pork chops. There is a supreme value, culturally, in the preparation of a "hot meal" for breakfast. When the Black child goes to school, a caring mother sees that she or he has had a "hot meal." Dinner is usually scheduled around noontime. The general meal consists of a "mess of greens" with pot licker (collards, turnips, cabbage, beet greens, or mustards seasoned with pork skins, fatback, or ham hocks), bread, potatoes, fresh-squeezed lemonade, maybe a meat, and a cobbler (maybe canned peaches or fresh apples, fresh wild fruits like huckleberries, blackberries, or dew berries). Desserts like bread puddings and homemade cookies are standard for the Black dinner. Supper is the evening meal, usually scheduled at about six to seven, where lighter food is served. This may include fruits, cut up in creamy milk, biscuits, ice cream, fried chicken, or creamed potatoes. Fresh buttermilk is regularly available to farm families; it is generally served as a dinner or supper drink. (For snacks between meals, I remember how in Alabama we enjoyed cornbread mashed up in bowls of butter milk; I also remember baked corn and fresh plums and peaches.) In retrospect, I realize that most of these meals were nutritious, although it appeared to be an economic selection at the time.

Throughout the world food has always been related to culture and economics. Recently, attention is devoted more and more to nutrition and "health foods." Many consumer purchases are made because health foods are "in." But many of the so-called health-food diets are nutritionally questionable, and they are chosen more often than not for their faddish cultural connotations. During slavery, however, nutrition was given prime consideration. Masters wanted slaves to be healthy and live longer in order to get maximum work from them.

> ... Negroes fed on three-quarters of a pound of bread and bacon are more prone to disease than if with less meat but with vegetables.[10]

The pro-vegetable attitudes (mess of greens) culturally inherited from Africa and reinforced by the masters still prevail among Blacks. In earlier history, the masters maintained a largely hedonistic attitude towards their own eating.

A casual relationship has been frequently hypothesized between the nutrition of Black foods and hypertension, the killer of Black folks. The apparent dietary practices of Blacks cause me to question the social malignancies that cause hypertension (stress due to economic and personal oppression, job deprivation, externally caused depression). It is well to note that diet may play a less significant role in killing Blacks than the oppressive conditions in American culture.

A combination of diet and exercise do, however, play a significant role in the body structure of Black women. The stereotyped Black female body appears on the screen in the form of Aunt Jemima—big bosom and a round, fat body. The ideal white female body is portrayed as thin and petite. Slimness, however, is not valued by middle-aged and older Black women. My mom worries about my slimness because at 5'4", I barely weigh 120, and I am a middle-aged woman. She asks me often, "Are you eating properly these days?" Maintaining my weight at 120 pounds is hard for me, because I was taught to enjoy eating and preparing food. If I ate the kind of food my mom prepares consistently, I would probably weigh 150 by now. I look forward to holidays at home, though, and I resort back to that joy-of-eating attitude which was prevalent in my childhood. Staying slim is difficult in a culture that values cooking and eating, and if anyone in the family is a "TV" watcher, you may find your house stocked with junk food, too. The pleasure of eating often results in the pain of being fat.

With so little pleasure in a socioeconomically oppressive environment, food, for Blacks, may become an escape. Blacks are prone to use such a psychological escape as a coping device. Stuffing the mouth with food (and other objects such as pacifiers, thumbs, cigarettes) has been acknowledged as a psychological expression of emptiness. Thus, excessive eating may be caused by emotional stress. Symbolically, there may be the experience of filling up the empty space by stuffing something into the mouth. Severely deprived and economically depressed Black women historically resorted to mouth-stuffing habits, an extension of the orality principle, believed to fill their emptiness. Such behaviors as eating clay and starches, stuffing their bottom lips with snuff, and stuffing their mouths with wine, liquors, and foods have been common among the socioeconomically deprived Black people of the South. Blacks seem to realize that, after all, the body and mind are destructible when oppression is a way of life, but the soul shall last externally.

Through her soul food preparation, the Black woman seeks to bring pleasure to her life and to her loved ones. It is one of her methods of taking away the painful realities of oppression, and introducing some pleasures in life. After all, short-term pleasure is better than a total life without any pleasure.

NOTES

1. Mendes, Helen. *The African Heritage Cookbook.* New York: Macmillan, 1971, p. 64.
2. Root, Waverly, and de Rochemont, Richard. *Eating in America.* New York: William Morrow & Co., 1976, p. 19.
3. Ibid., p. 39.
4. Ibid., p. 84.
5. Hess, John L., and Hess, Karen. *The Taste of America.* New York: Grossman Publishers, 1977, p. 71.
6. Ibid., p. 71.
7. Ibid.
8. Cleaver, Eldridge. *Soul on Ice.* New York: Dell Publishing Co., 1968, p. 29.
9. Ladner, Joyce. *Tomorrows' Tomorrow.* New York: Doubleday, 1972.
10. Root, Waverly, and deRochemont, Richard. *Eating in America.* New York: William Morrow and Co., 1976, p. 145.

SUGGESTED READINGS

Cleaver, Eldridge. *Soul on Ice.* New York: Dell Publishing Co., 1968.
Haley, Alex. *Roots.* New York: Dell Publishing Co., 1976.
Hess, John L., and Hess, Karen. *The Taste of America.* New York: Grossman Publishers, 1977.
Ladner, Joyce. *Tomorrow's Tomorrow.* New York: Doubleday, 1972.
Mendes, Helen. *The African Heritage Cookbook.* New York: Macmillan, 1971.
Root, Waverly, and deRochemont, Richard. *Eating in America.* New York: William Morrow and Co., 1976.
Spratt, Talmage, ed. *Black Light.* Hallmark Cards, Inc., 1973.

The Political Economy of Food

Commodification and Scarcity

The articles in this section demonstrate that food commodification is deeply implicated in perpetuating inequalities of gender, race, and class. Counihan's piece is a case study of how changes in production, distribution, and consumption of bread in Sardinia have weakened community relations. Allison describes how in the mundane quotidian act of preparing lunches for their children, Japanese women reproduce the cultural ideology of power. Mennell discusses the relationship between hunger, appetite, manners, and "the civilizing process." Goody ponders how changes in foodways towards the development of a "world cuisine" have affected local diets and food habits across the globe. Mintz shows how the rich controlled access to desirable, high-status sugar until it was produced in sufficient quantity to become a staple rather than a luxury. Fitchen describes the social conditions and cultural meanings of hunger in the United States to show how false our notions of hunger are and how they block social justice in food distribution. Van Esterik describes how the commodification of infant food through the international marketing of infant formula has had severe economic, cultural, and medical consequences. She also shows how women have been active in protesting the actions of transnational pharmaceutical and food companies promoting these products. Progress towards social justice, Lappé and Collins argue, can only come through a concerted effort on the part of social activists everywhere to end world hunger and bring about universal access to nutritious and adequate food.

21

Bread as World

Food Habits and Social Relations in Modernizing Sardinia

CAROLE COUNIHAN

This paper uses bread as a lens for analyzing contemporary social and economic change in the town of Bosa in the peripheral Italian region of Sardinia. Once men grew wheat and women baked bread together in their homes. The bread circulated through kin and friendship channels and was consumed communally. Now Bosans no longer grow wheat; they buy bakery bread which is distributed according to the principles of the market place, and they consume it ever more individualistically. The paper argues that Bosa's situation is characteristic of "modernization without development," and that this leads to an increasing atomization of social relations.

This paper is about contemporary social and economic change in the town of Bosa in the peripheral Italian region of Sardinia.[1] During the twentieth century Bosa has undergone an experience shared by many rural Mediterranean regions which Schneider, Schneider and Hansen (1972) have called "modernization without development," characterized by the stagnation of local production and the increasing emulation of Western industrial consumption patterns. These large-scale economic trends are accompanied by changes in social relations. One of these, this paper argues, is a process of individualization, where decisions and actions gradually become more independent of community ties. Description and analysis of the production, distribution and consumption of bread—the most important food in the Sardinian diet—illustrate these transformations in Bosan economy and society and enable consideration of their qualitative impact on the Bosan people.

The paper begins with the belief that human nature is a product of history and society (Geertz 1973; Gramsci 1957; Marx and Engels 1970; Sahlins 1976). Thus the extent to which a society is characterized by an individualistic conception of human nature varies a great deal. Individualism refers to a reduction in dependence on others for survival necessities and to an increasing autonomy of action and decision-making. The "Western conception of the person as a bounded, unique," autonomous individual is by no means universal (Geertz 1975:48). Anthropologists have consistently drawn our attention to the fact that members of foraging, tribal, and peasant societies are much more closely interdependent in behavior and ideology than citizens of modern Western states. They work together, continuously practice generalized reciprocal exchange, and conceptualize their identities not primarily as individuals, but as members of a group, a lineage, or a family (e.g., for Italy see Belmonte 1979; Banfield 1958; Pinna 1971).

Beginning with Marx, many social scientists have argued that individualism is created by the production system and market exchange typical of the capitalist economy (Gramsci 1957; Marx and Engels 1970; Sahlins 1972). Mauss's brilliant essay *The Gift* (1967) gives further support to this thesis. He shows not only the importance of gifts in tying people together in "archaic" societies, but also the fragmentation of social relations in the France of his day resulting from the increase in market exchange and the decline of gift-giving. This paper continues this line of inquiry. It describes changes in Sardinian bread habits, links them to large-scale economic processes, and assays through them the atomization of social relations.

THE CONTEXT

Geographical isolation, scarce natural resources, and low population density have made Sardinia traditionally one of the poorest regions of Italy (Counihan 1981:31–80). Up until the 1950s, malaria was endemic (Brown 1979; 1981), and population density was the lowest in the nation: 111 per square mile. In 1930 there were one million people and two million sheep on the island's 24,000 sq kms (Great Britain 1945:578,605). The population was concentrated in villages in the interior where extensive wheat cultivation and animal herding were the mainstays of the largely subsistence economy (Le Lannou 1941).

Since the Second World War, Sardinia has undergone massive cultural and economic changes fostered by governmental development measures like the *Cassa per il Mezzogiorno* (Fund for the South), the *Piano di Rinascita* (Plan of Rebirth) and the European Economic Community's Mansholt Memorandum (Di Giorgi and Moscati 1980; Graziani 1977; King 1975; Lelli 1975; Orlando et al. 1977). These measures have instigated industrial development in a few scattered areas—what Lelli (1975) calls "development poles"—and the capitalization of agriculture, principally in the rich Campidano valley. Subsistence agriculture has declined radically over much of the island. Animal-herding continues in the same small-scale and free-grazing style that it has been followed for centuries. The Sardinian standard of living as measured through education, hygiene and consumer goods has greatly improved (Musio 1969), and malaria has been eradicated (Brown 1979; 1981). This increased standard of living has been achieved through increases in available goods and cash. Consumer goods are largely imports, for Sardinian industry has concentrated on capital-intensive petrochemical conversions rather than on consumer products (Lelli 1975). People have greater access to cash largely through funds brought to Sardinia by vacationing tourists, by emigrants working in Northern Italy and Northern Europe, and by the State in the form of pensions, unemployment compensations, family assistance checks, and short-term public works projects. Sardinia has undergone modernization—cultural changes produced by the influx of models and practices "from already established centers," without development—"an autonomous and diversified economy on its own terms" (Schneider, Schneider and Hansen 1972:340).

Bosa is situated about two-thirds of the way up the west coast of Sardinia, on a small hill, about three kms inland on the north bank of the Temo river. Although it has never had more than the roughly 9000 inhabitants it has today, Bosa has always had an urban character that has differentiated it from its hinterland (Anfossi 1968) and from the small Sardinian villages of the mountainous interior and the plains (Angioni 1974; 1976; Barbiellini-Amidei and Bandinu 1976; Bodemann 1979; Mathias 1979; 1983; Musio 1969; Pinna 1971). It has highly stratified classes, has acted as a market and political center, and has had some rudimentary industry. Its unique ecological setting in the river valley near the coast surrounded by hills and highland plateau made possible a productive agro-pastoral and fishing economy. A small elite controlled most of the land, tools, animals, and fishing boats, as well as the machines of the mills, tanneries, and canneries. The majority of the population consisted of primary producers who were poor, dependent, and economically precarious, although the town itself was the scene of a certain economic bustle and relative cultural opulence the more notable today by their absence (Anfossi 1968; Counihan 1981). In the extent of its stratification, urbanity, and relative wealth, Bosa was somewhat unique in Sardinia, but its recent history has much in common with that of the rural Sardinian and Italian South (Davis 1977; Weingrod and Morin 1971).

In the past thirty years Bosa and its hinterland have been increasingly tied to the capitalist world system by sending emigrants to work in the industrial North of Italy and Europe and by serving as a growing market for imported consumer goods. Reliable figures on emigration do not exist, but in 1971 the government census listed 951 Bosans "absent for reasons of work"—11 percent of the total population, 36 percent of the employed. Agriculture, pastoralism, and fishing have declined dramatically in Bosa, from involving 52 percent of the employed in 1951 to 19 percent in 1971. During the same period, mining and manufacturing industry have increased only from 10 percent to 15 percent of the employed. There are no factories in Bosa: the nearest are the beer, cheese, and textile factories in Macomer, 27 kms away, which employ no more than 50 Bosans. None of the major "development poles" are within commuting distance, and even so they have never employed as many workers as was planned (Lelli 1975). Traditional crafts (cobbling, cooping, smithing, and tailoring) and semi-industrial activities (tanneries, canneries, olive and grain mills) are disappearing. The tertiary sector grew from 38 percent of the employed in 1951 to 67 percent in 1971. Most of the growth is in construction, tourism, and commerce, all of which are "precarious" in that they depend heavily on a prosperous economy for their own prosperity. In Bosa all are small in scale and "old-fashioned" in methods. Though exact figures do not exist, perhaps as many as three-quarters of the adult population receive government assistance of some sort (Brown 1979:282). Overall, work in Bosa is a patchwork quilt of various short-term jobs supplemented by small but steady infusions of cash from government pensions and assistance checks. This has made possible a visible increase in consumption.

Bosans have more and more goods produced in factories in faraway places. They wear the latest Italian fashions and eat imported, frozen, and processed foods. They get national and international culture in their living rooms through television, radio, magazines, and newspapers. Bosans are living through social and economic transformations that have much in common with those of other peripheral Italian regions (Davis 1973; 1977; Schneider and Schneider 1976). This paper will attempt to understand their impact on a particular sector of the Bosan people—those who were formerly primary producers—through a detailed study of changing relationships in the production, distribution, and consumption of bread.

BREAD IN SARDINIA

This paper treats bread as a "total social fact" (Mauss 1967). Like all foods, bread is a nexus of economic, political, aesthetic, social, symbolic, and health concerns. As traditionally the most important food in the Sardinian diet, bread is a particularly sensitive indicator of change.[2]

Bread has always been the *sine qua non* of food in Sardinia and all over Italy and Mediterranean Europe. Le Lannou (1941:288) estimates the daily diet of the Sardinian peasant in the 1930s as being 78 percent bread by weight (1200 gms of bread, 200 gms of cooked vegetables or legumes, 30 gms of cheese, and 100 gms of pasta). Other commentators on the Sardinian diet confirm this dominant role of bread over at least the last 150 years (Angioni 1976:34; Bodio 1879:200–206; Barbiellini-Amidei and Bandinu 1976:82–86; Cannas 1975; Chessa 1906:279–280; Delitala 1978:101; La Marmora 1839:243; Mathias 1983; Somogyi 1973). Although bread is still the most important food, its dietary significance has declined as other foods—especially meat, pasta, and cheese—have become more accessible (Counihan 1981:178–226).

As the most important food, bread was central to the Sardinian economy for centuries. The Romans allegedly exported Sardinian grain (Bouchier 1917:56). In recent times, durum wheat has been the island's principal crop though total hectares planted has dropped noticeably since the beginning of the century (see Table 1).

Table 1 Hectares of Durum Wheat

Year	Hectares Planted in Wheat	Source
1909	314,000	Le Lannou 1941:293–296
1924	138,000	Le Lannou 1941:293–296
1939	250,000	Le Lannou 1941:293–296
1950	243,000	King 1975:160
1970	112,700	King 1975:160
1976	94,907	Regione 1978: appendix[3]

The wheat-bread cycle was pivotal to the Bosan and Sardinian subsistence economy. Peasant men devoted much effort to growing grain, and peasant women to milling, sifting and baking it. The variety and beauty of the breads were astounding. There were three or four principal everyday types and tens of special, symbolic, and pictorial ones for holidays and rituals (see Cirese et al. 1977).

As the Sardinian staple food, bread was symbolic of life. Cambosu (in Cirese et al. 1977:40) reports the peasant proverb: *"Chie hat pane mai non morit"*—One who has bread never dies. In central Sardinia, security was *"pane in domu"*—bread in the home (Pinna 1971:86). Minimal well-being was expressed in the words "at least we have bread" *("pane nessi bi n'amus")* and poverty as "they do not even have bread" *("non b'ana mancu pane,"* [Barbiellini-Amidei and Bandinu 1976:83]). The outcome of the grain harvest was symbolic and determinant of the unfolding of life: *"Bellu laore, annu vonu; laore mezzanu, annu malu"*—"beautiful grain, good year, poor grain, bad year" (Barbiellini-Amidei and Bandinu 1976:80).[4]

The symbolic centrality of bread in Bosan society emerges in a story told me by an informant, Luisa Fois. In the first year of her marriage, she baked the traditional Easter breads out of her husband's wheat. One was a cross, representing Christ's crucifixion. This she gave to her husband to eat. But he said, "Don't you know that we have to eat the cross together, now that we are married. As we share our lives, so too must we share the cross, so that we bear life's burdens equally in the year ahead." Bread was a product of their union and its shared consumption reaffirmed their interdependence. Let us now consider changes in bread production, distribution, and consumption in Bosa and through them assess the individualization of social relations in a situation of modernization without development.[5]

BREAD PRODUCTION

Up until the 1960s wheat was a major crop in Bosa. Peasants cultivated grain principally on marginal soils—on the hillsides surrounding the town and north towards Capo Marragiu. In 1929, there were 393 hectares devoted to cereal cultivation (Catasto Agrario 1929:120) which had dwindled to 21 hectares in 1972 (ISTAT 1972).

Land ownership in Bosa has always been highly concentrated (see Table 2). Since the terrain is mostly hilly and mountainous, the large landowners have always used the *tancas* (large open tracts of land) mainly for grazing animals—sheep, goats, and some cattle and pigs. However, up until the 1960s, they also rented the land on long rotation cycles (every ten years or so) to landless peasants for growing wheat. This periodic cultivation of wheat improved the soil for pasture and provided landlords with extra income in the form of rents. Peasants generally worked the land alone or with a father or son. But the harvest and threshing were big jobs which involved reciprocal cooperation with other men and women.

Table 2 Land Tenure in Bosa

1929

Size of holding in hectares	#	%	Total hectares	%
0–3	342	64.9	445	3.9
3–10	127	24.1	654	5.6
10–20	12	2.3	160	1.4
20–50	3	0.6	86	0.7
Over 50	43	8.2	10230	88.2
TOTALS	527	100.1	11575	99.9

Source: Catasto Agrario 1929

1972

Size of holding in hectares	#	%	Total hectares	%
0–3	72	36.7	96	0.9
3–10	57	29.1	339	3.1
10–20	14	7.1	190	1.7
20–50	12	6.1	335	3.2
Over 50	41	20.9	9978	91.1
TOTALS	196	99.9	10958	100.0

Source: Catasto Agrario 1929
Source: ISTAT 1972

Men produced grain and brought it home; then it was the women's responsibility. They took it to the mill. Once milled, they sifted the flour into four different consistencies: *su poddine, sa simula, su chifalzu,* and *su fulfere* (see La Marmora 1839:241 for a similar description of nineteenth century Sardinia) which they used for different kinds of bread with different meanings and uses. The women organized the collective labor of baking, calling together neighbors, relatives or *comari*[6] to help, assistance which was later reciprocated when they did their own baking.

Women baked every ten to fourteen days. It was a long process, done entirely by hand. They used a sourdough leaven called *sa madriga* and made a stiff dough for the rounded figure-eight-shaped *palsiddas* and the flat, claw-shaped *ziccos,* and a sticky dough for the long, oval-shaped *covasas* with the characteristic hole in the middle. (For descriptions of baking in Sardinia see Angioni 1974:266–269; Cirese et al. 1977; Mathias 1983; and Satta 1979:67–68).

Women regularly worked together. In the work, they exchanged and intermingled their lives: their gossip, their skills, their pasts, their loves, and their losses. Barbiellini-Amidei and Bandinu (1976:83) say of central Sardinia, "The bread-makers are typical channels of social communication. . . . While they prepare the dough and bake the bread they make an X-ray of the town. Secrets are revealed; the women judge and absolve or condemn." By working together, Bosan women developed and acted out standards for "good bread," "good work," and a "good woman." Through the act of criticizing others *(fare la critica)* they collectively reaffirmed local norms and morality.

Informants told me that grain cultivation began to diminish in Bosa around 1960 and disappeared almost completely within six or seven years. This perplexing demise of a major crop becomes understandable in light of Italian agricultural policy and typifies modernization without development. There was a high protection of grain from Italian Unification in 1861 through the 1950s, about 60 percent higher than the world price for grain. In 1957 the government reduced the support prices for grain by L.100/quintal and in 1959 by L.500/quintal (Orlando et al. 1977:26–27). At the same time, government encouraged the mechanization and capitalization of agriculture by facilitating production and dissemination of tractors and fertilizer. In Sardinia the number of tractors increased from 993 in 1949 to over 7500 in 1971 (Brown 1979:302). Mechanization of agriculture stimulated industrial production in the North but reduced overall employment in agriculture by replacing workers with machines and causing increasing disparity in productivity between rich and poor zones (Orlando et al. 1977:21–30). Furthermore, salaried wages were rising, making wage labor ever more attractive than subsistence farming and encouraging emigration from marginal areas of the South to industrial centers of the North (Di Giorgi and Moscati 1980; Graziani 1977). The result of these trends was that many marginal and hilly areas where production per effort and acre was relatively low were abandoned. Bosan lands and many others in Sardinia were among them (Counihan 1981:64–80).

Specific events in Bosa exemplified these long-term socio-economic transformations. The first public bakery opened in Bosa in 1912. At first business was slow because housewives all baked their own bread and it was a source of *bilgonza* (shame) to buy it. But then with fascism, the centralized grain collection *(ammasso di grano),* and rationing in the war, people began to follow the habit of going to the bakery. After the Second World War there seems to have been a short-lived revival of home baking. But the steady abandonment of wheat cultivation in the early 1960s dealt a final blow to home baking which accompanied the demise of other home-processing activities like making tomato preserves, salting olives, or drying figs (Counihan 1981:127–159).

People no longer had their own grain and thus had to buy it. If one was buying grain, it was easier to get used to the idea of buying bread. The grain mills in Bosa closed, forcing a woman to go nine kms away to Tresnuraghes to mill flour. That involved considerable trouble and some expense. Women wondered why they should bother when more and more of their neighbors and relatives were buying bread and their own children were grumbling about stale bread. The shame of buying bread gradually disappeared and slowly all Bosan women stopped baking. Many have made the step irreversible by removing their ovens, spurred on by the desire to enlarge and "modernize" their houses.

Today there is not one woman left in Bosa who still bakes bread.[7] Bosans rely on five local bakers and three from the nearby villages of Suni, Tinnura and Montresta. The Bosan bakers produce from 80–500 kg per day and sell their bread in 27 retail stores throughout the town. Bread-making is the exclusive purview of male bakers and their salaried male assistants. It takes place on a larger scale, producing greater quantities with a wider dispersion than ever before. The culmination of this trend is the ironically named *"Pane della Nonna"*—"Grandmother's Bread," a sort of Italian-style Wonder Bread made by northern industry and sold presliced in cellophane packages with preservatives to facilitate long life on supermarket shelves.

In the past, the "domestic mode of production" (Sahlins 1972:41–99) characterized bread-making in Bosa. Husband and wife did complementary tasks which insured the harvesting of wheat and the baking of bread. Of necessity, they cooperated with people outside the family in the harvest, threshing and baking. Today, men and women still need each other for many things, but not to produce materially such a basic life necessity as bread. With a government pension or a salary, a man or woman can buy bread, obviating one form of dependence on each other. Furthermore, with the decline of subsistence production, Bosans have a less consistent need to rely on others outside the family. One of Pinna's (1971:94) informants expresses this change thus:

> In the old days, when we used to bake bread at home, we had to call on neighbors, and this was a form of dependence. Now that today we buy bread already made, this dependence has ended and we are all free in our own houses. And I am truly thankful for this progress, not so much for the sacrifice it cost to have to get out of bed at three in the morning to begin the job and to keep at it until four in the afternoon, but for the convenience of not having outsiders come into your house to help and go out of it to gossip about what they had seen, and even about what they had not seen. Now progress has brought the benefit that we all can be tranquil in our own houses. And this is a great comfort of the soul more than of the body.

This woman is expressing the trend of individualization: how she needs others less and is less subject to their moral scrutiny. She is culturally more autonomous because the incursion of new production methods has relieved her of her social dependence on others. Let us now look at how changes in distribution have contributed further to this trend.

BREAD DISTRIBUTION

Up until the Second World War, Bosa was a small but thriving center of maritime trade, exporting grain and other raw materials and importing consumer items. Today, Bosa is no longer a commercial center but is quite literally the end of the road. The town's failure to modernize its port, railroad or road systems has rendered it commercially obsolete. Intrepid entrepreneurs bring consumer goods to Bosa by truck over the long and winding road from Macomer, but with the decline in local production the town has little to export. This situation is typical of modernization without development (Counihan 1981:227–233).

There have also been changes in the market within Bosa. Shopping for food has become more important as subsistence production has declined. Women shop for bread and a few other small items daily in the small neighborhood stores that are centers of social relations. Increasingly, however, there is a trend away from reliance on these tiny, owner-operated neighborhood stores to bigger, self-service stores located in the center of town, staffed by salaried sales clerks, and owned by petty capitalists. Here women tend to buy greater quantities and hence to shop less often, decreasing their encounters with each other. Because these stores attract clients from all over town, shoppers are less intimate with each other than in the neighborhood stores. In the self-service stores, shopper and merchant interact only at the end of the shopping process at the cash register. Since the cashiers are hired help, they have no power to grant temporary credit as owner-operators frequently do. These debts serve to maintain an enduring relationship between merchant and client. The increasing reliance on the new, large self-service stores has resulted in an erosion of the sociability of shopping, both among shoppers and between shopper and storekeeper (Counihan 1981:242–266).

While market exchange has long governed commerce, a strict, incessant reciprocity has always been the basis of close and long-lasting social and economic relations in Bosa and all of Sardinia (Counihan 1981:233–242). There are many popular sayings like the following: *"Si cheres chi s'amore si mantenzat, prattu chi andet, prattu chi benzat"*—"If you want love to be maintained, for a plate that goes, let a plate come back" (Gallini 1973:60). Every social interaction is mediated by the giving and receiving of food and drink. One has an obligation to accept, just as the Sardinian has an obligation to give. A Bosan woman said, "My mother always said that if you were eating and a person came to the door, you must always offer him some food. And if you were baking bread, and a person came to the door, you must always give him a piece of freshly baked bread." Women sent gifts of fresh-baked bread to close relatives and friends. Cambosu (in Cirese et al. 1977:40) remembers, "My sisters waited for the first light of day and vied with each other in every season to bring to our closest relatives and dearest friends a gift of our bread. It was one of the most enjoyed gifts, and they reciprocated it every time that they lit their ovens."

Today, with the decline in local primary production and the increasing reliance on the market, the incessant mutual giving and receiving of foods slows and becomes less crucial to survival. Thus one of the most important forces in linking people together—reciprocal prestations—is fading away and with it goes people's interdependence. More and more they acquire goods through money in the formal market. Shopping becomes a concentrated economic act, aimed at the acquisition of food rather than of friendships on which to depend for food. Hence the changing distribution patterns accompanying modernization contribute further to the individualization seen occurring in production.

BREAD CONSUMPTION

Study of the mode of consumption reveals human relationships and basic values in a group. Commensality demonstrates fundamental social units. As Mauss (1967:55) says, "a man does not eat with his enemy." Meals are a central arena for the family in Italy, one of the domains through which domestic ties attain their strength (see especially Belmonte 1979). In rural Sardinia in the past there was an ironclad ethic and practice of consumption: daily consumption took place within the family and was parsimonious; festive consumption took place within society-at-large and was prodigal (Barbiellini-Amidei and Bandinu 1976:139, Gallini 1971). The rhythmic oscillation between these two different modes supported and mediated the individual's connection to the two most important social units—family and community.

Bosan meals involved consumption of a near-one's labor: one's father, brother, or son in the wheat; one's mother, sister, or daughter in the loaf. Consumption of bread reaffirmed the complementary of men and women and the nuclear family social structure, the basis of society, and locus of individual identity (Barbiellini-Amidei and Bandinu 1976:80–84). Consumption of bakery bread involves not this reaffirmation of Bosan society but rather participation in the broader world economy in which relationships are beyond Bosans' ability to retouch them.

The fundamental interdependence of consumption and the family and its current erosion are revealed in the case of a youth who died of hunger while I was living in Bosa, such an extraordinary event that it made the front page of the regional Sardinian newspaper *(L'Informatore del Lunedi,* 4/4/79:1):

> On the death certificate the doctor wrote "cardio-circulatory collapse," two words to veil a disturbing tragedy with a 23-year-old victim. I.C. died, more simply, of hunger. . . . Out front [of his house] only a few curious onlookers and neighbors: "He died like a dog, no one realized it." Did he live alone? "Yes, alone. His mother had been in a nursing home for years. He didn't have anyone else. . . . Alone, in that cave of a house, he lived worse than a dog."

I.C. did not die of poverty as so many did in the past; he had a government pension. Rather he died because the sustaining web of family and community had broken. Emigration and the decline of the domestic mode of production have made possible the existence of people like I.C., people who are totally alone.

Analysis of the evolution of redistributive feasts in Sardinia illustrates an erosion of community solidarity which was necessary to survival in the subsistence economy. In the past the collective *festa* (feastday) was the only legitimate locus of excessive and conspicuous consumption. "Exaggerated consumption was an exceptional act—festive and grandiose—to be carried out at an exceptional time and place" (Gallini 1971:11). As in many non-capitalist societies, excess consumption in the festa served to bring the community together, temporarily to obliterate social and economic differences, and to satiate hunger collectively, madly, and equally, at least for this one day (Mauss 1967, Turner 1969).

To a great extent, Bosan *feste* have lost this role. I observed it only in the Feast of Saint Anthony, January 17, the start of the Carnival period. People donated bread to the saint and the priest blessed it at the morning mass. Members of the saint's confraternity distributed it to the 200 or so faithful. People took it home to eat with family and neighbors for it was supposed to ward off stomach aches. Although this feastday contained some elements of a collective, redistributive feast, it was a minor, poorly attended event, peripheral to most people's lives.

The Feast of Saint Mark on March 25 in Tresnuraghes, a village nine kms from Bosa, was a much more significant feast of communal redistribution. Local shepherd families in this predominantly pastoral community offered sheep and oversaw cooking them in a gesture of thanks to Providence. Other families offered bread as thanksgiving or for favors desired. Hundreds of people, mostly Tresnuraghesi, but with a significant number of Bosans and outsiders, ate and drank to satiation.

The Feast of St. Mark is typical of the traditional Sardinian *feste,* theaters of excess consumption which served to solidify the social community (Angioni 1974:232–279; Gallini 1971). That *feste* of this sort persist today only in the less modernized areas of Sardinia (in Tresnuraghes but not in Bosa, for example) is an indication that their demise accompanies modernization. Tresnuraghesi expressed concern about the fate of this their most important *festa.* The growing presence of outsiders, who came not to participate in strengthening social community but only to acquire free food, threatened the meaning of the ritual and strained the economic ability of the community.

The ethic of consumption of the modern world market to which Bosa and Sardinia increasingly belong is very different from the one which traditionally animated Sardinian collective *feste.* In the past, "opulent consumption . . . was never an act to be carried out in private because this would have transformed it into secret guilt" (Gallini 1971:10). Today, "opulent consumption" increases and becomes a private, stratifying and individualizing act rather than a public, altruistic, and communalizing one. It becomes a mode of separating and differentiating people rather than bringing them together. People compete for prestige by buying and displaying high fashion clothes, color televisions, and all sorts of consumer goods. In the past, having more than others was dangerous, for it could only draw the evil eye and misfortune (Gallini 1973, 1981), but today it is a legitimate source of prestige (Barbiellini-Amidei and Bandinu 1976:139–145).

CONCLUSION

Anthropology's particular contribution to social science is to examine social phenomena holistically and to reveal the personal and qualitative side of large-scale economic and political processes studied more abstractly and quantitatively by other disciplines. This paper has attempted to fulfill anthropology's mandate by studying the nature of modernization and its effects on social relationships in rural Sardinia by using bread as a lens of analysis. Focusing on the production, distribution, and consumption of this most important food enables simultaneous examination of material changes in people's lives and their symbolic and social repercussions. Thus the paper demonstrates how the study of food habits can be an effective channel to attaining the holism central to anthropology.

Study of contemporary socio-economic changes in Bosa contributes to development studies by offering data from a backward region within an industrial nation. Thus it reveals the dynamics of internal underdevelopment. Furthermore, data on Sardinia are scarce in the literature and can increase our understanding of contemporary change in the Mediterranean region. Data from Bosa balance the strong concentration in Sardinian studies on small, mountainous, pastoral villages of the interior.

The findings presented here give support to the thesis of economic anthropology that the capitalist mode of production and exchange leads to an atomization of social relations. In subsistence wheat and bread production, men and women depend on each other for assistance and are unable to make a living without mutual exchange of labor and products. Social interdependence declines with the concentration of wheat production on capital-intensive farms and of bread production in a few bakeries operating with wage laborers to make profits. Bread acquisition takes place through increasingly impersonal money exchange. The continuous giving and receiving of bread and other foods so important to tying people together and ensuring their survival in the past fades away with the demise of subsistence production. Daily bread consumption used to be an affirmation of male and female complementarity and the integrity of the family unit. Although bread is still essential to meals, it has lost this particular symbolic meaning. Thus this study demonstrates that one effect of modernization without development is increasing autonomy of actions and identity.

The paper offers the reader the opportunity to assess the impact of individualization on Bosan people's lives. Pinna's central Sardinian informant offered one hint in her relief at the freedom gained through her productive independence, a freedom from grueling gossip. But one is tempted to ask another question: how does independence relate to the human need for others? Anthropology is based on the belief that the human is the social being. What happens to humanity if humans become increasingly separate, losing overriding communal ties? Finally, what does individualization signify in a class society? Bosans are gaining economic independence from each other, but they remain dependent on the state and the economic elite for pensions and jobs. Perhaps their independence would be more rewarding if it were accompanied by real control over the production, distribution and consumption of their survival neccessities. For, as they say, "one who has bread never dies."

NOTES

1. Fieldwork in Bosa from June 1978 through August 1979 was supported by a Fulbright-Hays Dissertation Research Grant, a Grant-in-Aid from Sigma XI, the Scientific Research Society of North America, and a University Fellowship from the University of Massachusetts. A trip to Bosa in February 1980 to make the film "Looking for Giolzi: Carnival and Anthropology in the Sardinian Town of Bosa" was made possible by the film's director, Stefano Silvestrini, and production manager, Sofia Manozzi. Research in Bosa in August 1982 was made possible in part by funds from Franklin and Marshall College. To these institutions and individuals I am most grateful. I thank the following people for commenting on earlier drafts of this paper: Peter Brown, Elizabeth Mathias, Nancy McDowell, James Taggart, Richard Ward, and several anonymous reviewers.
2. Bread seems to be expressive of the differential impact of change in different areas of Sardinia. In Bosa, bread-making has disappeared completely, whereas in the nearby mountainous village of Montresta almost every household regularly produces its own bread. In the central Sardinian mountain villages of Talana (Bodemann and Ostow 1982) and Esporlatu (Mathias 1983) many women still make bread, while this has declined in Tonara (Gallini 1981). Clearly many factors influence bread-making and this paper attempts to outline some of the important ones.
3. Production figures vary enormously from year to year, e.g., in 1912 production was 5.7 quintals/hectare, in 1911, 10 quintals/hectare, in 1970, 10.9 quintals/hectare, and in 1971, 15.6 quintals/hectare.
4. The expression *"a cercare pane migliore di quello di grano"*—"to look for bread better than that of wheat"—is a central and recurrent expression in Satta's (1979) penetrating ethno-

graphic novel about Nuoro, the isolated capital of the province which includes Bosa. The expression refers to the futility of challenging destiny, which is as immutable as the fact that bread is made of wheat.

5. I gathered data on contemporary Bosan food habits during 1978–1979 through participant-observation, interview, food logs, and questionnaires. Comparative data on past food habits come from the memories of older informants, travelers' and scholars' reports (Bodio 1879; Casalis 1835; Chessa 1906; Dessi' 1967; La Marmora 1839, 1860; Le Lannou 1941; Smyth 1828; Tyndale 1840; and Wagner 1928), and government statistics (ISTAT 1960; 1968). Cautious use of ethnographic analogy on contemporary data from areas more isolated than Bosa suggests a picture of how Bosa might have been in the past. See Angioni (1974; 1976). Barbiellini-Amidei and Bandinu (1976), Cambosu (1954), Cannas (1975), Delitala (1978), Satta (1979), and especially Cirese et al. (1977) and Mathias (1983).
6. *Comari* (female) and *compari* (male) are fictive kin. They are ordinarily the people who hold one's children at baptism and to whom one is thus linked by special obligation and respect.
7. I am grateful to Peter Brown for reminding me of the relevance of this *festa* to my paper.

WORKS CITED

Anfossi, Anna, 1968. Socialita'e organizzazione in Sardegna. Studio sulla zona di Oristano-Bosa-Macomer. Milano: Angeli.

Angioni, Giulio. 1974. Rapporti di produzione e cultura subalterna: contadini in Sardegna. Cagliari: EDES.

———. 1976. Sa laurera. Il lavoro contadino in Sardegna. Cagliari: EDES.

Banfield, Edward C. 1958. The moral basis of a backward society. Glencoe, IL: Free Press.

Barbiellini-Amidei, Gaspare and Bachisio Bandinu. 1976. Il re e' un feticcio. Romanzo di cose. Milano: Rizzoli.

Belmonte, Tom. 1979. The broken fountain. New York: Columbia University Press.

Bodemann, Y. Michael, and Robin Ostow. 1979. Telemula: Aspects of the micro-organization of backwardness in central Sardinia. Ph.D. Dissertation, Sociology, Brandeis University.

Bodemann, Y. Michael, and Robin Ostow. 1982. Personal Communication.

Bodio, Luigi. 1879. Sui contratti agrari e sulle condizioni materiali di vita dei contadini in diverse regioni d'Italia. Annali di Statistica, Ministero di Agricoltura, Industria e Commerico. Serie II, 8:125–206.

Bouchier, A. 1917. Sardinia in ancient times. Oxford: Blackwell.

Brown, Peter J. 1979. Cultural adaptations to endemic malaria and the socio-economic effects of malaria eradication. Ph.D. Dissertation, Anthropology, State University of New York at Stony Brook.

———. 1981. Cultural adaptations to endemic malaria in Sardinia. Medical Anthropology 4:3.

Cambosu, Salvatore. 1954. Miele amaro. Firenze: Valecchi.

Cannas, Marilena. 1975. La cucina dei sardi. 200 piatti caratteristici. Cagliari: EDES.

Casalis, G. 1834. Dizionario geografico-storico-statistico-commerciale degli stati di S.M. il Re diSardegna. Torino. "Bosa" Volume 2:526–546.

Catasto, Agrario. 1929. Compartimento della Sardegna. Volume 8, fascicolo 91:120.

Chessa, Federico. 1906. Le condizioni economiche e sociali dei contadini dell'agro di Sassari. Due monografie di famiglia. La Riforma Agraria, January-April.

Cirese, Alberto Maria, Enrica Delitala, Chiarella Rapallo, and Guilio Angioni. 1977. Pani tradizionali, arte efimera in Sardegna. Cagliari: EDES.

Counihan, Carole M. 1981. Food, culture and political economy. An investigation of changing lifestyles in the Sardinian town of Bosa. Ph.D. Dissertation, Anthropology, University of Massachusetts at Amherst.

Davis, John. 1973. Land and family in Pisticci. London: Athione.

———. 1977. People of the Mediterranean. An essay in comparative social anthropology. London: Routledge and Kegan Paul.

Delitala, Enrica. 1978. Come fare ricerca sul campo. Esempi di inchiesta sulla cultura subalterna in Sardegna. Cagliari: EDES.

Dessi', Giuseppe, ed. 1967. Scoperta della Sardegna. Antologia di testi di autori italiani e stranieri. Milano: Il Polifilo.

Di Giorgi, Umberto, and Roberto Moscati. 1980. The role of the state in the uneven spatial devel-

opment of Italy: The case of the Mezzogiorno. Review of Radical Political Economics 12: 3:50–63.
Gallini, Clara. 1971. Il consumo del sacro: feste lunghe in Sardegna. Bari: Laterza.
———. 1973. Dono e malocchio. Palermo: Flaccovio.
———. 1981. Intervista a Maria. Palermo: Sellerio.
Geertz, Clifford. 1973. The interpretation of cultures. Selected essays. N.Y.: Basic Books.
———. 1975. On the nature of anthropological understanding. American Scientist 63:47–53.
Gramsci, Antonio. 1957. The modern prince and other writings. N.Y.: International Press.
Graziani, A. 1977. Il messogiorno nell' economia italiana oggi. Inchiesta 29:3–18.
Great Britain Naval Intelligence Division. 1945. Italy. Geographical Handbook Series, B.R. 517. Volume 4.
ISTAT (Istituto Centrale Di Statistica). 1960. Indagine statistica sui bilanci di famiglie non agricole negli anni 1953–54. Annali di Statistica, 3,2. Roma.
———. 1968. Indagine statistica sui bilanci delle famiglie italiane, anni 1963–64. Annali di Statistica, 8, 21. Roma.
———. 1972. Secondo censimento generale dell'agricoltura 25/x/70, 2, 93. Roma.
King, Russell. 1975. Sardinia. Newton Abbot: David and Charles.
La Marmora, Alberto Ferrero Della. 1839. Voyage en Sardaigne. Paris: Arthus Bertrand. 2nd edition.
———. 1860. Intinéraire de l'ile de Sardaigne pour faire suite au voyage en cette contrée. Turin: Fratelli Brocca, 2 volumes.
Le Lannou, Maurice. 1941. Patres e paysans de la Sardaigne. Tours: Arrault et C.
Lelli, Marcello. 1975. Proletariato e ceti medi in Sardegna, una societa" dipendente. Bari: De Donato.
Lévi-Strauss, Claude. 1975. Tristes tropiques. New York: Atheneum.
Marx, Karl, and Frederick Engels. 1970. The German ideology. New York: International Press.
Mathias, Elizabeth Lay. 1979. Modernization and changing patterns in breastfeeding: The Sardinian case. In Breastfeeding and Food Policy in a Hungry World. Dana Raphael, ed., New York: Academic, pp. 75–79.
———. 1983. Sardinian born and bread. Natural History 1/83:54–62.
Mauss, Marcel. 1967. The gift. Forms and functions of exchange in archaic societies. New York: Norton.
Musio, Gavino. 1969. La cultura solitaria. Tradizione e acculturazione nella Sardegna arcaica. Bologna: Il Mulino.
Orlando, Giuseppe, Fabrizio De Filippis, and Mauro Mellano. 1977. Piano alimentare od politica agraria alternativa? Bologna: Il Mulino.
Pinna, Luca. 1971. La famiglia esclusiva, parentela e clientelismo in Sardegna. Bari: Laterza.
Regione Autonoma Della Sardegna. 1978. Convegno regionale sul piano agricolo-alimentare. Cagliari: Mulas.
Sahlins, Marshall. 1972. Stone age economics. New York: Aldine.
———. 1976. Culture and practical reason. Chicago: University of Chicago Press.
Satta, Salvatore. 1979. Il giorno del giudizio. Milano: Adelphi.
Schneider, Jane, and Peter Schneider. 1976. Culture and politics in western Sicily. New York: Academic.
Schneider, Peter, Jane Schneider, and Edward Hansen. 1972. Modernization and development: The role of regional elites and noncorporate groups in the European Mediterranean. Comparative Studies in Society and History 14: 3:328–350.
Smyth, William Henry. 1828. Sketch of the present state of the Island of Sardinia. London: Murray.
Somogyi, Stefano. 1973. L'alimentazione nell'Italia unita. Storia d'Italia, 5:1.
Turner, Victor. 1969. The ritual process: Structure and anti-structure. Chicago: Aldine.
Tyndale, John Warre. 1840. The island of Sardinia. London: Richard Bently. 3 volumes.
Wagner, Max Leopold. 1928. La vita rustica della Sardegna rispecchiata nella sua lingua. Valentino Martelli, trans. Cagliari: Soc. Ed. Italiana.
Weingrod, Alex, and Emma Morin. 1971. "Post-peasants": The character of contemporary Sardinian society. Comparative Studies in Society and History 13:3.

22

Japanese Mothers and *Obentōs*

The Lunch-Box as Ideological State Apparatus

ANNE ALLISON

Obentōs *are boxed lunches Japanese mothers make for their nursery school children. Following Japanese codes for food preparation—multiple courses that are aesthetically arranged—these lunches have a cultural order and meaning. Using the* obentō *as a school ritual and chore—it must be consumed in its entirety in the company of all the children—the nursery school also endows the* obentō *with ideological meanings. The child must eat the obentō; the mother must make an* obentō *the child will eat. Both mother and child are being judged; the subjectivities of both are being guided by the nursery school as an institution. It is up to the mother to make the ideological operation entrusted to the* obentō *by the state-linked institution of the nursery school, palatable and pleasant for her child, and appealing and pleasurable for her as a mother.*

INTRODUCTION

Japanese nursery school children, going off to school for the first time, carry with them a boxed lunch *(obentō)* prepared by their mothers at home. Customarily these *obentōs* are highly crafted elaborations of food: a multitude of miniature portions, artistically designed and precisely arranged, in a container that is sturdy and cute. Mothers tend to expend inordinate time and attention on these *obentōs* in efforts both to please their children and to affirm that they are good mothers. Children at nursery school are taught in turn that they must consume their entire meal according to school rituals.

Food in an *obentō* is an everyday practice of Japanese life. While its adoption at the nursery school level may seem only natural to Japanese and unremarkable to outsiders, I will argue in this article that the *obentō* is invested with a gendered state ideology. Overseen by the authorities of the nursery school, an institution which is linked to, if not directly monitored by, the state, the practice of the *obentō* situates the producer as a woman and mother, and the consumer as a child of a mother and a student of a school. Food in this context is neither casual nor arbitrary. Eaten quickly in its entirety by the student, the *obentō* must be fashioned by the mother so as to expedite this chore for the child. Both mother and child are being watched, judged, and constructed; and it is only through their joint effort that the goal can be accomplished.

I use Althusser's concept of the Ideological State Apparatus (1971) to frame my argument. I will briefly describe how food is coded as a cultural and aesthetic apparatus in Japan, and what authority the state holds over schools in Japanese society. Thus situating the parameters within which the *obentō* is regulated and structured in the nursery school setting, I will examine the practice both of making and eating *obentō* within the context of one nursery school in Tokyo. As an anthropologist and mother of a child who attended this school for fifteen months, my analysis is based on my observations, on discussions with other mothers, daily conversations and an interview with my son's teacher, examination of *obentō* magazines and cookbooks, participation in school rituals, outings, and Mothers' Association meetings, and the multifarious experiences of my son and myself as we faced the *obentō* process every day.

I conclude that *obentō* as a routine, task, and art form of nursery school culture are endowed with ideological and gendered meanings that the state indirectly manipulates. The manipulation is neither total nor totally coercive, however, and I argue that pleasure and creativity for both mother and child are also products of the *obentō*.

CULTURAL RITUAL AND STATE IDEOLOGY

As anthropologists have long understood, not only are the worlds we inhabit symbolically constructed, but also the constructions of our cultural symbols are endowed with, or have the potential for, power. How we see reality, in other words, is also how we live it. So the conventions by which we recognize our universe are also those by which each of us assumes our place and behavior within that universe. Culture is, in this sense, doubly constructive: constructing both the world for people and people for specific worlds.

The fact that culture is not necessarily innocent, and power not necessarily transparent, has been revealed by much theoretical work conducted both inside and outside the discipline of anthropology. The scholarship of the neo-Marxist Louis Althusser (1971), for example, has encouraged the conceptualization of power as a force which operates in ways that are subtle, disguised, and accepted as everyday social practice. Althusser differentiated between two major structures of power in modern capitalist societies. The first, he called (Repressive) State Apparatus (SA), which is power that the state wields and manages primarily through the threat of force. Here the state sanctions the usage of power and repression through such legitimized mechanisms as the law and police (1971: 143–5).

Contrasted with this is a second structure of power—Ideological State Apparatus(es) (ISA). These are institutions which have some overt function other than a political and/or administrative one: mass media, education, health and welfare, for example. More numerous, disparate, and functionally polymorphous than the SA, the ISA exert power not primarily through repression but through ideology. Designed and accepted as practices with another purpose—to educate (the school system), entertain (film industry), inform (news media), the ISA serve not only their stated objective but also an unsated one—that of indoctrinating people into seeing the world a certain way and of accepting certain identities as their own within that world (1971: 143–7).

While both structures of power operate simultaneously and complementarily, it is the ISA, according to Althusser, which in capitalist societies is the more influential of the two. Disguised and screened by another operation, the power of ideology in ISA can be both more far-reaching and insidious than the SA's power of coercion. Hidden in the movies we watch, the music we hear, the liquor we drink, the textbooks we read, it is overlooked because it is protected and its protection—or its alibi (Barthes 1957: 109–111)—allows the terms and relations of ideology to spill into and infiltrate our everyday lives.

A world of commodities, gender inequalities, and power differentials is seen not therefore in these terms but as a naturalized environment, one that makes sense because it has become our experience to live it and accept it in precisely this way. This commonsense acceptance of a particular world is the work of ideology, and it works by concealing the coercive and repressive elements of our everyday routines but also by making these routines of everyday familiar, desirable, and simply our own. This is the critical element of Althusser's notion of ideological power: ideology is so potent because it becomes not only ours but us—the terms and machinery by which we structure ourselves and identify who we are.

JAPANESE FOOD AS CULTURAL MYTH

An author in one *obentō* magazine, the type of medium-sized publication that, filled with glossy pictures of *obentōs* and ideas and recipes for successfully recreating them, sells in the bookstores across Japan, declares, "... the making of the *obentō* is the one most worrisome concern facing the mother of a child going off to school for the first time (*Shufunotomo* 1980: inside cover). Another *obentō* journal, this one heftier and packaged in the encyclopedic series of the prolific women's publishing firm, *Shufunotomo*, articulates the same social fact: "first-time *obentōs* are a strain on both parent and child" ("*hajimete no obentō wa, oya mo ko mo kinchoshimasu*") (*Shufunotomo* 1981: 55).

An outside observer might ask: What is the real source of worry over *obentō?* Is it the food itself or the entrance of the young child into school for the first time? Yet, as one look at a typical child's *obentō*—a small box packaged with a five- or six-course miniaturized meal whose pieces and parts are artistically arranged, perfectly cut, and neatly arranged—would immediately reveal, no food is "just" food in Japan. What is not so immediately apparent, however, is why a small child with limited appetite and perhaps scant interest in food is the recipient of a meal as elaborate and as elaborately prepared as any made for an entire family or invited guests?

Certainly, in Japan much attention is focused on the *obentō,* investing it with a significance far beyond that of the merely pragmatic, functional one of sustaining a child with nutritional foodstuffs. Since this investment beyond the pragmatic is true of any food prepared in Japan, it is helpful to examine culinary codes for food preparation that operate generally in the society before focusing on children's *obentōs*.

As has been remarked often about Japanese food, the key element is appearance. Food must be organized, reorganized, arranged, rearranged, stylized, and restylized to appear in a design that is visually attractive. Presentation is critical: not to the extent that taste and nutrition are displaced, as has been sometimes attributed to Japanese food, but to the degree that how food looks is at least as important as how it tastes and how good and sustaining it is for one's body.

As Donald Richie has pointed out in his eloquent and informative book *A Taste of Japan* (1983), presentational style is the guiding principle by which food is prepared in Japan, and the style is conditioned by a number of codes. One code is for smallness, separation, and fragmentation. Nothing large is allowed, so portions are all cut to be bite-sized, served in small amounts on tiny individual dishes, and arranged on a table (or on a tray, or in an *obentō* box) in an array of small, separate containers.[1] There is no one big dinner plate with three large portions of vegetable, starch, and meat as in American cuisine. Consequently the eye is pulled not toward one totalizing center but away to a multiplicity of de-centered parts.[2]

Visually, food substances are presented according to a structural principle not only of segmentation but also of opposition. Foods are broken or cut to make contrasts of color, texture, and shape. Foods are meant to oppose one another and clash; pink against green, roundish foods against angular ones, smooth substances next to rough ones. This oppositional code operates not only within and between the foodstuffs themselves, but also between the attributes of the food and those of the containers in or on which they are placed: a circular mound in a square dish, a bland-colored food set against a bright plate, a translucent sweet in a heavily textured bowl (Richie 1985: 40–41).

The container is as important as what is contained in Japanese cuisine, but it is really the containment that is stressed, that is, how food has been (re)constructed and (re)arranged from nature to appear, in both beauty and freshness, perfectly natural. This stylizing of nature is a third code by which presentation is directed; the injunction is not only to retain, as much as possible, the innate naturalness of ingredients—shopping daily so food is fresh and leaving much of it either raw or only minimally cooked—but also to recreate in prepared food the promise and appearance of being "natural." As Richie writes, ". . . the emphasis is on presentation of the natural rather than the natural itself. It is not what nature has wrought that excites admiration but what man has wrought with what nature has wrought" (1985: 11).

This naturalization of food is rendered through two main devices. One is by constantly hinting at and appropriating the nature that comes from outside—decorating food with season reminders, such as a maple leaf in the fall or a flower in the spring, serving in-season fruits and vegetables, and using season-coordinated dishes such as glassware in the summer and heavy pottery in the winter. The other device, to some degree the inverse of the first, is to accentuate and perfect the preparation process to such an extent that the food appears not only to be natural, but more nearly perfect than nature without human intervention ever could be. This is nature made artificial. Thus, by naturalization, nature is not only taken in by Japanese cuisine, but taken over.

It is this ability both to appropriate "real" nature (the maple leaf on the tray) and to stamp the human reconstruction of that nature as "natural" that lends Japanese food its potential for cultural and ideological manipulation. It is what Barthes calls a second-order myth (1957: 114–17): a language that has a function people accept as only pragmatic—the sending of roses to lovers, the consumption of wine with one's dinner, the cleaning up a mother does for her child—which is taken over by some interest or agenda to serve a different end—florists who can sell roses, liquor companies that can market wine, conservative politicians who campaign for a gendered division of labor with women kept at home. The first order of language ("language-object"), thus emptied of its original meaning, is converted into an empty form by which it can assume a new, additional, second order of signification ("metalanguage" or "second-order semiological system"). As Barthes point out, however, the primary meaning is never lost. Rather, it remains and stands as an alibi, the cover under which the second, politicized meaning can hide.

Roses sell better, for example, when lovers view them as a vehicle to express love rather than the means by which a company stays in business.

At one level, food is just food in Japan—the medium by which humans sustain their nature and health. Yet under and through this code of pragmatics, Japanese cuisine carries other meanings that in Barthes' terms are mythological. One of these is national identity: food being appropriated as a sign of the culture. To be Japanese is to eat Japanese food, as so many Japanese confirm when they travel to other countries and cite the greatest problem they encounter to be the absence of "real" Japanese food. Stated the other way around, rice is so symbolically central to Japanese culture (meals and *obentōs* often being assembled with rice as the core and all other dishes, multifarious as they may be, as mere compliments or side dishes) that Japanese say they can never feel full until they have consumed their rice at a particular meal or at least once during the day.[3]

Embedded within this insistence on eating Japanese food, thereby reconfirming one as a member of the culture, are the principles by which Japanese food is customarily prepared: perfection, labor, small distinguishable parts, opposing segments, beauty, and the stamp of nature. Overarching all these more detailed codings are two that guide the making and ideological appropriation of the nursery school *obentō* most directly: 1) there is an order to the food: a right way to do things, with everything in its place and each place coordinated with every other, and 2) the one who prepares the food takes on the responsibility of producing food to the standards of perfection and exactness that Japanese cuisine demands. Food may not be casual, in other words, nor the producer casual in her production. In these two rules is a message both about social order and the role gender plays in sustaining and nourishing that order.

SCHOOL, STATE, AND SUBJECTIVITY

In addition to language and second-order meanings I suggest that the rituals and routines surrounding *obentōs* in Japanese nursery schools present, as it were, a third order, manipulation. This order is a use of a currency already established—one that has already appropriated a language of utility (food feeds hunger) to express and implant cultural behaviors. State-guided schools borrow this coded apparatus: using the natural convenience and cover of food not only to code a cultural order, but also to socialize children and mothers into the gendered roles and subjectivities they are expected to assume in a political order desired and directed by the state.

In modern capitalist societies such as Japan, it is the school, according to Althusser, which assumes the primary role of ideological state apparatus. A greater segment of the population spends longer hours and more years here than in previous historical periods. Also education has now taken over from other institutions, such as religion, the pedagogical function being the major shaper and inculcator of knowledge for the society. Concurrently, as Althusser has pointed out for capitalist modernism (1971: 152, 156), there is the gradual replacement of repression by ideology as the prime mechanism for behavior enforcement. Influenced less by the threat of force and more by the devices that present and inform us of the world we live in and the subjectivities that world demands, knowledge and ideology become fused, and education emerges as the apparatus for pedagogical and ideological indoctrination.

In practice, as school teaches children how and what to think, it also shapes them for the roles and positions they will later assume as adult members of the society. How the social order is organized through vectors of gender, power, labor, and/or class, in other words, is not only as important a lesson as the basics of reading and writing, but is transmitted through and embedded in those classroom lessons. Knowledge thus is not only socially constructed, but also differentially acquired according to who one is or will be in the political society one will enter in later years. What precisely society requires in the way of workers, citizens, and parents will be the condition determining or influencing instruction in the schools.

This latter equation, of course, depends on two factors: 1) the convergence or divergence of different interests in what is desired as subjectivities, and 2) the power any particular interest, including that of the state, has in exerting its desires for subjects on or through the system of education. In the case of Japan, the state wields enormous control over the systematization of education. Through its Ministry of Education (Monbusho), one of the most powerful and influential ministries in the government, education is centralized and managed by a state bureaucracy that regulates almost every aspect of the educational process. On any given day, for example, what is taught in every public school follows the same curriculum, adheres to the same structure, and is informed by textbooks from the prescribed list. Teachers are nationally screened, school boards uniformly appointed (rather than elected), and students institutionally exhorted to obey teachers given their legal authority, for example, to write secret reports (*naishinsho*) that may obstruct a student's entrance into high school.[4]

The role of the state in Japanese education is not limited, however, to such extensive but codified authorities granted to the Ministry of Education. Even more powerful is the principle of the "*gakureki shakkai*" (lit., academic pedigree society), by which careers of adults are determined by the schools they attend as youth. A reflection and construction of the new economic order of post-war Japan,[5] school attendance has become the single most important determinant of who will achieve the most desirable positions in industry, government, and the professions. School attendance itself based on a single criterion: a system of entrance exams which determines entrance selection, and it is to this end—preparation for exams—that school, even at the nursery-school level, is increasingly oriented. Learning to follow directions, do as one is told, and "*ganbaru*" (Asanuma 1987) are social imperatives, sanctioned by the state, and taught in the schools.

NURSERY SCHOOL AND IDEOLOGICAL APPROPRIATION OF THE *OBENTŌ*

The nursery school stands outside the structure of compulsory education in Japan. Most nursery schools are private; and, though not compelled by the state, a greater proportion of the three- to six-year-old population of Japan attends pre-school than in any other industrialized nation (Tobin 1989; Hendry 1986; Boocock 1989).

Differentiated from the *hoikuen*, another pre-school institution with longer hours which is more like daycare than school,[6] the *yochien* (nursery school) is widely perceived as instructional, not necessarily in a formal curriculum but more in indoctrination to attitudes and structure of Japanese schooling. Children learn less about reading and writing than they do about how to become a Japanese student, and both parts of this formula—Japanese and student—are equally stressed. As Rohlen has written, "social order is generated" in the nursery school, first and foremost, by a system of routines (1989: 10, 21). Educational routines and rituals are therefore of heightened importance in *yochien*, for whereas these routines and rituals may be the format through which subjects are taught in higher grades, they are both form and subject in the *yochien.*

While the state (through its agency, the Ministry of Education) has no direct mandate over nursery-school attendance, its influence is nevertheless significant. First, authority over how the yochien is run is in the hands of the Ministry of Education. Second, most parents and teachers see the *yochien* as the first step to the system of compulsory education that starts in the first grade and is closely controlled by Monbusho. The principal of the *yochien* my son attended, for example, stated that he saw his main duty to be preparing children to enter more easily the rigors of public education soon to come. Third, the rules and patterns of "group living" (*shudanseikatsu*), a Japanese social ideal that is reiterated nationwide by political leaders, corporate management, and marriage counselors, is first introduced to the child in nursery school.[7]

The entry into nursery school marks a transition both away from home and into the "real world," which is generally judged to be difficult, even traumatic, for the Japanese child (Peak 1989). The *obentō* is intended to ease a child's discomfiture and to allow a child's mother to manufacture something of herself and the home to accompany the child as s/he moves into the potentially threatening outside world. Japanese use the cultural categories of *soto* and *uchi*; *soto* connotes the outside, which in being distanced and other, is dirty and hostile; and *uchi* identifies as clean and comfortable what is inside and familiar. The school falls initially and, to some degree, perpetually, into a category of *soto*. What is ultimately the definition and location of *uchi,* by contrast, is the home, where family and mother reside.[8] By producing something from the home, a mother both girds and goads her child to face what is inevitable in the world that lies beyond. This is the mother's role and her gift; by giving of herself and the home (which she both symbolically represents and in reality manages[9]), the *soto* of the school is, if not transformed into the *uchi* of the home, made more bearable by this sign of domestic and maternal hearth a child can bring to it.

The *obentō* is filled with the meaning of mother and home in a number of ways. The first is by sheer labor. Women spend what seems to be an inordinate amount of time on the production of this one item. As an experienced *obentō* maker, I can attest to the intense attention and energy devoted to this one chore. On the average, mothers spend 20–45 minutes every morning cooking, preparing, and assembling the contents of one *obentō* for one nursery school–aged child. In addition, the previous day they have planned, shopped, and often organized a supper meal with left-overs in mind for the next day's *obentō*. Frequently women[10] discuss *obentō* ideas with other mothers, scan *obentō* cookbooks or magazines for recipes, buy or make objects with which to decorate or contain (part of) the *obentō,* and perhaps make small food portions to freeze and retrieve for future *obentō.*[11]

Of course, effort alone does not necessarily produce a successful *obentō*. Casualness was never indulged, I observed, and even mothers with children who would eat anything prepared *obentōs* as elaborate as anyone else's. Such labor is intended for the child but also the mother: it is a sign of a woman's commitment as a mother and her inspiring her child to being similarly committed as a student. The *obentō* is thus a representation of what the mother is and what the child should become. A model for school is added to what is gift and reminder from home.

This equation is spelled out more precisely in a nursery school rule—all of the *obentō* must be eaten. Though on the face of it this is petty and mundane, the injunction is taken very seriously by nursery school teachers and is one not easily realized by very small children. The logic is that it is time for the child to meet certain expectations. One of the main agendas of the nursery school, after all, is to introduce and indoctrinate children into the patterns and rigors of Japanese education (Rohlen 1989; Sano 1989; Lewis 1989). And Japanese education, by all accounts, is not about fun (Duke 1986).

Learning is hard work with few choices or pleasures. Even *obentōs* from home stop once the child enters first grade.[12] The meals there are institutional: largely bland, unappealing, and prepared with only nutrition in mind. To ease a youngster into these upcoming (educational, social, disciplinary, culinary) routines, *yochien obentōs* are designed to be pleasing and personal. The *obentō* is also designed, however, as a test for the child. And the double meaning is not unintentional. A structure already filled with a signification of mother and home is then emptied to provide a new form: one now also written with the ideological demands of being a member of Japanese culture as well as a viable and successful Japanese in the realms of school and later work.

The exhortation to consume one's entire *obentō*[13] is articulated and enforced by the nursery school teacher. Making high drama out of eating by, for example, singing a song; collectively thanking Buddha (in the case of Buddhist nursery schools), one's mother for making the *obentō,* and one's father for providing the means to make the *obentō*; having two assigned class helpers pour the tea, the class eats together until everyone has finished. The teacher examines the children's *obentōs*, making sure the food is all consumed, and encouraging, sometimes scolding, children who are taking too long. Slow eaters do not fare well in this ritual, because they hold up the other students, who as a peer group also monitor a child's eating. My son often complained about a child whose slowness over food meant that the others were kept inside (rather than being allowed to play on the playground) for much of the lunch period.

Ultimately and officially, it is the teacher, however, whose role and authority it is to watch over food consumption and to judge the person consuming food. Her surveillance covers both the student and the mother, who in the matter of the *obentō* must work together. The child's job is to eat the food and the mother's to prepare it. Hence, the responsibility and execution of one's task is not only shared but conditioned by the other. My son's teacher would talk with me daily about the progress he was making finishing his *obentōs*. Although the overt subject of discussion was my child, most of what was said was directed to me: what I could do in order to get David to consume his lunch more easily.

The intensity of these talks struck me at the time as curious. We had just settled in Japan and David, a highly verbal child, was attending a foreign school in a foreign language he had not yet mastered; he was the only non-Japanese child in the school. Many of his behaviors during this time were disruptive: for example, he went up and down the line of children during morning exercises hitting each child on the head. Hamada-sensei (the teacher), however, chose to discuss the *obentōs*. I thought surely David's survival in and adjustment to this environment depended much more on other factors, such as learning Japanese. Yet it was the *obentō* that was discussed with such recall of detail ("David ate all his peas today, but not a single carrot until I asked him to do so three times") and seriousness that I assumed her attention was being misplaced. The manifest reference was to boxed lunches, but was not the latent reference to something else?[14]

Of course, there was another message for me and my child. It was an injunction to follow directions, obey rules, and accept the authority of the school system. All of the latter were embedded in and inculcated through certain rituals: the nursery school, as any school (except such nonconventional ones as Waldorf and Montessori) and practically any social or institutional practice in Japan, was so heavily ritualized and ritualistic that the very form of ritual took on a meaning and value in and of itself (Rohlen 1989: 21, 27–28). Both the school day and the school year of the nursery school were organized by these rituals. The day, apart from two free periods, for example, was broken by discrete routines—morning exercises, arts and crafts, gym instruction, singing—most of which were named and scheduled. The school year was also segmented into and marked by three annual events—sports day (*undokai*) in the fall, winter assembly (*seikatsu happyokai*) in December, and dance festival (*bon odori*) in the summer. Energy was galvanized by these rituals, which demanded a degree of order as well as a discipline and self-control that non-Japanese would find remarkable.

Significantly, David's teacher marked his successful integration into the school system by his mastery not of the language or other cultural skills, but of the school's daily routines—walking in line, brushing his teeth after eating, arriving at school early, eagerly participating in greeting and departure ceremonies, and completing all of his *obentō* on time. Not only had he adjusted to the school structure, but he had also become assimilated to the other children. Or, restated, what once had been externally enforced now became ideologically desirable; the everyday practices had moved from being alien (*soto*) to being familiar (*uchi*) to him, that is, from being someone else's to being his own. My American child had to become, in some sense, Japanese, and where his teacher recognized this Japaneseness was in the daily routines such as finishing his *obentō*. The lesson learned early, which David learned as well, is that not adhering to routines such as completing one's *obentō* on time results not only in admonishment from the teacher, but in rejection from the other students.

The nursery-school system differentiates between the child who does and the child who does not manage the multifarious and constant rituals of nursery school. And for those who do not manage, there is a penalty, which the child learns to either avoid or wish to avoid. Seeking the acceptance of his peers, the student develops the aptitude, willingness, and in the case of my son—whose outspokenness and individuality were the characteristics most noted in this culture—even the desire to conform to the highly ordered and structured practices of nursery-school life. As Althusser (1971) wrote about ideology: the mechanism works when and because ideas about the world and particular roles in that world that serve other (social, political, economic, state) agendas become familiar and one's own.

Rohlen makes a similar point: that what is taught and learned in nursery school is social order. Called *shudanseikatsu* or group life, it means organization into a group where a person's subjectivity is determined by group membership and not "the assumption of choice and rational self-interest" (1989: 30). A child learns in nursery school to be with others, think like others, and act in tandem with others. This lesson is taught primarily through the precision and constancy of basic routines: "Order is shaped gradually by repeated practice of selected daily tasks . . . that socialize the children to high degrees of neatness and uniformity" (p. 21). Yet a feeling of coerciveness is rarely experienced by the child when three principles of nursery-school instruction are in place: (1)

school routines are made "desirable and pleasant" (p. 30), (2) the teacher disguises her authority by trying to make the group the voice and unity of authority, and (3) the regimentation of the school is administered by an attitude of "intimacy" on the part of the teachers and administrators (p. 30). In short, when the desire and routines of the school are made into the desires and routines of the child, they are made acceptable.

MOTHERING AS GENDERED IDEOLOGICAL STATE APPARATUS

The rituals surrounding the *obentō*'s consumption in the school situate what ideological meanings the *obentō* transmits to the child. The process of production within the home, by contrast, organizes its somewhat different ideological package for the mother. While the two sets of meanings are intertwined, the mother is faced with different expectations in the preparation of the *obentō* than the child is in its consumption. At a pragmatic level the child must simply eat the lunch box, whereas the mother's job is far more complicated. The onus for her is getting the child to consume what she has made, and the general attitude is that this is far more the mother's responsibility (at this nursery school, transitional stage) than the child's. And this is no simple or easy task.

Much of what is written, advised, and discussed about the *obentō* has this aim explicitly in mind: that is making food in such a way as to facilitate the child's duty to eat it. One magazine advises:

> The first day of taking *obentō* is a worrisome thing for mother and *boku* (child[15]) too. Put in easy-to-eat foods that your child likes and is already used to and prepare this food in small portions. (*Shufunotomo* 1980:28)

Filled with pages of recipes, hints, pictures, and ideas, the magazine codes each page with "helpful" headings:

- First off, easy-to-eat is step one.
- Next is being able to consume the *obentō* without leaving anything behind.
- Make it in such a way for the child to become proficient in the use of chopsticks.
- Decorate and fill it with cute dreams (*kawairashi yume*).
- For older classes (*nencho*), make *obentō* filled with variety.
- Once he's become used to it, balance foods your child likes with those he dislikes.
- For kids who hate vegetables . . .
- For kids who hate fish . . .
- For kids who hate meat . . . (pp. 28–53)

Laced throughout cookbooks and other magazines devoted to *obentō*, the *obentō* guidelines issued by the school and sent home in the school flier every two weeks, and the words of Japanese mothers and teachers discussing *obentō*, are a number of principles: 1) food should be made easy to eat: portions cut or made small and manipulated with fingers or chopsticks, (child-size) spoons and forks, skewers, toothpicks, muffin tins, containers, 2) portions should be kept small so the *obentō* can be consumed quickly and without any left-overs, 3) food that a child does not yet like should be eventually added so as to remove fussiness (*sukikirai*) in food habits, 4) make the *obentō* pretty, cute, and visually changeable by presenting the food attractively and by adding non-food objects such as silver paper, foil, toothpick flags, paper napkins, cute handkerchiefs, and variously shaped containers for soy sauce and ketchup, and 5) design *obentō*-related items as much as possible by the mother's own hands including the *obentō* bag (*obentōfukuro*) in which the *obentō* is carried.

The strictures propounded by publications seem to be endless. In practice I found that visual appearance and appeal were stressed by the mothers. By contrast, the directive to use *obentō* as a training process—adding new foods and getting older children to use chopsticks and learn to tie the *furoshiki* [16]—was emphasized by those judging the *obentō* at the school. Where these two sets of concerns met was, of course, in the child's success or failure completing the *obentō*. Ultimately this outcome and the mother's role in it, was how the *obentō* was judged in my experience.

The aestheticization of the *obentō* is by far its most intriguing aspect for a cultural anthropologist. Aesthetic categories and codes that operate generally for Japanese cuisine are applied, though adjusted, to the nursery school format. Substances are many but petite, kept segmented and opposed, and manipulated intensively to achieve an appearance that often changes or disguises the food. As a mother insisted to me, the creation of a bear out of miniature hamburgers and rice, or a flower from an apple or peach, is meant to sustain a child's interest in the underlying food. Yet my child, at least, rarely noticed or appreciated the art I had so laboriously contrived. As for other children, I observed that even for those who ate with no obvious "fussiness," mothers' efforts to create food as style continued all year long.

Thus much of a woman's labor over *obentō* stems from some agenda other than that of getting the child to eat an entire lunch-box. The latter is certainly a consideration and it is the rationale as well as cover for women being scrutinized by the school's authority figure—the teacher. Yet two other factors are important. One is that the *obentō* is but one aspect of the far more expansive and continuous commitment a mother is expected to make for and to her child. "*Kyoiku mama*" (education mother) is the term given to a mother who executes her responsibility to oversee and manage the education of her children with excessive vigor. And yet this excess is not only demanded by the state even at the level of the nursery school; it is conventionally given by mothers. Mothers who manage the home and children, often in virtual absence of a husband/father, are considered the factor that may make or break a child as s/he advances towards that pivotal point of the entrance examinations.[17]

In this sense, just as the *obentō* is meant as a device to assist a child in the struggles of first adjusting to school, the mother's role is generally perceived as that of support, goad, and cushion for the child. She will perform endless tasks to assist in her child's study: sharpen pencils and make midnight snacks as the child studies, attend cram schools to verse herself in subjects her child is weak in, make inquiries as to what school is most appropriate for her child, and consult with her child's teachers. If the child succeeds, a mother is complimented; if the child fails, a mother is blamed.

Thus, at the nursery-school level, the mother starts her own preparation for this upcoming role. Yet the jobs and energies demanded of a nursery-school mother are, in themselves, surprisingly consuming. Just as the mother of an entering student is given a book listing all the pre-entry tasks she must complete—for example, making various bags and containers, affixing labels to all clothes in precisely the right place and of precisely the right size—she will be continually expected thereafter to attend Mothers' Association meetings, accompany children on field trips, wash her child's clothes and indoor shoes every week, add required items to her child's bag on a day's notice, and generally be available. Few mothers at the school my son attended could afford to work in even part-time or temporary jobs. Those women who did tended either to keep their outside work a secret or be reprimanded by a teacher for insufficient devotion to their child. Motherhood, in other words, is institutionalized through the child's school and such routines as making the *obentō* as a full-time, kept-at-home job.[18]

The second factor in a woman's devotion to over-elaborating her child's lunch box is that her experience doing this becomes a part of her and a statement, in some sense, of who she is. Marx writes that labor is the most "essential" aspect to our species-being and that the products we produce are the encapsulation of us and therefore our productivity (1970: 71–76). Likewise, women are what they are through the products they produce. An *obentō* therefore is not only a gift or test for a child, but a representation and product of the woman herself. Of course, the two ideologically converge, as has been stated already, but I would also suggest that there is a potential disjoining. I sensed that the women were laboring for themselves apart from the agenda the *obentō* was expected to fill at school. Or stated alternatively, in the role that females in Japan are highly pressured and encouraged to assume as domestic manager, mother, and wife, there is, besides the endless and onerous responsibilities, also an opportunity for play. Significantly, women find play and creativity not outside their social roles but within them.

Saying this is not to deny the constraints and surveillance under which Japanese women labor at their *obentō*. Like their children at school, they are watched not only by the teacher but by each other, and they perfect what they create, at least partially, so as to be confirmed as a good and dutiful mother in the eyes of other mothers. The enthusiasm with which they absorb this task, then, is like my son's acceptance and internalization of the nursery-school routines; no longer enforced from outside, it is adopted as one's own.

The making of the *obentō* is, I would thus argue, a double-edged sword for women. By relishing its creation (for all the intense labor expended, only once or twice did I hear a mother voice any complaint about this task), a woman is ensconcing herself in the ritualization and subjectivity (subjection) of being a mother in Japan. She is alienated in the sense that others will dictate, inspect, and manage her work. On the reverse side, however, it is precisely through this work that the woman expresses, identifies, and constitutes herself. As Althusser pointed out, ideology can never be totally abolished (1971: 170); the elaborations that women work on "natural" food produce an *obentō* that is creative and, to some degree, a fulfilling and personal statement of themselves.

Minami, an informant, revealed how both restrictive and pleasurable the daily rituals of motherhood can be. The mother of two children—one aged three and one a nursery-school student—Minami had been a professional opera singer before marrying at the relatively late age of 32. Now, her daily schedule was organized by routines associated with her child's nursery school: for example, making the *obentō,* taking her daughter to school and picking her up, attending Mothers' Association meetings, arranging daily play dates, and keeping the school uniform clean. While Minami wished to return to singing, if only on a part-time basis, she said that the demands of motherhood, particularly those imposed by her child's attendance at nursery school, frustrated this desire. Secretly snatching only minutes out of any day to practice, Minami missed singing and told me that being a mother in Japan means the exclusion of almost anything else.[19]

Despite this frustration, however, Minami did not behave like a frustrated woman. Rather she devoted to her mothering an energy, creativity, and intelligence I found to be standard in the Japanese mothers I knew. She planned special outings for her children at least two or three times a week, organized games that she knew they would like and would teach them cognitive skills, created her own stories and designed costumes for afternoon play, and shopped daily for the meals she prepared with her children's favorite foods in mind. Minami told me often that she wished she could sing more, but never once did she complain about her children, the chores of child-raising, or being a mother. The attentiveness displayed otherwise in her mothering was exemplified most fully in Minami's *obentōs*. No two were ever alike, each had at least four or five parts, and she kept trying out new ideas for both new foods and new designs. She took pride as well as pleasure in her *obentō* handicraft; but while Minami's *obentō* creativity was impressive, it was not unusual.

Examples of such extraordinary *obentō* creations from an *obentō* magazine include: 1) ("donut *obentō*"): two donuts, two wieners cut to look like a worm, two cut pieces of apple, two small cheese rolls, one hard-boiled egg made to look like a rabbit with leaf ears and pickle eyes and set in an aluminum muffin tin, cute paper napkin added, 2) (wiener doll *obentō*): a bed of rice with two doll creations made out of wiener parts (each consists of eight pieces comprising hat, hair, head, arms, body, legs), a line of pink ginger, a line of green parsley, paper flag of France added, 3) (vegetable flower and tulip *obentō*): a bed of rice laced with chopped hard-boiled egg, three tulip flowers made out of cut wieners with spinach precisely arranged as stem and leaves, a fruit salad with two raisins, three cooked peaches, three pieces of cooked apple, 4) (sweetheart doll *obentō—abekku ningyo no obentō*): in a two-section *obentō* box there are four rice balls on one side, each with a different center, on the other side are two dolls made of quail's eggs for heads, eyes and mouth added, bodies of cucumber, arranged as if lying down with two raw carrots for the pillow, covers made of one flower—cut cooked carrot, two pieces of ham, pieces of cooked spinach, and with different colored plastic skewers holding the dolls together (*Shufunotomo* 1980: 27, 30).

The impulse to work and re-work nature in these *obentō*s is most obvious perhaps in the strategies used to transform, shape, and/or disguise foods. Every mother I knew came up with her own repertoire of such techniques, and every *obentō* magazine or cookbook I examined offered a special section on these devices. It is important to keep in mind that these are treated as only flourishes: embellishments added to parts of an *obentō* composed of many parts. The following is a list from one magazine: lemon pieces made into butterflies, hard-boiled eggs into *daruma* (popular Japanese legendary figure of a monk without his eyes), sausage cut into flowers, a hard-boiled egg decorated as a baby, an apple piece cut into a leaf, a radish flaked into a flower, a cucumber cut like a flower, a *mikan* (nectarine orange) piece arranged into a basket, a boat with a sail made from a cucumber, skewered sausage, radish shaped like a mushroom, a quail egg flaked into a cherry, twisted *mikan* piece, sausage cut to become a crab, a patterned cucumber, a ribboned carrot, a flowered tomato, cabbage leaf flower, a potato cut to be a worm, a carrot designed as a red shoe, an apple cut to simulate a pineapple (pp. 57–60).

Nature is not only transformed but also supplemented by store-bought or mother-made objects which are precisely arranged in the *obentō*. The former come from an entire industry and commodification of the *obentō* process: complete racks or sections in stores selling *obentō* boxes, additional small containers, *obentō* bags, cups, chopsticks and utensil containers (all these with various cute characters or designs on the front), cloth and paper napkins, foil, aluminum tins, colored ribbon or string, plastic skewers, toothpicks with paper flags, and paper dividers. The latter are the objects mothers are encouraged and praised for making themselves: *obentō* bags, napkins, and handkerchiefs with appliqued designs or the child's name embroidered. These supplements to the food, the arrangement of the food, and the *obentō* box's dividing walls (removable and adjustable) furnish the order of the *obentō*. Everything appears crisp and neat with each part kept in its own place: two tiny hamburgers set firmly atop a bed of rice; vegetables in a separate compartment in the box; fruit arranged in a muffin tin.

How the specific forms of *obentō* artistry—for example, a wiener cut to look like a worm and set within a muffin tin—are encoded symbolically is a fascinating subject. Limited here by space, however, I will only offer initial suggestions. Arranging food into a scene recognizable by the child was an ideal mentioned by many mothers and cookbooks. Why those of animals, human beings, and other food forms (making a pineapple out of an apple, for example) predominate may have no other rationale than being familiar to children and easily re-produced by mothers. Yet it is also true that this tendency to use a trope of realism—casting food into realistic figures—is most prevalent in the meals Japanese prepare for their children. Mothers I knew created animals and faces in supper meals and/or *obentōs* made for other outings, yet their impulse to do this seemed not only heightened in the *obentō* that were sent to school but also played down in food prepared for other age groups.

What is consistent in Japanese cooking generally, as stated earlier, are the dual principles of manipulation and order. Food is manipulated into some other form than it assumes either naturally or upon being cooked: lines are put into mashed potatoes, carrots are flaked, wieners are twisted and sliced. Also, food is ordered by some human rather than natural principle; everything must have neat boundaries and be placed precisely so those boundaries do not merge. These two structures are the ones most important in shaping the nursery school *obentō* as well, and the inclination to design realistic imagery is primarily a means by which these other culinary codes are learned by and made pleasurable for the child. The simulacrum of a pineapple recreated from an apple therefore is less about seeing the pineapple in an apple (a particular form) and more about reconstructing the apple into something else (the process of transformation).

The intense labor, management, commodification, and attentiveness that goes into the making of an *obentō* laces it, however, with many and various meanings. Overarching all is the potential to aestheticize a certain social order, a social order that is coded (in cultural and culinary terms as Japanese. Not only is a mother making food more palatable to her nursery-school child, but she is creating food as a more aesthetic and pleasing social structure. The *obentō*'s message is that the world is constructed very precisely and that the role of any single Japanese in that world must be carried out with the same degree of precision. Production is demanding; and the producer must both keep within the borders of her/his role and work hard.

The message is also that it is women, not men, who are not only sustaining a child through food but carrying the ideological support of the culture that this food embeds.

No Japanese man I spoke with had or desired the experience of making a nursery-school *obentō* even once, and few were more than peripherally engaged in their children's education. The male is assigned a position in the outside world, where he labors at a job for money and is expected to be primarily identified by and committed to his place of work.[20] Helping in the management of home and the raising of children has not become an obvious male concern or interest in Japan, even as more and more women enter what was previously the male domain of work. Females have remained at and as the center of home in Japan, and this message too is explicitly transmitted in both the production and consumption of entirely female-produced *obentō*.

The state accrues benefits from this arrangement. With children depending on the labor women devote to their mothering to such a degree, and women being pressured as well as pleasurized in such routine maternal productions as making the *obentō*—both effects encouraged and promoted by institutional features of the educational system, which is heavily state-run and at least ideologically guided at even the nursery-school level—a gendered division of labor is firmly set in place. Labor from males, socialized to be compliant and hardworking, is more extractable when they have wives to rely on for almost all domestic and familial management. And females become a source of cheap labor, as they are increasingly forced to enter the labor market to pay domestic costs (including those vast debts incurred in educating children) yet are increasingly constrained to low-paying part-time jobs because of the domestic duties they must also bear almost totally as mothers.

Hence, not only do females, as mothers, operate within the ideological state apparatus of Japan's school system, which starts semi-officially with the nursery school, they also operate as an ideological state apparatus unto themselves. Motherhood *is* state ideology, working through children at home and at school and through such mother-imprinted labor that a child carries from home to school as the *obentō*. Hence the post-World War II conception of Japanese education as egalitarian, democratic, and with no agenda of or for gender differentiation, does not in practice stand up. Concealed within such cultural practices as culinary style and child-focused mothering is a worldview in which the position and behavior an adult will assume has everything to do with the anatomy she/he was born with.

At the end, however, I am left with one question. If motherhood is not only watched and manipulated by the state but made by it into a conduit for ideological indoctrination, could not women subvert the political order by redesigning *obentō*? Asking this question, a Japanese friend, upon reading this paper, recalled her own experiences. Though her mother had been conventional in most other respects, she made her children *obentōs* that did not conform to the prevailing conventions. Basic, simple, and rarely artistic, Sawa also noted, in this connection, that the lines of these *obentōs* resembled those by which she was generally raised: as gender-neutral, treated as a person not "just as a girl," and being allowed a margin to think for herself. Today she is an exceptionally independent woman who has created a life for herself in America, away from homeland and parents, almost entirely on her own. She loves Japanese food, but the plain *obentōs* her mother made for her as a child, she is newly appreciative of now, as an adult. The *obentōs* fed her, but did not keep her culturally or ideologically attached. For this, Sawa says today, she is glad.

NOTES

The fieldwork on which this article is based was supported by a Japan Foundation Postdoctoral Fellowship. I am grateful to Charles Piot for a thoughtful reading and useful suggestions for revision and to Jennifer Robertson for inviting my contribution to this issue. I would also like to thank Sawa Kurotani for her many ethnographic stories and input, and Phyllis Chock and two anonymous readers for the valuable contributions they made to revision of the manuscript.

1. As Dorinne Kondo has pointed out, however, these cuisinal principles may be conditioned by factors of both class and circumstance. Her *shitamachi* (more traditional area of Tokyo) informants, for example, adhered only casually to this coding and other Japanese she knew followed them more carefully when preparing food for guests rather than family and when eating outside rather than inside the home (Kondo 1990: 61–2).
2. Rice is often, if not always, included in a meal; and it may substantially as well as symbolically constitute the core of the meal. When served at a table it is put in a large pot or electric rice maker and will be spooned into a bowl, still no bigger or predominant than the many other containers from which a person eats. In an *obentō* rice may be in one, perhaps the largest, section of a multi-sectioned *obentō* box, yet it will be arranged with a variety of other foods. In a sense rice provides the syntactic and substantial center to a meal yet the presentation of the food rarely emphasizes this core. The rice bowl is refilled rather than heaped as in the preformed *obentō* box, and in the *obentō* rice is often embroidered, supplemented, and/or covered with other foodstuffs.
3. Japanese will both endure a high price for rice at home and resist American attempts to export rice to Japan in order to stay domestically self-sufficient in this national food *qua* cultural symbol. Rice is the only foodstuff in which the Japanese have retained self-sufficient production.
4. The primary sources on education used are Horio 1988; Duke 1986; Rohlen 1983; Cummings 1980.
5. Neither the state's role in overseeing education nor a system of standardized tests is a new development in post-World War II Japan. What is new is the national standardization of tests and, in this sense, the intensified role the state has thus assumed in overseeing them. See Dore (1965) and Horio (1988).
6. Boocock (1989) differs from Tobin *et al.* (1989) on this point and asserts that the institutional differences are insignificant. She describes extensively how both *yochien* and *hoikuen* are administered (*yoghien* are under the authority of Monbusho and *hoikuen* are under the authority of the Koseisho, the Ministry of Health and Welfare) and how both feed into the larger system of education. She emphasizes diversity: though certain trends are common amongst pre-schools, differences in teaching styles and philosophies are plentiful as well.
7. According to Rohlen (1989), families are incapable of indoctrinating the child into this social pattern of *shundanseikatsu* by their very structure and particularly by the relationship (of indulgence and dependence) between mother and child. For this reason and the importance placed on group structures in Japan, the nursery school's primary objective, argues Rohlen, is teaching children how to assimilate into groups. For further discussion of this point see also Peak 1989; Lewis 1989; Sano 1989; and the *Journal of Japanese Studies* issue [15(1)] devoted to Japanese pre-school education in which these articles, including Boocock's, are published.
8. For a succinct anthropological discussion of these concepts, see Hendry (1987: 39–41). For an architectural study of Japan's management and organization of space in terms of such cultural categories as *uchi* and *soto*, see Greenbie (1988).
9. Endless studies, reports, surveys, and narratives document the close tie between women and home, domesticity and femininity in Japan. A recent international survey conducted for a Japanese housing construction firm, for example, polled couples with working wives in three cities, finding that 97 percent (of those polled) in Tokyo prepared breakfast for their families almost daily (compared with 43 percent in New York and 34 percent in London); 70 percent shopped for groceries on a daily basis (3 percent in New York, 14 percent in London), and only 22 percent of them had husbands who assisted or were willing to assist with housework (62 percent in New York, 77 percent in London) (quoted in *Chicago Tribune*

1991). For a recent anthropological study of Japanese housewives in English, see Imamura (1987). Japanese sources include *Juristo zokan sogo tokushu* 1985; *Mirai shakan* 1979; *Ohirasori no seifu kenkyukai* 3.

10. My comments pertain directly, of course, to only the women I observed, interviewed, and interacted with at the one private nursery school serving middle-class families in urban Tokyo. The profusion of *obentō*-related materials in the press plus the revelations made to me by Japanese and observations made by other researchers in Japan (for example, Tobin 1989; Fallows 1990), however, substantiate this as a more general phenomenon.
11. To illustrate this preoccupation and consciousness: during the time my son was not eating all his *obentō*, many fellow mothers gave me suggestions, one mother lent me a magazine, my son's teacher gave me a full set of *obentō* cookbooks (one per season), and another mother gave me a set of small frozen-food portions she had made in advance for future *obentōs*.
12. My son's teacher, Hamada-sensei, cited this explicitly as one of the reasons why the *obentō* was such an important training device for nursery-school children. "Once they become *ichinensei* [first-graders], they'll be faced with a variety of food, prepared without elaboration or much spice, and will need to eat it within a delimited time period."
13. An anonymous reviewer questioned whether such emphasis placed on consumption of food in nursery school leads to food problems and anxieties in later years. Although I have heard that anorexia is now a phenomenon in Japan, I question its connection to nursery-school *obentōs*. Much of the meaning of the latter practice, as I interpret it, has to do with the interface between production and consumption, and its gender linkage comes from the production end (mothers making it) rather than the consumption end (children eating it). Hence, while control is taught through food, it is not a control linked primarily to females or bodily appearance, as anorexia may tend to be in this culture.
14. Fujita argues, from her experience as a working mother of a daycare (*hoikuen*) child, that the substance of these daily talks between teacher and mother is intentionally insignificant. Her interpretation is that the mother is not to be overly involved in nor too informed about matters of the school (1989).
15. "*Boku*" is a personal pronoun that males in Japan use as a familiar reference to themselves. Those in close relationships with males—mothers and wives, for example—can use *boku* to refer to their sons or husbands. Its use in this context is telling.
16. In the upper third grade of the nursery school (the *nencho* class; children aged five to six) that my son attended, children were ordered to bring their *obentō* with chopsticks rather than forks and spoons (considered easier to use) and in the traditional *furoshiki* (piece of cloth that enwraps items and is double-tied to close it) instead of the easier-to-manage *obentō* bags with drawstrings. Both *furoshiki* and chopsticks (*o-hashi*) are considered traditionally Japanese, and their usage marks not only greater effort and skills on the part of the children but their enculturation into being Japanese.
17. For the mother's role in the education of her child, see, for example, White (1987). For an analysis, by a Japanese, of the intense dependence on the mother that is created and cultivated in a child, see Doi (1971). For Japanese sources on the mother-child relationship and the ideology (some say pathology) of Japanese motherhood, see Yamamura (1971); Kawai (1976); Kyutoku (1981); *Sorifu seihonen taisaku honbuhen* (1981); *Kadeshobo shinsha* (1981). Fujita's account of the ideology of motherhood at the nursery-school level is particularly interesting in this connection (1989).
18. Women are entering the labor market in increasing numbers, yet the proportion who do so in the capacity of part-time workers (legally constituting as much as thirty-five hours per week but without the benefits accorded to full-time workers) has also increased. The choice of part-time over full-time employment has much to do with a woman's simultaneous and almost total responsibility for the domestic realm (Juristo 1985; see also Kondo 1990).
19. As Fujita (1989: 72–79) points out, working mothers are treated as a separate category of mothers, and nonworking mothers are expected, by definition, to be mothers full-time.
20. Nakane's much-quoted text on Japanese society states this male position in structuralist terms (1970). Though dated, see also Vogel (1963) and Rohlen (1974) for descriptions of the social roles for middle-class, urban Japanese males. For a succinct recent discussion of gender roles within the family, see Lock (1990).

REFERENCES

Althusser, Louis. 1971. *Ideology and ideological state apparatuses (Notes toward an investigation in Lenin and philosophy and other essays).* New York: Monthly Review Press.

Asanuma, Kaoru. 1987. *"Ganbari" no kozo (Structure of "Ganbari").* Tokyo: Kikkawa Kobunkan.

Barthes, Roland. 1957. *Mythologies.* Trans. Annette Lavers. New York: Noonday Press.

Boocock, Sarane Spence. 1989. Controlled diversity: An overview of the Japanese preschool system. *The Journal of Japanese Studies* 15(1): 41–65.

Chicago Tribune. 1991. Burdens of working wives weigh heavily in Japan. January 27, section 6, p. 7.

Cummings, William K. 1980. *Education and equality in Japan.* Princeton, NJ: Princeton University Press.

Doi, Takeo. 1971. *The anatomy of dependence: The key analysis of Japanese behavior.* Trans. John Becker. Tokyo: Kodansha International, Ltd.

Dore, Ronald P. 1965. *Education in Tokugawa Japan.* London: Routledge and Kegan Paul.

Duke, Benjamin. 1986. *The Japanese school: Lessons for industrial America.* New York: Praeger.

Fallows, Deborah. 1990. Japanese women. *National Geographic* 177(4): 52–83.

Fujita, Mariko. 1989. "It's all mother's fault": Childcare and the socialization of working mothers in Japan. *The Journal of Japanese Studies* 15(1): 67–91.

Greenbie, Barrie B. 1988. *Space and spirit in modern Japan.* New Haven, CT: Yale University Press.

Hendry, Joy. 1986. *Becoming Japanese: The world of the pre-school child.* Honolulu: University of Hawaii Press.

———. 1987. *Understanding Japanese society.* London: Croom Helm.

Horio, Teruhisa. 1988. *Educational thought and ideology in modern Japan: State authority and intellectual freedom.* Trans. Steven Platzer. Tokyo: University of Tokyo Press.

Imamura, Anne E. 1987. *Urban Japanese housewives: At home and in the community.* Honolulu: University of Hawaii Press.

Juristo zokan Sogotokushu. 1985. Josei no Gensai to Mirai (The present and future of women). 39.

Kadeshobo shinsha. 1981. *Hahaoya (Mother).* Tokyo: Kadeshobo shinsha.

Kawai, Jayao. 1976. *Bosei shakai nihon no Byori (The pathology of the mother society—Japan).* Tokyo: Chuo koronsha.

Kondo, Dorinne K. 1990. *Crafting selves: Power, gender, and discourses of identity in a Japanese workplace.* Chicago, IL: Univer sity of Chicago Press.

Kyutoku, Shigemori. 1981. *Bogenbyo (Disease rooted in motherhood).* Vol II. Tokyo: Sanma Kushuppan.

Lewis, Catherine C. 1989. From indulgence to internalization: Social control in the early school years. *Journal of Japanese Studies* 15(1): 139–157.

Lock, Margaret. 1990. Restoring order to the house of Japan. *The Wilson Quarterly* 14(4): 42–49.

Marx, Karl and Frederick Engels. 1970 (1947). *Economic and philosophic manuscripts*, ed. C. J. Arthur. New York: International Publishers.

Mirai shakan. 1979. Shufu to onna (Housewives and women). Kunitachishi Komininkan Shimindaigaku Semina—no Kiroku. Tokyo: Miraisha.

Mouer, Ross and Yoshio Sugimoto. 1986. *Images of Japanese society: A study in the social construction of reality.* London: Routledge and Kegan.

Nakane, Chie. 1970. *Japanese society.* Berkeley: University of California Press.

Ohirasori no Seifu kenkyukai. 1980. Katei kiban no jujitsu (The fullness of family foundations). (Ohirasori no Seifu kenkyukai—3). Tokyo: Okurasho Insatsukyoku.

Peak, Lois. 1989. Learning to become part of the group: The Japanese child's transition to preschool life. *The Journal of Japanese Studies* 15(1): 93–123.

Richie, Donald. 1985. *A taste of Japan: Food fact and fable, customs and etiquette, what the people eat.* Tokyo: Kodansha International Ltd.

Rohlen, Thomas P. 1974. *The harmony and strength: Japanese white-collar organization in anthropological perspective.* Berkeley: University of California Press.

————1983. *Japan's high schools.* Berkeley: University of California Press.

———1989. Order in Japanese society: attachment, authority, and routine. *The Journal of Japanese Studies* 15(1): 5–40.

Sano, Toshiyuki. 1989. Methods of social and socialization in Japanese day-care centers. *The Journal of Japanese Studies* 15(1): 125–138.

Shufunotomo Besutoserekushon shiri-zu. 1980. Obentō 500 sen. Tokyo: Shufunotomo Co., Ltd.

Shufunotomohyakka shiri-zu. 1981. 365 nichi no obentō hyakka. Tokyo: Shufunotomo Co.

Sorifu Seihonen Taisaku Honbuhen. 1981. Nihon no kodomo to hahaoya (Japanese mothers and children): kokusaihikaku (international comparisons). Tokyo: Sorifu Seishonen Taisaku Honbuhen.

Tobin, Joseph J., David Y. H. Wu, and Dana H. Davidson. 1989. *Preschool in three cultures: Japan, China, and the United States*. New Haven CT: Yale University Press.

Vogel, Erza. 1963. *Japan's new middle class: The salary man and his family in a Tokyo suburb*. Berkeley: University of California Press.

White, Merry. 1987. *The Japanese educational challenge: A commitment to children*. New York: Free Press.

Yamamura, Yoshiaki. 1971. *Nihonjin to haha: Bunka toshite no haha no kannen ni tsuite no kenkyu (The Japanese and mother: Research on the conceptualization of mother as culture)*. Tokyo: Toyo-shuppansha.

23

On the Civilizing of Appetite

STEPHEN MENNELL

Although *The Civilizing Process* has a great deal to say about the civilizing of table manners—how people ate—it says relatively little about what people ate and how much. Elias mentions in passing the well-known carnivorous bent of the medieval upper classes, in marked contrast to the largely leguminous and farinaceous diet of the peasants, and he discusses the gradual growth of feelings of repugnance towards the carving at table of large and recognizable carcasses (Elias, 1978: 117–22). Of appetite he says nothing. Yet the general thesis of *The Civilizing Process* is of course a powerful one, capable of wide application, and it gives general grounds for looking for evidence of a long-term process of the civilizing of appetite. Elias has demonstrated, not only in *The Civilizing Process* but in *The Court Society* and in many essays and lectures, how civilizing processes were manifested in changing taste in literature and the arts. The culinary arts are no exception, as I have tried to show in my book *All Manners of Food* (1985). In this particular paper, however, I am concerned less with changes in qualitative tastes in food than with the more difficult question of changes in the regulation of appetite in the quantitative sense. Is it not likely that the same long-term changes in the structure of societies which brought about changes in manners, in the expression of affect, and in the tension-balance of personalities would also be reflected in the patterning and expression of so basic a drive as appetite?

One aspect of the problem of control over appetite has been raised by Bryan Turner (1982a, 1982b, 1984) in his discussions of medical discourse about diet. Turner mentions Elias in passing, but his own theoretical orientation is derived from Foucault (especially *The Birth of the Clinic* and *Discipline and Punish)* and from Max Weber's views on rationalization in European culture and its roots in religion. In Foucault, however, as Turner pointed out, "the discourse appears to be almost sociologically disembodied" and "there is a pronounced reluctance to reduce systematic thought to interests, especially the economic interests of social groups, so that the growth of formal knowledge appears to be one which is immanent in discourse itself" (1982b: 257). Weber certainly is not vulnerable to that criticism, but in his case it is well to recall Goudsblom's (1977: 188–9) warning that although *The Protestant Ethic* will always stand as "a masterpiece of well-documented "interpretative understanding," it is hopelessly inconclusive when it comes to explaining the actual part played by Calvinism in the sociogenesis of capitalism. Attempts to extend notions of elective affinity into realms like medical writings on diet are likely to be even more inconclusive. Besides, medical opinion—and even the increasing power of the medical profession—are only small parts of the complex history of appetite and its control in European society. I therefore want to explore whether ideas derived from Elias and figurational sociology can help to make sense of that history, and to ask whether we can speak of the "civilizing of appetite."

HUNGER AND APPETITE

Appetite, it must be remembered, is not the same thing as hunger. Hunger is a body drive which recurs in all human beings in a reasonably regular cycle. Appetite for food, on the other hand, in the words of Daniel Cappon, a psychotherapist specializing in eating disorders is:

> basically a state of mind, an inner mental awareness of desire that is the setting for hunger. . . . An individual's appetite is his desire and inclination to eat, his interest in consuming food. Eating is what a person *does*. Appetite is what he *feels* like doing, mostly a psychological state. (1973: 21)

We tend to think of hunger and appetite as directly linked, but in fact, as Cappon argues, there is no simple relationship. The link between hunger and appetite is provided by what is sometimes referred to as the "appestat," by which is meant a *psychological,* not simply physiological, control mechanism regulating food intake. Just as a thermostat can be set too high or too low, so a person's "appestat" can be set too high or too low in relation to the physiological optimum range. Too high a setting, too much food intake, is a condition of "bulimia," likely to lead to excessive body weight; too low a setting represents the condition of "anorexia," leading to problems of underweight.

A person's "appestat" setting is determined not only by the underlying hunger drive, but also by often rather complex psychological processes in which social pressures can play a considerable part. Body image is a particularly notable element: how a person perceives his or her own body and its relation to what he or she perceives to be the socially approved body image. Today psychologists understand much more about the psychological problems which can lead individual people to have pathological "eating disorders" and body weights deviating from what is healthy.

But what about the regulation of appetite in the "normal" majority? Can that be studied according to the model provided by Elias in a long-term developmental perspective? Cappon (1973: 45) provides a clue that perhaps it can, when he argues that his patients with eating disorders are in some sense "immature" personalities, and that the normal mature individual today "is able to change his eating habits at will—when he eats, how long he lingers over a meal, what he eats, and the amount." In other words, Cappon is arguing that normal eating behavior involves a capacity for considerable self-control. Has this capacity developed over the long term in European society in the same way that Elias argues other facets of self-control have done?

THE APPETITE OF GARGANTUA

The celebrated banquets of the Middle Ages and Renaissance, known to us from literary sources like Rabelais and from numerous documents throughout Europe, give a misleading image of typical eating in that period. Not only did they involve just a small minority of society—even if we allow that servants and retainers received their share—but from the spectacular bills of fare it is difficult to work out how much each individual actually ate. For example, the menu for a feast given by the City of Paris for Catherine de Medici in 1549 (Franklin, 1887–1902: III, 93) lists twenty-four sorts of animals (mainly birds and other game, because butcher's meat was disdained for such grand occasions), many kinds of cakes and pastry, and a mere four vegetable dishes—but we do not know how many shared the food. At the feast for the enthronement of Archbishop Nevill at York in 1465 (Warner, 1791: 93ff.), a thousand sheep, two thousand pigs, two thousand geese, four thousand rabbits, fish and game by the hundred, numerous kinds of bird, and twelve porpoises and seals were eaten; but though we know the order of courses and even the seating plan for the most important guests, it is uncertain how many others took part, or indeed how long the feast lasted—it may have been several days. There is a little doubt that guests could if they wished eat as much as they could take. The number of dishes set before the diners on such great occasions was very large—for example, *Sir Gawain and the Green Knight* (c. 1400: 25) mentions twelve dishes between each pair of diners—but they did not necessarily finish them, for it is known that surplus from the high table generally found its way to lower tables and eventually to the poor. Whatever the uncertainties, however, there seems little doubt that prodigious feats of appetite were witnessed at these great feasts, which were at least symptomatic of great inequalities in the social distribution of nourishment. The great people who had the power to do so sometimes indulged in such banquets in times of widespread dearth,[1] which itself, as a sign of a relatively low level of identification with the sufferings of fellow men, marks in Elias's terms a relatively low point on the curve of civilizing processes.

In other ways, the great banquets are highly misleading as a guide to medieval patterns of appetite. Their social function can be understood more by analogy with the Kwakiutl potlatch than in relation to culinary taste and appetite (Mennell, 1985, chapter 3; Codere, 1950). Moreover they were untypical even of upper-class eating. They were high points of an oscillating dietary regime even for the courtiers and nobility. Even this élite did not eat like that all the time. Perhaps, unlike most people, they rarely went hungry, but they did not always enjoy the wide choice which (rather than the sophistication of the cooking) was the hallmark of the feast. The rhythm of the seasons and the hazards of the harvest impinged even on their diet; even they knew periods of frugality.[2] Breakfasts even in a royal household "would not now be regarded as extravagant in a day labourer's family," and on ordinary days dinners consisted of no more than two joints of meat, roast or boiled, or fish (Weber, 1973: 198). Robert Mandrou recognizes the significance of these fluctuations in the pattern of eating:

> without any doubt it was normal for all social classes to alternate between frugality and feasting. A consequence of the general insecurity where food was concerned, this oscillation imposed itself as a rite, some signs of which can still be found today. The festivals of the fraternities . . . in the towns and those of the harvest, vintage or St Martin's Day in the country were always occasions for fine living for a few hours at least—and with innumerable variations in the form it took, of course.[3] But these huge feasts, after which a man had to live on bread and water for months on end provided compensation, however meager, for ill-fortune, and were appreciated for that reason; the very precariousness of existence

> explained them. The virtue of thrift, of making one's resources spread evenly over a given period, cannot be conceived of without a certain margin of supply. . . . One other factor to be taken into account in explaining these "orgies" is the ever present dangers threatening the granary; what was the good of laying up large stocks if brigands or soldiers might come along the next day and carry them off? (Mandrou, 1975: 24)

This oscillation between fasting and feasting runs parallel to the extreme emotional volatility of medieval people noted by Elias, their ability to express emotion with greater freedom than today and to fluctuate quickly between extremes. And their sources are the same.

Mandrou, like Elias, Bloch (1961: I, 73 and II, 411) and Huizinga (1924: chapter 1) before him, notes this general psychological volatility but, curiously, relates it only indirectly to the insecurity of life in medieval and early modern Europe; he attributes it in large part to the physiological effects of inadequate and irregular feeding. "The effect of this chronic malnutrition was to produce in man the mentality of the hunted, with its superstitions, its sudden outbursts of anger and its hypersensitivity" (1975: 26). Such direct physiological effects of nutrition on psychology should perhaps not be entirely discounted, but they should equally not be overstressed; the suggestion merely adds one more complication to an already complex causal nexus. More important—as Mandrou himself seemed to see clearly when specifically discussing the fluctuation between feasting and fasting—is the link between the general precariousness and unpredictability of existence and its reflection in personality, beliefs and social behavior. Keith Thomas (1971) has emphasized the connection between the hazards of life in the sixteenth and seventeenth centuries and the prevalence of superstition and magical beliefs, which declined noticeably with the growing security of the late seventeenth and eighteenth centuries. But it is Norbert Elias who has traced most fully the general connection between the changing emotional economy of the personality and the gradually growing calculability of social existence brought about by long-term processes of change in the structure of societies.

The Civilizing Process presents a theory of state-formation and of the internal pacification of larger and larger territories which the growth of states involved. But Elias has also made it clear that state-formation is only one of several intertwining and interdependent long-term processes of social development which gradually increased the security and calculability of life in society. Internal pacification permitted the division of labour and growth of trade—eventually increasing the security of food supplies among many other things—which in turn provided the economic basis for further expansion of the territory and the internal regulative power of states. *The Civilizing Process* is also a study of the changing codes of manners and standards of social behavior which broadly accompanied these processes. Elias (1982: 233–4) gives a characteristically vivid illustration of the connection between these two aspects of his study. Travelling by road, he observes, was dangerous in medieval times, and it remains so today—but the nature of the danger has changed. The medieval traveller had to have the ability—temperamental as well as physical—to defend himself violently against violent attack. Today, the chief danger is from road accidents, and avoiding them depends to a great extent on high capacity for self-control in the expression of (and skill in warding off) aggression, whether in overt or in disguised form. And aggression is only one of the manifestations of affect over which people came gradually to be subject to increased pressures to exercise greater self-control. Not that the expression of feeling by people in the Middle Ages lacked all social patterning and control. There is no zero-point. But in the long-term the controls grew not just stronger but also more even.

Against this background, the oscillation between extremes of gluttonous gorging and enforced fasting seems all of a piece with other aspects of the medieval and early modern personality. I would therefore argue that it is connected not simply with the insecurity and unpredictability of food supplies alone, but also with the more general insecurity of conditions of life.

FAMINES AND OTHER HAZARDS

Life in medieval and early modern Europe certainly was by today's standards very insecure. Goubert (1960) speaks vividly of "steeples" of mortality, from the appearance of the suddenly soaring graphs of death-rates in the Beauvaisis in the seventeenth and early eighteenth centuries. Mortality among élites, whom one might expect to have been better fed, seems to have been just as high as among the mass of the population in Western Europe. This is especially well-documented for Britain. T. H. Hollingsworth's calculations (1977) of mortality in British peerage families since 1600 differ scarcely at all from those of Wrigley and Schofield (1981). As late as the third quarter of the seventeenth century, the life-expectancy of males at birth was only about thirty years. Mortality among ruling groups elsewhere in Europe was also high, "which makes it unlikely that they enjoyed appreciable advantages over the rest of the population" (Livi-Bacci, 1985: 98).

There were many causes other than dearth for the steeples of mortality which from time to time towered over localities, regions, or even whole countries. In towns, there were frequent disastrous fires, made worse in their consequences by organization inadequate to control them (Thomas, 1971: 15). Epidemic diseases including smallpox and plague periodically cut swathes through all ranks of society poor sanitation and hygiene—reflecting deficiencies in medical knowledge and technology as well as once more in social organization—played their part in this. And then there were wars and vagrancy. All these were in addition to crop failures, and they could interact in complex ways—war, for instance, not only killed people directly, but disrupted food supplies, led to increased vagrancy and helped to spread disease.

Not even in the worst times of famine is it thought that a great proportion of people actually starved to death. The general view is that hunger made many more people susceptible to disease, and that others who survived the immediate famine had their lifespans curtailed by the effects of hunger and malnutrition. Even this is in some dispute: Livi-Bacci (1985: 96) has pointed out "that the majority of cases of extraordinary and catastrophic mortality are independent of famine, hunger and starvation, and Watkins and Van de Walle (1985: 21) have contended that "the evidence linking malnutrition and mortality is surprisingly sparse and inadequate." Most historians would accept, however, at least that "even if many of the deaths in a famine period were due rather to disease than to outright starvation, nevertheless the sudden rise in death rates was sometimes associated with an abrupt fall in the availability of food, whatever the causes of this scarcity" (Watkins and Van de Walle, 1985: 17).

Famines, in any case, are not a simple function of crop failure. Sen's study of modern famines (1981) has already influenced historians' thinking about famines in the past. Sen shows that even in times of famine, food is available. People starve because of their inability to command food through "entitlement" relationships such as ownership, exchange, employment, and social security rights. In other words, the effects of crop failures have to be understood in terms of patterns of social interdependence. The breaking of the chain which linked crop failure to famines and famines to steeples of mortality is the story more of developing social organization's contribution to an increasing security of life than simply of increasingly reliable food production.

In medieval and early modern Europe, bad harvests and food shortages sometimes affected whole countries, even the whole continent at the same time. An example is the great European famine of 1315–17. Often, however, only a limited region was affected by harvest failure, though before authorities were able to organize the holding of sufficient stocks of grain, and before trade and transport were adequate to remedy local shortage, they could be serious enough.[4] Inadequate transport meant that food could not be moved, or could be moved only with difficulty, from surplus to deficit areas. Shortages led to panic buying, hoarding, and speculation, prices soaring and putting what food was available for sale quite beyond the means of the poor. Holding stocks could have helped to remedy this, but administrative difficulties defeated most governments before the late seventeenth or eighteenth century. The direct relationship between harvest failures and soaring rates of mortality only gradually disappeared from western Europe from the late seventeenth century onwards. By then, large grain stocks held for example at Amsterdam were helping to alleviate the effects of dearth not only in the Low Countries but in coastal and other areas of neighbouring countries accessible to trade. In the eighteenth century, food production increased markedly, but so did population. There was more food, though not necessarily greater consumption per capita. Food supplies, however, became gradually more reliable and shortages less frequent. After 1750, according to Braudel and Spooner (in Rich and Wilson, 1977: 396), only "suppressed" famines ('almost bearable ones') continued to occur in Western Europe, very largely because of improvements in trade and transport, the effects of which can be seen in the levelling out of food prices plotted (as on a weather map) across the continent. In England scarcity following crop failures no longer reached famine proportions by the first decade of the eighteenth century, though food prices rose very high and death rates were still noticeably up in years of bad harvests in the 1720s and 1740s. In France, the last full nation-wide famine was that of 1709–10, but regional dearths accompanied by rising mortality still happened as late as 1795–6 and 1810–12 (Cobb, 1970: 220–2).

Improved trade and transport were not altogether straightforward in their effects:

> the growth of trade, if it enabled the surplus of one region rather more often than before to relieve the dearth of another, also left a larger number of people at the mercy of market fluctuation, tended to depress or hold down real wages, and increase the gap between the rich and the poor. (Wernham, 1968: 5)

Furthermore, what Pelto and Pelto call the "delocalization" of food-use over the last few centuries—meaning "processes in which food varieties, production methods, and consumption patterns are disseminated throughout the world in an every-increasing and intensifying network of socio-economic and political interdependency" (1985: 309)—had a differential impact between centres and peripheries. In the industrialized countries it was eventually to bring about increased diversity for available foods and improved diets for lower as well as upper social ranks; in the less industrialized world, in contrast, the same process has led through commercialization to concentration in many regions on only a few cash crops with a concomitant reduction of food diversity. In a shorter-term period of transition, the same sort of contrast could be seen *within* the countries of Europe. This conflict between national markets and local needs was one reason why food riots were still common in eighteenth-century England and France (Tilly, 1975: 380–455; Rudé, 1964; Thompson, 1971; Cobb, 1970).

Another reason was more important: what could not immediately disappear with general famines was the fear of going hungry engendered by centuries of experience. Mandrou observes that one of the most characteristic features of early modern Europe was

> the obsession with starving to death, an obsession which varied in intensity according to locality and class, being stronger in the country than in the town, rate among the upper-classes and well-fed fighting men, and constant among the lower classes. (1961: 26–7)

The themes of starvation, child abandonment and outright cannibalism so common in European folklore are further evidence of the pervasive fear of food scarcity.[5] So equally, as Jacques Le Goff (1964) has argued, were the countering themes of the *mythes de ripaille* ("myths about having a good blow-out") found in early peasant folklore, becoming by the thirteenth century a literary theme in the French fable *Cocaigne* and the English poem *The Land of Cockaygne,* and the food miracles which multiplied around many saints. Both sets of themes, though superficially opposites, are signs of deep-rooted fears which could not disappear overnight. As late as 1828, notes Cobb (1970: 215), dearth was still being written about as a major threat to public order in France, because "the fear of dearth was permanent, especially at the lower levels of society, and it took very little at any time for this fear to become hysterical and to develop into the proportions of panic."

EXTERNAL CONSTRAINTS ON APPETITE: CHURCH, STATE AND DOCTORS

In these circumstances, self-control over appetite was scarcely a pressing problem for the vast majority of Europeans from medieval until relatively recent times. At first glance, in medieval and early modern Europe there might appear to be at least three sources of pressures towards self-control over appetite: first, the large number of fasts expected of the fervent Catholic; second, the sumptuary laws which apparently demonstrated the interest of states in suppressing gluttony; and, thirdly, medical opinion. I shall, however, argue that each of these represented a form only of external constraint (*Fremdzwang* was Elias's original word), and only very gradually did this come to be accompanied by a considerable measure of self-restraint *(Selbstzwang).*

Fasting

Fasting was in theory required on three days a week (Wednesday, Friday and Saturday), on the vigils for major saints days, for three days at each of the Quarter Days, and for the whole of Lent except Sundays. Strict fasting consisted essentially of eating only once in twenty-four hours, after Vespers, and as far as possible then eating only bread and water. But, of course, for all but the most ascetic, fish was permitted, as were vegetables, but wine as well as meat and any other animal product were excluded (Franklin, 1887–1902: VIII, 124ff.; Henisch, 1976: 28–50). As time passed, the Church made more and more exceptions, such as permitting eggs to be eaten on fast days—and made the requirements less stringent, but the rules were still in principle in force in Catholic countries in the late eighteenth century. After the Reformation, the Protestant churches generally disapproved of fasting on specific days as an integral part of Catholic ritual. In a characteristic compromise, the Elizabethan Church of England frowned on fasting as a form of display, though it allowed that at the discretion of individuals it could be a useful adjunct to prayer; and it adjured Christians to observe fasts decreed by law, not for religious but for political reasons:

> as when any realm in consideration of the maintenance of fisher-towns bordering upon the seas, and for the increase of fishermen, of whom do spring mariners to go upon the sea, to the furnishing of the Navy of the Realm, whereby not only commodities of other countries may be transported, but also may be a necessary defence to resist the invasion of the adversary. (Homilies, 1562: 300; cf. O'Hara-May, 1977: 122ff.)

Yet even when and where the Church's authority fully upheld the ritual of fasting, how much difference did it effectively make to how much people actually ate? The majority of people would have considered themselves fortunate if there was meat to eat as often as four days a week. Nor did the rules of fasting do anything to impede their enjoyment of the great binges which at times of plenty relieved the monotony and sparsity of their usual diet. As for the minority for whom plenty was not exceptional, they could eat sumptuously even on *jours maigres,* breaking not the letter but merely the spirit of the fasting rules. How little abstinence a dinner on a fish day might represent is suggested by the vigil dinner set before Sir Gawain on Christmas Eve:

> Several fine soups, seasoned lavishly
> Twice-fold, as is fitting, and fish of all kinds—
> Some baked in bread, some browned on coals,
> Some seethed, some stewed and savoured with spices,
> But always subtly sauced, and so the man liked it.
> The gentle knight generously judged it a feast,
> And often said so, while the servers spurred
> him on thus
> As he ate
> "This present penance do;
> It soon shall be offset." (1974: 54–5)

Much later, French courtly recipe books of the seventeenth and eighteenth centuries also show what could be achieved within the rules on *jours maigres*. In fact, the observance of fasting in the medieval and early modern period has all the hallmarks of *Fremdzwang* rather than *Selbstzwang*. That is to say, there is very little evidence of people having internalized the controls the rules embodied; few evidently felt any personal guilt or repugnance at breaking the rules. In any case, the prescribed fasts in their full severity were probably only ever observed in some religious orders.[6] And such exceptional instances of extreme abstinence are indeed a symptom of the unevenness of controls over eating. This general unevenness of controls is, according to Elias, typical of socially highly unequal societies, and Jack Goody has specifically pointed to fasting as characteristic of hierarchical societies:

> The other side of hierarchical cuisine was the extended notion of the fast, a rejection of food for religious, medical or moral reasons. . . . Abstinence and prohibition are widely recognized as ways of attaining grace in hierarchical societies such as Chin and India. . . . Such a philosophy of rejection could develop only within the context of hierarchical cuisine since abstention only exists in the wider context of indulgence. (1982: 116–17)

Very gradually there was to take place a process of development toward controls over appetite which, to use a phrase of Elias's, were both "more even and all-round"—meaning that individuals acquired the capacity typically to be able to exercise more consistent self-control, and that the controls came to apply more uniformly to people in all strata of society. But in this process, the teachings of the Church seem not to have played any very significant part.[7]

Sumptuary Laws

Perhaps more significant than the Church's teaching is that from the late Middle Ages onwards the secular authorities in England, France and other countries showed their concern to discourage over-elaborate banqueting by enacting sumptuary laws. That the problem was seen as one of social display, not of sheer physical appetite, can be seen from the fact that such laws often sought to control the clothes people wore as well as the food they ate (see Baldwin, 1926; Boucher d'Argis, 1765). The enactment of these laws is possibly a consequence of European society becoming somewhat more open. Enormous banquets were perhaps acceptable when given by feudal lords sharing their viands by custom and obligation with their followers and distributing remains to the poor, but were seen as excess and mere social display when copied by rising strata whose social obligations were ill-defined and dependents few. Not that sumptuary laws were ever effective. Like many other laws before the seventeenth century, the same law was often re-enacted at frequent intervals without ever being effectively enforced; the states simply did not have the power to enforce them. In France, a law of 1563 forbade even private families to have meals of more than three courses, and the number and type of dishes to constitute each course was also specified in detail. But very much the same law had to be re-enacted in 1565, 1567, 1572, 1577, 1590, 1591, and finally in 1629 (Franklin, 1887–1902: I, 102). In England, Archbishop Cranmer and his bishops agreed in 1541 on very detailed rules carefully grading the number of courses and number of dishes which the archbishops, bishops, deans, archdeacons and junior clergy might eat; but Cranmer appends a sad little memorandum "that this order was kept for two or three months, till, by the disusing of certain wilful persons, it came again to the old excess" (Combe, 1846: 491).

Medical Opinion

It would be equally incautious to overemphasize the influence of medical opinion, or of the rationalization of medical knowledge, in pressurizing people to exercise self-control over appetite. One of the major thrusts of *The Civilizing Process* is to demonstrate that

"'Rational understanding' is not the motor of the 'civilizing' of eating or of other behaviour" (Elias, 1978: I, 116; see also Goudsblom, 1979). Throughout the Middle Ages medical opinion, dominated by the views of the Salerno School, had favoured moderation in eating in the treatment of numerous illnesses. Doctors were certainly aware of the medical dangers of obesity, although they tended to interpret it as a result of inactivity and laziness rather than of overeating per se (O'Hara-May, 1977: 127). But medical opinion is and was brought to bear most effectively on the ill, and there is little evidence to suggest that their opinion had much effect on the daily eating habits of the normally healthy.

Although the social power of the medical profession was growing during the eighteenth century, it is too easy to follow Foucault in looking too hard for dramatic *ruptures*, and thus to exaggerate both the profession's power and the novelty of its opinions at this period. Certainly, as Jean-Paul Aron has shown (1961: 971–7), the notion of *régime alimentaire* began to be prominent in medical circles during the eighteenth century, and was reflected in the writings of Rousseau, who favored moderation and pure foods. Early in the century both in England and France a number of doctors advocated strict diets as a way to health. Bryan Turner has focused particularly on the writings of George Cheyne (1724, 1733), and a little earlier in France Philippe Hecquet (1709) propounded similar ideas in the famous controversy. Jones and Sonenscher (1983) describe how, later in the century, the diet of hospital inmates was the subject of conflict between doctors and nurses at the Hôtel-Dieu in Nîmes. The nursing sisters had traditionally seen their role as a charitable one and, aware that many illnesses had resulted from repeated subsistence crises, saw it as their duty to feed up the poor and needy ill. One of the doctor's at Nîmes complained bitterly against the overplentifulness of the patients' diet, which often impeded their recovery. "They are always afraid in this hospital that people will die of hunger . . . they always feed the sick too much." A colleague in neighbouring Montpelier in the 1760s documented how overfeeding by the sisters had led to patients' premature deaths, and "gave the impression that over-eating was one of the major causes of hospital mortality!" Significantly, the doctors in eighteenth-century Montpelier also launched an onslaught on the tradition of marking the hospital's patron saint's day with feasting.

All the same, it is unsound to pursue an explanation in terms of a few artificially isolated causal "factors." A figurational investigation looks first for the sorts of problems people encounter within the webs of social interdependence in which they are caught up. In this case, it is well to remember that the problem of appetite in relation to overabundant food had still scarcely arisen in the eighteenth century for the great majority of the people of western Europe; for them the most pressing external constraints on appetite were still the shortage or irregularity of food supplies. As for the minority for whom the problem had already arisen they had begun to show signs of adapting to it before any dramatic shift in medical opinion. (There is of course no reason why medical opinion, as one thread in a complex process of development, should not be both cause and effect in various ways and at different stages of the process.)

QUANTITY AND QUALITY

When food was scarce for most and supplies insecure and irregular for nearly all, the powerful distinguished themselves from their inferiors by the sheer quantities they ate: "those who could, gorged themselves; those who couldn't, aimed to" (Weber, 1973: 202). Evidence for this is found in one of the most detailed studies of diet in the late Middle

Ages, Stouff's study of Provence. Stouff shows that in various ecclesiastical communities, not only did those in the higher echelons eat proportionately more meat, fish and other proteins in relation to bread and wine than did their inferiors, they also ate a great deal more overall. In one case study typical of the general pattern,

> One conclusion must be drawn: in 1429 (and it appears to be equally true throughout the fifteenth century), the food intake of the Archbishop of Aries and the senior members of his household was too large, but relatively well balanced. (Stouff, 1970: 238)

In the sixteenth and seventeenth centuries, there were many who seem to have been noted more for their capacity than for their refinement of taste. Cathérine de Medici was celebrated for her appetite and frequent indigestion. Diarists at the court of Louis XIV have left graphic accounts of the great king's prodigious consumption. Nor does he appear to have been untypical of his court.[8]

Faint traces of the beginnings of pressures toward self-restraint in appetite can be seen a century earlier. In appetite as in so many other facets of the civilizing process, Montaigne is a good witness (Mennell, 1981). He reports that he himself has little self-restraint in eating, but bemoans the fact:

> if they preach abstinence once a dish is in front of me, they are wasting their time. . . . To eat greedily as I do, is not only harmful to health, and even to one's pleasure, but it is unmannerly into the bargain. So hurried am I that I often bite my tongue, and sometimes my fingers. . . . My greed leaves me no time for talk. (Montaigne, 1867: 445)

By the mid-eighteenth century extreme gluttony appears to have become the exception. Louis XVI, who saw off chicken, lamb cutlets, eggs, ham and a bottle and a half of wine before setting out to hunt, without it diminishing his appetite at dinner, appears to have been considered something of a throwback:

> By his appetite, and by his appetite alone did the unfortunate Louis XVI revive memories of Louis XIV. Like him, he did not bother himself with cookery, nor with any refinements; to him, always afraid of not having enough to eat, sheer quantity was more important than anything else; he did not eat, he stuffed himself, going as far as to incapacitate himself at his wedding dinner, scandalizing his grandfather [Louis XV]. (Gottschalk, 1939: 232)

Even in England, another famous trencherman of that time, Dr. Johnson, though of less exalted social rank, was also considered a coarse eater. Not only did he show so little sense of what was proper as to call for the boat containing the lobster sauce left over from the previous course and pour it over his plum-pudding (Piozzi, 1785), but he wolfed his food down in a shameful manner:

> When at table, Johnson was totally absorbed in the business of the moment; his looks seemed rivetted to his plate; nor would he, unless when in very high company, say one word, or even pay the least attention to what was said by others, till he had satisfied his appetite, which was so fierce and indulged with such intenseness that while in the act of eating, the veins of his forehead swelled and generally a strong perspiration was evident. To those whose

> sensations were delicate, this could not but be disgusting; and it was doubtless not very suitable to the character of a philosopher, who should be distinguished by self-command. (Boswell, 1791: I, 323)

Significantly, Boswell comments that everything about Johnson's character and manners was forcible and violent, and adds

> Johnson, though he could be rigidly *abstemious,* was not a *temperate* man either in eating or drinking. He could refrain, but he could not use moderately.

That sounds very much like a throwback to the mode of behavior typical of medieval and early modern Europe. But by the mid-eighteenth century it was no longer considered quite the right thing in the better circles. What changes were taking place?

The civilizing of appetite, if we may call it that, appears to have been partly related to the increasing security, regularity, reliability and variety of food supplies. But just as the civilizing of appetite was entangled with several other strands of the civilizing process including the transformation of table manners, so the improvement in food supplies was only one strand in a complex of developments within the social figuration which together exerted a compelling force over the way people behaved. The increased security of food supplies was made possible by the extension of trade, the progressive division of labour in a growing commercial economy, and also by the process of state-formation and internal pacification. Even a small improvement was enough to enable a small powerful minority to distinguish themselves from the lower ranks of society by the sheer quantities they ate and the regularity with which they ate them. As the improvement continued, somewhat wide segments of the better-off groups in society came to be able to copy the élite. The same structural processes, however, served not only to permit social emulation but positively to promote it. The longer chains of social interdependence produced by state-formation and the division of labour tended to tilt the balance of power little by little towards lower social groups, leading to increased pressure "from below" and to intensified social competition. The sumptuary laws, with their vain attempt to relate quantities eaten to social rank, seem symptomatic of that.

By the sixteenth or seventeenth centuries, for the nobility to eat quantitatively more would have been physically impossible.[9] That was one reason for increasing demands made upon the skill of the cook in making food more palatable; as a modern expert explains,

> A variety of studies demonstrates that hunger and palatability are substitutive for each other and algebraically additive in their effects. Equal amounts are eaten of a highly palatable food in a minimal state of hunger and even without hunger, and of a minimally palatable food in a state of hunger. Thus it is equally true to assume that hunger potentiates palatability and that palatability potentiates hunger in their common effect of eliciting eating. The consequence of this relationship is that the differential palatability of two foods decreases with increased hunger. (Le Magnen, 1972: 76)

Or, as Andrew Combe wrote in a nineteenth-century classic of dietetics,

> Appetite . . . may . . . be educated or trained to considerable deviations from the ordinary standard of quantity and quality . . . The most common source . . . of the errors into which we are apt to fall in taking appetite as our only guide, is unquestionably the *confounding of appetite with taste,* and continuing to eat for the gratification of the latter long after the former is satisfied. In fact, the whole science of a skilful cook is expended in producing this *willing* mistake on our part. (Combe, 1846: 29–30)

Here, then, is the psychological basis for the elaboration of cooking in an age of plenty. And the skills of cooks had another advantage: they could be applied not simply to stimulating the sated appetites of the glutton, but also to the invention and elaboration of an endless variety of ever more refined and delicate dishes; when the possibilities of quantitative consumption for the expression of social superiority had been exhausted, the qualitative possibilities were inexhaustible.

The links between the changing social figuration, changing patterns of social contest, the changing arts of the cook, and the civilizing of appetite are most clearly discernible, like so many facets of civilizing processes, in France. The development there of "court society" was particularly significant (Elias, 1982; Mennell, 1985, Ch. 5). The revenues, political power and social functions of the old *noblesse d'épée* were gradually declining, while those of the bourgeoisie and of the essentially bourgeois *noblesse de robe* were increasing. Parts of the old nobility acquired positions at court and became highly dependent on royal favour. They became in effect specialists in the arts of consumption, entrapped in a system of fine distinctions, status battles and competitive expenditure from which they could not escape because their whole social identity depended upon it. They were under constant pressure to differentiate themselves from the *robins,* the despised *noblesse compagnarde,* and the bourgeoisie. How was this reflected in eating?

The break with medieval cookery seems to have begun in the city courts of Renaissance Italy, but the leadership in matters of culinary innovation seems to have passed to France in the late sixteenth or early seventeenth centuries (see Mennell, 1985, Ch. 4). By early in the reign of Louis XIV, the beginnings of modern French cuisine are visible in the more refined techniques, the less exuberant use of ingredients, and the greater variety of dishes given in a book like La Varenne's *Le Cuisinier François* of 1651. Another period of rapid development followed in the next reign. The gluttony of Louis XIV and many of his courtiers was replaced by the delicate *soupers* for which the Regent was noted. Indeed the Regent himself, like several others among the high nobility, seems himself to have been an expert cook, and his mother the Princess Palatine implies that this was a part of *bon ton* which could be ranked with skills in other arts like music:

> My son knows how to cook; it is something he learned in Spain. He is a good musician, as all musicians recognize; he has composed two operas, which he had produced in his chambers and which had some merit, but he did not want them to be shown in public. (Orléans, 1855: I, 349–50)

The change of fashion during the eighteenth century away from quantitative display towards more varied and delicate ragouts is noted by Louis-Sebastien Mercier in 1783:

> In the last century, they used to serve huge pieces of meat, and pile them up in pyramids. These little dishes, costing ten times as much as one of those big ones, were not yet known. Delicate eating has been known for only half a century. The delicious cuisine of the reign of Louis XV was unknown even to Louis XIV. (1783: V, 597–8)

The sense of delicacy and pressures towards self-control are, as Elias has shown, closely interwoven. In eating it is the developing sense of delicacy which first becomes apparent, but that eventually becomes entangled with restraint. In the late sixteenth century Montaigne, who as we have already seen claimed to have little self-restraint over his own eating, also poked fun in his *essai* on "La vanité des paroles" at Cardinal Caraffa's Italian chef for the gravity with which he held forth on the propriety of courses and sauces, sequences of dishes and balances of flavours (Montaigne, 1967: 134–5). By the time of La Varenne, French cooks were at least as much concerned with such matters as their Italian forerunners. And only a couple of decades later, the next generation of French cookery writers spoke of La Varenne's meals and dishes as coarse and rustic. Molière mocks the seriousness with which these growing conventions were taken, and their social significance (see *Le Bourgeois Gentilhomme,* Act 4; L'Avare, Act 3). By the middle of the eighteenth century the first truly gastronomic controversies were taking place, in which defenders of old styles of cooking and eating railed against the preciousness, pretentiousness and over-developed sense of culinary propriety of the proponents of the *nouvelle cuisine.*

By then too, larger segments of the bourgeoisie were seeking to copy the courtly models of refined and delicate eating, and this probably gave increased impetus to the movement towards greater delicacy and self-restraint. The connections are complex. We have noted that courtly fashion moved towards the proliferation of small, delicate and costly dishes, and that knowledgeability and a sense of delicacy in matters of food became something of a mark of the courtier. Now a sense of delicacy implies a degree of restraint too, in so far as it involves discrimination and selection, the rejection as well as the acceptance of certain foods or combinations of foods, guided at least as much by social proprietaries as by individual fancies. No courtly gourmet would pour the lobster sauce over his plum pudding. But while the development of systems of fashionable preferences involves a degree of rationalization, what Elias calls "court-rationality" was antithetical to that of bourgeois economic rationality; lavish consumption was too closely part of the courtier's social identity for him to economize like a good bourgeois. While there is plenty of evidence that, in France at least, the bourgeoisie wanted in the eighteenth century to follow courtly models of eating, it is also clear that most did not have the resources to eat on such a lavish scale; they were therefore both under more pressure than the nobility to choose and select, and also more easily able to do so. The bourgeoisie was in many ways a more appropriate *couche* for the emergence of a body of gastronomic theorizing. Moreover, given that a fairly high degree of internal pacification and a measure of economic surplus are prerequisites for the development of the cultural syndrome of bourgeois rationality as a whole, it seems no coincidence that gastronomic theorizing as a genre first appeared during the period when the insecurity of food supplies ceased to be of catastrophic proportions, and burgeoned fully during the nineteenth century.[10] At any rate, when it did emerge, the theorists were indeed mainly members of the high bourgeoisie, and the themes of delicacy and self-restraint were prominent in their writings, the latter increasingly so as time went on.

GASTRONOMY AND MODERATION

Neither Grimod de la Reynière (1803–12) nor Brillat-Savarin (1826), the two most noted pioneers of gastronomy, entirely dismissed a large capacity as an epicurean virtue. But their writings emphasize the need for a discriminating palate and scorn as vulgar any merely quantitative display. They set the pattern for gastronomic writing in both France and England for the rest of the century. An Englishman strongly influenced by Grimod writes in 1822:

> Gluttony is, in fact, a mere effort of the appetite, of which the coarsest bolter of bacon in all Hampshire may equally boast with the most distinguished consumer of turtle in a Corporation: while Epicurism is the result of "that choicest gift of Heaven," a refined and discriminating taste: this is the peculiar attribute of the palate, that of the stomach. It is the happy combination of both these enviable qualities that constitutes that truly estimable character, the real epicure. He is not only endowed with a capacious stomach and an insatiable appetite, but with a delicate susceptibility in the organs of degustation, which enables him to appreciate the true relish of each ingredient in the most compound ragout, and to detect the slightest aberration of the cook; added to which advantages, he possesses a profound acquaintance with the rules of art in all the most approved schools of cookery, and an enlightened judgment on their several merits, matured by long and sedulous experience. (Sturgeon, 1822: 3–4)

A few decades later, in 1868, another writer bemoans England's lagging behind France in gastronomic *savoir-faire,* and now directly disparages the lack of discrimination masked by plenty:

> Not only our merchant princes, but our gentry and nobility, have merely a superficial knowledge of the science of cookery and the art of giving good dinners. Consider the barbarism implied in the popular phrase for ample hospitality! The table is described as groaning under the plenty of the host. (Jerrold, 1868: 5)

By the twentieth century, the theme of moderation was still more explicit. G. F. Scotson-Clark, in a book entitled *Eating without Fears* published in 1924, writes that

> Consuming large quantities of food is only a habit. What is often called a "healthy appetite" is nothing of the sort. The only people who should eat really large quantities of food are those whose regular daily life involves a vast amount of physical exercise—like the road-mender. (1924: 65)

And André L. Simon reiterates an argument prominent in his extensive writings between the 1930s and 1960s.

> There cannot be any intelligent choice nor real appreciation where there is excess. Gastronomy stands or falls by moderation. No gourmand and no glutton can be a gastronome. (1969: 94)

Gradually moderation became more clearly linked to questions of health as well as discrimination. Scotson-Clark says:

> Cookery plays such a large part in our life, it is really the fundamental basis of our life, our very existence, that it is foolish to belittle its importance. To take no interest in it is as bad for one's health as to take no interest in one's ablutions. An individual should cultivate his palate just as much as he should cultivate his brain. Good taste in food and wine is as necessary as good taste in art, literature and music, and the very fact of looking upon gastronomy as one of the arts will keep a man from becoming that most disgusting of creatures, a glutton . . .
>
> I am sure that moderation is the keynote of good health, and I contend that anyone can eat anything I mention in this book, without increasing his girth, and if taken in moderation he can reduce to normal weight. It is not necessary for one to deprive oneself of all the things one loves, for fear of getting too fat, but it is necessary to take an intelligent interest in the provender with which one intends to stoke the human furnace. (1924: 8–9)

At about the same time in France, Edouard de Pomiane, the medical doctor turned cookery writer, was developing similar themes in books (1922) and in the popular press. Although dieting for health and slimness became a prominent concern in mass-circulation publications like women's magazines (see Mennell, 1985, Ch. 9) only after the Second World War, the slim body-image had begun to appeal in higher social circles considerably earlier.

THE FEAR OF FATNESS

It would be interesting to know whether fatness was common and whether it carried any stigma in medieval and early modern Europe. The evidence is not entirely unambiguous. Kunzle (1982: 65) traces the ideal of the slender female figure as far back as courtly circles in the later Middle Ages, but it is easy to find literary evidence of plumpness being considered attractive. As for visual evidence, Jane O'Hara-May (1977: 127) argues that paintings show relatively few very fat people, and suggests that the frequent use of purges and the large amount of exercise which in this period even the wealthy could scarcely avoid tended to balance excessive intake. In contrast, Kristoff Glamann draws precisely opposite conclusions from portraits, and states that corporal bulk was in all ranks of society a source not of shame but of prestige.

> Eating made one handsome. A thin wife brought disgrace to a peasant. But of a plump wife it was said that "a man will love her and not begrudge the food she eats." Men too ought to be stout. That this ideal was not confined to the rustic world is plain from a glance at the magnificent amplitude of the human frame so abundantly depicted by the Renaissance painters. (Rich and Wilson, 1977: 195)

The contradictory conclusions about average girth in paintings point to the need for more systematic studies. But on the more general question of the prestige or otherwise of bodily bulk, the most likely conclusion is that while obesity which impeded health and activity was deplored (particularly by the doctors whom O'Hara-May is studying), a healthy stoutness was widely considered prestigious.

The problem and the fear of being overweight seems, not surprisingly, to have started towards the top of the social scale and progressed steadily downwards. The "magnificent amplitude of the human frame" which once constituted the cultural model in Europe—and still does in many societies where poverty is rife—was gradually replaced by the ideal of the slim figure. The changing standard of beauty among the upper strata can be seen around the time of the Romantic movement, when "for both women and men paleness, frailness, slenderness became the vogue" (Young, 1970: 16). Burnett (1966: 80) quotes some fairly abstemious diets recommended for well-to-do ladies at that period. Up to the end of the Edwardian era, as Dally and Gomez (1979: 25) point out, many successful men tended to be rather stout, but today there tend to be lower rates of obesity among the upper socio-economic groups.

Exactly when the ideal began to be reflected in an actual decline in typical body weights, and how the decline progressed down the social scale, is very difficult to demonstrate. Quite a lot of historical evidence is available about people's *heights* (Fogel et al., 1985) but, given the complexities of relating body weights to height, age and sex, let alone to social class, little in the way of time-series data over the long period required is available or likely to become available. An interesting clue is an article by Sir Francis Galton in *Nature,* 1884, comparing the weights of three generations of British noblemen among the customers of Berry Brothers, grocers and wine-merchants in St. James's, London, from the mid-eighteenth century to the late nineteenth (see Figure 1). This evidence is far from conclusive, but it does suggest that by the late nineteenth century men in the highest stratum of English society were no longer putting on weight so rapidly as young men as their fathers and grandfathers had done. They reached the same weight in the end, but possibly this is consistent with them having over-eaten slightly but persistently rather than indulging in dramatically excessive overeating.

Whatever happened to actual body weights, however, there is plenty of evidence of the worry the subject caused in the upper reaches of society. Gastronomic writers from Brillat-Savarin to Ali-Bab (1907) discussed obesity as a worry and affliction among gourmets. In the latter part of the nineteenth century, great innovating chefs such as Escoffier, Philéas

Figure 1 Mean Age—Weight of British Noblemen in Three Successive Generations

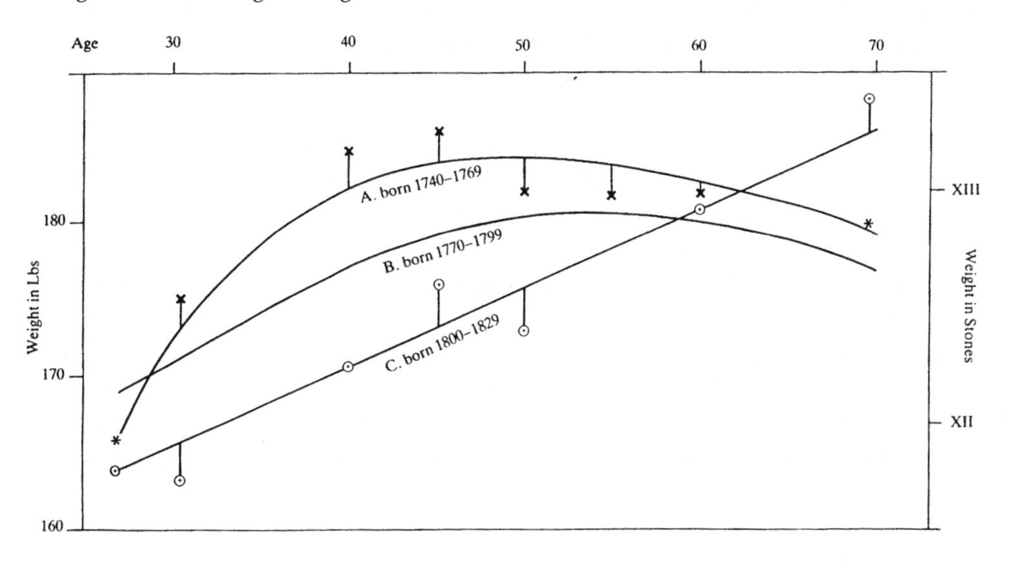

Gilbert and Prosper Montagné, cooking for a fashionable clientele, were beginning a trend towards simpler, lighter food and fewer courses. Yet at the same time, books were still being written on how to put on weight (for example, T. C. Duncan, *How to Become Plump,* 1878), and the cookery columns addressed to the lower middle classes (especially in England) emphasized the need to eat fat and heavy food for body-building (Mennell, 1985, Ch. 9). The upper and upper-middle classes often commented on the greed of servants:

> In towns we often observe the bad effects of overfeeding in your female servants recently arrived from the country. From being accustomed to constant exercise in the open air, and to the comparatively innutritious diet on which the labouring classes subsist, they pass all at once, with appetite, digestion and health in their fullest vigour, to the confinement of a house, to the impure atmosphere of a crowded city, and to a rich and stimulating diet. Appetite, still keen, is freely indulged; but waste being diminished, fulness is speedily induced . . . (Combe, 1846: 217)

And, at Buckingham Palace (no less), at the turn of the century:

> The plentiful meals of those days naturally enough encouraged greed, particularly among some of the servants. After a five-course breakfast those who visited the kitchens often slipped two or three hardboiled eggs into their pockets to help them last out the next few hours until it was time for morning tea. (Tschumi, 1954: 63)

It is hardly surprising if people drawn from ranks of society where the fear for centuries had been simply getting enough to eat did not immediately develop self-control when suddenly confronted with plentiful food.

Even at the present day, in the world's affluent societies the incidence of obesity is highest in the lower and poorer strata, in contrast to the countries of the Third World where it occurs only among the privileged few (Bruch, 1974: 14). Obviously the plentiful availability of food is a prerequisite for the development of obesity, but clinical evidence suggests that psychological pressures to overeat are often rooted in past hunger, perhaps in a previous generation. For instance, among the mothers of obese children in America,

> Many of these women had been poor immigrants who had suffered hunger during their early lives. They did not understand why anyone should object to a child's being big and fat, which to them indicated success and freedom from want. (Bruch, 1974: 15)

Conversely, cases of anorexia nervosa arise disproportionately among the more well-to-do strata. There may have been instances of this affliction, which is far more common among females than males, in earlier centuries. Neither classical literature nor the Bible, according to Dally and Gomez (1979: 1), contains any recognizable picture of anorexia nervosa, nor does it seem to have been known in the Middle Ages. A number of "miraculously fasting" girls are known to have excited attention from the sixteenth century onwards, and though several were probably frauds, some were possibly cases where psychological disturbance led to serious undereating (Morgan, 1977). The first reliable description of cases seems to have been by Morton in 1694, but the condition did not attract much medical interest and was probably not at all common until the latter half of the nineteenth century, when it was named by Sir William Gull (1874). Gull in England and E. C. Lasègue (1873) in France both gave clear accounts of it among their middle-class patients at that period. Today, it is a very familiar illness in Europe and North America. Again, there appears to be a clear connection with the reliable and plentiful availability of food: apparently anorexia nervosa is not reported from countries where there is still danger of widespread starvation or famine, nor among blacks and other underprivileged groups in the United States (Bruch, 1974: 13).

Anorexia nervosa and obesity can be regarded as similar if opposite disturbances of the normal patterns of self-control over appetite now normally expected and necessary in prosperous Western societies. Though the process may not yet be complete, in the course of the twentieth century the concern with weight-watching and slimming has gradually become more widespread in all ranks of society: its progress can be observed in cookery columns in popular magazines. For example, ever since the early 1950s, the French women's magazine *Elle* has had weekly columns giving menus and recipes with calorie counts, playing on and encouraging the reader's concern with her own weight and that of her family. A typical early instance is an article in *Elle*, February 1953, entitled "Unconscious Overeating can Threaten your Life," with a photograph of a slim girl in a swimsuit to illustrate the prevailing body image. The not-very-subliminal connection between self-control over the appetite, slimness, health, and sex appeal is one of the most salient themes in British as well as French mass-circulation women's magazines since the Second World War. Which is not to deny that persistent slight but definite overeating remains a characteristic problem among the populations of England, France and other Western industrial states. But a general anxiety to avoid obesity is very widespread, and the fitful extreme over-eating of an earlier era seems less common.

CONCLUSION

The process of the civilizing of appetite is in detail more complex than it has been possible to depict here. Very broadly speaking, the argument is that the increasing interdependence and more equal balances of power between social classes have been reflected in more equal distribution of foodstuffs, which in turn has been associated with somewhat greater similarity of cuisine, and also with less extreme differences between festival or banquet food and everyday eating; and that these changes have been accompanied by the growth of pressures towards more even and "all-round" self-controls over appetite. It has not, however, been a simple linear development; in fact there have been spurts and reversals, exceptions and sub-themes. For example, the seasonal rhythm and ritual of early modern eating, the observance of festivals and eating of festival fare, persisted longer in the countryside than in the towns,[11] where the special dishes of an earlier age have become the commonplace dishes of industrialized eating. At the opposite end of the social scale, a figure like King Edward VII might convince us that in the early twentieth century nothing had greatly changed since the carnivorous accomplishments of the medieval nobility. Yet within a couple of decades of his death, British royalty too was eating relatively abstemiously food which would not be very unfamiliar to most of their subjects (Magnus, 1964: 268–9; Tschumi, 1954). If it has not been possible here to pursue every detail and complexity of what I believe is one more example of a long-term civilizing process, I hope this paper has served to suggest the fruitfulness of applying figurational sociology—and in particular ideas derived from *The Civilizing Process*—to the development of eating and appetite.

NOTES

This paper is one of the products of research undertaken mainly in 1980–1 with the aid of a grant from the Nuffield Foundation. One earlier version was presented at a conference on "Civilization and Theories of Civilizing Processes" at the Zentrum für Interdisziplinäre Forschung, Bielefeld, 15–17 June 1984, and another forms part of my book *All Manners of Food* (1985). The argument has, however, been considerably clarified in the present paper.

1. Henry IV was one culprit (Franklin, 1887–1902: III, 115), and in February 1558, the Pope gave a banquet while people were dying of hunger in Rome (Weber, 1973: 194).
2. Although this is true, it should also be noted that historians no longer believe that noble households had to exist throughout the winter on salt meat following the "autumn slaughter" at Martinmas: some meat was salted, certainly, but some fresh meat was also generally available (see Dyer, 1983: 193). Again, however, the extent of autumn slaughter would probably be related to the success or failure of the harvest.
3. For more details of the cycle of feasts in medieval and early modern Europe, see Burke, 1978: 194–6; Coulton, 1926: 28–30; and Henisch, 1976: 50–1.
4. For a recent study of local famines in northwest England, see Appleby, 1978.
5. Nor should it be thought that fantasy was always necessarily very distant from reality: at least some incidents of cannibalism in time of famine seem reasonably well-authenticated—see Curschmann, 1900: 59–60.
6. And observed by no means in all. Accounts abound of monastic gluttony. See Alfred Gottschalk (1948: I, 343) who quotes St Bernard's denunciation of monks' gluttony.
7. For further discussion of religious views of gluttony in the medieval and early modern period, see Mennell, 1985: 29–30.
8. The Princess Palatine often describes the gorging of the French nobility, though the Duchesse de Berri's eating herself to death seems to have been even then considered an instance of a pathologically abnormal appetite (Orléans, 1855: I, 348; II, 54, 85, 131, 145).
9. Cf. Eli Hecksher on the Swedish nobility, cited by Glamann, in Rich and Wilson (1977: 195).
10. For a more adequate discussion of the social context of the emergence of gastronomes, see Mennell, 1985; 142–3, 265ff.
11. For a general impression supporting this point, see Thomas Hardy, 1874; Oyler, 1950; Guillaumin, 1905; for more scholarly evidence regarding France, see Tardieu, 1964; Claudian et al., 1969; Weber, 1977.

REFERENCES

Anon. (c. 1400) *Sir Gawain and the Green Knight*. Harmondsworth: Penguin, 1974.

Ali-Bab (Henri Babinski) (1907) *Gastronomie Pratique: Etudes Culinaries suivies du Traitement de l'Obésité des Gourmands*. Paris: Flammarion.

Appleby, A. B. (1978) *Famine in Tudor and Stuart England*. Liverpool: Liverpool University Press.

Aron, Jean-Paul (1961) "Biologie et alimentation au XVIIIe siècle et au debut du XIXe siècle," *Annales* E-S-C, 16 (2): 971–7.

Aron, Jean-Paul (1973) *Le Mangeur du 19e siècle*. Paris: Laffont.

Baldwin, F. E. (1926) *Sumptuary Legislation and Personal Regulation in England*. Baltimore: Johns Hopkins Press.

Bloch, Marc (1939–40) *Feudal Society*. 2 vols. London: Routledge & Kegan Paul, 1961.

Boswell, James (1791) *The Life of Samuel Johnson, LLD*. 2 vols. London: Odhams Press, n.d.

Boucher d'Argis, A.-G. (1765) "Lois somptuaires," in D. Diderot (ed.), *Encyclopédie*. Neufchâtel: Samuel Faulche, 1751–80, IX: 672–5.

Brillat-Savarin, J.-A. (1826) *La Physiologie du Goût*. Paris: A. Sautelet.

Bruch, Hilde (1974) *Eating Disorders: Obesity, Anorexia Nervosa and the Person Within*. London: Routledge & Kegan Paul.

Burke, Peter (1978) *Popular Culture in Early Modern Europe.* London: Temple Smith.
Burnett, John (1966) *Plenty and Want.* London: Nelson.
Cappon, Daniel (1973) *Eating, Loving and Dying: A Psychology of Appetites.* Toronto: University of Toronto Press.
Cheyne, George (1724) *An Essay of Health and Long Life.* London: G. Strachan.
Cheyne, George (1733) *The English Malady.* London: G. Strachan.
Claudian, J., Serville, Y. and Tremolières, F. (1969) "Enquête sur les facteurs de choix des aliments," *Bulletin de l'INSERM,* 24(5): 1277–1390.
Cobb, Richard (1970) *The Police and the People: French Popular Protest, 1789–1820.* Oxford: Clarendon Press.
Codere, Helen (1950) *Fighting with Property: A Study of Kwakiutl Potlatching and Warfare, 1792–1930.* Seattle: Washington University Press.
Combe, Andrew (1846) *The Physiology of Digestion, Considered with Relation to Dietetics.* Edinburgh: Maclachlan & Stewart.
Coulton, G. G. (1926) *The Medieval Village.* Cambridge: Cambridge University Press.
Curschmann, Fritz (1900) *Hungersnöte im Mittelalter.* Leipzig: B.G. Teubner.
Dally, P. and Gomez, J. (1979) *Anorexia Nervosa.* London: Heinemann.
Duncan, T. C. (1878) *How to Become Plump,* or *Talks on Physiological Feeding.* London.
Dyer, Christopher (1983) "English Diet in the Later Middle Ages," in T. H. Aston, P .R. Coss, C. Dyer and J. Thirsk (eds), *Social Relations and Ideas: Essays in Honour of R. H. Hilton,* pp. 191–216. Oxford: Past and Present Society.
Elias, Norbert (1939) *The Civilizing Process,* Vol. I: *The History of Manners.* Oxford: Basil Blackwell, 1978. Vol. II: *State Formation and Civilization.* Oxford: Basil Blackwell, 1982.
Fogel, R. W. et al. (1985) "Secular Changes in American and British Stature and Nutrition," in R. I. Rotberg and T. K. Rabb (eds), *Hunger and History,* pp. 247–83. Cambridge: Cambridge University Press.
Foucault, Michel (1973) *The Birth of the Clinic.* London: Tavistock.
Foucault, Michel (1977) *Discipline and Punish.* London: Allen Lane.
Franklin, Alfred (1887–1902) *Vie privée d'autrefois, 12e à 18e siècles.* 27 vols. Paris: Plon.
Galton, Sir Francis (1884) "The Weights of British Noblemen During the Last Three Generations," *Nature,* 29: 266–8.
Goody, Jack R. (1982) *Cooking, Cuisine and Class.* Cambridge: Cambridge University Press.
Gottschalk, Alfred (1939) "L'appetit de Louis XVI au Temple," *Grandgousier,* 6(4).
Gottschalk, Alfred (1948) *Histoire de l'alimentation et de la gastronomie, depuis la préhistoire jusqu'à nos jours.* 2 vols. Paris: Editions Hippocrate.
Goudsblom, Johan (1977) *Sociology in the Balance.* Oxford: Basil Blackwell.
Goudsblom, Johan (1979) "Zivilisation, Ansteckungsangst und Hygiene," in P. R. Gleichmann, J. Goudsblom and H. Korte (eds), *Materialien zu Norbert Elias's Zivilisationstheorie,* pp. 215–53. Frankfurt a.M.: Suhrkamp.
Grimod de la Reynière, A.-B.-L. (1803–12) *Almanach des Gourmands.* Paris.
Guillaumin, Emile (1905) *The Life of a Simple Man.* London: Selwyn & Blount, 1919.
Gull, W. W. (1874) "Apepsia Hysterica: Anorexia Nervosa," *Transactions of the Clinical Society of London, 7(2).*
Hardy, Thomas (1874) *Far from the Madding Crowd.* London.
Hecquet, P. (1709) *Traité des dispenses du carême, dans lequel on découvre la fausseté des prétextes qu'on apporte pour les obtenir, les rapports naturels des alimens maigres avec la nature de l'homme et par l'histoire, par l'analyse et par l'observation leur convenance avec la santé.* Paris: F. Fournier.
Henisch, B. A. (1976) *Fast and Feast: Food in Medieval Society.* University Park and London: Pennsylvania State University Press.
Hollingsworth, T. H. (1977) "Mortality in the British Peerage Families since 1600," *Population,* 32: 323–52.
Homilies (1562) *Certain Sermons or Homilies Appointed to be Read in Churches in the Time of Queen Elizabeth of Famous Memory.* London: George Wells, 1687.
Huizinga, Johan (1924) *The Waning of the Middle Ages.* Harmondsworth: Penguin 1972.
Jerrold, William B. (1868) *The Epicure's Yearbook and Table Companion.* London: Bradbury, Evans & Co.
Jones, C. D. H. and Sonenscher, M. (1983) "The Social Functions of the Hospital in Eighteenth-

Century France: The Case of the Hôtel-Dieu of Nîmes," *French Historical Studies,* 13(2): 172–214.

Kunzle, David (1982) *Fashion and Fetishism.* Totowa, N.J.: Rowman & Littlefield.

Lasègue, E.C. (1873) "De l'anorexie hysterique," *Archives générales médicales,* 2: 367.

La Varenne, François Pierre de (1651) *Le Cuisinier François.* Paris: chez Pierre David.

Le Goff, Jacques (1964) *La Civilisation de l'Occident Médiéval.* Paris: Arthaud.

Le Magnen, J. (1972) "Regulation of Food Intake," in F. Reichsmann (ed.), *Advances in Psychosomatic Medicine,* Vol. 7, *Hunger and Satiety in Health and Disease.* Basel: S. Karger: 73–90.

Livi-Bacci, Massimo (1985) "The Nutrition-Mortality Link in Past Times: A Comment," in R. I. Rotberg and T. K. Rabb (eds), *Hunger and History.* Cambridge: Cambridge University Press: 95–100.

Magnus, Philip (1964) *King Edward VII.* London: John Murray.

Mandrou, R. (1961) *Introduction to Modern France, 1500–1600.* London: Arnold, 1975.

Mennell, Stephen (1981a) "Montaigne, Civilization and Sixteen-Century European Society," in K. C. Cameron (ed), *Montaigne and His Age,* pp. 69–85. Exeter: University of Exeter.

Mennell, Stephen (1981b) *Lettre d'un pâtissier anglais et autres contributons à une polémique gastronomique du XVIIIe siècle.* Exeter: University of Exeter.

Mennell, Stephen (1985) *All Manners of Food: Eating and Taste in England and France from the Middle Ages to the Present.* Oxford: Basil Blackwell.

Montaigne, Michel de (1967) *Oeuvres Complètes.* Paris: Le Seuil.

Morgan, H. Gethin (1977) "Fasting Girls and Our Attitudes to Them," *British Medical Journal,* 2: 1652–5.

Morton, Richard (1964) *Phthisologia—or a Treatise of Consumptions.* London: Smith & Walford.

O'Hara-May, Jane (1977) *The Elizabethan Dyetary of Health.* Lawrence, Kansas: Coronado Press.

Orléans (1855) *Correspondance Complète de Madame la Duchesse d'Orléans.* 2 vols. Paris: Charpentier.

Oyler, Philip (1950) *The Generous Earth.* Harmondsworth: Penguin, 1961.

Pelto, G. H. and Pelto, P. J. (1985) "Diet and Delocalization: Dietary Changes since 1750," in R. I. Rotberg and T. K. Rabb (eds), *Hunger and History,* pp. 309–30. Cambridge: Cambridge University Press.

Piozzi, H. L. (1785) *Anecdotes of the Late Samuel Johnson.* London.

Pomiane, Edouard de (1922) *Bien manger pour bien vivre: essai de gastronomie théorique.* Paris: A. Michel.

Rich, E. E. and Wilson, C. H. (eds) (1977) *Cambridge Economic History of Europe,* Vol. V. Cambridge: Cambridge University Press.

Rudé, George (1964) *The Crowd in History.* London: Wiley.

Scotson-Clark, George F. (1924) *Eating Without Fears.* London: Jonathan Cape.

Sen, A. K. (1981) *Poverty and Famines.* Oxford: Oxford University Press.

Simon, André L. (1969) *In the Twilight.* London: Michael Joseph.

Stouff, Louis (1970) *Revitaillement et alimentation en Provence aux 14e et 15e siècles.* Paris: Mouton.

Sturgeon, Launcelot (1822) *Essays, Moral, Philosophical and Stomachical on the Important Science of Good-Living.* London: G. and B. Whittaker.

Tardieu, Suzanne (1964) *La Vie domestique dans le Mâconnais rurale pré-industriel.* Paris: Université de Paris.

Thomas, Keith (1971) *Religion and the Decline of Magic.* London: Weidenfeld and Nicolson.

Thompson, E. P. (1971) "The Moral Economy of the English Crowd in the Eighteenth Century," *Past and Present,* 50: 76–136.

Tilly, Charles (1975) "Food Supply and Public Order in Modern Europe," in Tilly (ed.), *The Formation of National States in Western Europe,* pp. 380–455. Princeton N.J.: Princeton University Press.

Tschumi, Gabriel (1954) *Royal Chef: Recollections of Life in Royal Households from Queen Victoria to Queen Mary.* London: W. Kimber.

Turner, Bryan S. (1982a) "The Discourse of Diet," *Theory, Culture & Society,* 1(1): 23–32.

Turner, Bryan S. (1982b) "The Government of the Body: Medical Regimens and the Rationalization of Diet," *British Journal of Sociology,* 33(2): 254–69.

Turner, Bryan S. (1984) *The Body and Society.* Oxford: Basil Blackwell.

Warner, Richard (ed.) (1791) *Antiquitates Culinariae*. London: R. Blamire.

Watkins, S. C. and van de Walle, E. (1985) "Nutrition, Mortality, and Population," in R. I. Rotberg and T. K. Rabb (eds), *Hunger and History*, pp. 7–28. Cambridge: Cambridge University Press.

Weber, Eugen (1973) *A Modern History of Europe*. London: Robert Hale.

Weber, Eugen (1977) *Peasants into Frenchmen*. London: Chatto and Windus.

Wernham, R. B. (ed.) (1968) *The Cambridge Modern History*, Vol. III. Cambridge: Cambridge University Press.

Wrigley, E. A. and Schofield, R. (1981) *The Population History of England, 1541–1871*. Cambridge: Cambridge University Press.

Young, James Harvey (1970) "Historical Aspects of Food Cultism and Nutrition Quackery," in G. Blix (ed.), *Food Cultism and Nutrition Quackery*, pp. 9–21. Uppsala: Symposia of the Swedish Nutrition Foundation, 8.

24

Industrial Food

Towards the Development of a World Cuisine

JACK GOODY

The British diet, claims the modern "Platine," went straight "from medieval barbarity to industrial decadence." With the general sentiment one has some sympathy. But as we have seen, medieval barbarity meant culinary differentiation, if not into something as grand as a sophisticated cuisine such as the French established in building on firm Italian foundations, at least into systems of supply, preparation, cooking, serving and consumption of food that resolutely set aside the high from the low. And "industrial decadence," whatever its consequences for the *haute cuisine* (larks' tongues are not promising ingredients for a mass cuisine, canned food is not always the best basis for a gourmet meal), has enormously improved, in quantity, quality and variety the diet (and usually the cuisine) of the urban working populations of the western world.[1] It is also making a significant impact on the rest of the world, initially on the productive processes, some of which have become geared to supplying those ingredients on a mass scale, and more recently on consumption itself, since the products of the industrial cuisine and of industrialised agriculture are now critical elements in the food supply of the Third World.

But before we consider the impact on the particular corner of the world on which we are concentrating, we need first to look at the world context in which those changes took place, at the rise of an industrial cuisine in the West. The immediate factors that made this possible were developments in four basic areas: (1) preserving; (2) mechanisation; (3) retailing (and wholesaling); and (4) transport. As we have seen, the preservation of food was a feature of relatively simple economics like those of northern Ghana. The drying of fish and meat enabled animal protein to be more widely distributed in time and space; the drying of vegetables such as ocro prolonged their use into the dry season when soup ingredients were scarce. The preservation of meat and vegetables, by drying, by pickling, by salting and in some regions by the use of ice, was characteristic of the domestic economy in early Europe.[2] With the developments in navigation that allowed the great sea voyages of the fifteenth century, the use of long-life foods became a matter of major importance; the navies and armies of Europe required considerable quantities of such products to feed their personnel. Werner Sombart has written of the revolution in salting at the end of the fifteenth century that permitted the feeding of sailors at sea. In the Mediterranean, salted fish and the ship's biscuit were already long-established (Braudel 1973: 132); in the Atlantic, much use was made of salted beef which came mainly from Ireland. The enormous catches of cod that arrived from Newfoundland by the end of the fifteenth century were mostly salted. Salt was much used by peasants to preserve food during the winter months.[3] Butter and vegetables were also preserved with salt, and until recently the French peasant placed part of the "family pig" in the salting tub, while the rest was made into sausages. But the importance of salt was not only dietary.[4] It was the hunger for salt, both for preserving, which became more common in eighteenth-century France, and for eating, that lay behind the peasant uprisings against the *gabelle*, the salt tax. Such taxes were an important source of revenue in Europe as in Asia, both to the merchants and to the governments; it was against such fiscal impositions, as well as against the alien government imposing them, that Gandhi led the famous march to the sea in British India.

Salting, of course, is only one method of preserving food. It is possible to pickle in vinegar as well as salt, and the production of vinegar was an important aspect of early industrial activity. Sugar was used to preserve fruit in forms such as marmalade and jam, as well as being used for coating ham and other meats. Spreading first from India and then from the eastern Mediterranean at the time of the Crusades, cane-sugar played an increasingly important part in the diets of Western Europe, a demand that led to the establishment of many of the slave plantations of the New World. Imports of sugar increased rapidly in the eighteenth century. It was the fact that supplies of cane-sugar were cut off from continental Europe during the Napoleonic wars that led to the fundamental invention embodied in the canning process, as well as to the use of the beet as a source of sugar, at the same time chicory developed as a substitute for the second of the trio of "junk foods," as Mintz has called them, that is, coffee (the third being tea). It was these "proletarian hunger-killers," to use another of Mintz's forceful phrases, that became such central elements of working-class diet in the nineteenth century and "played a crucial role in the linked contribution that Caribbean slaves, Indian peasants, and European urban proletarians were able to make to the growth of western civilization" (Mintz 1979: 60).

It was this general context of colonialism, overseas trade and long-lasting foods that saw the development of the great British Biscuit Industry. Its product owed much to the ship's biscuit which was known there at least as early as Shakespearean times and was manufactured by small bakeries situated around the many harbours of the kingdom. "Hard-tack" was essentially a substitute for bread (brown or white, depending on class, as had been the case since Roman times), which with ale, cheese and meat, was a basic feature of the diet of the common man (Drummond and Wilbraham 1958: 218). In the course of the eighteenth century the victualling authorities in certain of the king's dockyards such as Portsmouth set up their own large-scale bakeries "creating a human assembly line that economized each workman's movements to the utmost" (Corley 1976: 14). Despite these organisational developments, the fluctuation of demand caused by the various wars meant that dockyard production had to be supplemented by the work of contractors. The situation changed in 1833 when Thomas Grant of the Victualling Office invented steam machinery to mechanise certain of the processes, reducing labour costs, increasing output and improving the quality of the biscuits.

"Fancy biscuits" like "hard-tack" also had a long history, being employed for medicinal purposes as well as for the table, especially at festivals. The earliest proprietory brands were probably the Bath Oliver, invented by Dr. William Oliver (1695–1764) and the Abernethy, called after a doctor of that name (1764–1831). All these biscuits were initially made by hand, but mechanisation was applied to their manufacture not long after the technological changes had taken place in the dockyards. In the late 1830s a Quaker miller and baker of Carlisle, named Carr, designed machinery for cutting out and stamping biscuits. In 1841 George Palmer, another Quaker, went into partnership with his cousin Thomas Huntley who made biscuits in Reading.

The business that developed into Huntley and Palmers had been founded at a bakery in that town in 1822. Huntley's shop was opposite the Crown Hotel, a posting inn on the main London–Bath road. He had the idea of sending his delivery boy to sell biscuits to the waiting passengers. Their quality led customers to demand Huntley's produce from their grocers at home, opening up the market from a purely local one. So Huntley persuaded his son, who had been apprenticed to a Reading ironmonger and kept a shop nearby, to make tins and tinned boxes in order to keep the biscuits fresh. He also employed a traveller to collect orders for Abernethy, Oliver and other biscuits in the south of England, which he dispatched mainly by the canal system. When Palmer joined the firm, he immediately investigated the application of steam power to mixing the dough, to rolling and cutting, and to providing the oven with a continuous feed. These inventions subsequently led to the development of a whole secondary industry of specialised manufacturers of machinery for the trade, a development that helped to fuel the Industrial Revolution.

The sale of biscuits made rapid headway. The manufactured brands, Carrs, Huntley and Palmers, later Peek Freans, were distributed throughout the nation. In 1859, these firms sold 6 million lbs of their products. Changing eating habits in the shape of earlier breakfasts and later dinners led to a further increase in consumption, and by the late 1870s the figure had risen to 37 million lbs a year. Huntley and Palmers had become one of the forty most important companies in Britain, and within fifty years their biscuits were distributed not only throughout the nation but throughout the world. As with the early canning industry, much of the production of biscuits had first of all been directed to the needs of travellers, explorers and the armed forces. Such produce sustained sailors, traders and colonial officers overseas; only later did industrial production impinge upon the internal market in England or upon the local market overseas, eventually becoming part of the daily diet of the population.

PRESERVING: CANNING

The creation of a long-lasting cereal product, the biscuit, long pre-dated the Industrial Revolution, though its production and distribution were radically transformed by the course of those changes, making the biscuit an important element in the development of the industrial cuisine. But that cuisine was based in a large degree on two processes, the discovery of the techniques of canning and artificial freezing. The preserving of food in containers again dates back a long way, but the canning on which modern industry depends was invented by Nicolas Appert in response to an appeal of the Directoire in 1795 for contributions to solving the problems created by the war situation in France. During the Napoleonic wars France was cut off from its overseas supplies, and this separation stimulated the search for substitutes. At the same time the recruitment of a mass army of citizens raised in a very radical way the problem of supplying food for a large, mobile and non-productive element in the society; in 1811 Napoleon invaded Russia with over a million men. So the aim of the appeal was partly military, though the citation to Appert when he received the award refers to the advantages of the new invention for sea voyages, hospitals and the domestic economy (Bitting 1937).

This invention of "canning," what the English call bottling, was based on earlier practices and earlier devices, such as the "digester," a sort of pressure cooker, invented by Denis Papin in London in 1681, which provided John Evelyn, the diarist, with a "philosophical supper" (Cutting 1955: 5; Teuteberg in E. and R. Forster 1975: 88). A contemporary account of Appert's book in the *Edinburgh Review* (1814, vol. 45) calls the process "neither novel in principle, nor scarcely in any point of practice" (Bitting 1937: 38), and declares that "our fair country women . . . unless they have alike forgotten the example and precepts of their ancestors . . . must . . . be more or less acquainted with the methods" (p. 39). Nevertheless the author goes on to recognise the importance of Appert's contribution, especially as the ladies of 1814, having been relieved of various household tasks by the Industrial Revolution, tend to know too little about such things.

Appert had been a chef, and he worked out his new methods at his business near Paris. It was in 1804 that a series of public tests were made on his produce at Brest, and in the same year he opened his bottling factory at Massy, near Paris. Five years later he was awarded the prize of 12,000 francs by the committee that included Guy-Lussac and Parmentier, on condition that he deliver a description of the process, in 200 copies, printed at his own expense. This description he published in 1810 and proclaimed the use of his method of bottling as a general aid to domestic life. Entitled "Le livre de tous les ménages . . . ," the book gave instructions for bottling pot-au-feu, consommé, bouillon or pectoral jelly, fillet of beef, partridge, fresh eggs, milk, vegetables (including tomatoes or love apples, spinach, sorrel and petit pois), fruit and herbs. "No single discovery," declared Bitting, "has contributed more to modern food manufacturer nor to the general welfare of mankind" (1920: 13).

Nor did Appert stop there. Investing his prize money in production and research, he founded the house of Appert in 1812, produced bouillon cubes two years later and experimented with a number of other ideas, turning eventually to the use of the tin can to supplement that of the glass jar.

In England, where as much interest had been displayed in Appert's discoveries as in his own country, the tin can had been in use for some years. An English translation of Appert's book appeared in 1811, a second edition in 1812, and an American edition in the same year. Already in 1807 T. Saddington had been awarded a premium by the London Society for Arts for his work on bottling, and he probably learnt of Appert's process during his travels abroad. In 1810 Peter Durand and de Heine took out patents on the process, but in the former case it was adapted for preserving "food in vessels made of tin and other metals." The potentialities of these inventions aroused the interest of Bryan Donkin who was a partner in the firm of John Hall, founder of the Dartford Iron Works in 1785. Whether he acquired either of the earlier patents is not known, but he appreciated the potential value of Appert's discovery for his firm.[5] After various experiments, Donkin, in association with Hall and Gamble, set up a factory for canning food in metal containers in Blue Anchor Road, Bermondsey. The Navy immediately purchased supplies of "preserved provisions" to form part of their medical stores. These were used for numerous expeditions, by Ross in his voyage to the Far North in 1814, by the Russian von Kotzebue to the North-West Passage in 1815, by Parry to the Arctic in the same year. In 1831 Admiralty Regulations decreed that all ships should carry such provisions as part of their "medical comforts" (Drummond and Wilbraham 1958: 319). This was the prelude to their more general spread into the domestic economy which was still hindered by their great expense compared with other foods.

Glass containers continued to be used for most purposes and had a history dating back many years. At the end of the seventeenth century there were 37 glass-houses in London making only bottles, approximately three million a year, which were used mainly as containers for wine and medicine. The chemist Joseph Priestly gave the industry a further boost when early in the eighteenth century, he discovered how to make artificial mineral water which led to a flourishing new industry. By the end of the century, the Swiss, Jacob Schweppe, had set up a factory in London (Wright 1975: 46). But it was with Appert's invention that the glass container came into wide use for preserved foods.

From England the process spread to the United States, where bottles rather than cans were used. William Underwood, who had served his apprenticeship in pickling and preserving with a London house, left for New Orleans in 1817. By 1819 he had made his way to Boston, and in the following year he and one C. Mitchell founded a factory for bottling fruit which by 1821 was shipping goods to South America. Damsons, quinces, currants and cranberries were the main items preserved at the beginning, but the major part of the business had to do with pickles, ketchups, sauces, jellies and jams.

In England these preserved foods did not reach the shops until 1830 and were slow in selling because of the high price. In America too the local trade was initially poor, and most of Underwood's produce went abroad to India, Batavia, Hong Kong, Gibraltar, Manila, the West Indies and South America.[6] Much of it was marketed under an English label to counter the prejudice against American goods (Butterick, 1925). In 1828 Underwood was shipping preserved milk to South America and in 1835, having imported the seed from England, he started to bottle tomatoes, partly for export to Europe.[7] Up to this time the fruit was little known in the States and indeed was regarded as poisonous, even though it had been domesticated in the New World and taken from Mexico to Europe.[8]

In America the local trade developed with the shift from glass to the cheaper metal containers and with the immense boost to sales given by the Civil War (1861–5). Once again the demands of the army were of major significance. But the point of take-off had now been reached for society as a whole. At about the same time as Underwood started his factory in Boston, another English immigrant, Thomas Kensett, and his father-in-law, Ezra Doggett, set up a cannery in New York for salmon, lobsters and oysters. In 1825 Kensett took out a patent for tin cans but they did not become widely used until 1839.

Many of the pioneer factories in the States started with fish as the primary product, and fruit and vegetables as incidental (A. and K. Bitting, 1916: 14). In Europe the canning of sardines, that is, young pilchards, began in Nantes in the early 1820s. By 1836 Joseph Colin was producing 100,000 cans, and the industry spread along the coast of Brittany. But it was not until 1870 that a rapid expansion began. By 1880 50 million tins of sardines were being packed annually on the west coast of France, three million of which were exported to Britain. The world of industrial food had begun.

The canning of the other major object of fish packaging, the salmon, began about the same time in Aberdeen on a small scale. Others in Scotland followed, in an attempt to save the long haul of salmon, frozen and smoked, to the London market. The first large-scale salmon cannery was established at Cork, in the south of Ireland, in 1849 by Crosse and Blackwell. It was with the development of canneries on the Pacific coast in 1864 that large-scale production began in America.

The canning of meat was especially important for the army, being developed not only in the American Civil War but also by the Anglo-French forces in the Crimea. It continued to be of great military significance, and in the First World War the Germany Army were producing eight million cans of meat per month. The process was less important for the domestic market, especially after the advent of refrigeration techniques in the latter part of the nineteenth century, when frozen produce became widely available and was preferred by the consumers.

Condensed milk was another major product of the canning industry. In Britain Grimwade took out a patent for evaporated milk in 1847 and was supplying some to expeditions at an early date. In 1855 he took out another patent for powdered milk which could be reconstituted with water. A great improvement in milk processing was made possible by Borden's work in the United States. Borden, who had been stimulated by the needs of migrants in the Gold Rush to market pemmican and meat biscuits, applied for his patent in 1853 and his process was used for production not only in America but also by the Anglo-Swiss Company, later Nestlé's. Condensed milk became a major item of diet in Britain, and in 1924 over two million hundredweight of the product was imported, more than the total imports of tinned fruit and the combined imports of beef and fish.

Food was also processed by other techniques than canning, and some of the results played a prominent part in the new cuisine. Meat extract was developed by the Frenchmen, Proust and Parmentier, and after 1830 meat bouillon was boiled down to a stock soup, dried and sold as "bouillon bars" in pharmacies and for use on ships. Large-scale production became possible in 1857 as the result of Liebig's research on muscle meat, and in the following years factories were built in Fray Bentos, Uruguay, to process meat into a brown powder for shipping to the growing urban populations of Europe. Soon the process of manufacturing meat foods spread to Australia, New Zealand, Argentina and North America with the result that the soups and gravies of the English kitchen became dominated by the dehydrated products of the international meat industry.

PRESERVING: FREEZING

While the canning of food was the most significant step in the development of an industrial cuisine, other processes of preservation also played an important part, especially the artificial freezing of foodstuffs. In cold climates the technique had been practised since prehistoric times, and natural ice continued to be used until very recently. In the early nineteenth century the Russians were packing chickens in snow for consumption as occasion required, and frozen veal was being sent from Archangel to St Petersburg. In Scotland ice-houses had long been attached to richer homes, and the practice spread into the food trades. By the beginning of the nineteenth century every salmon fishery in Scotland was provided with an ice-house. The fish were packed in long boxes with pounded ice and dispatched to the London market (Bitting 1937: 29).[9]

In America natural ice was also packed in "refrigerators" from the beginning of the nineteenth century. Indeed, ice gathered from the Boston ponds became big business. Beginning in 1806 with the West Indies, Frederic Tudor developed a world-wide operation, which between 1836 and 1850 extended to every large port in South America and the Far East; within thirty years ice had become one of the great trading interests of the city of Boston, and one French source claimed that in 1847 the Asian trade of one house was almost equal in value to the whole of the Bordeaux wine harvest (Prentice 1950: 114ff.).

The demand of Asian countries for Boston ice had a history in the local usage of Chinese society where in pre-imperial times the feudal nobility had the prerogative of storing ice to keep sacrificial objects fresh. In later imperial times eunuchs supervised the cutting of blocks of ice from rivers and ponds, packing it in clean store and keeping it in trenches. The use of ice was not limited to the imperial house, though it helped to supply the court with fresh food. Mote remarks that "refrigerated shipping seems to have been 'taken for granted' in Ming times, long before we hear of such a development in Europe" (1977: 215).[10]

About the time of the development in the Boston trade, ice became widely used in America for transporting food on the railroad. It was in 1851 that the first refrigerated rail car brought butter from Ogdensburg, New York, to Boston, Massachusetts. Of greater significance was the ability to transport frozen meat from the Chicago stockyards to the urban centers of the East, an enterprise associated with the names of Armour and Swift, and providing a parallel development to the international shipment of meat to Europe from America and Australia.

With the development of rail transport around the middle of the century, fresh sea fish began to make its regular appearance in inland markets in England, leading to a decline in the popularity of salted and picked herrings, which had long been a staple food. Fish from rivers and dew ponds had long been available, sometimes being marketed alive by the fishmonger in tanks. Live fish were carried by the wives of sea-fishermen to towns as far inland as Coventry, being transported in brine on pack-horses (Davis 1966: 15). But the expense of such products meant that they were not available to the majority of the population, a situation that rail transport helped to change. Meanwhile in America a great deal of herring had been frozen by exposure since 1846, "providing the masses with a cheap and wholesome food" (Cutting 1955: 296).

In cooler climes and for longer distances the extensive use of refrigeration depended upon the development of artificial rather than natural ice. The newly exploited pastures of Australia and America produced an abundance of livestock, but the problem lay in getting meat to the industrial consumers. Shipping cattle live to Europe presented great difficulties until the length of the voyage had been reduced. Nevertheless American livestock was imported into Europe in appreciable numbers by the early 1870s. The Australian voyage was much longer and, although meat was now canned, its quality left much to be desired. In 1850, James Harrison, a Scottish immigrant to Australia, designed the first practical ice-making machine, using the evaporation and subsequent compression of ether. Ten years later, the French engineer Carré produced a much more efficient machine based on ammonia gas. But the problem of transporting Australian frozen meat to Europe was not solved until 1880 when the S.S. *Strathleven* brought a cargo from Melbourne to London. Meanwhile similar experiments were being carried on in other countries, and frozen meat reached the London market, probably from the United States, as early as 1872. In France, the engineer Charles Tellier had been working on the same problems since 1868, and after several unsuccessful attempts, the S.S. *Frigorifique* brought a cargo of meat from Argentina to Rouen in 1876. These developments led to a rapid diminution in the amounts of canned and salted meat being imported from those countries, just as the use of natural ice in Britain had earlier led to a decrease in the amount of salted and pickled fish.

Not only diet but cooking too responded in significant ways to technological changes. While domestic refrigeration had to wait until the following century, in the later part of the nineteenth century such machines were used in the new catering trades. It was these developments in freezing techniques that enabled Lyons tea-shops to serve their popular cucumber sandwiches all the year round. Started by the successful operators of a tobacco business in 1887, these shops followed the rise of the coffee public houses of the 1870s which had been organised by the temperance societies as an alternative to the pub. Out of these coffee houses developed two sizeable catering concerns, the first multiples of their kind, Lockharts and Pierce & Plenty. The growth of the commercial catering business was the counterpart of the decline of the domestic servant; in 1851 905,000 women in Britain were employed as domestic servants, plus 128,000 servant girls on farms; by the 1961 Census, there were no more than 103,000 resident domestic servants in England and Wales.[11]

Many of the other developments surrounding food in the nineteenth century had little to do with preservation *per se,* but rather with branding, packaging, advertising and marketing. In 1868 Fleischmann provided a standardised yeast for the American market; Heinz bottled horse-radish sauce in 1869. In England sauces, based on vinegar, featured early on in the commercial food business; their production was stimulated by the cheaper spices brought back by the East India Company in Elizabethan times, and the demand was later increased by the return of families from abroad. The well-known Worcester sauce, for example, was produced by the analytical chemists Lea and Perrins, who went into partnership in 1823 to run a pharmacy that also dealt in toiletries, cosmetics and groceries. They began to market their own medicinal products, and their first catalogue listed more than three hundred items, such as Essence of Sarsaparilla for scurvy and Taraxacum (dandelion coffee) for liver complaints. These products were soon in great demand from the new industrial towns of the Birmingham area as well as from abroad; travellers set off with a Lea and Perrins' medical chest, and this publicity led to orders from all over the world. However it was the invention of the sauce which exemplifies the rapid growth of prepared foods, the shift of focus from kitchen to factory, as well as the influence of overseas trade and overseas colonisation.

> Mr. Lea & Mr. Perrins were perfecting their medicines, hair lotions and marrow pomades when Marcus, Lord Sandys visited the shop in Worcester. Late Governor of Bengal, he had retired to his country estate in nearby Ombersley Court, and would be obliged if they would make up one of his favourite Indian sauces. They obliged. Having already arranged their own supplies of spices and dried fruits, from Asia and the Americas, they had the ingredients to hand. Scrupulously following his lordship's recipe, they made the required quantity, plus some for themselves. One taste was enough. The sauce was ghastly: an unpalatable, red hot, fire water. His lordship was entirely satisfied. The remainder however was consigned to a cellar below the shop, and there it stayed until the annual spring-cleaning and stocktaking.

It was on the point of being poured away when Mr. Lea and Mr. Perrins detected its appetising aroma. Tasting it once again they discovered it had matured into a rare and piquant sauce. The sauce was saved, more was made. Customers were persuaded to try the new Worcestershire Sauce, and did not need more persuasion: the sauce was an instant success. Sales rose. In 1842, Lea & Perrins sold 636 bottles. In 1845, a manufactory was set up in Bank Street, Worcester. Ten years later the yearly sales were up to 30,000 bottles of Worcestershire Sauce. Travellers covered Great Britain and there were agencies in Australia and the United States (Wright 1975: 31). The product still adorns the tables of cafés, restaurants and dining rooms the world over.

Another item of processed food which has substantially changed eating habits in many parts of the world, including the new bourgeoisie of Ghana, is the breakfast cereal. Initially these foods were developed in the United States to meet the needs of vegetarian groups like the Seventh Day Adventists, who were experimenting with cereal-based foods at Battle Creek, Michigan, in the 1850s. Dr John Kellogg was director of the "medical boarding house" at Battle Creek where he carried out research in the "dietary problem" and in the development of so-called "natural foods" (Deutch 1961). It was in the 1860s that he produced "Granola," the first ready-cooked cereal food, made from a mixture of oatmeal, wheat and maize baked in a slow oven until thoroughly dextrinised (Collins 1976: 31). It was his brother who later became the promoter of these foods.

The 1890s saw the invention of most of the basic types of pre-cooked cereal and manufacturing process—flaking, toasting, puffing and extrusion: Shredded Wheat appeared in 1892, and in 1898 Grape Nuts, invented by Charles W. Post, an ex-patient of Kellogg's who also produced Post Toasties. It was Post who pioneered the use of advertising techniques employed by the makers of patent medicines to market his products, "selling health foods to well people." Ever since this market has been heavily dependent upon massive publicity campaigns. After the First World War the American foods spread to Britain, where as elsewhere in the world the market is still dominated by more or less the same products. Pushing aside porridge and other breakfast foods, they owed their success to their ease of preparation, especially important for breakfast in households whose members are all working outside the home; but their widespread use was also due to the vigorous sales campaigns, later directed primarily towards children, to the general shift to lighter meals consistent with the changing nature of "work," and to rising real incomes which made it possible for people to buy "health" foods and so transform them into utility foods.[12]

MECHANISATION AND TRANSPORT

The use of machines in the production of industrial food has already been noted in the case of the biscuit industry. But it was equally important in the whole canning industry at three levels, in the mechanisation of the production of food, especially in agriculture, in the mechanisation of the preparation of the food, cleansing, peeling, podding, etc., and in the mechanisation of the canning itself.

When canned goods first reached the shops about 1830, they made little impact on domestic consumption on account of their price, even with the cheaper metal containers. A skilled man could fill only 50 or 60 cans a day; the cans themselves were made from rolled sheets of wrought iron; their lids were fitted by placing a sheet of metal across the top and then hammering it down at the sides, after which it was soldered on. The early cans were so heavy that they sometimes had to be fitted with rings to lift them, and they could only be opened with a hammer and chisel. During the course of the second half of the nineteenth century these obstacles were gradually overcome; in 1849 came a machine for pressing out the tops and bottoms, then a substitute for the soldering iron, and in 1876 the Howe machine produced a continuous stream of cans for sealing so that two men with assistants could produce 1,500 cans a day (Cummings 1941: 68). The methods of canning themselves improved in 1861 when the addition of calcium chloride to the water increased its temperature and cut down the boiling time required from five hours to thirty minutes, a time that was further reduced by the invention of the autoclave (a closed kettle). The other bottle-neck lay in the preparation of the food itself, a phase that also gradually became dominated by new machinery—washers, graders, peelers, corn huskers and cutters, bean snippers and filling machines. Once again a whole series of subsidiary industries arose to meet the new requirements.

It should be stressed that the manufacture of many processed and packaged foods did not require great advances in techniques of preservation but rather the adaptation of simple machinery for producing standard goods on a large scale. This happened not only with biscuits but with pasta. Possibly coming from China through Germany, this "typical" Italian dish was adopted in the fourteenth century.[13] It spread across the Atlantic with the influx of Italian immigrants late in the nineteenth century, and in 1890 it was manufactured on a large scale by Foulds, wrapped in a sanitary package and advertised as "Cleanly made by Americans." Mechanisation permitted the domestication and purification of foreign foods.

Not only production but distribution too was mechanised, involving a similar massive use of energy. The mechanisation of the process of distribution depended upon the development of a system of transport that could shift the very large quantities of goods involved in the ready-made market; in the United States this amounted to some 700 million cases of canned goods a year in the 1960s, each case including an average of 24 cans. The distribution of processed foods to a mass market was dependent upon the railway boom which in Britain marked the beginning of the second phase of industrialisation (roughly 1840–95). Following the little mania of 1835–7 came the big railway mania of 1845–7; during this period, from 1830, over 6,000 miles of railway were constructed in Britain (Hobsbawn 1968: 88). Here was an opportunity for investment, for employment and for the export of capital goods which laid the basis of working-class prosperity in the third quarter of the century, permitted the growth of the mass markets for preserved and processed foods, and built up the volume of imports from the Colonial (now the Third) World.

Critical to the growth of the overseas trade was the development of large cargo ships capable of transporting the raw materials to the metropolitan country in exchange for the mass export of manufactured goods. These various processes of mechanisation and transportation were essential to the preservation and distribution of food on a mass scale, and so to the industrialisation of the domestic diet of the new proletariat. But more immediately relevant to the domestic level were the social changes that took place in the organisation of the actual distribution of food to the household, since a whole series of agents now intervened between the producer and the consumer.

RETAILING

The changes in the English retailing trade were marked by two phases. The first was the shift from open market to closed shop which began in Elizabethan times, although the move to retail shops in the food trade was strongly resisted by many urban authorities. The second retail revolution occurred in the nineteenth century and was associated with industrialisation rather than with urbanisation itself; indeed it affected food and cooking in both town and country.

In medieval England market-places were areas for exchanging the products of the nearby countryside with those made by local craftsmen. Great efforts were made by the authorities to prevent the intervention of any middle men, except in the long-distance trades where their services could not be avoided. Shops for buying food scarcely existed; the town authorities forced the food trade into the street market for the purposes of control.[14] Frequent regulations were made against "forestalling" the market by buying goods outside it, against "regrating" those goods, that is, selling them at a higher price, and against "engrossing," or hoarding. At the same time an attempt was made to control the quality and the price;[15] control lay partly in the hands of the occupational associations and partly in the hands of the collective authority of the corporation. In Chester, butchers and bakers were made to take a public oath that the food they supply to their fellow-citizens shall be wholesome and fair in price.[16] Such regulations also included rules about the use of standard weights and measures, instructing men "not to sell by aime of hand."

It was impossible completely to control the marketing of food, even at this time. In addition to "foreigner" or "stranger" markets (i.e., those catering for country folk) there was a certain amount of hawking. But the move to establishing food shops only developed with the growth of suburban London, the main markets being too far away for their inhabitants. By this time the cornchandler and a few other traders were beginning to act as wholesalers with fixed premises. Nevertheless it was well into the eighteenth century before the big open markets ceased to be the normal place to buy food (Davis 1966: 74).

Apart from the shift from market to shop, little had changed in London as late as 1777, when the country was on the threshold of the great changes in the processing and sale of food that industrialisation brought. In that year, James Fitch, the son of an Essex farmer who, like others in a period of depression following the great agricultural changes of the earliest part of the century, came to London to enter the food trade.

> The only retailers allowed to trade were those who had been apprenticed to a freeman of the City, and who, on completion of their apprenticeship, had themselves taken up the freedom of a City Company or Guild. The Lord Mayor and his Court were still virtual dictators of trade within the City walls and for up to seven miles outside. Price control was exercised from the Guildhall, and it was an offence of offer goods for sale other than in the markets. No one was allowed to own more than one shop selling poultry, butter or eggs. (Keevil 1972: 2)

Control was exercised by the great City companies, and freeman engaged in selling produce had to belong to the appropriate Livery Company, Butchers', Poulterers', Bakers', Fruiterers' or Grocers.'

It was the last of these, the grocers, who were the key to later developments. Originally one of the minor food trades, grocery overtook all the others. For many years grocers were associated with the import of foreign goods. In the fifteenth century they had been general merchants dealing in most goods except fresh food and clothing. But in London they gradually concentrated on the non-perishable items of food arriving in increasing quantities from the Mediterranean, the Far East and the New World. The English (London) Company of Grocers was made up of merchants dealing in spices, dried fruit and similar commodities which they imported or bought in bulk ("gross," *grossier* in French is a wholesaler), and sold in small quantities. Later they added tea, coffee, cocoa and sugar, all of them initially "luxury" goods.[17] One of the most important commodities was sugar, first imported at a high price from Arabia and India but then brought in much more cheaply first from the Canaries and then from the Caribbean, one of the "junk foods" that were key components of colonial plantations and working-class diets.

The grocer was distinguished from the provision merchant who dealt in butter, cheese and bacon, and the distinction obtained until very recently in the larger shops where there was a "provision" counter dealing in these items; the British housewife herself referred to these central foods as "provisions" and to the dry goods as "groceries."

It was the grocer dealing in dry, imported goods who led the second retailing revolution. Authors such as Davis (1966) and Jefferys (1954) have considered this great development in retailing (as distinct from wholesaling) to lie in the growth of multiples, of shops that were organised in branches along national lines. Since this growth was based upon the rise of working-class incomes, it is not surprising that the first such organisation was that of the Rochdale Pioneers whose cooperative was begun in 1844—not the first of its kind but the first commercial success. In 1856, at the request of loyal customers, it opened the first of its many branches. At the same time the cooperative experimented in vertical integration by starting the Wholesale Society in 1855 which moved from purchase and distribution directly into production. It was not until some time later that private firms entered into the same field.[18] Indeed, the great boom came in the last twenty years of the century, when Coop membership rose from half a million in 1881 to three million in 1914.

One of the first trades to develop a network of chain stores was footwear which had 300 shops in 1875, rising to 2,600 in 1900. Butchers started later, with 10 shops in 1880 and 2,000 in 1900, while grocery branches jumped from 27 to 3,444 over the same period (Hobsbawm 1968: 131). In 1872 Thomas Lipton started his grocery shop in Glasgow; twenty-six years later there were 245 branches scattered all over the kingdom. Selling a limited number of cheaper goods, the new multiples in turn influenced the trade of the old-fashioned grocer who now had to deal with the appearance of "an entirely new style of commodity in the form of manufactured foods" (Davis 1966: 284)—tinned goods, jams, powders for custards, grains and so forth. Just as imported goods became cheaper with the new developments in transport, so too manufactured goods and items packaged before sale came to dominate the market. These products were generally branded goods, "sold" before sale by national advertising.

Advertising in the modern sense was a critical factor in these developments and had begun with printing itself; already in 1479 Caxton printed an advertisement for books from his press. The first newspaper advertisement appeared in 1625, and by the mid-eighteenth century Dr Johnson was complaining of their ubiquity. But the major development of the business came in the nineteenth century with the advent of the wholesale trade which itself derived from the industrialisation of production, as well as with the coming of the mass newspaper dependent upon the rotary press. Large-scale manufacture brought with it an increased gap between producer and consumer so that some new way of communication was required. During the fifty years from 1853 the quality of soap bought in Britain increased fourfold, with the main companies, Levers', Pears', and Hudson's, competing by means of national advertising campaigns of which the use of Sir John Millais' painting "Bubbles" was the most famous example. In America, N.W. Ayer and Son, Inc., the first modern agency, was founded in Philadelphia in 1869, acting as an intermediary between producer and media, and making possible the more complex advertising campaigns which not only enables products to reach a wider market but which to some extent create that market, as in the case of breakfast cereals.

The role of advertising in promoting the Hovis loaf is an interesting example of this process. This patent bread has been the product leader in brown bread since 1890 when it had probably been adapted from a loaf invented by an American vegetarian, Graham, in the 1840s. The problem was to overcome the popular prejudice, in existence since Roman times at least, in favour of white bread.[19]

> In the early years the goal was to acquire a sound reputation with a public that was sceptical of patent foods and wary of adulteration. Thus early advertisements made considerable play of royal patronage, of awards and diplomas for quality and purity, and of the need to beware of cheap imitations. (Collins 1976: 30)

The relics of the system of royal patronage and international diplomas employed by the industry remain with us to this day. They were essential parts of the initial legitimation of processed foods, just as the advertisement and the grocery trade were essential aspects of their distribution.

This whole process led to a considerable degree of homogenisation of food consumption and was dependent upon the effective increase in demand from the "working class," which now had no direct access to foodstuffs, to primary production. Because of this mass demand, mass importation and mass manufacture, grocery, formerly one of the minor food trades, became by far the most important. "The vast majority of consumers of all income groups drink the same brand of tea, and smoke the same cigarettes, and their children eat the same cornflakes just as they wear the same clothes and watch the same television sets" (Davis 1966: 84). Differences in income, class and status have to manifest themselves in other ways.

While some historians see the second retailing revolution as developing with the growth of the multiples, Blackman would place it earlier in the century when processed foods and mass imports began to make an impact on the market as a result of changes in technology that were linked with the new demands of the industrial workers. Cheaper West Indian sugar and less expensive Indian teas became essential items in the improved diets of the working class (when they were employed) in the latter half of the nineteenth century. In the 1860s grocers added other new lines, processed foods, including cornflours, baking powders and dried soups, such as Symingtons.' As we have seen, many of these "processed" foods were not the results of changing techniques of food preservation so much as the advent of national instead of local products, such as soap or "patented" branded foods consisting of established items broken down, packaged and sold through public advertising campaigns. Blackman notes that at this time one grocer in Sheffield was buying dried peas, oatmeal and groats from Symingtons' Steam Mills at Market Harborough, and mustard, cocoa, chicory and other commodities from forty different firms including starch and blue from J. & J. Coleman's, the mustard manufacturers who had a dramatic rise from the time in 1854 when they purchased a windmill in Lincolnshire, before moving to Norwich. In the United States too the national canning industry took off in the mid-1860s when Blue Label canned foods, founded in 1858, started advertising nationally, though items like Borden's condensed milk (1857), Burham and Morrill's sweet corn (c. 1850), Burnett's vanilla essence (1847) and various brands of soap were already available.

By 1880 a grocer in Hull was buying Wotherspoon's cornflour, Brown and Polson's brand of the same commodity, Symingtons' pea flour, Goodall's custard and egg powders, several brands of tinned milk, including Nestlé's, tinned fruit, Crosse & Blackwell jams, and many other items which are still household names. It was the technical revolution of mass-producing and semi-processing foodstuffs in common use together with the increased volume of trade in tea and sugar that now brought the grocer into focus as "the most important food trader for regular family purchases" (Blackman 1976: 151).

ADULTERATION UNDER THE NEW DISPENSATION

Complaints against the adulteration of food are as old as the sale of foodstuffs itself. In Athens protests about the quality of wine led to the appointment of inspectors to control its quality. In Rome wines from Gaul were already accused of adulteration, and local bakers were said to add "white earth" to their bread.[20]

Adulteration is a feature of the growth of urban society, or rather of urban or rural society that is divorced from primary production. The agro-towns of West Africa were not so divorced, while even many of the rural inhabitants of modern England have little or nothing to do with the land. With the growth of a distinct town life in England in the centuries after the Norman Conquest, an increasing number of merchants, artisans and shopkeepers had to rely on others for their supply of food, and it was these non-food producers who were the targets (and sometimes the perpetrators) of adulteration. The quality and price of bread and ale was controlled as early as 1266 and continued to be so for more than five hundred years.

It was that "most revolutionary social change" of the first half of the nineteenth century, the rapid growth of towns, and especially the industrial towns of the midlands and north of England (Burnett 1966: 28), which was based on the development of manufacturing industry in the previous century, that made the adulteration of food a major social problem. Protests against impure food had already taken a literary form by the middle of the eighteenth century and were mainly aimed at millers, bakers and brewers. In 1757 "My Friend," a physician, published a work entitled *Poison Detected;* the next year saw the appearance of *Lying Detected* by Emmanuel Collins, attacking the general trend of this literature, as well as a work by Henry Jackson entitled *An Essay on Bread . . . to which is added an Appendix; explaining the vile practices committed in adulterating wines, cider, etc.* In a later period of scarcity and high prices, we find *The Crying Frauds of London Markets, proving their Deadly Influence upon the Two Pillars of Life,*

Bread and Porter, by the author of the *Cutting Butcher's Appeal* (1795). But it was the work of Frederick Accum in 1820, *A Treatise on Adulterations of Food and Culinary Poisons, etc.*, that had the greatest influence on the public since he was a respected analytical chemist and a professor at the Surrey Institution. He gave widespread publicity both to the methods adopted and to named individuals until he had to flee to Berlin, possibly as a result of a "a deliberate conspiracy of vested interests" (Burnett 1966: 77).

The adulteration of food continued to be a problem for the industrialisation of cooking, particularly in these early days. While Accum's departure from the country led to a temporary neglect of his work, the fight continued. The main analytic contribution was a series of reports made by Dr Hassall between 1851 and 1855, published collectively in the latter year. Hassall recounts that within months of coming to live in London in 1850 he saw that "there was something wrong in the state of most of the articles of consumption commonly sold" (1855: xxxvii). So he examined a range of items sold at grocers' shops (coffee, cocoa, mustard, sauces, preserved goods, prepared flour) as well as butter, bread, beer and gin. Some passed the test; many failed. To take a typical example, 22 out of 50 samples of arrow root were adulterated; one variety advertised on the label as:

Walker's
Arrow-root

sold in packages, "2d the quarter pound," elicited the comment (most were equally lapidary) "Consists entirely of potato-flour" (1855: 41). Much of the produce was packaged in the stores in which it was sold. Some had been wrapped and even produced elsewhere, being labelled with the maker's name. These names were publicised by Hassall so that the brand became a mark of quality, or lack of it. Names like Frys and Cadbury's already appear in the cocoa trade (pp. 264–5), Crosse & Blackwell and Fortnum & Mason's in the sauce trade, J. & J. Coleman for mustard (p. 131). Each of these well-known firms was indicted for selling adulterated products and no doubt took steps to improve the quality. On the other hand a positive recommendation for a branded product was clearly an important aspect of publicity as well as of quality control, as for example in Hassall's conclusion: "That Borden's Patent Meat Biscuit was in a perfectly sound state, and that there is much reason to regard it as a valuable article of diet in the provisioning of ships, garrisons, etc." (1855: xx). Firms soon began to employ the label as a certification.

> Vicker's Genuine Russian Isinglass for invalids and culinary use. . . . Purchasers who are desirous of protecting themselves from the *adulteration* which is now extensively practised are recommended to ask for "Vicker's Genuine Russian Isinglass," in *sealed packets*.

The samples that were tested by Hassall sometimes indicate the foreign provenance of their constituents or recipes. India had an obvious influence on the appearance of chutneys as well as on the fact that three varieties of "King of Oude" Sauce were on the market as well as an "India Soy," consisting of burnt treacle. Hassall tested seven tomato sauces, six of which were adulterated, including two from France, one of which was from Maille ("very much of the red earth," he comments).

The combined contribution of public medical testing, branded goods and widespread advertising brought adulteration under control at the same time as creating a national cuisine, at least as far as processed ingredients and prepared foods were concerned. At the same time the pattern of the grocery trade changed radically. For now the shopkeeper was no longer the one who selected and certified the product; that was done by the producer and packager, by the name and the advertisement. Regional tastes continued to be important, as Allen (1968) has pointed out. But these comprised only a small component of a largely nationalised, even internationalised, repertoire.

Given these developments in retailing, the move towards self-service, even automated service, was the next major step. Consumer services become of less and less importance; small special shops tend to vanish while large general stores prosper. But not altogether. In some European countries the tendency is less marked. So it is in rural areas where the owner-managed general store persists. In towns, smaller shops often specialise in new products, in second-hand and antique goods, filling the spatial interstices created by the supermarkets, the department stores and the discount houses, while market stalls arise in unexpected places to sell objects of craft manufacture or local produce. But in general the larger stores offer lower prices, wider choices and the impersonality of selection that a socially mobile population often appears to prefer.

The effect of these changes on the diet and cuisine was enormous. A great deal of domestic work was now done before the food ever entered the kitchen. Many foods were already partly or fully processed, and even sold in a ready-to-eat form. Consequently not only have the ingredients become standardised but a number of the dishes as well, at least in many homes in England and America where only the festive occasion, either in the house or at the restaurant, requires the food to make some claim to be "home-made." While Ghana is far from attaining this extreme condition, the industrialisation of food has begun to affect the country not only as a supplier (essentially of cocoa) but also as a consumer. Within a relatively short space of time tinned sardines, condensed milk, tomato paste and cartons of lump sugar have become standard features of the small markets throughout Anglophone West Africa. A drain on the limited resources of foreign exchange, the demand for which is continuously expanding while cocoa production remains static, the absence of these items causes hardship and complaint; these industrial foods of the West have now become incorporated in the meals of the Third World.

NOTES

1. For an example of an attempt to "improve" the diet of the urban proletariat the reader is referred to the booklet of Charles Elmé Francatelli, late maître d'hôtel and chief cook to Her Majesty the Queen, entitled "A Plain Cookery Book for the Working Classes' (1852, reprinted by the Scholar Press, London, 1977). The first entry, Boiled Beef, is described as "an economical dinner, especially where there are many mouths to feed," and, "as children do not require much meat when they have pudding" there should be "enough left to help out the next day's dinner, with potatoes." Such fare contrasts with that of the poorer classes in France and many parts of Europe who "very seldom taste meat in any form" (p. 47). The staples of later English cuisine are already there: Toad in the Hole, Meat Pie, Sausage Roll, Bread Pudding, Rice Pudding, Cocky Leaky, Irish Stew, Bubble and Squeak, Jugged Hare, Fish Curry, Boiled Bacon and Cabbage, Tapioca Pudding, Brown and Polson Pudding, Blancmange, Stewed Prunes, Welsh Rarebit—the list reads like the roll of honour of school catering, and one is surprised to find it reinforced by a cook in royal employ and with such continental credentials as his name and title imply. However, the working classes are also taught how

to bake their own bread, brew their own beer, to cure hams, and to make A Pudding of Small Birds. Their potential benefactors are also instructed on how to prepare "economical and substantial soup for distribution to the poor."

2. On the early use of dried food for soldiers and travellers in China, see Yü 1977: 75. Chang points out the many ways in which China made use of preservation techniques, by smoking, salting, sugaring, steeping, pickling, drying, soaking in many kinds of soy sauces, etc. For example, meat (either raw or cooked) was pickled or made into a sauce. It also seems that even human flesh was pickled and some famous historical personages ended up in the sauce jar (Chang 1977: 34). In Turkey dried meat was used by soldiers in the field (Braudel 1973: 134); elsewhere such food was intended for the poor, as with the *carne do sol* of Brazil and the *charque* of Argentina.
3. On the processing of fish see the valuable book, *Fish Saving,* by F. L. Cutting (1955), and for a fascinating catalogue of North Atlantic fish as food, see Alan Davidson's *North Atlantic Seafood* (1979). Both works contain useful bibliographic information.
4. Butter, cheese, cream and, in Asia, yoghurt were ways of preserving milk. Throughout Eurasia, eggs were preserved by a variety of methods. Butter was salted for preservation, not only for internal use but also for sale. At Isigny in Normandy the inhabitants benefited from the right of "franc salé" which enabled them to set their butter and export it to the capital as early as the twelfth century (Segalen 1980: 88).
5. According to Drummond and Wilbraham (1939). Other authors however are sure that it was Durand's patent that was used by Donkin and Hall. Keevil mentions the figure of £1,000 for the purchase of the invention (1972: 6) and Durand is even referred to as the "father of the tin can." However iron containers coated with tin were already used by the Dutch to preserve salmon, a forerunner of the sardine process (Cutting 1955: 86). Appert himself visited London in 1814 to try and place orders for his bottled goods, and found the English industry more advanced in certain ways.
6. The contribution of canned goods to the running of the colonial regime, or rather to the maintenance of the metropolitican cuisine in foreign parts, is illustrated by Mrs Boyle's account of her journey to Wenchi in the interior of the Gold Coast in the middle of the First World War. She remarks that a large number of the boxes that had to be headloaded from railhead at Kumasi contained "food supplies of all kinds, provided in those days for so many Colonial officers by Fortnum & Mason, and ranging from soap and candles to tinned peaches, butter, sausages and so on—in fact enough to stock a District Officer's house for at least nine or ten months" (1968: 3).
7. Despite early contributions to the technology of canning, the British industry remained small until the 1930s. Quantities of canned food were imported before the First World War, mainly from America; even in the field of canned vegetables, the share of the home market in 1924 was only 5.1 percent (Johnston 1976: 173).
8. See Doudiet (1975) on coastal Maine cooking from 1760.
9. Drummond and Wilbraham report the case of a London fishmonger who "as early as 1820" used ice to bring Scotch salmon to London (1957: 308–9).
10. On the making and storing of ice among wealthy families in the Indian city of Lucknow, see Sharar 1975: 168.
11. See L. A. Coser, "Domestic Servants: the obsolescence of an occupational role" in *Greedy Institutions* (New York, 1974). According to Hobsbawm (1968: 131), the figure rose to 1.4 million by 1871, of which 90,000 were female cooks and not many more were housemaids. See also Keevil's remark on the growth of the firm of Fitch Lovell:

    Prosperity, high wages and inflation have also had their effect on the food trade. The well-off can no longer obtain domestic servants to cook for them. Attracted by high wages more and more women take jobs outside their homes, so a demand has been created for convenience foods and labour-saving gadgets. Nowadays the typical domestic servant is in a factory making washing machines, the typical shop assistant working in a cannery. A whole new food manufacturing industry has sprung up (1972: 9).
12. Indicative of this change, Collins points out, was the change of names. Post Toasties was formerly "Elijah's Manna," and the Kellogg Co. was the "Sanitas Nut Food Company." "Most breakfast cereals were originally marketed as 'natural,' 'biologic' or, in the case of Grape Nuts, 'brain" foods' (1976: 41).

13. See Anderson 1977: 338; but also Root 1971.
14. A London regulation, revoked in the fourteenth century, ran "Let no baker sell bread in his own house or before his own oven, but let him have a basket with his bread in the King's market" (Davis 1966: 24).
15. See for example the entries from the Chester Mayors' Books reprinted by Furnivall (1897: lxiii).
16. In Chester in 1591 the oath was taken by 33 Butchers and 30 Bakers. The Butchers' oath included the statement that "all such your victuall that you shall utter and sell, to poore and Riche, at reasonable prices" (Furnivall 1897: 153).
17. In German they were originally *Kolonialwarenhändler* (dealers in colonial produce).
18. On the impact of multiples on the retail trade itself, see Keevil's account of the growth of Fitch Lovell:

    Already by 1900 the growth of multiple shops was well under way and in the 1930s great integrated companies like Allied Suppliers, and Unigate began to appear. These companies expanded rapidly, as did multiple shops in every field, helped by the low value of property and low rents. The standardisation of packaging made possible mass advertising of branded food, selling it long before it reached the shop. Traditional wholesaling was becoming an expensive luxury and far-sighted firms like Fitch Lovell diversified out of it as quickly as possible. (1972: 8)
19. McCance and Widdowson note that the distinction between brown and white breads goes back at least to Roman times when "white bread . . . was certainly one of the class distinctions" (1956: 6). Indeed in eighteenth-century France white bread was customarily offered to the master and brown to the servants, even when eating at the same table (Flandrin 1979: 105). But the milling and preparation of bread was another, perhaps more basic ground, for class distinction, since both water and wind mills were at first the property of the manor or of the monastery. In Norman times and later, the miller rented the mill, and the other tenants were expected to bring their grain to be ground. Such relations lasted well into the seventeenth century. Like the mill, the bakehouse was also the property of the lord; serfs were compelled to bring their meal or dough to be baked there for a fixed charge, an imposition that they tried to avoid whenever possible for there was never the same need to use the bakehouse as the mill (McCance and Widdowson 1956: 13–14). All monasteries had their own bakehouses until the dissolution under Henry VIII. This basic food of the mass of the British population has shifted from domestic production in the early Middle Ages, to the lord's bakehouse under feudalism, to the local bakery under early capitalism, to the concentration of production in a few firms—four controlling about 70 per cent of bread production, according to Collins (1976: 18)—and its distribution through that contemporary descendant of the grocer's shop, the supermarket.

    The Chinese preference for the less beneficial white rice is also associated with status, partly because the milling was more expensive and partly because it stored better (Anderson 1977: 345). The same preference is found in India and the Middle East.
20. See Vehling 1936: 33 for a comment on Roman adulteration of food.

REFERENCES

Anderson, E. N. and M. L. 1977. Modern China: South. In K. C. Chang, ed., *Food in Chinese Culture*. New Haven.

Bitting, A.W. and K.G. 1916. *Canning and How to Use Canned Foods*. Washington, D.C.

Bitting, A.W. 1937. *Appertizing: Or, The Art of Canning; Its History and Development*. San Francisco.

Bitting, K.G. 1920. Introduction to N. Appert, *The Book for All Households*. Chicago.

Blackman, J. 1976. The Corner Shop: The Development of the Grocery Trade and General Provisions Trade. In D. Oddy and D. Miller, eds., *The Making of the Modern British Diet*. London.

Boyle, L. 1968. *Diary of a Colonial Officer's Wife*. Oxford.

Braudel, F. 1973. *Capitalism and Material Life, 1400–1800*. London.

Burnett, J. 1966. *Plenty and Want: A Social History of Diet in England from 1815 to the Present Day*. London.

Butterick. 1925. *The Story of the Pantry Shelf, An Outline of Grocery Specialties*. New York.

Chang, K. C., ed. 1977. *Food in Chinese Culture: Anthropological and Historical Perspectives.* New Haven.
Collins, E. J. T. 1976. The "Consumer Revolution and the Growth of Factory Foods: Changing Patterns of Bread and Cereal-Eating in Britain in the Twentieth Century. " In D. J. Oddy and D. Miller, eds., *The Making of the Modern British Diet*. London.
Corley, T. A. B. 1976. Nutrition, Technology and the Growth of the British Biscuit Industry, 1820–1900. In D. J. Oddy and D. Miller, eds., *The Making of the Modern British Diet*. London.
Coser, L. A. 1974. *Greedy Institutions*. New York.
Cummings, R. O. 1941. *The American and His Food*. Chicago, 2nd ed.
Cutting, C. L. 1955. *Fish Saving: A History of Fish Processing from Ancient to Modern Times.* London.
Davidson, A. 1979. *North Atlantic Seafood*. London
Davis, D. 1966. *Fairs, Shops and Supermarkets: A History of English Shopping*. Toronto.
Deutch, R. M. 1961. *The Nuts among the Berries*. New York.
Doudiet, E. W. 1975. Coastal Maine Cooking: Foods and Equipment from 1970. In M. L. Arnott, ed. *Gastronomy: The Anthropology of Food and Food Habits*. The Hague.
Drummond, J. C. and A. Wilbraham. 1939. *The Englishmen's Food: A History of Five Centuries of English Diet*. London.
Flandrin, J-L. 1979. *Families in Former Times: Kinship, Household and Sexuality*. Cambridge.
Francatelli, C. E. 1852. *A Plain Cookery Book for the Working Classes*. London.
Furnivall, F. J. 1897. *Child-marriages, Divorces, and Ratifications. . . .* London.
Hassall, A. H. 1855. *Food and Its Adulteration*. London.
Hobsbawm, E. J. 1968. *Industry and Empire*. New York.
Jefferys, J. B. 1954. *Retail Trading in Britain 1880–1950*. London.
Johnston, J. P. 1976. The Development of the Food-Canning Industry in Britain during the Inter-War Period. In D. J. Oddy and D. Miller, eds., *The Making of the Modern British Diet*. London.
Keevil, A. 1972. *The Story of Fitch-Lovell 1784–1970*. London.
McCance, R. A. and E. M. Widdowson. 1956. *Breads Brown and White*. London.
Mintz, S. 1979. Time, Sugar, and Sweetness. *Marxist Perspectives*, 2: 56–73.
Mote, F. W. 1977. Yuan and Ming. In K. C. Chang, ed., *Food in Chinese Culture*. New Haven.
Prentice, E. P. 1950. *Progress: An Episode in the History of Hunger?* New York.
Root, W. 1971. *The Food of Italy*. New York.
Segalen, M. 1980. *Mari et femme dans la société paysanne*. Paris.
Sharar, A. H. 1975. *Lucknow: The Last Phase of an Oriental Culture*. London.
Teuteborg, H. J. 1975. The General Relationship between Diet and Industrialization. In E. and R. Forster, eds., *European Diet from Pre-Industrial to Modern Times*. New York.
Vehling, J. D. trans. 1977. *Apicus: Cooking and Dining in Imperial Rome*. New York.
Wright, L. 1975. *The Road from Aston Cross: An Industrial History, 1875–1975*. Leamington Spa.
Yü, Y-S. 1977. Han. In K. C. Chang, ed., *Food in Chinese Culture*. New Haven.

25

Time, Sugar, and Sweetness

SIDNEY W. MINTZ

Food and eating as subjects of serious inquiry have engaged anthropology from its very beginnings. Varieties of foods and modes of preparation have always evoked the attention, sometimes horrified, of observant travelers, particularly when the processing techniques (e.g., chewing and spitting to encourage fermentation) and the substances (e.g., live larvae, insects, the contents of animal intestines, rotten eggs) have been foreign to their experience and eating habits. At the same time, repeated demonstrations of the intimate relationship between ingestion and sociality among living peoples of all sorts, as well as the importance attributed to it in classic literary accounts, including the Bible, have led to active reflection about the nature of the links that connect them. Long before students of Native America had invented "culture areas," or students of the Old World had formulated evolutionary stages for pastoralism or semiagriculture, W. Robertson Smith had set forth elegantly the concept of commensality and had sought to explain the food prohibitions of the ancient Semites.[1] But food and eating were studies for the most part in their more unusual aspects—food prohibitions and taboos, cannibalism, the consumption of unfamiliar and distasteful items—rather than as everyday and essential features of the life of all humankind.

Food and eating are now becoming actively of interest to anthropologists once more, and in certain new ways. An awakened concern with resources, including variant forms of energy and the relative costs of their trade-offs—the perception of real finitudes that may not always respond to higher prices with increased production—seems to have made some anthropological relativism stylish, and has led to the rediscovery of a treasure-trove of old ideas, mostly bad, about natural, healthful, and energy-saving foods. Interest in the everyday life of everyday people and in categories of the oppressed—women, slaves, serfs, Untouchables, "racial" minorities, as well as those who simply work with their hands—has led, among other things, to interest in women's work, slave food, and discriminations and exclusions. (It is surely no accident that the best early anthropological studies of food should have come from the pens of women, Audrey Richards and Rosemary Firth.[3]) What is more, the upsurge of interest in meaning among anthropologists has also reenlivened the study of any subject matter that can be treated by seeing the patterned relationships between substances and human groups as forms of communication.

While these and other anthropological trends are resulting in the appearance of much provocative and imaginative scholarship, the anthropology of food and eating remains poorly demarcated, so that there ought still to be room for speculative inquiry. Here, I shall suggest some topics for a study of which the skills of anthropology and history might be usefully combined; and I shall raise questions about the relationship between production and consumption, with respect to some specific ingestible, for some specific time period, in order to see if light may be thrown on what foods mean to those who consume them.

During and after the so-called Age of Discovery and the beginning of the incorporation of Asia, Africa, and the New World within the sphere of European power, Europe experienced a deluge of new substances, including foods, some of them similar to items they then supplemented or supplanted, others not readily comparable to prior dietary components. Among the new items were many imports from the New World, including maize, potatoes, tomatoes, the so-called "hot" peppers (*Capsicum annuum, Capsicum frutescens,* etc.), fruits like the papaya, and the food and beverage base called chocolate or cacao.

Two of what came to rank among the most important post-Columbian introductions, however, did not originate in the New World, but in the non-European Old World: tea and coffee. And one item that originated in the Old World and was already known to Europeans, the sugar cane, was diffused to the New World, where it became, especially after the seventeenth century, an important crop and the source of sugar, molasses, and rum for Europe itself. Sugar, the ingestible of special interest here, cannot easily be discussed without reference to other foods, for it partly supplemented, partly supplanted, alternatives. Moreover, the character of its uses, its association with other items, and, it can be argued, the ways it was perceived, changed greatly over time. Since its uses, interlaced with those of many other substances, expressed or embodied certain continuing changes in the consuming society itself, it would be neither feasible nor convincing to study sugar in isolation. Sweetness is a "taste," sugar a product of seemingly infinite uses and functions; but the foods that satisfy a taste for sweetness vary immensely. Thus, a host of problems arise.

Until the seventeenth century, ordinary folk in Northern Europe secured sweetness in food mostly from honey and from fruit. Lévi-Strauss is quite right to emphasize the "natural" character of honey,[4] for he has in mind the manner of its production. Sugar, molasses, and rum made from the sugar cane require advanced technical processes. Sugar can be extracted from many sources, such as the sugar palm, the sugar beet, and all fruits, but the white granulated product familiar today, which represents the highest technical achievement in sugar processing, is made from sugar cane and sugar beet. The sugar-beet extraction process was developed late, but sugar-cane processing is ancient. When the Europeans came to know the product we call sugar, it was cane sugar. And though we know sugar cane was grown in South Asia at least as early as the fourth century B.C., definite evidence of processing—of boiling, clarification and crystallization—dates from almost a millennium later.

Even so, sugar crudely similar to the modern product was being produced on the southern littoral of the Mediterranean Sea by the eighth century A.D., and thereafter on Mediterranean islands and in Spain as well. During those centuries it remained costly, prized, and less a food than a medicine. It appears to have been regarded much as were spices, and its special place in contemporary European tastes—counterpoised, so to speak, against bitter, sour, and salt, as the opposite of them all—would not be achieved until much later. Those who dealt in imported spices dealt in sugar as well. By the thirteenth century English monarchs had grown fond of sugar, most of it probably from the Eastern Mediterranean. In 1226 Henry III appealed to the Mayor of Winchester to obtain for him three pounds of Alexandrine sugar, if possible; the famous fair near Winchester made it an entrepôt of exotic imports. By 1243, when ordering the purchase of spices at Sandwich for the royal household, Henry III included 300 pounds of *zucre de Roche* (presumably, white sugar). By the end of that century the court was consuming several tons of sugar a year, and early in the fourteenth century a full cargo of sugar reached Britain from Venice. The inventory of a fifteenth-century chapman in York—by which time sugar was beginning to reach England from the Atlantic plantation islands of Spain and Portugal—included not only cinnamon, saffron, ginger, and galingale, but also sugar and "casson sugar." By that time, it appears, sugar had entered into the tastes and recipe books of the rich; and the two fifteenth-century cookbooks edited by Thomas Austin[5] contain many sugar recipes, employing several different kinds of sugar.

Although there is no generally reliable source upon which we can base confident estimates of sugar consumption in Great Britain before the eighteenth century—or even for long after—there is no doubt that it rose spectacularly, in spite of occasional dips and troughs. One authority estimates that English sugar consumption increased about fourfold in the last four decades of the seventeenth century. Consumption trebled again during the first four decades of the eighteenth century; then more than doubled again from 1741–1745 to 1771–1775. If only one-half of the imports were retained in 1663, then English and Welsh consumption increased about twenty times, in the period 1663–1775. Since population increased only from four and one-half million to seven and one-half million, the per capita increase in sugar consumption appears dramatic.[6] By the end of the eighteenth century average annual per capita consumption stood at thirteen pounds. Interesting, then, that the nineteenth century showed equally impressive increases—the more so, when substantial consumption at the start of the nineteenth century is taken into account—and the twentieth century showed no remission until the last decade or so. Present consumption levels in Britain, and in certain other North European countries, are high enough to be nearly unbelievable, much as they are in the United States.

Sugar consumption in Great Britain rose together with the consumption of other tropical ingestibles, though at differing rates for different regions, groups, and classes. France never became the sugar or tea consumer that Britain became, though coffee was more successful in France than in Britain. Yet, the general spread of these substances through the Western world since the seventeenth century has been one of the truly important economic and cultural phenomena of the modern age. These were, it seems, the first edible luxuries to become proletarian commonplaces; they were surely the first luxuries to become regarded as necessities by vast masses of people who had not produced them; and they were probably the first substances to become the basis of advertising campaigns to increase consumption. In all of these ways, they, particularly sugar, have remained unmistakably modern.

Not long ago, economists and geographers, not to mention occasional anthropologists, were in the habit of referring to sugar, tea, coffee, cocoa, and like products as "dessert crops." A more misleading misnomer is hard to imagine, for these were among the most important commodities of the eighteenth- and nineteenth-century world, and my own name for them is somewhat nastier:

> Almost insignificant in Europe's diet before the thirteenth century, sugar gradually changed from a medicine for royalty into a preservative and confectionery ingredient and, finally, into a basic commodity. By the seventeenth century, sugar was becoming a staple in European cities; soon, even the poor knew sugar and prized it. As a relatively cheap source of quick energy, sugar was valuable more as a substitute for food than as a food itself; in western Europe it probably supplanted other food in proletarian diets. In urban centres, it became the perfect accompaniment to tea, and West Indian sugar production kept perfect pace with Indian tea production. Together with other plantation products such as coffee, rum and tobacco, sugar formed part of a complex of "proletarian hunger-killers," and played a crucial role in the linked contribution that Caribbean slaves, Indian peasants, and European urban proletarians were able to make to the growth of western civilization.[7]

If allowance is made for hyperbole, it remains true that these substances, not even known for the most part by ordinary people in Europe before about 1650, had become by 1800 common items of ingestion for members of privileged classes in much of Western Europe—though decidedly not in all—and, well before 1900, were viewed as daily necessities by all classes.

Though research by chemists and physiologists on these substances continues apace, some general statements about them are probably safe. Coffee and tea are stimulants without calories or other food value. Rum and tobacco are both probably best described as drugs, one very high in caloric yield, and the other without any food value at all, though apparently having the effect at times of reducing hunger. Sugar, consisting of about 99.9 percent pure sucrose, is, together with salt, the purest chemical substance human beings ingest and is often labeled "empty calories" by physicians and nutritionists. From a nutritional perspective, all are, in short, rather unusual substances. With the exception of tea, these hunger-killers or "drug foods" destined for European markets were mostly produced in the tropical Americas from the sixteenth century onward until the nineteenth century; and most of them continue to be produced there in substantial amounts. What, one may ask, was the three-hundred-year relationship between the systems of production of these commodities, their political and economic geography, and the steady increase in demand for them?

Though remote from his principal concerns, Marx considered the plantations of the New World among "the chief momenta of primitive accumulation":[8]

> Freedom and slavery constitute an antagonism. . . . We are not dealing with the indirect slavery, the slavery of the proletariat, but with direct slavery, the slavery of the black races in Surinam, in Brazil, in the Southern States of North America. Direct slavery is as much the pivot of our industrialism today as machinery, credit, etc. Without slavery, no cotton; without cotton, no modern industry. Slavery has given their value to the colonies; the colonies have created world trade; world trade is the necessary condition of large-scale machine industry. Before the traffic in Negroes began, the colonies only supplied the Old World with very few products and made no visible change in the face of the earth. Thus slavery is an economic category of the highest importance.[9]

These and similar assertions have been taken up by many scholars, most notably, Eric Williams, who develops the theme in his famous study, *Capitalism and Slavery* (1944). In recent years a lively controversy has developed over the precise contribution of the West India plantations to capitalist growth in the metropolises, particularly Britain. The potential contribution of the plantations has been viewed in two principal ways: fairly direct capital transfers of plantation profits to European banks for reinvestment; and the demand created by the needs of the plantations for such metropolitan products as machinery, cloth, torture instruments, and other industrial commodities. Disputes continue about both of these potential sources of gain to metropolitan capital, at least about their aggregate effect. But there is a third potential contribution, which at the moment amounts only to a hunch: Possibly, European enterprise accumulated considerable savings by the provision of low-cost foods and food substitutes to European working classes. Even if not, an attractive argument may be made that Europeans consumed more and more of these products simply because they were so good to consume. But it hardly seems fair to stop the questions precisely where they might fruitfully begin. Of the items enumerated, it seems likely that sweet things will prove most persuasively "natural" for human consumption—if the word dare be used at all. Hence, a few comments on sweetness may be in order.

Claude Lévi-Strauss in his remarkable *From Honey to Ashes* (1973), writes of the stingless bees of the Tropical Forest and of the astoundingly sweet honeys they produce, which, he says,

> have a richness and subtlety difficult to describe to those who have never tasted them, and indeed can seem almost unbearably exquisite in flavour. A delight more piercing than any normally afforded by taste or smell breaks down the boundaries of sensibility, and blurs its registers, so much so that the eater of honey wonders whether he is savouring a delicacy or burning with the fire of love.[10]

I shall resist an inclination here to rhapsodize about music, sausage, flowers, love and revenge, and the way languages everywhere seem to employ the idiom of sweetness to describe them—and so much else—but only in order to suggest a more important point. The general position on sweetness appears to be that our hominid capacity to identify it had some positive evolutionary significance—that it enabled omnivores to locate and use suitable plant nutrients in the environment. There is no doubt at all that this capacity, which presumably works if the eating experience is coupled with what nutritionists call "a hedonic tone," is everywhere heavily overladen with culturally specific preferences. Indeed, we know well that ingestibles with all four of the principal "tastes"—salt, sweet, sour, and bitter—figure importantly in many if not most cuisines, even if a good argument can be made for the evolutionary value of a capacity to taste sweetness.

Overlaid preferences can run against what appears to be "natural," as well as with it. Sugar-cane cultivation and sugar production flourished in Syria from the seventh century to the sixteenth, and it was there, after the First Crusade, that north Europeans got their first sustained taste of sugar. But the Syrian industry disappeared during the sixteenth century, apparently suppressed by the Turks, who, according to Iban Battuta, "regard as shameful the use of sugar houses." Since no innate predisposition, by itself, explains much about human behavior, and since innate predispositions rarely get studied before social learning occurs—though there is at least some evidence that fetal behavior is intensified by the presence of sucrose, while human newborns apparently show a clear preference for sweetened liquids—how much to weigh the possible significance of a "natural" preference remains moot. For the moment, let it suffice that, whether there exists a natural craving for sweetness, few are the world's peoples who respond negatively to sugar, whatever their prior experience, and countless those who have reacted to it with intensified craving and enthusiasm.

Before Britons had sugar, they had honey. Honey was a common ingredient in prescriptions; in time, sugar supplanted it in many or most of them. (The term "treacle," which came to mean molasses in English usage, originally meant a medical antidote composed of many ingredients, including honey. That it should have come to mean molasses and naught else suggests, in a minor way, how sugar and its byproducts overcame and supplanted honey in most regards.) Honey had also been used as a preservative of sorts; sugar turned out to be much better and, eventually, cheaper. At the time of the marriage of Henry IV and Joan of Navarre (1403), their wedding banquet included among its many courses. "Perys in syrippe." "Almost the only way of preserving fruit," write Drummond and Wilbraham, "was to boil it in syrup and flavour it heavily with spices."[11] Such syrup can be made by supersaturating water with sugar by boiling; spices can be added during the preparation. Microorganisms that spoil fruit in the absence of sugar can be controlled by 70 percent sugar solutions, which draw off water from their cells and kill them by dehydration. Sugar is a superior preservative medium—by far.

Honey also provided the basis of such alcohol drinks as mead, metheglin, and hypomel. Sugar used with wine and fruit to make hypocras became an important alternative to these drinks; ciders and other fermented fruit drinks made with English fruit and West Indian sugars represented another; and rum manufactured from molasses represented an important third. Here again, sugar soon bested honey.

The uses of spices raises different issues. Until nearly the end of the seventeenth century, a yearly shortage of cattle fodder in Western Europe resulted in heavy fall butchering and the preservation of large quantities of meat by salting, pickling, and other methods. Though some writers consider the emphasis on spices and the spice trade in explanation of European exploration excessive, this much of the received wisdom, at least, seems well founded. Such spices were often used to flavor meat, not simply to conceal its taste; nearly all were of tropical or subtropical origin (e.g., nutmeg, mace, ginger, pepper, coriander, cardamom, turmeric—saffron is an important exception among others). Like these rare flavorings, sugar was a condiment, a preservative, and a medicine; like them, it was sold by Grocers (*Grossarii)* who garbled (mixed) their precious wares, and was dispensed by apothecaries, who used them in medicines. Sugar was employed, as were spices, with cooked meats, sometimes combined with fruits. Such foods still provide a festive element in modern Western cuisine: ham, goose, the use of crab apples and pineapple slices, coating with brown sugar, spiking with cloves. These uses are evidence of the obvious: that holidays preserve better what ordinary days may lose—just as familial crises reveal the nature of the family in ways that ordinary days do not. Much as the spices of holiday cookies—ginger, mace, cinnamon—suggest the past, so too do the brown sugar, molasses, and cloves of the holiday ham. More than just a hearkening to the past, however, such practices may speak to some of the more common ways that fruit was preserved and meat flavored at an earlier time.

Thus, the uses and functions of sugar are many and interesting. Sugar was a medicine, but it also disguised the bitter taste of other medicines by sweetening. It was a sweetener, which, by 1700, was sweetening tea, chocolate, and coffee, all of them bitter and all of them stimulants. It was a food, rich in calories if little else, though less refined sugars and molasses, far commoner in past centuries, possessed some slight additional food value. It was a preservative, which, when eaten with what it preserved, both made it sweeter and increased its caloric content. Its byproduct molasses (treacle) yielded rum, beyond serving as a food itself. For long, the poorest people ate more treacle than sugar; treacle even turns up in the budget of the English almshouses. Nor is this list by any means complete, for sugar turns out to be a flavor-enhancer, often in rather unexpected ways. Rather than a series of successive replacements, these new and varied uses intersect, overlap, are added on rather than lost or supplanted. Other substances may be eliminated or supplanted; sugar is not. And while there are medical concerns voiced in the historical record, it appears that no one considered sugar sinful, whatever they may have thought of the systems of labor that produced it or its effects on dentition. It may well be that, among all of the "dessert crops," it alone was never perceived as an instrument of the Devil.[12]

By the end of the seventeenth century sugar had become an English food, even if still costly and a delicacy. When Edmund Verney went up to Trinity College, Oxford in 1685, his father packed in his trunk for him eighteen oranges, six lemons, three pounds of brown sugar, one pound of powdered white sugar in quarter-pound bags, one pound of brown sugar candy, one-quarter pound of white sugar candy, one pound of "pickt Raisons, good for a cough," and four nutmegs.[13] If the seventeenth century was the century in which sugar changed in Britain from luxury and medicine to necessity and food, an additional statistic may help to underline this transformation. Elizabeth Boody Schumpeter has divided her overseas trade statistics for England into nine groups, of which "groceries," including tea, coffee, sugar, rice, pepper, and other tropical products, is most important. Richard Sheridan points out that in 1700 this group comprised 16.9 percent of all imports by official value; in 1800 it comprised 34.9 percent. The most prominent grocery items were brown sugar and molasses, making up by official value two-thirds of the group in 1700 and two-fifths in 1800. During the same century tea ranked next: The amount imported rose, during that hundred years, from 167,000 pounds to 23 *million* pounds.[14]

The economic and political forces that underlay and supported the remarkable concentration of interest in the West India and East India trade between the seventeenth and nineteenth centuries cannot be discussed here. But it may be enough to note Eric Hobsbawm's admirably succinct summary of the shift of the centers of expansion to the north of Europe, from the seventeenth century onward:

> The shift was not merely geographical, but structural. The new kind of relationship between the "advanced" areas and the rest of the world, unlike the old, tended constantly to intensify and widen the flows of commerce. The powerful, growing and accelerating current of overseas trade which swept the infant industries of Europe with it—which, in fact, sometimes actually *created* them—was hardly conceivable without this change. It rested on three things: in Europe, the rise of a market for overseas products for everyday use, whose market could be expanded as they became available in larger quantities and more cheaply; and overseas the creation of economic systems for producing such goods (such as, for instance,

slave-operated plantations) and the conquest of colonies designed to serve the economic advantage of their European owners.[15]

So remarkably does this statement illuminate the history of sugar—and other "dessert crops"—between 1650 and 1900 that it is almost as if it had been written with sugar in mind. But the argument must be developed to lay bare the relationships between demand and supply, between production and consumption, between urban proletarians in the metropolis and African slaves in the colonies. Precisely how demand "arises"; precisely how supply "stimulates" demand even while filling it—and yielding a profit besides; precisely how "demand" is transformed into the ritual of daily necessity and even into images of daily decency: These are questions, not answers. That mothers' milk is sweet can give rise to many imaginative constructions, but it should be clear by now that the so-called English sweet tooth probably needs—and deserves—more than either Freud or evolutionary predispositions in order to be convincingly explained.

One of Bess Lomax's better-known songs in this country is "Drill, ye Tarriers, Drill."[16] Its chorus goes:

And drill, ye tarriers, drill,
Drill, ye tarriers, drill,
It's work all day for the sugar in your tay,
Down behind the railway . . .

As such, perhaps it has no particular significance. But the last two verses, separated and followed by that chorus, are more pointed:

Now our new foreman was Gene McCann,
By God, he was a blamey man.
Last week a premature blast went off
And a mile in the air went Big Jim Goff.
Next time pay day comes around,
Jim Goff a dollar short was found.
When asked what for, came this reply,
You're docked for the time you was up in the sky.

The period during which so many new ingestibles became encysted within European diet was also the period when the factory system took root, flourished, and spread. The precise relationships between the emergence of the industrial workday and the substances under consideration remain unclear. But a few guesses may be permissible. Massive increases in consumption of the drug-food complex occurred during the eighteenth and nineteenth centuries. There also appears to have been some sequence of uses in the case of sugar; and there seems no doubt that there were changes in the use, by class, of sugar and these other products over time, much as the substances in association with which sugar was used also changed. Although these are the fundamentals upon which further research might be based, except for the first (the overall increases in consumption) none may be considered demonstrated or proved. Yet, they are so general and obvious that it would be surprising if any turned out to be wrong. Plainly, the more important questions lie concealed behind such assertions. An example may help.

To some degree it could be argued that sugar, which seems to have begun as a medicine in England and then soon became a preservative, much later changed from being a direct-use product into an indirect-use product, reverting in some curious way to an earlier function but on a wholly different scale. In 1403, pears in syrup were served at the feast following the marriage of Henry IV to Joan of Navarre. Nearly two centuries later, we learn from the household book of Lord Middleton, at Woollaton Hall, Nottinghamshire, of the purchase of two pounds and one ounce of "marmalade" at the astronomical price of 5s. 3d., which, say Drummond and Wilbraham, "shows what a luxury such imported preserved fruits were."[17]

Only the privileged few could enjoy these luxuries even in the sixteenth century in England. In subsequent centuries, however, the combination of sugars and fruit became more common, and the cost of jams, jellies, marmalades, and preserved fruits declined. These changes accompanied many other dietary changes, such as the development of ready-made (store-bought) bread, the gradual replacement of milk-drinking by tea-drinking, a sharp decline in the preparation of oatmeal—especially important in Scotland—and a decrease in the use of butter. Just how such changes took place and the nature of their interrelationship require considerable detailed study. But factory production of jams and the increasing use of store-bought (and factory-made) bread plainly go along with the decline in butter use; it seems likely that the replacement of milk with tea and sugar are also connected. All such changes mark the decline of home-prepared food. These observations do not add up to a lament over the passage of some bucolic perfection, and people have certainly been eating what is now fashionably called "junk food" for a very long time. Yet, it is true that the changes mentioned fit well with a reduction in the time which must be spent in the kitchen or in obtaining foodstuffs, and that they have eased the transition to the taking of more and more meals outside the home. "Only in the worst cases," writes Angeliki Torode of the mid-nineteenth-century English working class, "would a mother hesitate to open her jam jar, because her children ate more bread if there was jam on it."[18] The replacement of oatmeal by bread hurt working-class nutrition; so, presumably, did the other changes, including the replacement of butter by jam. Sugar continues to be used in tea—and in coffee, which never became a lower-class staple in England—but its use in tea is direct, its use in jam indirect. Jam, when produced on a factory basis and consumed with bread, provides an efficient, calorie-high and relatively cheap means of feeding people quickly, wherever they are. It fits well with changes in the rhythm of effort, the organisation of the family, and, perhaps, with new ideas about the relationship between ingestion and time.

"What is wanted," wrote Lindsay, a nutritionist of the early twentieth century, about Glasgow, "is a partial return to the national dish of porridge and milk, in place of tea, bread and jam, which have so universally replaced it in the towns, and which are replacing it even in the rural districts."[19] But why, asks R. H. Campbell, the author of the article in which Lindsay is cited, did people fail to retain the more satisfactory yet cheap diet of the rural areas?"[20] Investigators in Glasgow found a ready answer: "When it becomes a question of using the ready cooked bread or the uncooked oatmeal, laziness decides which, and the family suffers." In the city of Dundee, home of famous jams and marmalades, other investigators made an additional observation: The composition of the family diet appears to change sharply when the housewife goes to work. There, it was noted that such time-consuming practices as broth-making and oatmeal-cooking dropped out of domestic cuisine. Bread consumption increases; Campbell cites a statistic for the nineteenth century indicating that one family of seven ate an average of fifty-six pounds of bread per week.[21] Jam goes with bread. The place of laziness in these changes in diet remains to be established; the place of a higher value on women's labor—labor, say, in jam factories (though women worked mainly in jute factories in Dundee)—may matter more.

The rise of industrial production and the introduction of enormous quantities of new ingestibles occurred during the same centuries in Britain. The relationship between these phenomena is, on one level, fairly straightforward: As people produced less and less of their own food, they ate more and more food produced by others, elsewhere. As they spent more and more time away from farm and home, the kinds of foods they ate changed. Those changes reflected changing availabilities of a kind. But the availabilities themselves were functions of economic and political forces remote from the consumers and not at all understood as "forces." People were certainly not compelled to eat the specific foods they ate. But the range of foods they came to eat, and the way they came to see foods and eating, inevitably conformed well with other, vaster changes in the character of daily life—changes over which they plainly had no direct control.

E. P. Thompson has provided an illuminating overview of how industry changed for working people the meaning—nay, the very perception—of the day, of time itself, and of self within time: "If men are to meet both the demands of a highly-synchronized automated industry, and of greatly enlarged areas of 'free time,' they must somehow combine in a new synthesis elements of the old, and the new, finding an imagery based neither upon the seasons nor upon the market but upon human occasions."[22] It is the special character of the substances described here that, like sugar, they provide calories without nutrition; or, like coffee and tea, neither nutrition nor calories, but stimulus to greater effort, or, like tobacco and alcohol, respite from reality. Their study might enable one to see better how an "imagery based . . . upon human occasions" can take shape partly by employing such substances, but not always with much success. Perhaps high tea can one day become a cozy cuppa; perhaps the afternoon sherry can find its equivalent in the grog shop. But a great amount of manufactured sweetness may eventually lubricate only poorly, or even partly take the place of, human relations on all occasions.

The coffee break, which almost always features coffee or tea, frequently sugar, and commonly tobacco, must have had its equivalent before the industrial system arose, just as it has its equivalent outside that system today. I have been accused of seeing an inextricable connection between capitalism and coffee-drinking or sugar use; but coffee and sugar are too seductive, and capitalism too flexible, for the connection to be more than one out of many. It is not that the drug-food habits of the English working classes are the consequence of long-term conspiracies to wreck their nutrition or to make them addicted. But if the changing consumption patterns are the result of class domination, its particular nature and the forms that it has taken require both documentation and specification. What were the ways in which, over time, the changing occupational and class structure of English society was accompanied by, and reflected in, changes in the uses of particular ingestibles? How did those ingestibles come to occupy the paramount place they do in English consumption? Within these processes were, first, innovations and imitations; later, there were ritualizations as well, expressing that imagery based upon human occasions to which Thompson refers. But an understanding of those processes, of those meanings, cannot go forward, I believe, without first understanding how the production of the substances was so brilliantly separated by the workings of the world economy from so-called meanings of the substances themselves.

I have suggested that political and economic "forces" underlay the availabilities of such items as sugar; that these substances gradually percolated downward through the class structure; and that this percolation, in turn, probably fit together social occasion and substance in accord with new conceptions of work and time. And probably, the less privileged and the poorer imitated those above them in the class system. Yet, if one accepts this idea uncritically, it might appear to obviate the research itself. But such "imitation" is, surely, immeasurably more complicated than a bald assertion makes it seem. My research to date is uncovering the ways in which a modern notion of advertising and early conceptions of a large clientele—a mass market, or "target audience" for a mass market—arose, perhaps particularly in connection with sweet things and what I have labeled here "drug-foods." How direct appeals, combined with some tendency on the part of working people to mimic the consumption norms of those more privileged than they, can combine to influence "demand" may turn out to be a significant part of what is meant by meaning, in the history of such foods as sugar.

As anthropologists turn back to the study of food and eating and pursue their interest in meaning, they display a stronger tendency to look at food in its message-bearing, symbolic form. This has resulted in an enlivening of the discipline, as well as in attracting the admiration and attention of scholars in kindred fields. Such development is surely all to the good. But for one interested in history, there is reason to wonder why so few anthropological studies have dealt with long-term changes in such things as food preferences and consumption patterns, to which historians and economic historians have paid much more attention. In part, the relative lack of anthropological interest may be owing to the romanticism of an anthropology once resolutely reluctant to study anything not "primitive." But it appears also to stem from a readiness to look upon symbolic structures as timeless representations of meaning.

Hence, we confront difficult questions about what we take "meaning" to mean and within what limits of space and time we choose to define what things mean. No answers will be ventured here. But if time is defined as outside the sphere of meaning in which we are interested, then certain categories of meaning will remain and may then be considered adequate and complete. In practice, and for the immediate subject-matter, the structure of meaning would in effect be made coterminous with the political economy. For the substances of concern here—plantation products, tropical products, slave products, imported from afar, detached from their producers—the search for meaning can then be confined within convenient boundaries: the boundaries of consumption.

But if one is interested in the world economy created by capitalism from the sixteenth century onward, and in the relationships between the core of that economy and its subsidiary but interdependent outer sectors, then the structure of meaning will not be coterminous with the metropolitan heartland. If one thinks of modern societies as composed of different groups, vertebrated by institutional arrangements for the distribution and maintenance of power, and divided by class interests as well as by perceptions, values, and attitudes, then there cannot be a single system of meaning for a class-divided society. And if one thinks that meanings arise, then the separation of how goods are produced from how they are consumed, the separation of colony from metropolis, and the separation of proletarian from slave (the splitting in two of the world economy that spawned them both in their modern form) are unjustified and spurious.

Such substances as sugar are, from the point of view of the metropolis, raw materials, until systems of symbolic extrusion and transformation can operate upon them. But those systems do not bring them forth or make them available; such availabilities are differently determined. To find out what these substances come to mean is to reunite their availabilities with their uses—in space and in time.

For some time now anthropology has been struggling uncomfortably with the recognition that so-called primitive society is not what it used to be—if, indeed, it ever was. Betrayed by its own romanticism, it has sought to discover new subject-matters by imputations of a certain sort—as if pimps constituted the best equivalent of "the primitive" available for study. Without meaning to impugn in the least the scientific value of such research, I suggest that there is a much more mundane modernity equally in need of study, some of it reposing on supermarket shelves. Anthropological interest in things—material objects—is old and highly respectable. When Alfred Kroeber referred to "the fundamental thing about culture . . . the way in which men relate themselves to one another by relating themslves to their cultural material . . . ,"[23] he meant objects as well as ideas. Studies of the everyday in modern life, of the changing character of such humble matters as food, viewed from the perspective of production and consumption, use and function, and concerned with the emergence and variation of meaning, might be one way to try to renovate a discipline now dangerously close to losing its purpose.

NOTES

Versions of this paper were presented during the past few years at the University of Minnesota, Bryn Mawr College, Rice University, Wellesley College, Cornell University, the University of Pennsylvania, and at Johns Hopkins University's Seminar in Atlantic History and Culture. In radically modified form, these materials also formed part of my 1979 Christian Gauss Lectures at Princeton University. I benefited from comments by participants at all of these presentations, and from criticisms from other friends, including Carol Breckinridge, Carol Heim, and Professors Fred Damon, Nancy Dorian, Eugene Genovese, Jane Goodale, Richard Macksey, Kenneth Sharpe, and William Sturtevant.

1. W. Robertson Smith, *Lectures on the Religion of the Semites* (New York, 1889).
2. Audrey I. Richards, *Hunger and Work in a Savage Tribe: A Functional Study of Nutrition Among the Southern Bantu* (London, 1932); *Land, Labour and Diet in Northern Rhodesia: An Economic Study of the Bemba Tribe* (London, 1939).
3. Rosemary Firth, *Housekeeping Among Malay Peasants* (London, 1943).
4. Claude Lévi-Strauss, *From Honey to Ashes* (New York, 1973).
5. Thomas Austin, *Two Fifteenth-Century Cookbooks* (London, 1888).
6. Richard Sheridan, *Sugar and Slavery* (Baltimore, 1974).
7. Sidney W. Mintz, "The Caribbean as a Socio-cultural area," *Cahiers d'Histoire Mondiale,* IX (1966), 916–941.
8. Karl Marx, *Capital* (New York, 1939), I, 738.
9. Karl Marx to P. V. Annenkov, Dec. 28, 1846, *Karl Marx to Friedrich Engels: Selected Works* (New York, 1968).
10. Lévi-Strauss, *From Honey to Ashes,* 52.
11. J. C. Drummond and Anne Wilbraham, *The Englishman's Food* (London, 1958), 58.
12. I am indebted to Professor Jane Goodale of Bryn Mawr College, who first suggested to me that I investigate this possibility.
13. Drummond and Wilbraham, *Englishmen's Food,* 111.
14. Sheridan, *Sugar and Slavery,* 19–20. Statistics on tea are somewhat troublesome. Smuggling

was common, and figures on exports are not always reliable. That the increases in consumption were staggering during the eighteenth century, however, is not open to argument. See Elizabeth Schumpeter, *English Overseas Trade Statistics,* 1697–1808 (Oxford, 1960).

15. Eric Hobsbawm, *Industry and Empire* (London, 1968), 52.
16. See A. Lomax, *The Folk Songs of North America* (Garden City, N.Y., 1975).
17. Drummond and Wilbraham, *Englishmen's Food,* 54.
18. Angeliki Torode, "Trends in Fruit Consumption," in T. C. Barker, J. C . McKenzie, and John Yudkin, eds., *Our Changing Fare* (London, 1966), 122.
19. R. H. Campbell, "Diet in Scotland: An Example of Regional Variation," in ibid., 57.
20. Ibid.
21. Ibid., 58.
22. E. P. Thompson, "Time, Work Discipline and Industrial Capitalism," *Past and Present,* no. 38 (1967), 96.
23. Alfred Kroeber, *Anthropology* (New York, 1948), 68.

26

The Politics of Breastfeeding

An Advocacy Perspective

PENNY VAN ESTERIK[1]

INTRODUCTION

Politics may be defined as the practice of prudent, shrewd and judicious policy. Politics is about power. How, then, can politics have anything to do with breastfeeding? When health, profits and the empowerment of women are at stake, how could politics not be involved? Extraordinary changes in the way power is allocated in the world would be necessary for breastfeeding to flourish in this world. Many people believe such changes are impossible to make, that we have "advanced" too far into industrial capitalism to ever retreat into natural infant feeding regimes not based on profits. But even state policies influencing infant feeding practices can change, particularly when people begin to ask some very basic questions about child survival.

Advocacy on behalf of breastfeeding is incomplete and probably ineffective unless accompanied by a politically informed analysis of the obstacles to breastfeeding. These obstacles include the marketing practices of infant formula manufacturers, physician-dominated medical systems, and the relationship between industry and health professionals. This relationship has resulted in widespread misinformation about breastfeeding, including false claims of the equivalence between breastmilk and artificial substitutes, and the devaluing of women's knowledge about the management of breastfeeding.

The purpose of this paper is to trace the development of infant feeding as a public policy issue over the last few decades, to examine the role of non-governmental groups (NGOs) in influencing public policy, and to place breastfeeding within the advocacy debates on the promotion of commercial breastmilk substitutes, with the modest goal of putting the voices of industry critics more directly into discussions of the politics of breastfeeding. The paper concludes with a call for anthropologists to include advocacy discourses as a valid addition to other modes of understanding and interpretation.

THE DEVELOPMENT OF THE CONTROVERSY

Throughout history, women have substituted animal milks or wetnursing for maternal breastfeeding. This is, however, the first time in history when infants lived through these experiments long enough for others to measure the impacts on their health. This is also the first time that huge industries have promoted certain options for women, and profited from mothers' decisions not to breastfeed or to supplement breastmilk with a commercial product. It is this historical and economic fact that requires us to place breastfeeding in a broad political context.

An early presentation on the problem of bottle feeding may be traced to a Rotary Club address made by Dr. Cicely Williams in Singapore in 1939 entitled "Milk and Murder." She argued that the increased morbidity and mortality seen in Singapore infants was directly attributable to the increase in bottle feeding with inappropriate breastmilk substitutes, and the decline of breastfeeding. And she dared to call this murder—not something that happens to poor people over there, but murder. Her words:

> If you are legal purists, you may wish me to change the title of this address to *Milk and Manslaughter*. But if your lives were embittered as mine is, by seeing day after day this massacre of the innocents by unsuitable feeding, then I believe you would feel as I do that misguided propaganda on infant feeding should be punished as the most criminal form of sedition, and that these deaths should be regarded as murder. (Williams, 1986:70)

Although conditions in other cities in the developing world may have been similar or worse than in Singapore, the voices of warning and reproach were hesitant, isolated, and easily ignored. Conditions in many inner city and Native communities in North America today way be little improved over the conditions Williams found in Singapore in 1939.

Occasionally, reports from missionaries and health workers would confirm the devastating effects of bottle feeding on infant morbidity and mortality. But these were single voices and never stimulated a social movement. And it was easy to assume that the "problem" was "over there" and thus was irrelevant to promotional practices of infant food manufacturers in developed industrial countries. Only recently has the full extent of the dangers of commercial infant formula been acknowledged or publicized (Cunningham, Jelliffe, and Jelliffe, 1991; Palmer, 1993:306–312; Walker, 1993).

From the 1930s, the promotion of breastmilk substitutes steadily increased, particularly in developed countries. In North America, competition between American pharmaceutical companies and the depression reduced the number of companies producing infant formula to three large firms—Abbott (Ross), Bristol-Myers (Mead-Johnson) and American Home Products (Wyeth) (cf. Apple, 1980). Food companies like Nestlé were already producing baby foods before the turn of the century. Both food and drug-based companies producing infant formula expanded their markets during the post–World War II baby boom, as breastfeeding halved between 1946 and 1956 in America, dropping to 25 percent at hospital discharge in 1967 (Minchin, 1985:216). By that time, the birth rate in industrialized countries had dropped, and companies sought new markets in the rapidly modernizing cities of developing countries. As industry magazines reported "Bad News in Babyland" as births declined in the sixties in North America, their sales in developing countries increased, with only isolated and occasional protests from health professionals and consumer groups.

Other points of resistance to the increasing collaboration between infant formula manufacturers and health professions in North America came from mothers who wanted to breastfeed their infants and met with resistance or lack of support from the medical profession. These voices of resistance were not raised against the infant formula industry nor against the medical profession *per se*. Rather, they took the form of mother-to-mother support groups. The prime example is La Leche League, a group founded in 1956 in Chicago by breastfeeding mothers. The founding of La Leche League represented women's growing dissatisfaction with physician-directed bottle feeding regimes. While mother to mother support groups in some countries have lent support to infant food industry critics, it is important to remember that since its inception, La Leche League never directed its energies outward against infant formula companies, but rather inward toward the nursing couple. Only in recent years has the linkage been made between advocacy groups oriented towards consumer protests and mother support groups.

One phrase in a speech in 1968 by Dr. Derrick Jelliffe caught the attention of a much wider audience. He labelled the results of the commercial promotion of artificial infant feeding as "commerciogenic malnutrition." Like "Milk and Murder," this phrase grabbed headlines and became the focus for advocacy writing. By the mid-1970s, publications like the *New Internationalist* (1973) were bringing the problem to public attention. Reports such as Muller's *The Baby Killer* (1974) and the version by a Swiss group called *Nestlé Totet Babys* (Nestlé Kills Babies) prompted responses from Nestlé. In 1974, Nestlé filed libel charges in a Swiss court for five million dollars against the Third World Action Group for their publication *Nestlé Kills Babies*, leading to a widely publicized trial. The judge found the members of the group guilty of libel and fined members a nominal sum, but clearly recognized publicly the immoral and unethical conduct of Nestlé in the promotion of their infant feeding products. The libel suit and these popular publications provided focal points around which public opinion gradually developed, strengthening the efforts of advocacy groups in two complementary directions, the organization of a consumer boycott and drafting a code to regulate the promotion of baby foods (bottles, teats, and all breastmilk substitutes, not just infant formula).

STRATEGY FOR CHANGE: CONSUMER BOYCOTTS

Since the mid-1970s, a broad range of people from all walks of life, in many different parts of the world, have participated in a public debate known as the infant formula controversy, the baby food scandal or the breast-bottle debate. In North America, one catalyst for the "back to the breast" movement and a resurgence of interest in breastfeeding was a consumer movement organized by grass roots advocacy groups that drew attention to how the existence and advertising of commercial infant formula affected women's perceptions of their breasts, breastmilk and breastfeeding. They demonstrated that there was a direct and specifiable link between changes in infant feeding practices and the promotion of commercial infant formula in developing countries. The participation of ordinary people in North America in this debate was mostly through the direct action of a consumer boycott. Without the social mobilization of the consumer boycott, the work to promote a code for the marketing of breastmilk substitutes would not have been as effective.

Both boycott groups and promoters of a code to regulate the way infant formula was being promoted and marketed argued that the decline in initiation rates and the duration of breastfeeding could be linked to the expanding promotion of breastmilk substitutes, usually by multinational food and drug corporations, and to bottle feeding generally. The boycott against Nestlé's products, and eventually those of other infant formula manufacturers, generated the largest support of any grass roots consumer movement in North America, and its impact is still being felt in industry, governments, and citizen's action groups around the world. Women were the primary supporters of the boycott against Nestlé and other manufacturers of infant formula, although the movement in North America was strongly male dominated. Nevertheless, many women gained experience in analyzing the relations between corporate power and public health through their experience of working on the boycott campaign.

The groups that took on the task of challenging the infant formula companies were for the most part small, underfunded and in many cases ran on voluntary labor. While they were not the only people to recognize the problems of bottle feeding, they were the first to effectively mobilize to challenge the industries promoting it. Their success against the forces ranged against them, including powerful governments and multinational corporations, is a study in the power of co-operative networking. The importance of these small, non-governmental groups cannot be overstressed.

IBFAN (the International Baby Food Action Network) is a single-issue network of extraordinarily dedicated people—flexible, non-hierarchical, decentralized and international in organization (Allain, 1991). IBFAN works to promote breastfeeding worldwide, eliminate irresponsible marketing of infant foods, bottles and teats, advocate implementation of the WHO/UNICEF International Code of Marketing of Breastmilk Substitutes, and monitor company compliance with the Code.

In North America and Europe, advocacy groups also formed around the issue—most notably the Interfaith Centre for Corporate Responsibility (ICCR), the Infant Formula Action Coalition (INFACT) in Canada and the United States, the Baby Milk Action Group in Britain, the Geneva Infant Feeding Association (GIFA), and the many groups in developing countries that formed part of the IBFAN network. Throughout the late 1970s and early 1980s, these groups provided evidence of the unethical marketing of infant formula in their communities. This evidence was critically important in convincing delegates to the World Health Assembly (WHA, the meetings of the World Health Organization) that a regulatory code of industry practices was necessary.

The New York based ICCR, formed in 1974, monitored multinational corporations, provided information to church groups on responsible corporate investments, and publicized cases such as the lawsuit filed by the Sisters of the Precious Blood against Bristol-Myers in 1976 for misleading stockholders about their infant formula marketing practices. Although the lawsuit was dismissed, information about the marketing of breastmilk substitutes circulated in church basements among groups interested in Third World development and justice issues, bringing a new constituency into the movement. Public education on the promotion of breastmilk substitutes often featured the 1975 film, *Bottle Babies*, a vivid portrayal of the tragic effects of bottle feeding in Kenya.

In 1977, several action networks began the campaign to boycott Nestlé products in North America. The American INFACT (now called Action for Corporate Accountability) grew out of a student group at the University of Minnesota, while the Canadian. INFACT groups developed around justice ministries of the Anglican and United Churches, first in Victoria, British Columbia. These groups were linked together through IBFAN to represent the views of coalition members at international health policy meetings such as the World Health Assembly.

It was through these groups that the general public in North America was made aware of the infant formula controversy (or the breast-bottle controversy) through an increasingly sophisticated campaign involving public debates, newsletters, radio and T.V. shows, petitions, demonstrations, posters, buttons, and the first consumer boycott of Nestlé's products, which ended in 1984.

The advocacy position as defined by the boycott groups is quite straightforward. It argues that the makers of infant formula should not be promoting infant formula and bottle feeding in developing countries where breastfeeding is prevalent and the technology for adequate use of infant formula is absent. Advocacy groups claim that multinational corporations (like Nestlé), in their search for new markets, launched massive and unethical campaigns directed toward medical personnel and consumers that encouraged mothers in developing countries to abandon breastfeeding for a more expensive, inconvenient, technologically complex, and potentially dangerous method of infant feeding—infant formula from bottles. For poor women who have insufficient cash for infant formula, bottles, sterilization, equipment, fuel, or refrigerators; who have no regular access to safe, pure drinking water; and who may be unable to read and comprehend instructions for infant formula preparation, the results are tragic. Misuse of infant formula is a major cause of malnutrition and the cycles of gastroenteritis, diarrhea and dehydration that lead eventually to death. Advocacy groups place part of the blame for this "commerciogenic malnutrition" on the multinational companies promoting infant formula.

The boycott groups have never advocated a ban on the sale of infant formula, although some have advocated its "demarketing" (Post, 1985). Nor were women to be pressured to breastfeed against their will, although their critics represented their aims in this light. "Better to bottle feed with love than breastfeed with reluctance" is a cliché cited by many different people convinced that protecting mothers from feelings of guilt for not breastfeeding is more important than removing obstacles to breastfeeding. The intentions of INFACT and other boycott groups are clearly stated in their demands:

1. An immediate halt to all promotion of infant formula.
2. An end to direct product promotion to the consumer, including mass media promotion and direct promotion through posters, calendars, baby care literature, baby shows, wrist bands, baby bottles, and other inducements.
3. An end to the use of company "milk nurses."
4. An end to the distribution of free samples and supplies of infant formula to hospitals, clinics, and homes of new mothers.
5. An end to promotion of infant formula to the health professions and through health care institutions.

The infant formula companies responded to the boycott groups by modifying their advertising to the public, but they were slow to meet all INFACT demands and certainly never met the spirit of the demands, namely, to stop promoting their products. They simply promoted new products such as follow-on milks for toddlers, developed new marketing strategies, and hired public relations firms to answer their critics and to improve their corporate image.

Nestlé's efforts were concentrated on trying to improve their tarnished public image by hiring a prestigious public relations firm, sending clergy glossy publications about their contributions to infant health, and generally discrediting their critics as being merely uninformed opponents of the free enterprise system (Chetley, 1986: 46, 53). Companies such as Nestlé continue their efforts to buy social respectability by sponsoring events at international medical and nutrition conferences, and events celebrating the Canadian Year of the Family, for example, in addition to funding research on infant feeding.

Food boycotts have been a successful tool for social mobilization. Like all mass action social movements, the rhetoric used by advocacy groups oversimplifies the issue and seldom provides the statistically significant evidence that both the infant formula industry and medical journals call for (cf. Gerlach, 1980). But that is the nature of advocacy communication used by all social mobilization groups. At one level of analysis, the issue is both clear and simple; it is only complicated by the many obstacles ranged against breastfeeding. Nevertheless, the words and sentiments voiced in the original advocacy documents still ring clear today, as North American hospitals continue to make lucrative deals with infant formula companies.

STRATEGY FOR CHANGE: CODE WORK

Another parallel stream of activities for advocacy groups concerned lobbying and attending drafting sessions on the development of a code to regulate the marketing of breastmilk substitutes. Health professionals called for establishing policy guidelines on infant feeding through United Nation groups such as the Protein-Calorie Advisory Group. In 1979, WHO and UNICEF hosted an international meeting to develop an international code regulating the marketing of breastmilk substitutes. That meeting enabled nine infant formula companies to form the International Council for Infant Food Industries (ICIFI) (Palmer, 1993:237), and to lobby UN agencies for guidelines least damaging to their profits. The code was drafted with the cooperation and consent of the infant formula industry and is very much a compromise, a minimal standard rather than the ideal.

North American advocacy groups in IBFAN "... had to divide their very scarce resources and energy between running a boycott of Nestlé and the expensive periodic visits to Geneva for the Code drafting sessions" (Allain, 1991:10). Work in the United States to document abusive marketing practices of infant formula companies was brought to a head in 1978 by the Congressional Hearings on the Marketing and Promotion of Infant Formula in the Developing Nations chaired by Edward Kennedy. During the hearings, Ballarin, a manager of Nestlé's Brazilian operations, claimed—to the amazement of the hearing—that the boycott and the campaign against the infant formula companies was really an "attack on the free world's economic system," led by "a worldwide church organization with the stated purpose of undermining the free enterprise system" (United States Congress, 1978:127).

In May of 1981, the World Health Assembly adopted a non-binding recommendation in the form of the WHO/UNICEF Code for the Marketing of Breastmilk Substitutes with a vote of 118 for, 3 abstentions, and one negative vote. The negative vote was cast by the United States, in spite of the fact that it was the United States Senate that had proposed the idea of a Marketing Code and had initiated and actively participated in the drafting process. The American delegate to the WHA had been an enthusiast for the Marketing Code until shortly before the vote, when direct orders from his government ordered him to vote against its adoption. The Reagan White House had responded to direct lobbying from the infant formula industry (Chetley, 1986). The delegate who was ordered to reverse his nation's stance did so, and then resigned his post.

The Marketing Code is not a code of ethics but a set of rules for industry, health workers and governments to regulate the promotion of baby foods through marketing. It covers bottles, teats and all breastmilk substitutes, not just infant formula.

The code includes these provisions:

- No advertising of any of these products to the public.
- No free samples to mothers.
- No promotion of products in health care facilities, including the distribution of free or low-cost supplies.
- No company sales representatives to advise mothers.
- No gifts or personal samples to health workers.
- No words or pictures idealizing artificial feeding, or pictures of infants on labels of infant milk containers.
- Information to health workers should be scientific and factual.
- All information on artificial infant feeding, including that on labels, should explain the benefits of breastfeeding, and the costs and hazards associated with artificial feeding.
- Unsuitable products, such as sweetened condensed milk, should not be promoted for babies.
- Manufacturers and distributors should comply with the Code's provisions even if countries have not adopted laws or other measures. (World Alliance for Breastfeeding Action, 1994)

AFTER THE CODE

Following the establishment of the Code, Nestlé and other infant formula companies publicly released special instructions to its marketing personnel to comply with the Code, and asked the International Boycott Committee, a subgroup of IBFAN groups who were working on the boycott, to call it off. However, the boycott continued until 1984 when some means of monitoring company compliance with the Code could be established, and WHO member countries could draft national codes.

The advocacy groups, in the absence of national machinery, continued their monitoring role, recording and publicizing noncompliance of the Code (IBFAN, 1991). WHO and UNICEF have never monitored Code compliance, although they occasionally have taken individual companies to task. UNICEF's executive board extracted a promise that manufacturers would end all free supplies of infant formula to hospitals by the end of 1992. This was not done.

In the Philippines, a law banning free supplies was passed, but was evaded by the company tactic of invoicing for milk supplies and not bothering to collect payment. In the face of this and other flagrant violations, a second boycott against Nestlé and American Home Products in the United States, and Nestlé and Milupa in Germany was launched in 1988 by groups who were part of the IBFAN network. To date (1994) the second Nestlé boycott has spread to 14 countries and is most active in Europe and Latin America. However, it has never gained momentum in North America, ironically, because it could never live up to the success of the first boycott.

Disagreements about Code interpretations were to be referred to WHO. By the late 1980s, it was clear that Nestlé and other baby food companies had diverted part of their marketing budgets from public promotion into expanding the tactic of placing large quantities of free or low-cost milk in maternity facilities. Because of the inadequacy of medical training in breastfeeding management, health officials use these supplies for routine bottle feeding of newborns, which sabotages the successful establishment of breastfeeding.

In 1986, a World Health Resolution was adopted that acknowledged the detrimental effect of free or low-cost supplies and clarified the relevant Articles in the Code by banning such supplies. According to the resolution, free or low-cost supplies of infant formula were not to be given to hospitals. If supplies were donated to an infant, they were to be continued for as long as the infant required the milk. Hospitals that needed small quantities of infant formula for exceptional cases could buy them through the normal procurement channels. Thus, free supplies could no longer be used as sales inducements. Most of the major companies who were giving free supplies ignored the resolution, arguing that they would only stop distributing free supplies if governments brought in laws against them. However, the Code states that "Independently of other measures" manufacturers and distributors should take steps to ensure their conduct at every level conforms to the principles and aims of the Code.

At the World Health Assembly meeting in May of 1994, advocacy groups' successful lobbying reminded delegates that free and low-cost supplies of infant formula are marketing devices pure and simple, and not charity, a point made in 1989 by the Nigerian Minister of Health during the WHA. A few European countries including Ireland and Italy, and most forcefully, the United States delegation tried to defeat the resolution to end free supplies. But their efforts were thwarted by a block of African delegates and a very effective Iranian delegate who made it clear that the American position was the industry position as advocated by the International Association of Infant Food Manufacturers (IFM, the successor to ICIFI). The meeting ended with a consensus to withdraw all amendments and support the original text proposed by WHO's Executive Board to end donations of infant formula to all parts of the health care system worldwide. Once again, the question remains how such resolutions can be implemented and monitored. No doubt the advocacy groups will take up the challenge, or at least ensure that the issue does not quietly disappear from the world's conscience.

For all their rhetoric, and what some have decried their so-called confrontational tactics, the advocacy groups deserve great credit for bringing about what decades of clinical observations alone failed to accomplish: public awareness and concern about the dangers of breastmilk substitutes. This struggle for corporate accountability is often recounted in development education workshops as well as marketing classes (Post, 1985). For the first time, non-governmental organizations like INFACT, IBFAN and ICCR had a direct role in the deliberations at WHO and UNICEF in 1979 and in subsequent meetings regarding infant feeding policy. Chetley points out that in spite of industry's concerns about the "scientific integrity" of allowing popular organizations, mother's groups and consumer groups to participate, delegates to the international meetings were impressed with the contributions of the non-governmental organizations (1986:65–69). It is the NGOs that keep alive the underlying concern about corporate responsibility, human rights, and infant feeding as a justice issue.

THE UNHOLY ALLIANCE

Allain refers to the "unholy alliance" (1991:15) between the medical profession and the baby food companies. Certainly the medical profession and medical associations followed rather than led the advocacy groups in their criticism of industry. Although there was resistance by some doctors to the promotion of commercial baby foods, only occasional voices of protest were heard from health professionals in the 1950s and 1960s, as infant feeding became more completely medicalized.

In the United States, continuing efforts by doctors like Derrick Jelliffe and Michael Latham continually brought the issue of breastfeeding and promotional practices of industry to the attention of health organizations. Internationally, the advocacy groups turned a number of physicians into more outspoken public advocates for breastfeeding, stimulating a medical consensus on the value of breastfeeding. But many university and medical school research projects on infant nutrition are funded by industry money. Doctors are beginning to speak out against practices in their own hospitals, but they may be criticized by the medical establishment for doing so. As researchers are increasingly being warned (Margolis, 1991), there is no such thing as a free lunch, nor do people bite the hand that feeds them.

In 1986, an international group of doctors at an IBFAN Conference in Thailand drafted and endorsed the "Doctors' Declaration on Breastfeeding." Among several statements of commitment to pro-breastfeeding practices the Declaration promises not to permit the use of feeding bottles, teats or pacifiers in hospitals nor to accept personal funding from an infant food company. UNICEF's Physician's Pledge simply asks doctors to do their part "to protect, promote and support breastfeeding and to work to end the free and low-cost distribution of breastmilk substitutes in our health care systems" (UNICEF news release, July 22, 1994). While individual health professionals around the world work to change practices in their hospitals, hospitals and medical research still depend on industry support.

BREASTFEEDING IN THE NINETIES

In 1990, a global initiative sponsored by a number of bilateral and multilateral agencies resulted in the adoption of the Innocenti Declaration, which reads in part:

> As a global goal for optimal maternal and child health and nutrition, all women should be enabled to practice exclusive breastfeeding and all infants should be fed exclusively on breast milk from birth to 4–6 months of age. Thereafter, children should continue to be breastfed, while receiving appropriate and adequate complementary foods, for up to two years of age or beyond . . . Efforts should be made to increase women's confidence in their ability to breastfeed. Such empowerment involves the removal of constraints and influences that manipulate perceptions and behavior towards breastfeeding, often by subtle and indirect means. This requires sensitivity, continued vigilance, and a responsive and comprehensive communications strategy involving all media and addressed to all levels of society. Furthermore, obstacles to breastfeeding within the health system, the workplace and the community must be eliminated" (Innocenti Declaration, 1991:271–272)[2]

This carefully worded statement is nothing less than a challenge to change the priorities of the modern world. The language stresses the empowerment of women rather than their duty to breastfeed, a change that should bring more advocates for women's health to support breastfeeding.

Later in 1990, UNICEF convened a meeting to review progress on breastfeeding programs and concluded that if the Innocenti Declaration were ever to be implemented, work would have to be done by NGOs rather than governments alone. This led to the formation of an umbrella group called the World Alliance for Breastfeeding Action (WABA). WABA is a global network of organizations and individuals who are actively working to eliminate obstacles to breastfeeding and to act on the Innocenti Declaration. The groups include those who approach problems of breastfeeding from different perspectives—from consumer advocates to mother support groups and lactation consultants.

As part of their social mobilization efforts to gain public support for implementing the Innocenti Declaration, WABA sponsors World Breastfeeding Week (August 1–7) to pull together the efforts of all breastfeeding advocates, governments, and the public. The first campaign, in 1992, focused on hospital practices, and was called the Baby Friendly Hospital Initiative (BFHI). This campaign established steps that hospitals should take to support breastfeeding and to implement the Innocenti Declaration, and was based on the WHO/UNICEF statement, Ten Steps to Successful Breastfeeding. By 1994, over 1000 hospitals worldwide had been approved as baby-friendly (BFHI Progress Report, April, 1994). The second campaign, in 1993, tackled the problem of developing Mother-Friendly Workplaces, where breastfeeding and work could be combined. The complexity of the integration of women's productive and reproductive work, and the relevant cultural and policy issues have been explored elsewhere (Van Esterik, 1992). In 1994, attention returns to implementing the Code for the Marketing of Breastmilk Substitutes in all countries to meet the goals of the Innocenti Declaration.

BROADENING SUPPORT FOR BREASTFEEDING

The public appeal of the infant formula controversy was that it was presented as a simple, solvable problem. People in North America were attracted to the campaign because it put many of their unspoken concerns about the power of multinational corporations into a clear, concrete example of exploitative behavior that could be acted upon. For some boycott groups, the solution to the problem of bottle feeding with infant formula was for multinational infant formula manufacturers to stop promoting infant formula in developing countries. When the companies agreed to abide by the conditions of the WHO/UNICEF Code and the boycott was lifted, this marked the end of the campaign, a victory of small grass roots organizations over huge corporations. As with other social movements, it was hard to sustain interest in the issue after a "victory" had been declared. But the advocacy groups and most breastfeeding supporters recognized that infant feeding decisions are not related to marketing abuse alone; rather, the issue was embedded in a set of problems that require rethinking broader questions about the status of women, corporate power over the food supply, poverty and environmental issues.

For example, the implications of bottle feeding have not been explored from an environmental perspective with the exception of the position paper by Radford (1992). The ecology and environmental justice movements have been slow to recognize breastfeeding as part of sustainable development and breastmilk as a unique under-utilized natural resource. The report of the world Commission on the Environment and Development, *Our Common Future* (1987), made no reference to nurturance or infant feeding, although economy, population, human resources, food security, energy, and industry are all discussed as part of sustainable development.

Sustainability refers to courses of action that continue without damaging the environment and causing their own obsolescence. A sustainable infant feeding policy must consider the impact of decisions a number of years in the future, rather than simply examining conditions at the present. If we compare breastfeeding and bottle feeding as modes of infant feeding, each has very different implications for sustainability; breastmilk is a renewable resource, a living product that increases in supply as demand increases. It reinforces continuity with women's natural reproductive phases and is a highly individualized process, adapting itself to the needs of infant and mother. The infant is actively empowered and "controls" its food supply.

By contrast, the bottle feeding mode—most commonly associated with infant formula even in developing countries—is a prime example of using a non-renewable resource that uses even more non-renewable resources to produce and to prepare. It puts demands on fuel supplies and produces solid wastes—for every 3 million bottle-fed babies, 450-million tins are discarded. It is a standardized product that does not take into consideration individual needs (although in practice it is not really standardized; it is commonly adulterated in its industrial production with insect parts, rat hairs, iron filings and accidental excesses of chlorine and aluminum, or adjusted by the preparer to individualize it by the addition of herbs and sugar). The bottle-fed infant is passive, controlled by others, and becomes a dependent consumer from birth.

The issue of environmental pollution is critically important. A sustainable development policy for infant feeding must take careful note of the fact that women's capacity to breastfeed successfully is often a gauge for judging when our capacity to adapt to environmental stresses—air and water pollution, environmental toxins, radiation—has been overstrained. But women are not canaries or cows or machines. Breastfeeding promotion that treats women as merely milk producers is bound to fail, and the issue itself will be rejected by women's groups. Hence, an ongoing task for advocacy groups is to reposition breastfeeding so that it can be productively placed within the agendas of other advocacy groups, particularly environmental and feminist groups (cf. Van Esterik, 1994).

ADVOCACY AND ANTHROPOLOGY

Food advocacy is tiring work, and most anthropologists working in academic settings are not full-time activists or breastfeeding counsellors. Yet many of us have been drawn to research on breastfeeding by our personal experiences of mothering or by witnessing the commercial exploitation of women in different countries. Advocacy lessons are personal lessons because they require each and every one of us to put our values on the line—even occasionally to suspend academic canons of reserve and non-involvement, and respond emotionally to things we feel strongly about. In the study of breastfeeding, there is a convergence of different ways of knowing—a convergence of scientific knowledge, experimental knowledge, and experiential knowledge of generations of women, with moral and emotional values that all support action to support, protect and promote breastfeeding. Few areas of research in anthropology encourage such integration. Further, advocacy lessons are never far from us, as advocacy action permeates different parts of our lives and links diverse causes—from the women's movement to environmental concerns.

In this climate of reflexive anthropology and the increasing responsibility that the profession as a whole is taking in human rights debates, it is important that we clarify our relation to advocacy discourse and action as professional anthropologists and as citizens. Anthropology has a long history of applied work, but more recent and more problematic is the commitment of individual anthropologists to advocacy work (cf. Harries-Jones, 1985). But advocacy anthropology is still suspect to some in the profession. Advocacy refers to the act of interceding for or speaking on behalf of another person or group (Van Esterik, 1986), or promoting one course of action over another. This takes us beyond presentations, analyses, and discussion of evidence to recommend particular alternatives. Advocacy work draws some anthropologists into taking action with regard to well-defined goals that may best be implemented outside of academic settings. What has made this position acceptable in anthropology? First, the increasing numbers of anthropologists who have become involved in "causes" such as the rights of indigenous peoples, famine, AIDS, and women's rights, have made such commitment more visible within the profession. At the same time, the increasing involvement of indigenous peoples and special interest groups in advocating on their own behalf has resulted in anthropologists working with or for these groups.

Second, these individual and collective initiatives occurred at the same time as theoretical work arguing that there is no such thing as "scientific objectivity," and that many past examples of applied anthropology were both paternalistic and supportive of the status quo.

Third, feminist anthropology's epistemological stance on the lack of separation between theory and action justifies and even requires advocacy stances. Feminist methodology calls for explicit statements of the positionality of the author. The feminist axiom "the personal is political" breaks down opposition between "emotional advocacy action" and "cool, detached scientific reasoning," and accepts experience and emotion as valid guides to moral stands. This is particularly true of food activism, where the line between objective and participatory approaches to food is blurred. But as advocacy groups remind us, it is politics that determine whose truth is heard.

Finally, the recent involvement of all branches of anthropology in human rights debates, the theme of the 1994 meetings of the American Anthropological Association, requires rethinking the relation between advocacy and anthropology.

Advocacy for breastfeeding is one enormous anthropology lesson. Breastfeeding is simultaneously biologically and culturally constructed, deeply embedded in social relations, and yet cannot be understood without reference to varying levels of analysis including individual, household, community, institutional and world industrial capitalism. As much a part of self and identity as political economy; as personal as skin and as impersonal as the audit sheets of international multinational corporations, breastfeeding research requires a synthesis of multiple methods and theoretical approaches. At a time when anthropology hovers on the brink of self-reflexive nihilism and fragmentation on the one hand, and greater involvement in studying global change, internationalism, and public policy on the other (cf. Givens and Tucker, 1994), breastfeeding provides a challenging focus for holistic, biocultural, interdisciplinary research.

NOTES

1. The most comprehensive history of the controversy is Andrew Chetley's *The Politics of Baby Foods* (1986) and *Baby Milk: Destruction of a World Resource* from the Catholic Institute for International Relations (1993). I also review the history in my book, *Beyond the Breast-Bottle Controversy* (1989), and am using this opportunity to update that discussion. Here, I trace the development of the controversy highlighting the focal points of the movement up to and including 1994. This update has benefited from the views and writings of G. Palmer.
2. "The Innocenti Declaration was produced and adopted by participants at the WHO/UNICEF

policy-makers' meeting on 'Breastfeeding in the 1990s: A Global Initiative,' co-sponsored by the United States Agency for International Development (A.I.D.) and the Swedish International Development Authority (SIDA), held at the Spedale degli Innocenti, Florence, Italy, on 30 July–1 August 1990. The Declaration reflects the content of the original background document for the meeting and the views expressed in group and plenary sessions. (Innocenti Declaration, 1991:273)

3. The detailed reports on the infractions of the Marketing Code by infant formula companies are mostly in "fugitive literature"—letters, newspaper advertisements, and brief reports in low budget newsletters in many languages. The violations are most accurately reflected in the "SOCs," red and blue folders published by IBFAN since 1988, documenting the State of the Code, by country and by company. The advocacy groups in individual countries are the best sources for these records, particularly ACTION for Corporate Responsibility in the United States, INFACT Canada, IBFAN in Penang, Malaysia, GIFA in Geneva, Switzerland, and the Baby Milk Action Group in Britain. Their files are treasure troves for studying a social movement, but are not easily made to conform to academic standards of citation.

REFERENCES

Allain, A. 1991. IBFAN: On the cutting edge. *Development Dialogue* offprint, April:1–36, Uppsala, Sweden.

Apple, R. 1980. To be used only under the direction of a physician: Commercial infant feeding and medical practice, 1870–1940. *Bulletin of the History of Medicine* 54:402–417.

Baby Friendly Hospital Initiative. 1994. *Progress Report*. See endnote 3.

Catholic Institute for International Relations. 1993. *Baby Milk: Destruction of a World Resource* London:

Chetley, A. 1986. *The Politics of Baby Food*. London: Frances Pinter.

Cunningham, A. S., D. B. Jelliffe, and E. F. P. Jelliffe. 1991. Breast-feeding and health in the 1980s: A global epidemiologic review. *Journal of Pediatrics* 118(5):659–666.

Fildes, V. 1986. *Breasts, Bottles and Babies*. Edinburgh: Edinburgh University Press.

Gerlach, L. P. 1980. The flea and the elephant: Infant formula controversy. *Transaction* 17(6): 51–57.

Givens, D., and R. Tucker. 1994. Sociocultural anthropology: The next 25 years. *Anthropology Newsletter* 35 (4):1.

Harries-Jones, P. 1985. From cultural translator to advocate: Changing circles of interpretation. In *Advocacy and Anthropology*, edited by R. Paine, pp. 224–248. St. John's, Newfoundland: Institute of Social and Economic Research.

IBFAN. 1991. *Breaking the Rules*. Penang. See endnote 3.

INFACT, Canada. 1982. See endnote 3.

Innocenti Declaration. 1991. Innocenti declaration: On the protection, promotion and support of breastfeeding. *Ecology of Food and Nutrition* 26:271–273.

Margolis, L. H. 1991. The ethics of accepting gifts from pharmaceutical companies. *Pediatrics* 88(6):1233–1237.

Minchin, M. 1985. *Breastfeeding Matters*. Sydney, Australia: George Allen and Unwin.

Muller, M. 1974. *The Baby Killer*. London: War on Want.

Palmer, G. 1993. *The Politics of Breastfeeding*. London: Pandora Press.

Post, J. 1985. Assessing the Nestlé boycott: Corporate accountability and human rights. *California Management Review* 27(2):113–131.

Radford, A. 1992. The *Ecological Impact of Bottle Feeding*. WABA Activity Sheet #1. Penang, Malaysia.

Sussman, G. D. 1982. *Selling Mother's Milk: The Wet-Nursing Business in France, 1715–1914*. Urbana, IL: University of Illinois Press.

UNICEF News Release. 1994. See endnote 3.

United States Congress. 1978. Marketing and promotion of infant formula in the developing nations. Washington, D.C.: U. S. Government Printing Office.

Van Esterik, P. 1986. Confronting advocacy confronting anthropology. In *Advocacy and Anthropology*, edited by R. Paine, pp. 59–77. St. John's, Newfoundland: Institute for Social and Economic Research.

———. 1989. *Beyond the Breast-Bottle Controversy*. New Brunswick, N.J.: Rutgers University Press.

———. 1992. *Women, Work and Breastfeeding*. Cornell International Nutrition Monograph No. 23. Ithaca, New York.

———. 1994. Breastfeeding and feminism. *International Journal of Gynecology and Obstetrics*. Vol. 47, Suppl. pp. S.41–54.

Walker, M. 1993. A fresh look at the risks of artificial infant feeding. *J. Hum. Lact.* 9(2):97–107.

Williams, C. 1986. Milk and murder. In *Primary Health Care Pioneer: The Selected Works of Dr. Cicely Williams*, edited by N. Baumslag, pp. 66–70. Geneva, Switzerland: World Federation of Public Health Associations.

World Alliance for Breastfeeding Action. 1994. *Protect Breastfeeding: Making the Code Work*. World Alliance for Breastfeeding Action, Action Folder. Available through La Leche League International.

World Commission on Environment and Development. 1987. *Our Common Future*. New York: Oxford Press.

27

Hunger, Malnutrition, and Poverty in the Contemporary United States

Some Observations on Their Social and Cultural Context

JANET M. FITCHEN

That malnutrition and hunger exist in the contemporary United States seems unbelievable to people in other nations who assume that Americans can have whatever they want in life. Even within the United States, most people are not aware of domestic hunger or else believe that government programs and volunteer efforts must surely be taking care of any hunger that does exist here. And, to some extent, the focus of American public attention on "Third World hunger" and the enthusiasm for mass media events to raise money for famine relief divert attention from hunger and malnutrition at home. After all, the television pictures of distended bellies, matchstick legs, and gaunt faces are from Ethiopia, not the United States.

In the United States, hunger can go unnoticed because there is little overt begging for food and little obvious starvation. In fact, people who are poor enough to qualify for government-issued food stamps may be seen in grocery stores purchasing not only basic, inexpensive staples but also such widely popular items as frozen pizza, potato chips, soda pop, prepared desserts, and sometimes a beef steak. With such purchases, low-income people may be seeking to satisfy subjective as well as metabolic aspects of eating, perhaps attempting to convert their perceived hunger into a sense of well-being or to affirm that they can live like other Americans. But in so doing, they may inadvertently be transforming their hunger into malnutrition and also hiding their hunger from public awareness. The public, observing such "luxury" items among the grocery purchases of the poor, concludes that if poor people can eat steak, then they must be neither very poor nor very hungry. And so the problem of hunger receives little serious public attention.

But the diet of Americans who are poor should be compared to the nutritional status and eating patterns of the rest of the American population rather than to the condition of starving refugees in Africa. Although hunger anywhere is fundamentally a metabolic phenomenon with absolute dimensions, it also has important cognitive and relativistic attributes. Hunger in America, like the poverty that spawns it, must be understood in relation to the standards of living and eating in the surrounding society. In this article I argue that hunger is a significant problem in the United States today, in both its physical and its cognitive forms; and also that malnutrition, which is a more purely physical condition involving insufficient nutrients for growth and health, is a serious problem. I emphasize the need to understand the cultural aspect of hunger.

Hunger in the United States is not the result of insufficient foodstuffs for the total population: in fact, national food production is at unprecedently high levels, and farm surpluses continue to be a problem for the economy. It is, instead, a matter of some people regularly having inadequate access to sufficient food. And the recent increase in hunger outlined below is not solely a result of the policies and priorities of the present federal administration. True, the problem has been exacerbated as government programs providing food assistance to needy people have been cut back, altered, or eliminated. But simply to blame the current administration is to fail to understand the complexity of the situation.

To provide increased understanding of what hunger means in the American context and why this society has tolerated the continued existence of hunger and malnutrition in the midst of affluence, I look more deeply at the cultural dimensions of hunger, of eating, and of food assistance programs. I look not only at the people who go chronically underfed but also at the cultural context in which their hunger is embedded. The first part of the article presents the case that hunger and malnutrition do indeed exist in the United States, that they are closely associated with poverty, and that currently they are growing more prevalent. The main part of the article presents ethnographic data on the food and eating patterns of low-income people, suggesting that these patterns result both from the economic constraints of poverty and from the fact that the poor, despite their limited economic resources, follow many dominant American cultural ideas and practices. I indicate how these two sets of factors together shape eating patterns that may actually exacerbate malnourishment. In the following section I summarize prevailing American cultural assumptions about poverty and the poor, showing that these assumptions generate societal ideas about how the poor should eat. I then demonstrate that the same assumptions shape governmental policy and programs for food assistance. I conclude with the suggestion that in America there is a strong cultural belief, enshrined in government food assistance programs, that the poor should eat differently from other Americans because they are different: the poor should not buy steak.

RECENT HISTORY OF HUNGER AND MALNUTRITION IN AMERICA

Hunger Is Closely Associated with Poverty

A brief overview of recent trends of hunger and poverty at the national level underscores the correlation between the two phenomena. In the 1950s it was widely assumed that the poverty and hunger of the Depression era had been totally eradicated by a combination of federal New Deal programs, the economic stimulus of World War II, and post-war economic growth. But in the early 1960s serious poverty was "discovered" in the midst of America's affluence: it was estimated that between 22 percent and 25 percent of the U.S. population was living in poverty.[1] The John F. Kennedy campaign and presidency gave public recognition to the extent and severity of poverty and to the hunger and malnutrition associated with it. In 1967, when a group of U.S. senators and teams of physicians toured some of the newly discovered pockets of poverty, they found staggering evidence of hunger and malnutrition. These forays into the underclass led to books and television documentaries that stirred public opinion and to congressional testimony that galvanized the government into action.[2] From the mid-1960s through the mid-1970s, a massive effort was undertaken to combat poverty through community action programs, job training, various compensatory educational programs (such as Head Start), regional development schemes, and so forth.

While the "war on poverty" was attacking some of the underlying causes of poverty that contributed to hunger and malnutrition, new food assistance programs were also developed to attack hunger directly. The commodity distribution program, which provided handouts of such commodities as lard, milk powder, cheese, and dried beans from federal stockpiles to those of the poor who could get to the pick-up stations, was expanded to reach more people with more food items. Later the program was replaced by nationwide food stamps as a more effective form of assistance.[3] In 1969 President Richard Nixon, responding to the growing tide of public and congressional concern, declared to the nation that he would work "to put an end to hunger in America for all time." In a spirit of real commitment that surmounted entrenched opposition (see Kotz, 1969), Congress made adjustments in the new food stamp program to extend its benefits to a greater percentage of the needy. The federal program to provide free school lunches was expanded, advertised, and more adequately funded; and free breakfasts became available in many schools. A home-based nutrition education program was set up at the national level and implemented in all states, tailoring nutrition lessons to the exigencies of low-income living.[4] Extra food assistance was made available to poor people with particular needs, such as pregnant or lactating mothers, infants, and small children.[5]

As a result of the combined effort to combat hunger directly and to attack the underlying problem of poverty, and with a generally strong economy, both poverty and hunger decreased significantly. The poverty rate fell dramatically from its 1959 level of approximately 25 percent, nearly 40 million people, to a 1979 low of just under 12 percent (using a constant definition of poverty with dollar levels adjusted for inflation).[6] In 1977, when two Senate committees and a team of physicians restudied the same poverty pockets they had visited a decade earlier, they found real improvements in nutrition levels. Nationwide surveys conducted by the United States Department of Agriculture also found that although deficiencies of calories, calcium, iron, and vitamin C were still more common in low-income populations than in the general population, there was substantial improvement from 1966 to 1977 (Physician Task Force on Hunger in America, 1985: 68–69).

Recently, however, the figures on poverty, hunger, and malnutrition have deteriorated again. Nationwide, poverty rose to 13 percent in 1980 and to 15.2 percent in 1984—the highest level in two decades. Although by late 1985 the poverty rate had fallen back down to 14 percent, 33.1 million Americans were officially classified as poor. And there is little to suggest that the poverty rate will once again decline to the 12 percent level of 1979. The position of the poor relative to the rest of the society has also deteriorated recently, as is evidenced by the fact that the official poverty level for a family of four in 1985, $10,989, was almost the same as the median national income in that same year,

$11,013 (Pear, 1986). Figures on the percentage of people below the officially defined poverty line, moreover, give no indication of how far below that income level people actually are; most observers report that many people are further below the line now than they were a decade ago. The poor are getting poorer.

If the poor are more numerous and poorer now than when this decade began, they are hungrier and less well nourished also, as two recent studies show. One was coordinated by the Harvard School of Public Health and compiled under the title *Hunger in America: The Growing Epidemic* (Physician Task Force on Hunger in America, 1985); the other, compiled by Public Voice for Food and Health Policy, is entitled *Rising Poverty, Declining Health: The Nutritional Status of the Rural Poor* (Shotland, 1986). The poor are less well fed now not only because of reduced purchasing power (due to increased food costs relative to wages and welfare benefit levels) but also because fewer of them are currently receiving government food assistance and those who do get assistance are receiving less of it. As a result of more stringent eligibility guidelines for the food stamp program, which is the major form of federal food assistance, the number of households receiving food stamps decreased by 6.1 percent from 1983 to 1984 (from 21,073,000 to 19,778,000 people)—while the poverty rate dropped a mere 0.8 percent (Physician Task Force on Hunger in America, 1985: 99). (Even among those who are potentially eligible, only about 60 percent actually participate in the program.) For those who do receive food stamps, the monthly allotments have been reduced. The stamp allotment has never been intended to cover all household food needs, but generally it has been thought to fill the gap between food a household can afford to purchase on its own and food it should have to approach minimum daily requirements for all of its members. At present, the food stamp allotments fill a smaller part of that gap: the average bonus level for a four-person household in 1984 was only $147 (ibid.:91). Within the poverty population, specific groups who are at high risk for malnutrition now get less assistance than previously. For example, the number of children receiving free and reduced-price school lunches has decreased by 12 percent since 1980, according to United States Department of Agriculture figures (ibid.:98). Since most poor people lack food reserves, either in their bodies or in their cupboards, their being dropped from a government food assistance program or having benefit levels reduced is likely to have serious consequences for nutritional well-being.

Ample evidence of the recent increase in hunger and malnutrition appears throughout the Harvard study, including disturbing reports from pediatricians around the country. For example, doctors in Chicago report cases of marasmus (protein-calorie deficiency) and kwashiorkor (protein deficiency) reminiscent of the findings in the 1960s (ibid.:49–50). Children from low-income families continue to have high rates of anemia: in Minneapolis, 13 percent of the infants and 21 percent of the children among families applying for supplementary food assistance were determined by the Health Department to be anemic (ibid.:78).

Evidence of another kind comes from across the nation where in the last few years churches, volunteer groups, and local governments have found it necessary to set up soup kitchens, food banks, and other emergency food give-aways to supplement federal food assistance programs. Despite the rapid growth of such facilities, the number of people coming to each of them rose remarkably. In 1983 a random sample survey of 181 emergency food programs in the United States found that one-third of them had experienced at least a 100 percent increase in one year in the number of people coming for food (ibid.:9). In the Boston area food pantries served about 13,000 people monthly in 1982, nearly 30,000 in 1984 (ibid.:14). In Alabama, the Birmingham Community Kitchens that had served only 1,200 meals in 1980 served over 130,000 in 1984 (ibid.:25). A recently completed survey in New York State indicated dramatic rises in the number of people regularly turning to such food sources, with 63 percent of the programs reporting an increase in people served from 1984 to 1985 alone and only 5.5 percent reporting a decrease (Cornell University and New York State Department of Health, 1985). It can safely be assumed that most of the people who are turning increasingly to these food sources do so because they cannot afford to eat sufficiently well on their own—because they are hungry.

Corroborating evidence of the worsening problem of hunger comes from human-service workers in various states and programs. Nutrition educators conducting lessons in low-income homes increasingly observe empty cupboards and refrigerators and see more families lacking food or resources for the next day's meals. Teachers in Head Start have told me that children are eating voraciously at school on Monday mornings. And volunteers who operate soup kitchens have felt it necessary to give their Friday diners an extra bag of food for the weekend.

Hunger Is Unevenly Distributed in the Population

Within the national figures, those population groups, geographic regions, and age ranges most likely to fall below the poverty line (such as blacks, Hispanics, Indians, members of households headed by women, and children) are also most at risk for being hungry and malnourished. Children, for example, are disproportionately hungry or malnourished due to poverty. With the poverty rate at about 15 percent in 1983, more than 20 percent of all children under eighteen were living in households below the poverty line (O'Hare, 1985:17). Probably at least 15 percent of all children in the United States routinely experience sufficient hunger and malnutrition to cause such problems as anemia and lowered resistance to infection. If malnutrition occurs early enough, is severe enough, or persists for long periods, many of these children may also suffer long-term consequences such as impaired brain development and stunted body growth.

The South as a region is vulnerable to poverty-related hunger; and blacks as a group are particularly vulnerable. Predictably, therefore, in Mississippi there is again, or still, a serious problem of hunger in the black population (Physician Task Force on Hunger in America, 1985:18–28). In many of that state's counties, 50 percent to 75 percent of the black population falls below the poverty line (Mississippi Research and Development Center, 1981). However, only slightly over half of the state's households potentially eligible for food stamps are receiving them. Low levels of welfare benefit and a 6 percent sales tax on food exacerbate the hunger problem among black Mississippians.

Closely associated with both poverty and hunger is a high infant mortality rate (number of children per thousand who die before reaching age one). Among some populations, especially low-income blacks, infant mortality has risen recently despite technological advances in neonatal care. For American blacks as a whole, the infant mortality rate is roughly equal to the national rate in Costa Rica; and the disparity between infant mortality rates for blacks and whites in the United States, currently at a ratio of 2 to 1, continues to grow. In Pittsburgh, Pennsylvania, a city recently dubbed in a national survey one of the best places in America to live, the infant mortality rate among blacks is higher than in any other city in the United States, and nearly three times the rate for whites in the city (Physician Task Force on Hunger in America, 1985:72). The major causes of infant mortality in the United States are premature birth and low birth weight, which are closely associated with malnutrition of mothers, especially young mothers. Studies have shown that making supplemental foods available to pregnant women decreases the infant mortality rate by as much as 22 percent.[7] Even for those infants of low birth weight who do survive, as for all other infants, inadequate nutrition may lead to or exacerbate mental, developmental, and physical limitations that cannot be erased later. Born of an undernourished mother and undernourished during its first years of life, a child is likely to grow up to repeat the cycle: hungry because poor, and poor because hungry.

EATING PATTERNS OF THE POOR

In any human population, hunger is embedded in the larger context of eating, and so to understand hunger we need to understand eating. Eating is not simply a matter of ingesting calories, proteins, vitamins, and minerals. And just as eating is a culturally shaped act, so too is hunger culturally defined and invested with meanings that may outweigh its metabolic or nutritional aspects.

The cultural dimensions of hunger among poor Americans can be elucidated by the use of ethnographic research methods, which are particularly well suited for examining such food-related patterns as food preferences, frequency, quantity, and regularity of eating, distribution of food within the household, attitudes about foods, and social interactions associated with food. Several years of participant-observation research among rural poor people living in pockets of poverty in upstate New York have given me ample opportunity to witness food-related activities, including shopping and cooking as well as eating, within the context of everyday life (Fitchen, 1981). My research focused on over forty families living in several rural depressed neighborhoods, with twenty families studied quite intensively. During frequent, unscheduled, drop-in visits in homes, I listened to many food-related conversations and observed innumerable interactions involving food.

I have since supplemented this ethnographic research about hunger and foodways in one poverty setting with both first-hand and indirect observation in other regions of the nation. Particular insight has come from conducting training workshops for EFNEP, the Expanded Food and Nutrition Education Program that is operated nationwide by Co-operative Extension to bring nutrition information to low-income homemakers. From discussions with hundreds of EFNEP staff who conduct lessons in the homes of low-income families, I have learned about nutritional conditions and food patterns among the poor of various regions and ethnic and racial backgrounds, for example, southwestern Hispanics, urban blacks in the Northeast and deep South, rural whites in the Northwest, Indians in the upper Midwest, and Samoans in Hawaii. The opportunity to accompany EFNEP outreach educators in several states as they conducted lessons in the homes of their low-income clients has enabled me to observe mothers and children interacting over food and to listen to women talk about their food buying and preparation. These observations, though brief, have added breadth to the case study research and have confirmed that the eating patterns observed in the smaller sample can indeed be generalized. Additional insight has come from years of interaction with various food assistance and antipoverty programs at the federal, state, and local levels.

Low-income people probably vary as much as any other population segment in the United States in terms of individual and family food behaviors. But there are also some important similarities, so we can generalize about eating patterns of the poor without doing violence to variations among individuals and between groups.

Eating Patterns Are Shaped by the Constraints of Poverty

Poverty obviously affects the total amount of money a household can spend on food. Although the amount spent by a poor household on food may be considerably smaller than what a similar-size household with more available money spends, the poorer household is likely to spend a greater percentage of its income on food. Some low-income people are able to supplement their food supply from noncash sources, such as fishing and hunting, vegetable gardening and gathering wild fruits, raising chickens or other animals, and of course food stamps, free school lunches, and other forms of food assistance. Most poor people also obtain additional food by using their social resources, for example, by trading services and goods with relatives and neighbors to obtain food or food stamps. But for most poor Americans, as in the population as a whole, the majority of food is purchased (cf. Whitehead in Douglas, 1984:112).

Poverty also affects the amount of food that poor people can obtain for their money or their food stamps. Because of the constraints of poverty, the foods purchased by America's poor often cost more than the same foods purchased by more affluent people. Most poor people generally lack the surplus cash needed to take advantage of sale prices or to buy in quantity. Furthermore, inadequate storage space and refrigeration at home necessitate frequent trips to the store and smaller-scale purchases, usually at higher unit prices. Limited cash for public transportation (where it is available) and the lack of private cars restrict some people to shopping in small neighborhood stores rather than in more distant supermarkets with lower prices. From the inner cities of the eastern seaboard to the small towns on southwestern Indian reservations, food may cost more for those who are poor.

The economic exigencies of poverty also determine the types of foods that people eat most frequently. Interesting (and tasty) variations certainly exist in the foods preferred by different poverty populations; but in their menu combinations, in cuts of meat used (ibid.:118), and in modes of preparation (Goode et al. in Douglas, 1984:148) the inescapable constraints of poverty tend to override ethnic and regional differences. Indeed, I have been struck by overall similarities in the diets of America's poor, from Maine to Hawaii, from Mississippi to Alaska. By and large, poverty diets nationwide appear to be excessive in starches, fats, and sugars while being deficient in any or all of meats and other proteins, vegetables and fruits, and milk products. Particular dietary excesses and deficiencies vary considerably, as do particular preferences within any single good group. For example, the main starch may be rice, potatoes, tortillas, or Indian fry bread, according to ethnic preference, but whatever the case the preferred starch constitutes the bulk of a diet that may achieve satiety but also produces malnutrition. As a consequence of these dietary similarities, diet-related health problems are also similar across different poverty populations. For example, obesity and adult-onset diabetes, both common in many low-income populations, have been reported as epidemic among the Zuñi Indians, whose diet consists of an abundance of fats, sugar, and fast foods (Peterson, 1986). In Hawaii and American Samoa, low-income people are purchasing such high-fat, imported, and expensive starches as potato chips, a trend that EFNEP nutrition educators hope to reverse by stimulating new interest in the native, traditional taro, which is both cheaper and nutritionally better.

A common eating pattern in poverty populations across the nation is a marked periodicity in food consumption levels. Both purchase and consumption peak immediately after the paycheck, welfare check, or food stamps arrive. (Grocery stores may be especially crowded with customers using food stamps just after the first of each month.) Quantity and variety of food consumed subsequently taper down and level off, then take a nosedive in the last few days before the next check or stamp allotment. One mother I observed in Boston fried potatoes—with no accompaniment—for dinner for her four children on three successive nights at the end of one month. Even preschool children in the rural households I studied were fully aware of the check-to-check cycle of food availability: while examining bare cupboards, they eagerly listed treats they would ask for on payday. These periodic reductions in availability of food can create problems of both hunger and malnutrition, and EFNEP nutrition educators, who see this pattern frequently among their low-income clients, are attempting to help families stretch money, food stamps, and food to avoid the end-of-month hunger periods. There is some evidence that families receiving welfare, food stamps, and food assistance maintain relatively constant, if inadequate, nutrient intake throughout the month (Emmons, 1986).

The amount of food consumed also drops from time to time, when decreased income or unexpected expenses place additional strains on household finances. The unreliability of household income is only one source of fluctuations in food expenditures. Food is a nonfixed cost, and money set aside for food may go to the bill collector instead: one woman called this "eating light to pay for the lights." Human service workers, from nutrition educators to budget counselors, report that money allocated for food is often diverted to other uses when expenses increase unexpectedly, when debts become more pressing, or when income drops. Even personal family pleasure may sometimes come ahead of eating well, as long as the children are not complaining much.

The amount and quality of food that people eat is also frequently reduced by the necessity to stretch the household's food supply to feed extra people temporarily eating in or staying at the home. This sharing of food has been reported with increasing frequency in the last few years by nutrition educators in EFNEP in all states, and especially among black, Indian, and Hispanic populations. Although food sharing may temporarily diminish the food intake of the host household, it is an important coping strategy found in many poverty populations: it enables people to get through tight times by maintaining a system of reciprocity, an informal security network (Stack, 1974; Fitchen, 1981:106). Young children in families I observed were explicitly encouraged to share food and praised for food generosity; it was common to see an elementary school child bring home to share with a younger sibling a pocket full of cookies or candies from a party at school.

Another eating pattern found in diverse populations of low-income people is that food consumption is unevenly distributed within households. Although I collected no quantified food intake data in my case study to prove the point, observations during mealtimes and conversations with women indicated this tendency. When the cost of feeding a family must be eked out of a small income, always competing with other urgent needs, some individuals may go seriously underfed. Although the man of the household may receive ample quantities of food, for example, an older relative living in the household, perhaps an otherwise homeless person, may receive an insufficient amount of food. Some individual children routinely have insufficient access to food, through differential size or order of serving, outright denial of certain foods, or parental failure to accommodate a child's particular food needs. The vulnerable children are often those who occupy problematic positions within the household, for example children of a previous marriage or handicapped children. Some children may eat very little because of chronic untreated health problems, such as bad teeth, gastroenteritis, or anemia; but even if the health problem is addressed, the portions served may not be sufficiently increased as the child's health improves.

Even more common than the vulnerable child pattern is the phenomenon of the wife-mother who shortchanges her own food needs. In many poor households the woman eats only starches without any of the meats or vegetables she serves to the rest of the family. For herself, she may scrape the pot or lick the spoon and take whatever her children leave uneaten on their plates, but basically she eats just plain macaroni—or potatoes, or fry bread, or tortillas and beans. This pattern, which I observed during my field research, is recognized by EFNEP workers as one of the most common causes of nutritional problems among low-income women throughout the nation. It shows up clearly in the 24-hour food recalls, in which women enrolled in EFNEP are asked to record their own food intake. Many of these women proudly report that they have fixed a certain nutritious recipe for their children but admit that they did not eat any themselves. Although American women of other socioeconomic levels may also place the food needs and wants of family ahead of their own (see Whitehead in Douglas, 1984:126), "sacrificing for the sake of the children" has more deleterious nutritional consequences for poor women. It undoubtedly contributes to the observably high incidence of obesity, poor dental health, and generally low nutritional status of low-income women as compared to their non-poor counterparts and even to other members of their families.[8]

The perpetual condition of limited financial resources also affects when, where, and with whom people eat. Shortage of chairs, plates, or forks may mean that in many households meals are not taken with all members of the household assembled in one place or at one time. Some young children I observed were given or helped themselves to food off and on throughout the day: a nursing bottle filled with milk, juice, or soda pop or other artificially flavored sweet drink; a bowl of dry cereal; a peanut butter and jelly sandwich; soup eaten directly out of a can; doughnuts or a bag of potato chips. When a greater proportion of food is taken as snacks rather than meals, the result is apt to be a less well balanced diet, perhaps seriously so. In some homes, however, women went to considerable effort to cook meals, especially the evening dinner, and some were excellent cooks and ingenious at devising substitutes for ingredients or equipment they lacked. (One woman made her own bread, baking it in coffee cans in an oven that had been retrieved from the local garbage dump, propping the oven door shut with a stick.) But when financial or other problems overwhelmed these women, mealtime eating ceased to be planned, organized, or nutritionally balanced. During fieldwork, I was often able to gain a quick estimate of the current state of family life by observing or hearing about eating. For example, one woman in my sample was very conscious of nutritional needs and normally made a real effort to feed her family as well as possible. But during periods of parental depression or alcoholism, or marital violence, often brought on by money problems, each child helped herself from whatever happened to be in the cupboards.

Poverty also shapes the eating patterns of the poor by generating anxieties that center on food. Many adults, in recounting their childhood, cited periods when "there was no food in the house and no way to get any" and when "all we had for supper in those years was boiled potatoes and the water they were cooked in—we called it potato soup." The sense of deprivation engendered by such food shortages in early childhood seems to leave a lifelong sensitivity to the problem of having sufficient and desirable food. It may cause some adults to eat beyond the point of satiety or metabolic need, to overeat regularly and become obese. Food anxiety often carries over to the next generation too. Many of the mothers in my study consciously linked recollections of food deprivation in childhood and a present desire to give their children whatever foods they request, including soda pop and potato chips, so that the children might never feel denied or deprived. Thus for people living a whole lifetime in poverty, as for many of America's poor, the memory of childhood hunger in one generation may be a factor leading to malnutrition in the next.

For all poor people, the constraints of having to feed a family on an inadequate budget are exacerbated by the fact that hunger is cognitive as well as metabolic. The necessity of keeping children reasonably satisfied despite the shortage of money may take its toll on nutritious eating. In one common management strategy a mother responds to her children's complaints about being hungry by giving them a food item that is not only filling but also desired and liked. A package of frosting-covered cupcakes (high in desirability, sugar, and cost, but low in needed nutrients) may quickly pacify a child, thereby addressing the perceived and expressed hunger of the moment and allowing the mother to turn to other demands on her time and attention. Repeated reliance on this strategy, however, may lead to long-term nutritional deficit for the child. But the child's fussing now is more pressing and immediate and cannot be ignored; malnourishment, on the other hand, is delayed, is not so readily apparent, and has a less clear cause. Consciously as well as unconsciously, mothers may be dealing with the problem of inadequate access to food by trading off between hunger in the present and malnutrition in the future. In responding to cognitive hunger, they are inadvertently contributing to physical malnutrition.

Food may also occupy people's attention more when obtaining enough of it is problematic. Some young children I observed during my studies of rural poverty-stricken families seemed preoccupied with food: when not actually eating or begging a parent for something to eat, they would stand for whole minutes at a time just looking at whatever foods were still in the cupboard or refrigerator. One such child seemed simply to be reassuring himself that there was something left to eat. Perhaps this visual reassurance is part of the reason, along with limited kitchen storage space, why many food items always remain out on the kitchen table: bottles of ketchup, jars of instant coffee, bread, jam, peanut butter. (Perhaps it is also a literal manifestation of a saying one hears frequently among low-income people: "We may be poor, but at least we have managed to keep a roof over our heads and food on the table.") Some babies cling to the learned security of a nursing bottle until they are three or four; some carry the bottle around the house, hide it, and return to it periodically throughout the day.

Food is the source or the center of considerable interpersonal friction in many low-income homes. Many of the young children's temper tantrums I observed were connected with a demand for food. Minor disputes between mothers and children often revolved around the child's demands for something to eat and the mother's refusal—followed frequently by her capitulation. In many homes, arguments between spouses or lovers began over expenditures for food or the selection and preparation of foods. The most frequent form of this argument that I observed or heard about was a man castigating a woman for wasting "his" money on the purchase of some "unnecessary" food such as fruit juice, fruit, or vegetables. Food was sometimes used as a weapon (both literally and figuratively) in marital quarreling: a man who throws a plate full of food at his wife, a woman who punishes her man by preparing a dinner that she knows he dislikes. Where such food disagreements were common in the home, some children developed strong negative associations with all food and eating and ate very little even when food was available; other children, even in the same family, reacted in the opposite way, with an insatiable, voracious appetite.

Just as eating takes up a large portion of the budget and thoughts of low-income people, so also it occupies a large part of their actions. During the vast majority of my home visits in various poverty-stricken communities around the nation, at least one person in the household was eating. But it is not sheer physical hunger or metabolic need alone that fastens people's attention so much on food. Food and eating are enmeshed in feelings about self, interpersonal relationships, and dreams for the future, and these in turn are shaped by the surrounding culture.

Eating Patterns Are Shaped by General American Culture

As Mary Douglas (1984:3) has said, "unlike livestock, humans make some choices that are not governed by physiological processes. They choose what to eat, when and how often to eat, in what order, and with whom." As in any society, so in America, the definitions of acceptable and preferred foods are largely cultural. Contemporary food preferences that lean towards finger foods, fun foods, snack foods, and fast and convenient foods express basic American cultural values (Jerome, 1969). Low-income people express their membership in the society and their adherence to its dominant values through many of the same food choices that characterize the rest of the population. And so they, too, desire and purchase foods with these characteristics. Like most other Americans, poor people want to exercise "freedom of choice" in their food selection. (This is one reason why food stamps or vouchers for purchasing food at the grocery store are generally preferred to commodities.) In exercising this freedom, poor people select not only for price but also for desirability and therefore often purchase heavily advertised, status-invested foods "seen on television." Hence among the poor, as for the nation as a whole, diets may be high in processed foods, in sugars and fats, and in the category loosely termed "junk food."

The effect of junk foods on the nutritional status of the poor is probably worse than it is on the affluent. The well-to-do can afford both junk food and nutritious food; the poor can seldom afford both. The low-income parent who frequently succumbs to children's constant pleading for potato chips, soda pop, cheese-flavored puffs, and creme-filled cupcakes has no money left over for milk or carrots or apples. From the corner convenience store in inner-city Boston to the street vans parked by public housing projects in Honolulu, low-income parents have difficulty denying their children these advertised, desired products that are high in cost but low in nutritional value. One important factor contributing to the purchase of these products for children is the low self-image and sense of failure of the parent, a problem that seems pervasive in the American poverty population. Mothers with low self-esteem report that they have difficulty saying no to their children's demands for junk foods, even when they know that these foods are economically costly and nutritionally detrimental. (The association between poverty and low self-image is attested to by the responses of EFNEP outreach educators to a questionnaire about problems affecting the homemakers they teach. With over a thousand responses received from different regions of the country, I have found that the vast majority ranked "low self-esteem" as number one, two, or three in prevalence.) Single low-income mothers have told me that they are particularly vulnerable to buying treats for their children as an attempt to make up for the fact that the child has only one parent.

Foods and drink are as important to social interaction among the poor as they are in other segments of the American population—or any population. Some items, such as coffee, are more social than dietary. During field research, I was offered—and accepted—a cup of coffee almost every time I entered a home. The cost to the financially strapped family was outweighed in their minds by the importance of making this gesture of hospitality, this culturally prescribed presentation of themselves. In many homes, coffee was also consumed in great quantities whenever a family was going through some calamity and relatives and neighbors were dropping in to discuss events.

Poor people also reflect general American cultural patterns in their use of foods for celebrating. In rural poor households, I found, people were fully aware of the food choices and food consumption patterns appropriate for holidays and rites of passage. One woman started in September to use food stamps to purchase one Thanksgiving item each week: a can of cranberry sauce, a can of pumpkin, and so on. A mother on a very limited budget made a child's birthday more special by purchasing a "real" birthday cake, with icing and lettering on it, rather than buying the ingredients and making it herself at a lower cost. Women regarded the scrimping before and after these celebrations as preferable to the sense of deprivation felt when such socially prescribed foods of celebration are missing.

If people ate entirely on the basis of rational appraisal of nutritive value relative to dollars spent, then poor people could be convinced to ignore the preferences of their society. But poor people cling to and may even exaggerate dominant American food preferences because, despite their poverty, they are American—by culture if not by riches. And so, to the detriment of their own nutritional well-being, they may spend their food stamps and their scarce money on foods that are both expensive and nutritionally inferior. Despite what they may learn about nutritional needs and smart shopping, low-income people will continue to purchase convenience foods, snack foods, holiday foods, and status foods because they continue to classify themselves first of all as Americans and only second as poor Americans. If they could only accomplish it, most poor people would eat their way into the middle class.

CULTURAL ASPECTS OF THE RESPONSE TO HUNGER IN THE UNITED STATES

Dominant Cultural Values Shape Societal Attitudes about the Poor and What They Should Eat

Dominant American culture not only influences the foods poor people eat; it also influences the way the nonpoor think about eating, about poverty, and about what the poor should eat. To a considerable extent Americans are unaware of the impact of culture on what they eat. When asked, most claim that their food selections are purely a matter of individual preference, a claim that fits well with the cultural emphasis on individualism and individual choice. Most are equally unaware of the multiple functions served by their own food behaviors. Because of this general oversight, the public seriously underestimates the cultural and psychosocial factors shaping food patterns of people who are poor; and so there is ample room for societal attitudes about poverty to shape ideas about what poor people should eat.

The American cultural system rests on a belief that in this land of opportunity, the individual can and should shape his or her own destiny. One corollary is a pervasive conviction that the poor are casualties not of society but of their own shortcomings. Poverty is a condition of individuals, not of society: people are poor, the assumption is, because they are lazy and won't work and because they spend their money foolishly, purchasing only for immediate gratification, with no care for the future. It is an article of faith that the opportunity to escape poverty exists; the responsibility of poor people is to seize that opportunity and pull themselves up.

These beliefs about poverty and poor people shape dominant societal beliefs about what and how poor people should eat. Because the poor have little money, they should eat rationally on a cost/benefit basis, where costs are measured only in dollars and benefits solely in terms of nutrition. If poor people would eat this way, then surely there would not be malnutrition and hunger in America. By this reasoning, just as poverty is the fault of the poor, so hunger and malnutrition are the fault of the hungry and malnourished. The proposed solution, then, is that poor people should change their eating habits:

a. People who are poor should eat only the basics, consume only the foods needed by the body to maintain growth and health. (In fact, the official federal definition of poverty, and the determination of the poverty level, is based on a calculation involving the cost of a minimally adequate diet, actually 80 percent of it, assuming no expenditure for any foods other than minimal daily requirements.)

b. Poor people should not waste their money buying convenience foods. (Since the poor are too lazy to work, it is thought, they have plenty of time on their hands. They should economize by spending their unused time making meals from scratch rather than purchasing the convenience foods that busy working people eat.)

c. People who are poor should forego the favored, advertised, status foods, purchasing only cheap substitutes. (If people remain poor because they seek instant gratification, then to escape poverty they should learn to postpone pleasures for the future.)

These common public attitudes about poverty and what the poor should eat reveal a complete lack of awareness of the cultural aspects of eating. And they have important consequences, for they blend with our attitudes about food to shape the governmental response to hunger and malnutrition.

Cultural Attitudes about Poverty Shape Food Assistance Programs

The inadequacy of government food assistance programs goes deeper than the political conservatism of the present administration and the current constraints of budget deficits. The inadequacy exists—and is tolerated by the public—because of the ascendant myth that any individual who really wants to can overcome poverty and the accompanying assumption that if the poor are hungry and malnourished it must be of their own doing; they eat wrongly. Another deep and long-standing reason for the failure to guarantee decent, reliable food assistance to anyone who needs it may be a culturally generated fear that guaranteed food assistance (like guaranteed income) would create recipients dependent on government help. In a society that strongly favors independence and individual effort, aversion to creating dependence carries sufficient political appeal to outweigh concern that government food assistance is insufficient to prevent many children from growing up badly nourished, unhealthy, and poorly developed. When concern does rise, the more culturally appropriate solution involves voluntary efforts rather than governmental assistance.

Government assistance programs for the needy also reflect the fact that food is used instrumentally in our society. Early in life we learn that food is given or withheld at the discretion of the donor; food is a means by which we are controlled and can control others; food is used to reward and punish. So it is culturally appropriate that government food assistance is politically more palatable than the provision of a guaranteed minimum income, since the former contains an element of control over recipients. Food programs give assistance on their own terms: the donor determines what is given, when and where it is given, to whom, and under what conditions.

Government food assistance is neither shaped by nor solely conducted for the benefit of the people in need of food. In general, the interests and political power of the food industry and the agricultural sector of the economy exert a strong influence on food assistance programs. The food stamp program, the major form of food assistance, is entirely under the supervision of the United States Department of Agriculture (USDA) and subject to congressional and lobbying interests mostly representing agricultural producers, processors, and distributors—not the poor.[9] The USDA controls other food assistance programs as well, including the free and reduced-price school lunches and nutrition education for the poor. However, the poor are not the USDA's major constituency, and their needs do not claim top priority. Among the major food assistance programs, only the Women, Infants, and Children program is under a federal department primarily charged with people's well-being, the Department of Health and Human Services.

Some government food assistance programs also contain an element of "feeding the poor our leftovers." Consider, for example, the commodity distributions to the poor, also administered by the USDA. Such a program was used in the 1950s and early 1960s but was phased out in the late 1960s when it was decided that food stamps would better assist the poor in obtaining food. In the early 1980s, however, the commodity distribution program was reinstated as a supplement to food stamps, ostensibly in response to concern about reports of increasing hunger in America. The new version of the food give-away program started with cheese—more accurately, with blocks of pasteurized processed cheese food—and was subsequently expanded to include other commodities with seasonal and regional variations, depending on availability and deliveries. For example, in 1985 in an upstate New York county, families who proved or declared their poverty could receive every other month, on a first-come-first-served basis, the following goods: two five-pound blocks of processed cheese, two one-pound blocks of butter, and a choice of any two of the following items: 10 lbs. cornmeal, 4 lbs. powdered milk, 3 lbs. honey, 5 lbs. flour, or 2 lbs. rice. Many low-income people have responded positively to the give-aways, standing in lines at Salvation Army centers, churches, community halls, fire stations, and hospitals all across the nation to receive their generically packaged handouts.[10] (There are also nongovernmental food give-aways, many organized by church groups, in which excess or unsold foods donated by manufacturers and distributors are given to poor people.)

The foodstuffs given out by the government are not exactly scraps from the tables of the affluent, but they are clearly the leftovers from the food production industry. They represent the overproduction that threatens to bring down the price received by the producer/processor, the "surplus" purchased by the federal government to keep it off the market. The real objectives of the commodity distribution program, it appears, are to reduce the supply reaching the market, to dispose of government-owned surplus commodities in a way that obviates long-term storage costs, and to reduce embarrassment over surplus in the face of reportedly growing hunger. Although the public may believe that food distribution is designed, funded, and carried out solely for the altruistic purpose of reducing hunger, one could easily argue that the beneficiaries include not only the hungry but also the well-fed.

CONCLUSION: THE RICH BUY STEAK AND THE POOR GET CHEESE

Hunger, eating, and helping hungry people to eat—each is a cultural as well as a metabolic phenomenon. Food is culturally embedded for both the affluent and the poor. Through food patterns, all Americans enact cultural values and conduct interpersonal relationships. Cultural preferences, group identity, social interactions, and psychological needs all shape food-related behavior, no matter what the income level. Poor people are also subject to the same general, systemwide economic and social factors that tend to prevent most Americans from getting maximum nourishment per dollar. Because Americans are captive consumers in a total food production system that emphasizes profits rather than national nutritional well-being (Silverstein, 1984), and because all Americans buy foods to satisfy a variety of needs beside caloric and nutritional ones (Counihan, 1985), few Americans are as well nourished as they could afford to be.

The problem is significantly worse for low-income people, however, because for them there is no cushion of good health to tide them over periods of inadequate eating. The chronically poor have little opportunity to eat both the nutritious foods and also the heavily advertised, status-invested, and culturally preferred foods available to the upper classes. If they opt only for nutritionally "sensible" and cheaper foods, and forego the foods "that everyone else eats," their perceived sense of deprivation will be as genuine and gnawing as are the pangs of an empty stomach. But if they give in to the pressures of advertising, to the desire to eat like other people, and to the pleading of their children, then they will remain nutritionally and financially shortchanged—and also criticized by societal opinion for their "wasteful spending." It is a difficult set of choices, choices that have to be made every day but provide no clear way to win.

Because poverty-stricken Americans are influenced by dominant cultural values as well as by financial exigencies and metabolic requirements, a low-income person may occasionally purchase a beef steak. When other shoppers and grocery store employees observe such a purchase, behavioral conventions may restrict immediate reaction to a stare, but subsequent comments are uniformly negative. That the grocery cart is filled mostly with macaroni, potatoes, and bread may go unnoticed. The image of a poor person purchasing a steak sticks in the public mind as undeniable proof that in America the poor are living in luxury. If the poor are hungry in this land, it is thought to be their own fault, a result of their unwise spending.

The purchase of a steak by a poor person represents more than just an "unwise" use of money, however, and it carries significance far beyond the act itself. It is also culturally "inappropriate" behavior, for it violates an implicit cultural statement about the way things are or ought to be: the rich buy steak and the poor get cheese. This cultural statement actually encompasses three significant oppositions. The rich are the opposite of the poor. Buying food is the opposite of getting or being given food. And steak and cheese are familiar opposites: a thick piece of red meat is an ultimate desired food for Americans, while ordinary, processed cheese, such as the government gives to the poor, is a well-known cheap substitute for meat. Rich—poor; buy—get; steak—cheese. Each of these three pairs is a culturally recognized opposition, and the first term of each pair should be combined with only the first term of the other pairs. Rich is appropriately associated with buy and with steak. Poor is appropriately associated with getting and with cheese.

Thus it is culturally fitting that the government gives handouts of cheese to the poor: by defining what they shall eat and how they shall obtain it, the "cheese give-away" also defines who the poor are. When a poor person buys a steak, she or he is committing a symbolic inversion, performing an action associated with the rich and acquiring a food appropriate only to the rich. If the poor person gets the steak with food stamps rather than with cash, the purchase further violates what is thought to be appropriate because it violates the notion of how certain categories of food should be obtained: luxury foods should not be obtained with food stamps. But whether the steak is obtained with money or stamps, steak for the poor is a notable transgression because it violates the idea that the poor are different from the rest of us. It mocks our sense of societal order that demands separation of rich and poor.

NOTES

Adapted from a paper presented at the annual meeting of the Northeastern Anthropological Association, Buffalo, N.Y., March 1986, as part of a symposium, "Feast or Famine," organized by Carole Counihan.

1. Michael Harrington stirred public and governmental concern with his *The Other America* (1962), which documented the high incidence of poverty. He stressed both the uneven distribution of poverty in the population and its invisibility. Journalist Nick Kotz examined the politics of hunger and food programs (1969). The widely read works of anthropologist Oscar Lewis called attention to the intergenerational persistence of poverty. The untiring efforts of philanthropist-activist Robert Choate, of psychiatrist-writer Robert Coles (1969), and of political leaders such as senators George McGovern and Robert Kennedy contributed greatly to public awareness and demand for government action. Private foundations, such as the Field Foundation, were also instrumental in funding research and publicizing findings.
2. Investigations into the problem of hunger and malnutrition around the nation were sponsored or directly carried out by the Senate Select Committee on Nutrition and Human Needs, the Senate Subcommittee on Employment, Manpower, and Poverty, and the Citizens' Board of Inquiry (the last funded by the Field Foundation). In 1968 the powerful CBS television documentary *Hunger in America,* which was based on these investigations, brought action from Washington—despite sharp criticism from some in government.
3. The Food Stamp Program provides coupons that can be used like money in purchasing any foods (but no pet foods and no nonedible household supplies). The value of the coupons allotted is determined by household income and size. Booklets of coupons are issued monthly

to eligible families. Originally participating households paid a set amount each month, calculated according to their income, and received an amount of stamps equal to payment plus the bonus level (based on income and the household size). Subsequently the cash payment has been eliminated, so households receive stamps equivalent to the bonus level alone.

4. EFNEP, the Expanded Food and Nutrition Education Program, has been active in all states for 15 years as part of Co-operative Extension (which transmits information about agricultural and household topics from the state land-grant colleges to the public). EFNEP uses the extension mode of education to carry information about food and nutrition directly to a low-income audience. Through EFNEP low-income people (usually women) receive training in food and nutrition and then conduct individualized lessons in the homes of program participants. Lessons include not only nutrition needs and nutrient values but also food selection, preparation, and handling. Funded by a combination of federal, state, and local money, the program has recently been jeopardized by a threat of complete withdrawal of federal money.
5. This nationwide program, available in most counties of most states, is the Special Supplementary Food Program for Women, Infants, and Children, commonly known as WIC. Participating women are issued vouchers that are redeemable only on specified foods, such as milk, cheese, infant formula, infant cereals, and fruit juices.
6. The official definition of poverty used by the federal government to count the poor, and by most federal, state, and local programs to determine eligibility and benefit levels for assistance, is a straight income definition adjusted to the number of people in the household. It is based on the cost of obtaining food that would provide only about 80 percent of minimum daily dietary requirements and assumes that people need three times this much income to meet minimally adequate nutritional levels and other basic needs as well. The poverty level thus calculated is adjusted periodically to reflect the cost of living. In 1970 the poverty level for a family of 4, for example, was about $4,000; during 15 years of inflation it has been adjusted upward, reaching $10,989 for the same size household in 1985. All 4-person households with less than that level of income are officially defined as poor.
7. The Women, Infants, and Children program, WIC, has perpetually been underfunded and threatened with reduction in funds, which will mean longer waiting periods to obtain WIC benefits. For a single mother with little money to spend on food, a wait of just a few months before receiving WIC coupons for supplemental foods can have serious effects on her infant's well-being. For the pregnant woman, delay in obtaining WIC may affect the outcome of the pregnancy (premature birth, low birth weight). In many states the WIC program has lacked sufficient federal funding and staff to certify all those who apply. In New York State, as of March 1985, 16,300 low-income women who were pregnant and/or had young children were on the waiting list to obtain WIC benefits. Some states have recently had to reduce costs by dropping the eligibility age limit for children from six to three or below. In New Mexico only 28 percent of eligible infants and children could be served by WIC. Figures from Physician Task Force on Hunger in America, 1985.
8. Roe, 1973, in a study conducted on rural and urban women in the same region, reports a high number of medical complaints and chronic ailments, as well as obesity, among the low-income women to whom her team administered questionnaires, physical examinations, and laboratory tests.
9. See Kotz, 1969. When the Food Stamp program began, participants could not use their stamps to purchase any imported items. When it was seen that this regulation meant no coffee, tea, or bananas for the poor, the ruling was changed to allow imports if no domestic product were available. Thus tinned corned beef from Argentina could not be purchased with food stamps even when cheaper than its domestic counterpart. Even this restriction was subsequently dropped, but agricultural interests still prevail in shaping this program. Hence food stamps cannot be used to purchase essential kitchen cleaning supplies that might contribute to better health of the nation's poor: only food items are allowed.
10. Some products given out in the current distribution are not necessarily nutritionally advisable for all recipients. "Pasteurized process cheese food" is not an allowable purchase on WIC vouchers because this product is felt to be inadvisable for pregnant/lactating mothers and small children. The elderly have been cautioned about their free cheese because of its sodium content. Nutritionists warn recipients that honey should not be given to babies; but with the ending of federal honey subsidies, this commodity will no longer be distributed.

REFERENCES

Coles, Robert. 1969. *Still Hungry in America.* New York: New American Library.

Cornell University and New York State Department of Health. 1985. *Joint Report on Emergency Food Relief in New York State.* Ithaca, N.Y.: Cornell University, Division of Nutritional Sciences.

Counihan, Carole M. 1985. "What Does It Mean to Be Fat, Thin, and Female in the United States? A Review Essay." *Food and Foodways* 1:77–94.

Douglas, Mary. 1984. *Food in the Social Order: Studies of Food and Festivities in Three American Communities.* New York: Russell Sage Foundation.

Emmons, Lillian. 1986. "Food Procurement and the Nutritional Adequacy of Diets in Low-Income Families." *Journal of the American Dietetic Association* 86:1684–93.

Fitchen, Janet M. 1981. *Poverty in Rural America: A Case Study.* Boulder: Westview.

Harrington, Michael. 1962. *The Other America: Poverty in the United States.* New York: Macmillan.

Jerome, Norge W. 1969. "American Culture and Food Habits: Communicating through Food in the U.S.A." In Jacqueline Dupont, ed., *Dimensions of Nutrition,* pp. 223–34. Fort Collins: Colorado Dietetic Association.

Kotz, Nick. 1969. *Let Them Eat Promises: The Politics of Hunger in America.* Englewood Cliffs, N.J.: Prentice-Hall.

Lewis, Oscar. 1961. *The Children of Sanchez.* New York: Random House.

——. 1966. "The Culture of Poverty." *Scientific American* 215:19–25.

Mississippi Research and Development Center. 1981. *Handbook of Selected Data for Mississippi.* Jackson: State of Mississippi.

O'Hare, William P. 1985. "Poverty in America: Trends and New Patterns." *Population Bulletin* 40, 3.

Pear, Robert. 1986. "Poverty Rate Shows Slight Drop for '85, Census Bureau Says," *New York,* 27 August.

Peterson, Iver. 1986. "Surge in Indians' Diabetes Linked to Their History." *New York Times,* 18 February.

Physician Task Force on Hunger in America. 1985. *Hunger in America: The Growing Epidemic.* Boston: Harvard University School of Public Health.

Roe, Daphne A., and Kathleen R. Eickwort. 1973. "Health and Nutritional Status of Working and Non-Working Mothers in Poverty Groups." Report prepared for the U.S. Department of Labor, Manpower Administration. Ithaca, N.Y.: Cornell University, School of Nutrition.

Shotland, Jeffrey. 1986. *Rising Poverty, Declining Health: The Nutritional Status of the Rural Poor.* Washington, D.C.: Public Voice for Food and Health Policy.

Silverstein, Brett. 1984. *Fed Up: The Food Forces That Make You Fat, Sick, and Poor.* Boston: South End.

Stack, Carol B. 1974. *All Our Kin: Strategies for Survival in a Black Community.* New York: Harper & Row.

28

Beyond the Myths of Hunger

What We Can Do?

FRANCES MOORE LAPPÉ AND JOSEPH COLLINS

Some approaches to world hunger elicit our guilt (that we have so much) or our fear (that they will take it from us). Others imply impossible tradeoffs. Do we protect the environment *or* grow needed food? Do we seek a just *or* an efficient good system? Do we choose freedom *or* the elimination of hunger?

But our search for the roots of hunger has led us to a number of positive principles that neither place our deeply held values in conflict nor pit our interests against those of the hungry. We offer the following principles as working hypotheses, not to be carved in stone but to be tested through experience:

- Since hunger results from human choices, not inexorable natural forces, the goal of ending hunger is obtainable. It is no more utopian than the goal of abolishing slavery was not all that long ago.
- While slowing population growth in itself cannot end hunger, the very changes necessary to end hunger—the democratization of economic life, especially the empowerment of women—are key to reducing birth rates so that the human population can come into balance with the rest of the natural world.
- Ending hunger does not necessitate destroying our environment. On the contrary, it requires protecting it by using agricultural methods that are both ecologically sustainable and within the reach of the poor.
- Greater fairness does not undercut the production of needed food. The only path to increased production that can end hunger is to devise food systems in which those who do the work have a greater say and reap a greater reward.
- We need not fear the advance of the poor in the third world. Their increased well-being can enhance our own.

These and other liberating principles point to possibilities for narrowing the unfortunate rifts we sometimes observe among those concerned about the environment, rapid population growth, and world hunger.

GIVING CHANGE A CHANCE

Elsewhere we explained why most U.S. foreign aid actually sends our tax dollars to work against the hungry. But the question remained, if not promoting more U.S. aid, what *is* our responsibility to the hungry?

We responded that the most important step Americans can take to end hunger is to remove U.S. support—financial, diplomatic, and military—from regimes determined to resist the changes necessary to end hunger.

Even many Americans who agree with our approach to the problem of hunger may balk at this recommendation. "But, we can't do that! If we don't support those regimes, the Soviet Union will fill the vacuum. Nothing new will be allowed to emerge; things will only be worse." How often we have heard this!

We have thought long and hard about this fear. We understand it. We have tried to think through exactly what choices we have. Aren't there really only two? On the one hand, we can allow our government to continue on its present course—blocking change. Or we can give change a chance.

Where does the first choice lead?

Two quite different countries have come to symbolize for us the logical consequences of this course.

The first is Guatemala. In the early 1950s, the U.S. government abetted the overthrow of an elected government attempting to carry out a modest land reform. Over many years, U.S. military and economic aid strengthened the grip of governments responsible for imprisoning, torturing, and murdering tens of thousands of Indian peasants (and making many more into refugees), virtually all opposition leaders, and hundreds of churchworkers—that is, anyone seeking political and economic reforms.

Guatemala has perhaps the worst human rights record in all Latin America. In 1984, the respected human rights organization Americas Watch called Guatemala a "nation of prisoners."[1] That same year the Guatemalan military permitted the election of a civilian as president, but terror against the poor and other dissidents persists. So well entrenched are the oligarchy and their military that many observers doubt the elected government will be able to enact reforms addressing Guatemala's appalling poverty and hunger. And even if the coming years were to bring reform, several decades would be required to undo the damage wrought with U.S. backing. The Philippines, El Salvador, Zaire, Chile, Haiti, South Africa, Paraguay, Indonesia—we could use these and several other countries to make the same point.

Cuba represents an equally predictable consequence of the same course—a policy based on blocking change. Historically, the United States supported corrupt, authoritarian regimes in Cuba even though they perpetuated misery and hunger for many Cubans. When Fidel Castro's government threatened to nationalize a U.S.-owned oil refinery—as decades earlier Mexico had nationalized its oil fields—the United States retaliated with hostilities that continue to this day. Along with multiple failed attempts to assassinate or overthrow Castro, the United States has used all its power to isolate Cuba internationally: trade embargoes, travel restrictions, and lobbying against aid by international lenders. The United States even imposes its policy of fear on its allies, refusing to import goods containing Cuban-made parts.[2]

If U.S. policymakers fear the emergence of a Soviet satellite near our borders, no policy could have been better designed to turn that fear into reality. And if their concern is for political freedom in Cuba, the U.S. government's unrelenting hostility and repeated attempts at subversion help create in Cuba an environment *least* likely to allow the flowering of civil liberties.[3]

Guatemala and Cuba represent the outcome of one choice. Fortunately, there is another choice. Primarily, it would entail our government's obeying the law—both U.S. laws and U.S.-signed international treaties that forbid supporting governments notorious for their human rights violations. It would mean an end to covert and overt operations to "destabilize" societies where reforms necessary to end hunger are under way.

Americans are told that following such a course would pave the road for Soviet satellites throughout the third world, with Cuba cited as proof. But as we have just pointed out, developments in Cuba are in part the outcome of policies based on U.S. hostility to change, *not* on an alternative course.

Years of study about and experience in numerous third world nations have led us to predict a different outcome if the United States were to change its course. Our prediction is based on the observation that any nation that has for decades, even centuries, been under the control of elites beholden to foreign interests will above all yearn for sovereignty. Such movements for change will want to do it *their* way—if they are given the chance. The last thing they will want to become is a puppet of a foreign power. And domestically, they will likely seek to avoid becoming a carbon copy of either dominant model—U.S.-style capitalism or Soviet-style statism.

Our close-up observation of Nicaragua over the last seven years has strongly confirmed our hunch.[4] Looking at the pattern of Nicaragua's aid and trade ties with other countries, we have been struck by the new government's efforts to avoid dependency on any one power bloc. In 1984, most of the value of Nicaragua's imports came from Latin America, Western Europe, and the United States. About one quarter came from the Eastern bloc. In 1984, only 6 percent of Nicaragua's exports went to socialist countries. In loans to Nicaragua, a similar pattern emerges. Between 1979 and 1984, of the almost $3 billion in loans made to Nicaragua, nearly two-thirds came from other Latin American countries, multilateral lending institutions like the World Bank, and Western European countries, while about a quarter came from the Eastern bloc. Only as Western sources of aid have cut back, in large part in response to U.S. pressure to isolate Nicaragua, has the share of its loans from the Eastern bloc increased, reaching 60 percent in 1984.[5]

Nicaragua's domestic economic policies also confirm our sense that third world movements for change will seek to break loose from *both* dominant economic models. About 60 percent of Nicaragua's economy (and over three-quarters of its farmland) is in private hands, and its experiments in political participation include forms tried in neither East nor West.[6]

Americans have been told that Nicaragua, like Cuba, is a direct threat to our own security. But can anyone seriously believe either of these tiny countries could harm the United States? Since the 1962 Cuban missile crisis, the United States has made clear that it would not tolerate weapons installations near our borders that might threaten our security. Satellite-gathered intelligence allows us to be certain that prohibition is not violated. Rather than a threat, both Nicaragua and Cuba could contribute to the U.S. economy if the United States established trading ties, as we now have with China and in certain fields with the Soviet Union.

But if the U.S. government continues its hostility to change, we may be deprived of knowing the full possibilities of economic and political change in the interests of the majority. Any society attacked by a much more powerful enemy will find it difficult to allow free debate or to invest its scarce resources in an alternative development path.

A RELEVANT EXAMPLE

If both common sense and historical experience suggest that third world peoples, if allowed to do so, will want to chart new paths, what do Americans have to offer?

We Americans have always thought of our country as a beacon of hope for the world's oppressed. But as we travel throughout the third world, we sense a change. We fear our example is becoming increasingly irrelevant to the poor majority abroad.

While our government extolls the virtues of democracy and freedom, America's present version of these two values appears unrelated to the concerns of the hungry—food, access to land, and jobs. Our government praises third world elections as creating democracies, but most of the hungry people in the world today live in countries—India, Brazil, El Salvador, Pakistan, Sudan, Egypt, Indonesia—where there have been elections, yet the majority of people find themselves no better able to meet their needs.

Even more directly stated, if amid our nation's fantastic food bounty, poor American children are stunted by malnutrition, what example of hope do we offer to children in the third world? If, with an unparalleled industrial and service economy, millions go without work even during a period of economic growth and millions more work full time yet remain in poverty, what hope do we offer the impoverished and jobless in the third world?

We fear the answer is very little as long as Americans' understanding of democracy and freedom fails to address the most central concerns of the poor.

This realization suggests that we can contribute toward ending world hunger not only by helping to remove obstacles in the way of change in the third world but also by what we do right here at home. In the preceding chapter, we quoted philosopher Henry Shue who argues that subsistence rights—what we call economic rights—are just as central to freedom as is the right to security from physical assault.

We would only add that until we expand our understanding of democracy and freedom to include economic rights—a job for all those able to work and income with dignity for those not able—the United States can't be an example of hope in the eyes of the world's poor. Moreover, unless we so enlarge our understanding of democracy here at home, we doubt our government's capacity to understand or tolerate attempts for such change in the third world.

BEYOND ECONOMIC DOGMA

What would be required to expand our understanding of freedom and democracy, necessary both to end hunger here and to allow our nation to open the way to change in the third world?

First and foremost, a willingness to challenge the grip of economic dogma. In the opening essay of this book, we pinpointed what we see as the root of hunger—the antidemocratic concentration of power over economic resources, especially land and food.

But why have we allowed such concentration of power to continue, even at the price of untold human suffering? We began by answering that myths block our understanding. Here we want to probe deeper. We believe the answer lies in our imposed and self-imposed powerlessness before economic dogma.

Seventeenth-century intellectual breakthroughs forced us to relinquish the comforting notion of an interventionist God who would put the human house aright. And what a frightening void we then faced! Running from the weighty implication—that indeed *human beings* are responsible for society-inflicted suffering—we've desperately sought a substitute concept. We've longed for overriding laws we could place above human control, thus relieving us of moral responsibility.

With Newton's discovery of laws governing the physical world and with Darwin's parallel discovery in the realm of nature, we became convinced that there must indeed be laws governing the social world.

And we thought we had found them! Here we'll mention two such "absolutes" that relate most directly to the causes of hunger. Though they be human creations, our society has made them sacred.

The first is the market. Who can deny that the market is a handy device for distributing goods? As we stated elsewhere, any society that has attempted to do away with the market has run up against serious stumbling blocks. But once transformed into dogma, this useful device can become the cause of great suffering. As such, we are made blind to even the most obvious shortcomings of the market—its ability to respond only to the demands of wealth, not to the needs of people, its inability to register the real resource costs of production, and its inherent tendency to concentrate power in ever fewer hands.

Facing up to these shortcomings does not mean that we throw out the market in favor of another dogma, such as top-down state control. It means that we approach the market as a useful device, asking ourselves, under what circumstances can the market serve our values? We have set forth the very simple proposition that *the more widely purchasing power is distributed, the more the market will respond to actual human needs.*

But within a market system in which everything—land, food, human skills—is bought and sold with no restrictions, how can we work toward a more equal distribution of buying power? The answer is we cannot. Yet if we agree that tossing out the market would be foolish, what do we do?

In answering this question, we face the second major stumbling block posed by the prevailing economic dogma, the notion of unlimited private control over productive property.

Taken as economic dogma, the right to unlimited private control over productive property allows many Americans to accept as fair and inevitable the accelerating consolidation of our own farmland in fewer hands and the displacement of owner-operated farms, just as we have seen in much of the third world. In Iowa, a symbol of family-farm America, more than half the land is now controlled by absentee landlords, not working farmers.[7] Similarly, we accept the accelerating concentration of corporate power.

Although many Americans believe that the right to unlimited private control over productive property is the essence of the American way, this was certainly not the vision of many of our nation's founders, as we pointed out in the preceding chapter. In their view, property could serve liberty only when ownership was widely dispersed, and the right to property was valid only when it served society's interests. This view was widely held well into the nineteenth century. "Until after the Civil War, indeed, the assumption was widespread that a corporate charter was a privilege to be granted . . . for purposes clearly in the public interest," writes historian Alan Trachtenberg."

But by 1986, Ford executive Robert A. Lutz could declare without apology that his "primordial duty" is to his shareholders, while lamenting that his company's investment decisions meant the loss of tens of thousands of jobs.[9] Lutz seemed unaware that the notion that a corporation is responsible to its shareholders, but not to its workers nor to the larger society, is in fact a very *new* idea.

More accurately, Lutz's view is the revival of a once-discarded idea. When our nation's founders rejected monarchy their cry was no taxation without representation. It was a demand for the accountability of governing structures. Applied to the much-altered economic world of the twentieth century, their demand seems especially appropriate vis-à-vis our major corporations. Corporations now "can have more impact on the lives of more people than the government of many a town, city, province, state," notes Yale political scientist Robert A. Dahl.[10] Thus today's claim by corporations of an unfettered right to allocate wealth we all helped to create may be closer to the concept of the divine right of kings than it is to the principles of democracy.

Ownership with Responsibility

Working against hunger requires a fundamental rethinking of the meaning of ownership, certainly when applied to the productive resources on which all humanity depends. Such effort would be a first step in breaking free of the constraints of dogma.

In this rethinking, we believe Americans would be well served by going back to our roots, to the concept of property-cum-responsibility held by our nation's founders and to that of the original claimants to these soils, the American Indian nations. Because the community endures beyond the lifetime of any one individual, the Indian concept of community tenure carried within it an obligation to future generations as well.[11]

Indeed, we see a worldwide movement toward the rethinking of ownership already under way. In this rethinking, ownership of productive resources, instead of an absolute to be placed above other values, becomes a cluster of rights and responsibilities at the service of our deepest values. It is neither the rigid capitalist concept of unlimited private ownership nor the rigid statist concept of public ownership.

Where do we see movement toward such rethinking? In 1982, we visited one of the most productive industrial complexes in Europe: Mondragon, in the Basque region of Spain. Here some 100 enterprises—including a banking system, technical training school, and social services—are owned and governed by the people who work there. This noncapitalist, nonstatist form of ownership results in very different priorities and values. During the recession of the early 1980s, for example, when Spain suffered 15 percent unemployment, virtually no one in Mondragon was laid off. Worker-owners were retrained to meet the needs of the changing economy.[12]

We can detect a values-first approach to ownership in the third world too. In Nicaragua's pragmatic agrarian reform the goal is not the elimination of private property; indeed many more landowners are being generated by the reform. The keystone is attaching an obligation to the right to own farmland. Since this resource is essential and finite, every owner is obliged to use it efficiently so as to benefit society. Land left idle or grossly underproducing is taken away and given free of charge to families with no land. The concept of ownership is thus protected, but not above a higher value—life itself, the right of all human beings to eat.

Do these examples sound far away, irrelevant, even alien to our own experience? Then consider the recent decision of Nebraskans on this very question of farmland ownership. A few years ago, they amended their state's constitution so that only working farmers and their families can own farmland. Corporations like Prudential Insurance that had been speculating in Nebraska farmland could buy no more. In their support for this amendment, Nebraskans put the value of dispersed ownership in family farm agriculture above the notion of anyone's absolute right to buy whatever their dollars can pay for.[13]

We introduced our discussion of property rights in response to the question, what would be required to achieve such a dispersion of economic power that the market could actually reflect human needs rather than the demand for wealth? Part of the answer, we have suggested, lies in rethinking property rights as a device to serve higher values, not as ends in themselves. But an additional approach is worthy of consideration.

Just Too Important

Price fluctuations in a market economy can be troublesome for the consumer, but in the special case of food, such variation can be catastrophic. For this reason, and because movement toward fairer distribution of income takes time, some societies have simply decided that what is necessary to life itself should not be left to the vagaries of the market.

As we have mentioned in earlier chapters, a number of both capitalist and noncapitalist societies—as vastly different as Sweden and China—have decided that wholesale food prices are too vital to everyone's well-being to be left to the uncertainties of the market. Health care is equally essential to life. Thus some third world societies and all Western industrialized societies except the United States have also concluded that health care should not be distributed by the market, that is, to those who can afford to pay for it, but should be a citizen's right.

These examples are hardly the final word. We present them as signs of growing courage to confront the rigid "isms," courage to put one's deepest values first and judge economic policies according to how they serve those values—not the other way around.

WHAT CAN WE DO?
DOWN TO THE MOST PERSONAL QUESTION

Believing in the possibility of ending hunger means believing in the possibility of real change.

Ironically, the greatest stumbling block of all is the notion held by many Americans that in the United States we have achieved the best that can be—no matter how flawed it may appear. Why is this ironic? Because as Americans we have a very different heritage. Near his death, the father of the Constitution, James Madison, said of our newborn nation, "[America] has been useful in proving things before held impossible."[14] Thus the belief that indeed something new is always possible should be our very birthright.

But how is it possible to believe that those who are poor and downtrodden—those who have so much working against them—can construct better lives? Observing ourselves and others, we've come to appreciate how hard it is to believe that others can change unless we experience change ourselves.

With this realization, the crisis of world hunger becomes the personal question, how can I use my new knowledge to change myself so that I can contribute to ending hunger? The answer lies in dozens of often mundane choices we make *every day.*

These choices determine whether we are helping to end world hunger or to perpetuate it. Only as we make our choices conscious do we become less and less victims of the world handed to us, and more and more its creators. The more we consciously align our life choices with the vision of the world we are working toward, the more powerful we become. We are more convincing to ourselves and more convincing to others.

How do we begin?

A first step is getting alternative sources of information. As we hope to have demonstrated, as long as we only get world news from television and the mainstream press, our vision will remain clouded by myths. That's why the resource section at the end of the book includes a list of useful periodicals that continually challenge prevailing dogma. Without a variety of independent sources, we can't fulfill our role as citizens to help reshape our government's definition of our national interest and its policies toward the third world.

Then we must put that new knowledge to use. We are all educators—we teach friends, coworkers, and family. With greater confidence born of greater knowledge, we can speak up effectively when others repeat self-defeating myths. Letters to the editor, letters to our representatives, letters to corporate decision makers—they all count too.

Perhaps the most important step, however, in determining whether we will be part of the solution to world hunger, is the choice of a career path. The challenge is to think through just how we apply our skills in jobs that confront, rather than accept, a status quo in which hunger and poverty are inevitable.

To have a real choice of career path or to contemplate involvement in social change, we also have to decide what level of material wealth we need for happiness. Millions of Americans are discovering the emptiness of our society's pervasive myth that material possessions are the key to satisfying lives. They are learning that the *less* they need, the more freedom of choice they have in where to work, where to live, in learning experiences.

In every community in America, people go hungry and lack shelter. Through our churches, community groups, trade unions, and local government, we can help address immediate needs and participate in generating a new understanding of democracy—not as a vote one casts every few years but as active participation in community planning for more and better jobs, affordable housing, and environmental protection. Working to elect officials committed to addressing the roots of hunger is essential to such change.

Where and how we spend our money—or don't spend it—is also a vote for the kind of world we want to create. For example, in most communities we can now choose to shop at food stores that offer less-processed and less wastefully packaged foods, stores managed by the workers themselves, instead of conglomerate-controlled supermarkets. And we can choose to redirect our consumer dollars in support of specific product boycotts, such as the successful boycott of Nestlé that alerted the world to the crisis of infant deaths caused by the corporate promotion of infant formula in the third world, or the boycott of Campbell Soup that brought the company to the negotiating table with a Midwest farmworkers organization.

We can take responsibility for the invisible role our savings play when we put them in the bank. Instead of allowing our savings to be invested in weapons manufacturing, nuclear power, or South Africa, we can use our savings to support our values. Socially responsible investment funds have been created in recent years that use criteria of fairness and environmental protection, along with monetary return, in deciding where to put our money.[15]

But little is possible by oneself. We need others to push us and to console us when we are overwhelmed by the enormity of the problems we face. The points we make about the myth of the passive poor apply equally well to "passive" North Americans. We, too, need the example of others. *Community Is Possible* by Harry Boyte[16] and *Helping Ourselves* by Bruce Stokes[17] are just two books offering inspiring glimpses into local initiatives for change in America.

Actually *going* to the third world ourselves can profoundly alter our perceptions. A superficial tourist's view might confirm one's despair; but making the effort to meet those working for change, we can discern tremendous energy and hope. And looking back to the United States from abroad, we gain new insights on the role of our government. Today several nonprofit groups and travel agencies offer study-tours to selected third world countries. Individuals with specialized skills can consider actually living for awhile in the third world, offering their services to locally organized initiatives.

At the end of this book, we have included a selected list of some of the organizations working at a number of levels: all are part of the growth in understanding necessary to end hunger.

THE ESSENTIAL INGREDIENT

Our capacity to help end world hunger is infinite, for the roots of hunger touch every aspect of our lives—where we work, what we teach our children, how we fulfill our role as citizens, where we shop and save. But whether we seize these possibilities depends in large measure on a single ingredient. You might expect us to suggest that the needed ingredient is compassion—compassion for the millions who go hungry today. As we have pointed out, compassion is indeed a profoundly motivating emotion. It comes, however, relatively easy. Our ability to put ourselves in the shoes of others makes us truly human. Some even say it's in our genes and that we deny our innate compassion only at great peril to our own emotional well-being. There is another ingredient that's harder to come by. It is moral courage.

At a time when the old "isms" are ever more clearly failing, many cling even more tenaciously to them. So it takes courage to cry out, "The emperor wears no clothes! The world is awash in food, and all of this suffering is the result of human decisions!"

To be part of the answer to world hunger means being willing to take risks, risks many of us find more frightening than physical danger. We have to risk being embarrassed or dismissed by friends or teachers as we speak out against deeply ingrained but false understandings of the world. It takes courage to ask people to think critically about ideas so taken for granted as to be like the air they breathe.

And there is another risk—the risk of being wrong. For part of letting go of old frameworks means grappling with new ideas and new approaches. Rather than fearing mistakes, courage requires that we continually test new concepts as we learn more of the world—ever willing to admit error, correct our course, and move forward.

But from where does such courage come?

Surely from the same root as our compassion, from learning to trust that which our society so often discounts—our innate moral sensibilities, our deepest emotional intuitions about our connectedness to others' well-being. Only on this firm ground will we have the courage to challenge all dogma, demanding that the value of human life be paramount. Only with this new confidence will we stop twisting our values so that economic dogma might remain intact while millions of our fellow human beings starve amid ever greater abundance.

NOTES

1. Americas Watch, *A Nation of Prisoners* (New York: Americas Watch, 1984). For further documentation of human rights abuses perpetrated by the Guatemalan government, see U.S. Department of State, *Country Reports on Human Rights Practices for 1984*, Report submitted to the U.S. Senate Committee on Foreign Relations, February 1985, 541–57.
2. Jonathan Kwitny, *Endless Enemies: The Making of an Unfriendly World* (New York: Cogdon and Weed, 1985), chapter 15; and Warren Hinckle and William Turner, *The Fish is Red: The Story of the Secret War Against Castro* (New York: Harper and Row, 1981).
3. Most evaluations of Cuban society are so one-sided as to make informed, reasonable discussion impossible. Our institute's study—*No Free Lunch: Food and Revolution in Cuba* by Medea Benjamin, Joseph Collins, and Michael Scott (New York; Grove Press/Food First Books, 1986)—offers a penetrating, critical review of the Cuban food system, discussing both positive and negative lessons.
4. Joseph Collins with Frances Moore Lappé, Nick Allen, and Paul Rice, *Nicaragua: What Difference Could a Revolution Make?* (New York: Grove Press/Food First Books, 1986); and Frances Moore Lappé and Joseph Collins, *Now We Can Speak: A Journey through the New Nicaragua* (San Francisco: Food First Books, 1982).
5. Central American Historical Institute, Intercultural Center, Georgetown University, Washington, D.C., *Update* 5 (27 February 1986): 6–7.
6. For ongoing information about developments in Nicaragua, Re monthly *Envio* and periodic *Updates* from the Central American Historical Institute listed above.
7. *Iowa Land Ownership Survey: Preliminary Report on Land Tenure and Ownership in 47 Counties* (Des Moines: Farmers Unions, 1982), 3, 10.
8. Robert Bellah et al., *Habits of the Heart* (Berkeley: University of California Press, 1985), 289.
9. *Business Week* (3 March 1986): 62.
10. Robert A. Dahl, *Dilemmas of Pluralist Democracy: Autonomy Versus Control* (New Haven: Yale University Press, 1982), 184. For a thoughtful discussion on the issues raised here, see Robert A. Dahl, *A Preface to Economic Democracy* (Berkeley: University of California Press, 1985).
11. For a useful discussion, see William Cronon, *Changes in the Land: Indians, Colonists, and the Ecology, of New England* (New York: Hill and Wang, 1983).
12. Based upon interviews by Frances Moore Lappé, September 1982. For a general introduction to the Mondragon experience, see Hank Thomas and Chris Logan, *Mondragon: An Economic Analysis* (London: Allen & Unwin, 1981).
13. For information on Nebraska's constitutional amendment, write to the Center for Rural Affairs, Box 405, Walthill, Nebraska 68067.
14. Robert A. Dahl, *Democracy in the United States*. 3rd edition (Boston: Houghton Mifflin Co., 1981), 32.
15. For more information about one such investment fund, write to Working Assets, 230 California St., San Francisco CA 94111.
16. Harry C. Boyte, *Community Is Possible* (New York: Harper and Row, 1984). See also Boyte's new book, coauthored with Sara M. Evans, *Free Spaces: The Sources of Democratic Change in America* (New York: Harper and Row, 1986), for a study of democratic movements in America.
17. Bruce Stokes, *Helping Ourselves: Local Solutions to Global Problems* (New York: W.W. Norton, 1980).

Permissions

M. F. K. FISHER, "Foreword" from The Gastronomical Me, p. 353.Reprinted with permission of Macmillan General Reference USA, a Division of Simon and Schuster Inc., from The Art of Eating by M.F.K. Fisher. Copyright c 1943 by M.F.K. Fisher. Copyright renewed c 1971 by M.F.K. Fisher.

MARGARET MEAD, "The Changing Significance of Food." Reprinted from American Scientist 58 (March-April 1970), pp. 176-181. Reprinted with permission of the Science Research Society.

ROLAND BARTHES, "Toward a Psychosociology of Contemporary Food Consumption," published as "Vers une psycho-sociologie de l'alimentation moderne." Reprinted from Annales: Economies, Societes, Civilisations no. 5 (September-October 1961), pp. 977-986.

CLAUDE LÉVI-STRAUSS, "The Culinary Triangle." Reprinted from The Origin of Table Manners: Introduction to a Science of Mythology, Volume 3 (1968), by Claude Levi-Strauss, pp. 471-495. Copyright c 1968 by Librairie Plon; English Translation Copyright c 1978 by Jonathan Cape Ltd., and Harper and Row, Publishers, Inc.. Reprinted by permission of HarperCollins Publishers, Inc..

MARY DOUGLAS, "Deciphering a Meal." Reprinted from Implicit Meanings (1975) by Mary Douglas, pp. 249-275. Reprinted with permission of the author and the publisher.

JEAN SOLER, "The Semiotics of Food in the Bible." Reprinted from ESC, July-August 1973, pp. 943-955. Reprinted with permission of the Johns Hopkins University Press.

MARVIN HARRIS, "The Abominable Pig." Reprinted from Good to Eat: Riddles of Food and Culture by Marvin Harris, pp. 67-81. Reprinted with permission of the author.

E.N. ANDERSON, "Traditional Medical Values of Food." Reprinted from The Food of China (1988), by E.N. Anderson, pp. 187-198. Reprinted with permission of the Yale University Press.

ANNA MEIGS, "Food as a Cultural Construction." Reprinted from Food and Foodways 2 (1988), pp. 341-359. Reprinted with permission of the Gordon and Breach Publishing Group.

ANNA FREUD, "The Psychoanalytic Study of Infantile Feeding Disturbances." Reprinted from The Psychoanalytic Study of the Child, Volume 2 by Anna Freud, pp 119-132. Reprinted with permission of International Universities Press, Inc..

DOROTHY SHACK, "Nutritional Processes and Personality Development among the Gurage of Ethiopia." Reprinted from Ethnology 8:3 (1969), pp. 292-300. Reprinted with permission of Ethnology.

WILLIAM SHACK, "Hunger, Anxiety, and Ritual: Deprivation and Spirit Possession among the Gurage of Ethiopia." Reprinted from Man 6:1 (March 1971), pp. 30-43. Reprinted with permission of the Royal Anthropological Institute of Great Britain and Ireland.

CAROLINE WALKER BYNUM, "Fast, Feast, and Flesh: The Religious Significance of Food to Medieval Women." Reprinted from Representations, no. 11 (Summer 1984), pp. 1-25. Copyright c 1985 by the Regents of the University of California. Reprinted with permission of the University of California Press.

JOAN JACOBS BRUMBERG, "The Appetite as Voice." Reprinted from Fasting Girls: The Emergence of Anorexia Nervosa as a Modern Disease (1988) by Joan Jacobs Brumberg, pp. 164-188. Copyright c 1988 by the President and Fellows of Harvard College. Reprinted with permission of the Harvard University Press, Cambridge, MA.

MARJORIE DEVAULT, "Conflict and Deference." Reprinted from Feeding the Family by Marjorie DeVault, pp. 227-243. Reprinted with permission of the University of Chicago Press.

HORTENSE POWDERMAKER, "An Anthropological Approach to the Problem of Obesity." Reprinted from Bulletin of the New York Academy of Medicine 36:5 (May, 1960).

HILDE BRUCH, "Body Image and Self-Awareness." Reprinted from Eating Disorders (1973) by Hilde Bruch, pp. 87-105. Copyright c 1973 by Basic Books, Inc.. Reprinted with permission of BasicBooks, a division of HarperCollins Publishers, Inc..

SUSAN BORDO, "Anorexia Nervosa: Psychopathology as the Crystallization of Culture." Reprinted from Unbearable Weight: Feminism, Western Culture and the Body (1993) by Susan Bordo, pp. 139-164. Copyright c 1993 by the Regents of the University of California. Reprinted with permission of the University of California Press.

EMILY MASSARA, "Que Gordita." Reprinted from Que Gordita (1989), Chapter 8. Reprinted with permission of the AMS Press.

ELISA J. SOBO, "The Sweetness of Fat: Health, Procreation, and Sociability in Rural Jamaica." Reprinted from Many Mirrors: Body Image and Social Relations (1994), ed. Nicole Sault, pp. 132-154. Copyright c 1994 by Rutgers, The State University. Reprinted with permission of the Rutgers University Press, New Brunswick, New Jersey.

MARVALENE H. HUGHES (formerly STYLES), "Soul, Black Women, and Food." Reprinted from A Woman's Conflict: The Special Relationship between Women and Food (1980), ed. Jane Rachel Kaplan, pp. 161-176. Reprinted with permission of the author.

CAROLE COUNIHAN, "Bread as World: Food Habits and Social Relations in Modernizing Sardinia." Reprinted from Anthropological Quarterly 57:2 (April 1984), pp. 47-59. Reprinted with permission of the Catholic University of America Press.

ANNE ALLISON, "Japanese Mothers and Obentos: The Lunch Box as Ideological State Apparatus." Reprinted from Anthropological Quarterly 64:4 (October 1991), pp. 195-208. Reprinted with permission of the Catholic University of America Press.

STEPHEN MENNELL, "On the Civilizing of Appetite." Reprinted from All Manners of Food by Stephen Mennell, pp. 20-39. Reprinted with permission of Blackwell Publishers.

JACK GOODY, "Industrial Food: Towards the Development of a World Cuisine."

SIDNEY W. MINTZ, "Time, Sugar, and Sweetness." Reprinted with permission of the author.

PENNY VAN ESTERIK, "The Politics of Breastfeeding: An Advocacy Perspective." Reprinted from Breastfeeding: Biocultural Perspectives (1995), eds Patricia Stuart-Macadam and Katherine A. Dettwyler, chapter 6. Copyright c 1995 Walter de Gruyter, Inc.. Reprinted with permission of Aldine de Gruyter.

JANET M. FITCHEN, "Hunger, Malnutrition, and Poverty in the Contemporary United States: Some Observations on Their Social and Cultural Context." Reprinted from Food and Foodways (1988), pp. 309-333. Reprinted with permission of Gordon and Breach Publishers.

FRANCES MOORE LAPPÉ AND JOSEPH COLLINS, "Beyond the Myths of Hunger: What We Can Do." Reprinted from World Hunger: Twelve Myths (1986), by Frances Moore Lappe and Joseph Collins, pp. 137-149. Reprinted with permission of the Grove Press.

Contributors

Anne Allison is associate professor in the Department of Cultural Anthropology at Duke University. She is currently working on Japanese superheroes in popular culture and how they are exported to markets around the world. Her research interests are gender, sexuality, popular culture, mass culture, transnationalism, and Japan. Her recent publications are *Nightwork: Sexuality, Pleasure, and Corporate Masculinity in a Tokyo Hostess Club* and *Permitted and Prohibited Desires: Mothers, Comics, and Censorship in Japan.*

E. N. Anderson is the author of *The Food of China* and *Ecologies of the Heart* and the editor of *Bird of Paradox: The Unpublished Writings of Wilson Duff.* His research interests are cultural ecology, political ecology, traditional ecological knowledge, foodways, China, and the Yucatec Maya.

Roland Barthes, teacher and writer, was appointed in 1976 as the first person to hold the Chair of Literary Semiology at the Collège de France. He pioneered the semiologic interpretation of foodways in articles including the preface to Brillat-Savarin's *Physiology of Taste.* He also authored the antiautobiography *Roland Barthes by Roland Barthes.*

Susan Bordo is Professor of Philosophy and Singletary Chair of Humanities at the University of Kentucky. She is the author of *The Flight to Objectivity: Essays on Cartesianism and Culture* and *Unbearable Weight: Feminism, Western Culture, and the Body* (a *New York Times* Notable Book for 1993). Her forthcoming books are *In Plato's Cave: Essays on Reality and Illusion in Contemporary American Culture* and *My Father's Body and Other Unexplored Regions of Sex, Masculinity, and the Male Body.*

One of the world's pioneering authorities on the diagnosis and treatment of eating disorders, **Hilde Bruch** received an M.D. from the University of Freiburg in 1929 and was Professor of Psychiatry at Baylor College of Medicine in Houston. She has authored scores of journal articles and several books, including *Eating Disorders: Obesity, Anorexia Nervosa, and the Person Within; The Golden Cage: The Enigma of Anorexia Nervosa;* and the posthumously published *Conversations with Anorexics.*

Joan Jacobs Brumberg is a historian who teaches at Cornell University. She is the author of *Fasting Girls: The Emergence of Anorexia Nervosa as a Modern Disease* and *The Body Project: An Intimate History of Americans Girls.* She has held fellowships from the Guggenheim and Rockefeller Foundations as well as the National Endowment for the Humanities.

Professor of History at Columbia University, **Caroline Walker Bynum** has pioneered the feminist study of foodways in history with *Fragmentation and Redemption: Essays on Gender and the Human Body in Medieval Religion* and *Holy Feast and Holy Fast: The Religious Significance of Food to Medieval Women.*

Co-founder with Frances Moore Lappé of Food First: The Institute for Food and Development Policy, **Joseph Collins** is an expert on world hunger. He is the author of many books, including *Food First* (with Lappe), *What Difference Could a Revolution Make: Food and Farming in the New Nicaragua,* and *Agrarian Reform and Counter-Reform in Cuba.*

Carole Counihan has been pursuing the anthropological study of food and culture for two decades through fieldwork in the United States and Italy. She is Associate Professor of Anthropology and Women's Studies at Millersville University in Pennsylvania and associate editor of the scholarly journal *Food and Foodways*. She has published in diverse anthropological journals and is writing a book on food, culture, and gender in Florence, Italy.

Marjorie DeVault teaches Sociology and Women's Studies at Syracuse University. She has written a groundbreaking feminist study of women's food work in the United States, entitled *Feeding the Family: The Gendered Nature of Caring Work,* and many articles on women's work.

Mary Douglas recently retired from teaching anthropology at Cambridge University and has held a number of other academic appointments, including the Avalon Foundation Chair at Northwestern University. She is a trailblazer in the anthropological study of food, and her books include *Constructive Drinking; Purity and Danger;* and *Food in the Social Order.*

M. F. K. Fisher is a prolific essayist and memoirist whose writings center on the pleasures of cooking and eating, principally in California and Southern France. She writes about food with passion and brilliance, encapsulating the complex human relations centering around food. She is author of over sixteen volumes, including *The Art of Eating* and an acclaimed translation of Brillat-Savarin's *The Physiology of Taste*. She spent the last years of her life in Glen Ellen, California, and died in 1992.

Janet M. Fitchen chaired the Department of Anthropology at Ithaca College for many years before her appointment to the Department of Rural Sociology at Cornell University. Her books include *Endangered Spaces, Enduring Places: Change, Identity and Survival in Rural America* and *Poverty in Rural America: A Case Study*. Her recent untimely death occurred during the planning stages of this book, and we mourn her passing with regret and affection.

Youngest child and intellectual heir of Sigmund Freud, **Anna Freud** is renowned for her interpretations and applications of psychoanalytic theory, especially to children. In a few brilliant articles, she established a definitive psychoanalytic interpretation of eating and gluttony. Her writings are collected in the eight volumes of *The Writings of Anna Freud.*

Jack Goody, formerly on the faculty of anthropology at St. John's College, Cambridge University, is the author of *Cooking, Cuisine and Class: A Study in Comparative Sociology*, *Production and Reproduction*, and *The Culture of Flowers*.

Professor of Anthropology at the University of Florida, **Marvin Harris** pioneered the cultural materialist interpretation of seemingly quirky human food habits in *Good to Eat: Riddles of Food and Culture* and co-edited *Food in Evolution* with Eric Ross. He is the author of many other books, including *Cultural Materialism*; *Cows, Pigs, Wars and Witches*; and *The Rise of Anthropological Theory*.

Marvalene H. Hughes currently serves as the sixth resident of California State University, Stanislaus, one of the 23 campuses of the CSU system. Previously, she served as Vice President for Student Affairs / Vice Provost and Professor of Educational Psychology at the University of Minnesota, Twin Cities, as well as system-wide vice president for all University of Minnesota campuses. Dr. Hughes studied at Tuskegee University, NYU, and Columbia University and earned her Ph.D. at Florida State University. She has pursued postdoctoral study at three Harvard University institutes and attended management seminars at the University of California, San Diego.

An authority on world hunger and participatory democracy, **Frances Moore Lappé** is the author of many books, including *Diet for a Small Planet; Rediscovering America's Values;* and *Food First* (with Joseph Collins). With Joseph Collins, she co-founded Food First: The Institute for Food and Development Policy in San Francisco, and in 1990 she co-founded the Institute of the Arts of Democracy with Paul Martin Du Bois.

Claude Lévi-Strauss is a French cultural anthropologist best known for applying structuralism to the study of symbolism in mythology. He has held the Chair of Social Anthropology at the Collège de France since 1959. His books include *Totemism; The Raw and the Cooked; The Origin of Table Manners*; and *The Story of the Lynx*.

Emily Massara has always had an interest in applying cultural principles to practical issues. She worked in the weight-management field for many years. Among her publications is "A Method of Quantifying Cultural Ideals of Beauty and the Obese" (*International Journal of Obesity*), co-authored with Albert J. Strunkard, M.D. Currently she records life histories using a "life portrait" approach that she developed in her capacity as Director of Community Relations at Wills Eye Hospital's Geriatric Psychiatry Program in Philadelphia.

Margaret Mead was a cultural anthropologist and curator of ethnology at the American Museum of Natural History in New York. She carried out extensive fieldwork in Oceania, and during World War II she was Executive Secretary of the Committee on Food Habits of the National Research Council. Some of her many books are *Coming of Age in Samoa; Growing Up in New Guinea; Sex and Temperament in Three Societies;* and *Male and Female*.

Anna Meigs is associate professor of anthropology at Macalester College. She is the author of *Food, Sex, and Pollution: A New Guinea Highlands Religion*. She is currently working on cultural constructions of race.

Stephen Mennell is a professor of sociology at Monash University in Australia. He is the author of *All Manners of Food* and co-editor of *The Sociology of Food, Eating, Diet and Culture.*

Sidney W. Mintz teaches anthropology at Johns Hopkins University. He has written widely about the Caribbean, and his food analyses include *Sweetness and Power* and the new *Tasting Food: Tasting Freedom.*

Hortense Powdermaker was a cultural anthropologist with a Ph.D. from the University of London. Her diverse and fascinating fieldwork in the Pacific, Mississippi, Hollywood, and Zambia is described in her pathbreaking piece of reflexive anthropology, *Stranger and Friend,* originally published in 1966.

Dorothy N. Shack carried out research on Gurage childrearing practices from 1962–65. She was an instructor in psychology at the former Haile Sellassie I University, and retired in 1989 as a school psychologist in the Berkeley Unified School District.

William A. Shack, Professor Emeritus of Anthropology at the University of California, Berkeley, first carried out field research among the Gurage of Ethiopia in 1957–59. He is the author of *The Gurage: A People of the Ensete Culture* and numerous other publications.

Elisa J. Sobo is a sociocultural anthropologist specializing in health. She teaches in the Department of Anthropology at the University of California, San Diego. Her recent books are *Choosing Unsafe Sex: AIDS-Risk Denial and Disadvantaged Women* and *One Blood: The Jamaican Body,* about traditional Jamaican ethnophysiological and ethnomedical beliefs and practices.

Jean Soler taught French, Latin, and Greek in Algeria and France for five years before entering France's Ministry of Foreign Affiars, where he served as Director of the French Cultural Center in Poland (1965–68); Cultural Councilor for the French Embassy in Israel (1968–73, 1989–93), Iran (1973–77), and Belgium (1977–80); and Director of Cultural Affairs for the Provence–Côte d'Azur Region (1981–85). His influential anthropological studies of Hebrew dietary restrictions first earned him the respect of Claude Lévi-Strauss in 1973.

Penny Van Esterik is a cultural anthropologist teaching at York University in Toronto, Canada. She has long-term interests in infant-feeding practices, food advocacy, and Thai food. Her books include *Beyond the Breast-Bottle Controversy; Women, Work, and Breastfeeding*; and *Taking Refuge: Lao Buddhists in North America.* She coordinates the Taskforce on Breastfeeding and Women's Work for the World Alliance for Breastfeeding Action (WABA).

Index